THE COLLECTIVIZATION OF AGRICULTURE IN COMMUNIST EASTERN EUROPE

THE COLLECTIVIZATION OF AGRICULTURE IN COMMUNIST EASTERN EUROPE

Comparison and Entanglements

Edited by
CONSTANTIN IORDACHI
and
ARND BAUERKÄMPER

CEU PRESS

Central European University Press
Budapest–New York

Published in 2014 by

Central European University Press

An imprint of the
Central European University Limited Liability Company
Nádor utca 11, H-1051 Budapest, Hungary
Tel: +36-1-327-3138 or 327-3000
Fax: +36-1-327-3183
E-mail: ceupress@ceu.hu
Website: www.ceupress.com

224 West 57th Street, New York NY 10019, USA
Tel: +1-212-547-6932
Fax: +1-646-557-2416
E-mail: martin.greenwald@opensocietyfoundations.org

Published with the support of
the *Bundesstiftung zur Aufarbeitung der SED-Diktatur,* Berlin

BUNDESSTIFTUNG
AUFARBEITUNG

ISBN 978-615-5225-63-5

LIBRARY OF CONGRESS CATALOGING-IN-PUBLICATION DATA

The collectivization of agriculture in communist Eastern Europe : comparison and entanglements / edited by Constantin Iordachi and Arnd Bauerkämper.
pages cm
Includes bibliographical references and index.
ISBN 978-6155225635 (hardbound)
1. Collectivization of agriculture—Europe, Eastern—History. I. Iordachi, Constantin. II. Bauerkämper, Arnd.

HD1493.E852C65 2013
338.1'847—dc23

2012048156

Printed in the USA

Table of Contents

II. Land Collectivization in Central Europe

III. Land Collectivization in Southeastern Europe

IV. Axes of Differentiation: Center and Periphery, "Class Struggle," Social and Ethnic Cleavages

V. Appendix

List of Maps

List of Tables

Introduction

The Collectivization of Agriculture in Eastern Europe: Entanglements and Transnational Comparisons

ARND BAUERKÄMPER and CONSTANTIN IORDACHI

The sweeping economic and sociopolitical changes set into motion by the collapse of communist dictatorships and the end of the Cold War have significantly transformed Eastern European societies. In the last two decades, numerous scholars in various disciplines have attempted to understand the nature of these changes and assess the societal impact of Eastern European integration into European and Euro-Atlantic security and sociopolitical organizations. These research efforts have mostly focused on the transformation of legal-political systems, on mechanisms of elite (re)production, and on processes of privatization and restructuring of property relations. Much less attention has been devoted to changes in rural transformation areas, despite the fact that the peasantry continues to be an important segment of the population in all countries in the region. Moreover, when attention has been paid to post-communist rural communities, researchers tended to adopt a short-term perspective, focusing mostly on the issue of decollectivization, reconstitution of private ownership over land, and integration into the European market-based agricultural policies. At the same time, historical perspectives on the agrarian question under communism, in general, and the establishment and history of collective farms, in particular, are still scarce.

Despite its profound, long-term sociopolitical implications, the collectivization of agriculture in Eastern Europe has remained a relatively under-researched field. In this volume, we posit that, in order to understand the communist and post-communist transformation of Eastern European societies, scholars need to pay attention to the status and evolution of the peasantry and to employ a historical perspective by exploring the communist agricultural policies and their enduring economic and social implications.[1] To contribute to this analytical effort, the volume explores the interrelated campaigns of agricultural collectivization in the USSR during the interwar and postwar periods and in the communist dictatorships established in the Soviet-dominated Eastern Europe after World War II.

In line with recent works on the topic, the volume addresses collectivization in its full complexity—both as a core aspect of the establishment of communist regimes in the Soviet Union and in Eastern Europe and as an essential dimension of the profound transformation of political and state institutions, as well as socioeconomic relations and everyday life.[2] Although the contributors to this volume focus mostly on individual case studies, they address a unitary set of analytical issues, relating to: 1) the crystallization of a Soviet model of collectivization; 2) its imposition in Eastern Europe, as well as its adaptation to local conditions; 3) spatial differences in the collectivization campaigns, particularly with regard to the relationship between center and peripheral regions; 4) the temporal dynamics of collectivization in rural societies, marked by periods of acceleration, temporization, or stagnation; and 5) types of collectivization and socialist agricultural systems. Based on novel primary sources, the volume provides a reappraisal of the theoretical implications and research agenda of studies on collectivization in Eastern Europe.

The collection combines detailed studies of collectivization campaigns in individual states of Eastern Europe with issue-oriented comparative essays. In order to provide a solid empirical ground for trans- or cross-national comparisons, all studies make extensive use of archival materials; they also provide comprehensive bibliographies and useful chronologies of the collectivization process in each country, thus facilitating novel comparative insights into this process at the regional level. Along with providing an up-to-date overview of recent research in the field, the contributors in this volume also promote new approaches to the topic, combining historical comparisons with studies of transfers and transnational entanglements.[3]

---

1 For a comparative illustration of the deep, long-term impact of collectivization patterns on post-communist agrarian developments in two Hungarian villages situated in Romania and Hungary, see Thelen, "Violence and Social (Dis)Continuity."

2 On this point, see Iordachi and Dobrincu, eds., *Transforming Peasants*, 2.

3 Transnational approaches first emerged in the 1970s in the field of international relations, and later in world and global history. According to a pioneering definition advanced by Robert O. Keohane and Joseph S. Nye, "transnational relations" describe "contacts, coalitions and interactions across state boundaries

In line with recent theoretical and methodological innovations such as "shared," "connected," or "relational" history and *histoire croisée,* the volume aims to shift the analytical research focus from isolated national case studies to multiple forms of connectedness, entanglements, and transfers within the Soviet "camp," at various levels. To this end, the national case studies included in the volume are approached from a relational perspective, the contributors highlighting transnational exchanges in the process of Eastern Europe's Sovietization in general, and in collectivization campaigns in particular, in synchronic as well as diachronic historical perspectives.[4]

Methodologically, transnational approaches have the merit of enabling researchers to transcend the "national paradigm," highlighting the nation-states' position in a wider geographical context and their connections to regional or global developments. As such, transnational history paves the way for multiple, symmetrical, or asymmetrical comparisons and more complex, multilateral studies of cross-border exchange. Applied to the field of collectivization, this research perspective might lead to a full-fledged *histoire croisée* of collectivization campaigns in Eastern Europe. According to Michael Werner and Bénédicte Zimmermann, *histoire croisée* seeks to "investigate relational configurations that are active and asymmetrical, as well as the labile and evolving nature of things and situations, to scrutinize not only novelty but also change." *Histoire croisée* approaches underscore "relational, interactive, and process-oriented dimensions" that give rise to a "multiplicity of possible intercrossings," challenging scholars to continually alter their analytical perspectives.[5]

---

that are not controlled by the central foreign policy organs of government." See Keohane and Nye, "Transnational Relations," 331. For a more recent perspective on transnationalism, see Robinson, "Beyond Nation-State Paradigms." On the permutations of the term "transnational," first coined by the German linguist Georg Curtius in 1862, see Saunier, "Transnational," 1047.

4 Transnational approaches have been only recently applied to the history of communist regimes in Eastern Europe. Discussing its potential impact on communist studies, Patricia Calvin argues that transnationalism should be "best understood not as fostering bounded networks, but as creating honeycombs, a structure that sustains and gives shapes to the identities of nation-states, international and local institutions, and particular social and geographic spaces. A honeycomb binds, but it also contains hollowed-out spaces where organizations, individuals and ideas can wither away to be replaced by new groups, people and innovations." Calvin, "Defining Transnationalism," 421.

5 Werner and Zimmermann, "Beyond Comparison," 38; on the conceptual implications for the history of modern Europe, see Bauerkämper "Europe;" Bauerkämper, "Wege."

In this introduction, we first evaluate the state of the research on agricultural collectivization in Eastern Europe and emphasize the importance of studying this process from a comparative and transnational perspective. The second part of the introduction highlights this volume's specific contribution to the existing scholarship on this topic and identifies new avenues of future research.

Between National History, Regional Comparisons, and Entangled Histories

Scholarship on collectivization in Eastern Europe has been profoundly shaped by the sharp ideological debates that took place during the Cold War over the nature of communist dictatorships established in the region after World War II. Due to political restrictions imposed on academic research, the great majority of local works published on the topic in Eastern Europe refrained from tackling controversial aspects in the collectivization campaigns. Instead, they presented the socialist transformation of the agriculture as a major step towards the creation of the envisaged utopian communist society in the countryside, and dogmatically praised the heroism of the peasantry, who allegedly responded to the Communist Party's call and prevailed upon the regime's enemies in rural areas. Outside of the Soviet camp, in contrast to this apologetic "official" literature, Western scholars interpreted collectivization as a form of "agrarian despotism,"[6] a direct result of the forceful and oppressive campaign of "Sovietization," and as a typical Leninist response by communist elites to economic dependence and the inequities of the capitalist system.[7] Research on collectivization thus generally suffered from ideological constraints in the East, while in the West it was hampered by a lack of access to primary archival or oral sources. In addition, due to political interdictions, these streams of literature developed in relative isolation from each

---

[6] Lewin, *The Soviet Century*, esp. 143–49, 311. Departing critically from the concept of "Oriental despotism," Lewin proposes the concept of "agrarian despotism" as a more accurate label for the Soviet "non-modern modernizer state" during the Stalin era, which "reinvigorated" an old system of tight agrarian control by "means of a breakneck industrialization."

[7] Jowitt, *New World Disorder*, esp. 32–39.

other, with limited academic interaction and cross-references except for counter-propaganda polemics.[8]

After the collapse of communist regimes in 1989–91, research on collectivization has experienced a relative upsurge, in East and West alike. On the one hand, political restrictions have been lifted, enabling local historians to tackle research topics that had been political taboos; on the other hand, scholars have gradually gained access to new archival funds, being thus able to pursue more in-depth studies of collectivization at the local and national levels.

The first and arguably most dramatic campaign of collectivization, the one that occurred in the USSR in the late 1920s and early 1930s, has been subject to substantial research.[9] Based on new documentary evidence, recent investigations have covered all the major stages in the collectivization campaign, from the Soviet leadership's major ideological dilemmas and debates over the course of collectivization to the impact of Joseph Stalin's and Vyacheslav Molotov's political initiatives and decrees on wholesale collectivization in late 1929 and early 1930, Stalin's abrupt criticism of the direction taken by the creation of collective farms at the local level (see Stalin's article "Dizzy with Success," published in *Pravda* on March 2, 1930), and the violent acceleration of the collectivization campaign, leading to full collectivization in the mid-1930s. In addition, other studies integrated collectivization in a broader perspective by addressing the preconditions of the collectivization of agriculture, the emergence of industrialized agricultural production since the 1960s, and the variations in the collective farming systems under state socialism.[10]

The role of collectivization in sustaining forced industrialization, and more generally, the socioeconomic and demographic impact of collectivization generated ample historiographical debates.[11] Most historians highlighted

---

8 See, for example, the West Germany response to the triumphalist official propaganda of the GDR regime on the completion of collectivization campaign: *The Compulsory Collectivization*, 1961.

9 For earlier studies of collectivization in the USSR, see Ladejinsky, "Collectivization of Agriculture;" Lewin, "The Immediate Background;" Narkiewicz, "Stalin, War Communism, and Collectivization;" Lewin, *Russian Peasants*, and Davies, *Socialist Offensive*.

10 Francisco, Laird, and Laird, eds., *Agricultural Policies*. For the GDR, see Last, *After the "Socialist Spring*;" Last, "Rural Functionaries."

11 See Sirc, "Economics of Collectivization;" Fallenbuchl, "Collectivization and Economic Development;" Holland, "Soviet Agriculture."

the link between collectivization and the procurement campaigns that the Bolshevik regime adapted from Tsarist rule in order to secure food provision and—even more important—to promote the ambitious industrialization goals that were mandated in Stalin's First Five-year Plan (enacted in 1928). They argued that the high rate of accumulation of resources and industrial investment in the 1930s could not be sustained without a tight state control over agriculture and the extraction of agricultural surplus.[12] Other historians posited that industrialization would have still been possible without collectivization, but not at such intensive levels; they also argued that the mobilization of additional labor and not surplus grain was central to the industrialization efforts.[13] Few historians argue that industrialization could have been even within the confines of the New Economic Policies of the mid-1920s.[14]

Drawing on previously inaccessible archival sources, recent works have also provided important new insights into the destructive dynamics of Soviet collectivization in the countryside at the federal or regional level.[15] First, a new wave of scholarship has shed new light on the organized Stalinist terror against the *kulaks*, who were dispossessed, murdered or deported to labor camps, frequently in Siberia—*dekulakization* thus becoming a cudgel of collectivization.[16] Due to the rapid and unforeseeable changes in agricultural policy and vagueness in defining the term *kulak*, rural committees arrested, dispossessed and deported rich peasants, in order to meet the quotas their superiors had imposed on them. Second, new studies have also explored the forced requisition campaigns in 1932–33 that resulted in hunger and mass starvation, especially in the Ukraine in 1932–33.[17] Overall, based on new archival evidence, historians

---

[12] Wheatcroft, Davies, and Cooper, "Soviet Industrialization Reconsidered."

[13] Allen, "Imposition et mobilisation du surplus."

[14] Millar, "Mass Collectivization;" Millar, "Collectivization and Its Consequences."

[15] On regional studies of collectivization and *dekulakization* in Soviet Union, see Olcott, "Collectivization Drive;" Hughes, *Stalinism*; Pianciola, "Collectivization Famine."

[16] See mainly Getty et al., "Victims;" Viola, *Peasant Rebels*; Viola, *Role of the OGPU*; Viola et al., eds., *War Against.*

[17] See Conquest, *Harvest of Sorrow*; Tucker, *Stalin in Power*, 119–45; Economakis, "Soviet Interpretations of Collectivization;" Coombe, *Cultural Life*. For the debate in Ukrainian studies over the famine, and its relation to the politics of history, see, selectively Dolot, *Execution by Hunger*; Davies and Wheatcroft, *The Years of Hunger*; Luciuk, *Holodomor*; and Kasianov, *Danse macabre*. For other case studies on the relation between famine and collectivization, see Pianciola and Finnel, "Famine in the Steppe."

have been able to re-evaluate the human losses and the long-term demographic costs generated by the liquidation of the kulaks and the famine of 1932–33, an issue which led to a grand debate over excess collectivization deaths in the USSR.[18]

A major turn in studies of collectivization in the Soviet Union has been the shift of focus from political to social and then to cultural history, and from macro to micro perspectives of research. Thus, if earlier works on collectivization in the Soviet Union focused mostly on violent state policies and their impact, new scholarship concentrates on state–peasant relations and local-level politics.[19] Using new methods provided by cultural history and anthropology, recent works on the Soviet Union have also dealt with aspects of everyday life in the countryside during the collectivization campaign and with peasants' variegated responses to the Soviet agricultural policies.[20] On the one hand, based on declassified Soviet archives, new studies have further explored strategies of peasant resistance against collectivization, pointing out that anti-collectivization riots were the most violent and sustained revolts against the Soviet state after the Russian Civil War, joined by men as well as women.[21] On this

---

[18] See Rosefielde, "New Demographic Evidence;" Wheatcroft, "New Demographic Evidence;" Rosefielde, "New Demographic Evidence," 1985b; Conquest, Cohen, and Wheatcroft. "New Demographic Evidence," 1986. Based on the difference between the population figures provided by official Soviet censuses in 1926 and 1937, Conquest "conservatively" evaluates the death toll generated by collectivization and famine at circa 12 million. Among these, 3.5 million peasants died as a result of the campaign of *dekulakization*, 1.5 died as a result of the Kazakh "catastrophe," around 5 million died in the Ukrainian famine, 1 million died in the famine in North Caucasus, and one million died due to famine elsewhere. See Conquest, *The Harvest of Sorrow*, 299–307, esp. 306. In a more recent article, Livi-Bacci evaluated the death toll caused by forced collectivization within a range varying between 6 and 13 million people. See Livi-Bacci, "On the Human Costs."

[19] Conquest, *Harvest of Sorrow*; Fitzpatrick, *Stalin's Peasants*; Davies, *Crisis and Progress*; Viola, "Campaign to Eliminate;" Viola, *Best Sons of the Fatherland*; Viola, *Peasant Rebels*; Viola, *Role of the OGPU*; Viola, et al., eds., *War Against*; Viola, ed., *Unknown Gulag*, 13–88, 183–98, 201–22, 245–47.

[20] See, in particular, Merl, *Bauernunter Stalin*; Fitzpatrick, *Stalin's Peasants*; Fitzpatrick, *Everyday Stalinism*, 115–38, 218–27; Hindus, *Red bread*; Storella and Sokolov, *The Voice of the People*. More generally on recent scholarship, see Beer, "Modernity and Resistance."

[21] Viola, *Peasant Rebels*; see also Reese, "Red Army Opposition;" and McDonald, "A Peasant Rebellion." For the gender dimension of anti-collectivization resistance see Viola, "Bab'i Bunty."

basis, they pointed out that terror pervaded everyday life in rural communities in the early 1930s, leading to new conflicts that impeded cooperation in the newly founded collectives. On the other hand, new studies have underscored the fact that the lower strata of the peasantry acted as the state's allies in the collectivization campaign, and that many villagers were able to adapt to the new conditions created by collectivization and even attain some immediate benefits and economic interests, thus highlighting issues of state–peasant "collaboration" that were taboo in the Cold War type of discourse. The new scholarship is thus able to provide a more informed but at the same time more nuanced account of the collectivization campaign.

In addition to the case of the USSR, research on collectivization has also focused, although to a lesser extent, on various East European countries.[22] As historians have stressed, this process was related to long-term shifts from nineteenth-century liberal traditions of property to national-collective conceptions. Moreover, institutions and actors associated with land ownership (for instance property registry and solicitors) had been weak intermediary instances as early as the interwar years.[23] Historians, social scientists and anthropologists have paid attention to the transfer of the Soviet model of collectivization to the newly occupied territories in the West, such as the Baltic States,[24] Moldavia,[25] or Western Ukraine after 1945.[26] The bulk of the existing research has nevertheless concentrated on the third and more arduous wave of collectivization campaigns in the Central and Southeastern European countries that had been forcibly integrated into the Soviet camp after World War II: Albania,[27] Bulgaria,[28]

---

22 For early synthetic views on collectivization in the region, see Spulber, "Eastern Europe;" Tomasevich, "Agriculture in Eastern Europe;" Sanders, "Collectivization of Agriculture;" Katona, "Collectivization of Agriculture in Central Europe."

23 Müller, "Bodeneigentum."

24 Labsvirs, "The Effect of Collectivization;" Taagepera, "Soviet Collectivization;" Skultans, "The Expropriated Harvest;" Girnius, "Collectivization of Lithuanian;" Kõll, *Village and the Class War.*

25 Gribincea, *Agricultural.*

26 Marples, "The *Kulaks*;" Marples, "Toward a Thematic;" Graziosi, *The Great Soviet Peasant War.*

27 Sjöberg, *Rural Change*, 1991; Vickers and Pettifer, *Albania.*

28 Creed, *Domesticating Revolution*; Josifov, *Totalitarnoto nasiliev*; Gruev, *Preorani slogove.*

Yugoslavia,[29] Romania,[30] Hungary,[31] Czechoslovakia,[32] the German Democratic Republic,[33] and Poland.[34]

Among these, the cases of Yugoslavia and Poland, where collectivization was abandoned in 1953 and 1956, respectively, have been of special interest. In the case of Yugoslavia, Melissa Bukovoy argued that unrest among the peasants, who had supported the communist partisans in the Second World War, forced Yugoslav communists to reconfigure their agrarian policies, contributing to the party's retreat from collectivization in 1953.[35] Like the Yugoslav communists, unwilling to pay the heavy political price of the forced agricultural transformation, the reformist leadership of the Communist Party in Poland also abandoned collectivization, even if at a later, intermediate stage, following a decision taken of the Ninth Plenum of the Polish Communist Party's Central Committee (October 29, 1956).[36]

Among the countries that pursued full collectivization, the historiography of collectivization in the GDR and Romania, in particular, exemplify the successive changes in interpreting collectivization before and after the fall of communism in 1989, in a joint endeavor of local and Western scholars. The GDR was the last country in the Eastern block to initiate the collectivization campaign, but also among the first ones to complete it, in 1960. Capitalizing on this success, the Socialist Unity Party (SED) glorified the allegedly "voluntary" process of collectivization as a genuine "revolution from below." According to the official party interpretation,

---

29 A. V. S., "Agriculture in Yugoslavia;" V. M., "Collectivization in Yugoslav;" Tochitch, "Collectivization in Yugoslavia;" Bokovoy, *Peasants and Communists.*

30 Kideckel, "Socialist Transformation;" Kideckel, *Solitude*; Cătănuş and Roske, eds., *Colectivizarea agriculturii.*

31 Balassa, "Collectivization in Hungarian;" Hollos, "Ideology and Economics;" Hollos and Maday, eds., *New Hungarian Peasants*; Swain, *Collective Farms*; Swain, "Collective farms;" Rév, "Advantages of Being;" Lampland, *Object of Labor*; Swain, "From Kolkhoz;" Thelen, "Violence and Social."

32 Miller, "Collectivization;" Jech, *Soumrak selského stavu*; Jech, *Kolektivizace.*

33 Langenhan, "Halte Dich fern;" Bauerkämper, "Collectivization and Memory;" Bauerkämper, *Ländliche Gesellschaft*; Buechler and Buechler, eds., *Agriculture*; Schöne, *Frühling.*

34 J. C., "The Peasant in Poland Today;" Korboński, *Politics of Socialist*; Robakowski, *Społeczno-polityczne problemy*; Próchniak, *Kolektywizacja*; Jarosz, *Politykawładz*; Jarosz, "Polish Peasants;" Jarosz, *Polacy a stalinizm*; Dobieszewski, *Kolektywizacja wsi polskiej*; Miernik, *Opórchłopów*; Markiewicz, *Kolektywizacja.*

35 Bokovoy, *Peasants and Communists.*

36 Korboński, *Politics of Socialist.*

German peasants had taken up the legacy of the failed Peasants' War of 1524–25 and significantly contributed to the disempowerment of the feudal large landowners, securing the victory of "socialism" in the countryside. Working under official political tutelage, East German historians refrained from investigating the dramatic impact of forced collectivization on agriculture and rural society, concentrating instead on the "progressive" role of the supposedly omniscient party, praised as the "vanguard of the proletariat." It was only in the 1980s that scholarship half-heartedly abandoned the Leninist doctrine on collectivization and shifted its focus to the Politburo's agricultural policies or to state–peasant relations. In this stage, scholars of collectivization focused especially on the role of particular groups, such as the refugees, or on the process, as well as the impact of collectivization on social relations, topics dealt with in particular in ethnological and cultural studies.[37]

The collapse of the dictatorship of the Socialist Unity Party and the reunification of Germany in 1989–90 has profoundly changed scholarly interpretations of collectivization. Research on this topic, and on the history of the socialist regime in the GDR, in general, has benefited significantly from the input of West German historians. Instead of propagandistically glorifying the role of the party in the collectivization campaign, new studies have reconstructed the relationship between the state and sections of rural society, and assessed the long-term impact of collectivization on everyday life and the natural environment.[38] They have explored the paramount roles played by rural elites and police surveillance in the collectivization campaign, but have also emphasized the limits of party rule and state power in the countryside. The legacy of collectivization in the individual and collective memories of local villagers has also been examined. Not least, historiography has demolished the myth of the "revolution from below" by exposing the meticulous preparation of the two waves of collectivization in 1952–53 and 1959–60, when the process was officially finalized. Gradually, the previous fixation on political history has given way to social and cultural-historical perspectives.[39]

---

[37] Bauerkämper, "Von der Polithistorie;" Bauerkämper, "The Industrialization of Agriculture."

[38] Buechler and Buechler, *Contesting Agriculture*; Bauerkämper, "Vertreibung als Exklusion;" Bauerkämper, *Ländliche Gesellschaft*; Schöne, *Frühling*; Langenhan, "Halte Dich fern."

[39] For a brief overview of this literature, see Schöne, "Landwirtschaft und ländliche Gesellschaft."

The historiography on collectivization in Romania largely conforms to this pattern as well.[40] If the GDR was the first collectivized country in the Soviet bloc, Romania, with its large agricultural sector, was among the very last countries to complete collectivization. The few studies on collectivization published in communist Romania were either heavily censored or sheer propaganda,[41] with more insightful work being published on the evolution of the socialist collective farms after the end of the collectivization process.[42] For many decades, significant research on collectivization could only be conducted in foreign academic settings, mainly in the United States and Western Europe. Pioneering foreign studies of collectivization mostly focused on the political repression accompanying the processes of the communist takeover of rural society. In the late 1950s and early 1960s, in an effort to account for the consolidation of the communist regime in Romania, scholarship began to devote attention not only to state repressive actions in the countryside, but also to the evolution of the "peasant question" and rural relations, in their complexity.[43]

In the 1970s, the research agenda on communist Romania shifted toward the cultural policies of the Ceauşescu regime. Since the end of the 1970s, the study of the communism in Romania was substantially enriched by a new generation of anthropologists who conducted fieldwork in Central and Eastern Europe.[44] These scholars promoted an interdis-

---

40 For an ampler overview, see Iordachi and Dobrincu, "Introduction," in Iordachi and Dobrincu eds, *Transforming Peasants*, esp. 3–8.

41 Popescu-Puţuri, et al., eds., *Din lupta PCR*; Feneşan, *Dezvoltarea*; Hartia and Dulea, *Constanţa*; Biji, ed., *Dezvoltarea economică*.

42 See the work of sociologist Mihail Cernea, *Sociologia Cooperativei*. For extensive comments on the importance of Cernea's work, mostly in relation to the changing socio-economic role of the extended family in rural areas in communist period, see Jowitt, *New World Disorder*, 33–34, 37.

43 Drawing upon the scholarly tradition created by David Mitrany during the interwar period (see mostly *Land and the Peasant* and *Marx Against*) Henry Roberts authored a highly perceptive book on rural relations—*Rumania*. For other pioneering works on collectivization in Romania, see Jowitt, *Leninist Response*; and Gilberg, "The Costly Experiment."

44 For a comprehensive account of research methodology and the political implications of field research in communist Romania, see Kideckel and Sampson, "Fieldwork in Romania;" Sampson and Kideckel, "Anthropologists Going." For the evolution of anthropological studies of Eastern European societies, see Halpern and Kideckel, "Anthropology of Eastern Europe." For the post-1989 convergence of the agendas of foreign and local Romanian studies, see Iordachi and Trencsényi, "In Search of a Usable Past."

ciplinary research agenda focusing on the agrarian question, social relations in rural communities, and the collectivization of agriculture. David Kideckel's works, in particular, provided path-breaking anthropological perspectives on collectivization combining micro with macro perspectives.[45] Challenging the view that collectivization was a unitary, centralized, and homogeneous process, Kideckel used anthropological theories about "complex societies" in order to map out local power relations and villagers' individual and collective strategies for resisting the socialist regime's policies.

The liberalization of academic studies in Romania after 1989 stimulated research on the history of the communist period.[46] However, in the first stage, the rule of Gheorghe Gheorghiu-Dej enjoyed much less scholarly attention than Ceauşescu's post-1965 regime, the few existing studies on the former period being published relatively recently.[47] In addition, early works on collectivization were hampered by methodological limitations.[48] A first strand of scholarship consisted of regionally focused studies of rural issues, which routinely included references to the history of collectivization.[49] Several valuable edited volumes have also been published, containing documents about the political and legislative aspects of collectivization,[50] or on the local campaigns in specific regions.[51] Gradually, a new, more complex strand of scholarship emerged, both in approach and methodology. For instance, a volume by Alina Mungiu-Pippidi and Gérard Althabe investigated comparatively the history of two villages that represent markedly different "symbolic" case studies in the

---

45 See Kideckel, "Dialectic of Rural;" Kideckel, *Agricultural Cooperativism*; Kideckel, "Socialist Transformation;" Kideckel, *The Solitude of Collectivism*.

46 For synthetic views of the communist period, see Deletant, *România sub regimul*; and mostly Tismăneanu, *Stalinism*. For issue-oriented research, Kligman, *Politics of Duplicity*.

47 For groundbreaking works, see Tismăneanu, *Fantomalui*; Tănase, *Elite şi societate*; and Deletant, *Communist Terror*.

48 Tănăsescu, *Colectivizarea*.

49 Recent scholarship includes Sándor Oláh's study of collectivization and resistance techniques deployed by ethnic Szekler villagers from Valea Homoroadelor (Oláh, *Csendes csatatér*), a short monograph on collectivization in Dobrogea between 1945 and 1957, and a number of Ph.D. dissertations (e.g.: Avram, *Procesul de colectivizare*).

50 Roske, ed., *Dosarul*; Cătănuş and Roske, eds., *Colectivizarea*; Iancu et al., eds., *Colectivizarea*.

51 Damian et al., eds., *Colectivizarea*; Dobeş et al., eds., *Colectivizarea în Maramureş*.

Romanian context—Scorniceşti (Ceauşescu's birthplace) and Nucşoara (a hotbed of armed resistance to the regime during early communism).[52] The history of collectivization has also been touched upon by scholars who have conducted research on the restitution of land rights in 1990–91.[53]

To date, the most comprehensive investigation of collectivization in Romania has been undertaken by Gail Kligman and Katherine Verdery in a large-scale, collaborative research project that resulted in several studies.[54] Unlike other works on the subject, these studies encompassed the entire history of the collectivization campaign, between 1949 and 1962, and covered the full territory of Romania. Research concentrated on four major themes elaborated by Kligman and Verdery: 1) the effect of collectivization on property relations; 2) the impact of collectivization on collective and individual identities in the rural world; 3) the creation of the party-state; and 4) the process of shaping and reshaping historical memory. The project set a comparative conceptual framework as well as an interdisciplinary methodological approach that can serve as a model for conducting research on collectivization at a broader regional level.

In sum, recent works on collectivization have undoubtedly added to our understanding of this complex transnational phenomenon. Overall, however, despite the abundance of recent studies on collectivization in East European countries, upon closer scrutiny the academic literature on the topic appears to be rather fragmentary and uneven. With few but notable exceptions, the collectivization of agriculture in Eastern Europe has been generally approached as an appendix of broader social and political transformations and not as a sociopolitical process that is, in itself, worthy of scholarly attention.[55] Moreover, until recently, scholarship has concentrated on the political history of the collectivization campaigns, on the decision-making process at the higher echelons of power, and on the issue of state repression and peasants' "heroic resistance" to collectivization. At the same time, the social and cultural dimensions of collec-

---

52 Mungiu-Pippidi and Althabe, *Secera şi buldozerul*; Mungiu, *Tale of Two Villages*.

53 For an example, see Cartwright, *Return of the Peasant*. For the most comprehensive analysis of decollectivization in post-communist Transylvania and its profound effects on rural communities, see Verdery, *Vanishing Hectare*, 2003.

54 For research results, see Dobrincu and Iordachi, eds., *Ţărănimea şi puterea*; Iordachi and Dobrincu, eds., *Transforming Peasants*; and especially Kligman and Verdery, *Peasants under Siege*, the most comprehensive study on the topic available to date.

55 For a notable exception, see Jowitt, *Leninist Response*; and *The New World Disorder*.

tivization have been largely neglected, despite the fact that this process fundamentally challenged not only entrenched modes of production but also traditional rural values and lifestyles. In addition, scholars have often treated both the party-state and the peasantry as unitary and monolithic actors, thus denying historical agency and subjectivity to individual or collective local actors. While collectivization was described as a "war" between the peasants and the communist regime,[56] few studies have dealt with the differentiated response to collectivization within various historical regions or social strata of the peasantry, ranging from ardent resistance and unequal negotiations to accommodation and collaboration.[57] Similarly, historical scholarship has often underrated conflicts as well as forms of communal solidarity among peasant communities. Moreover, agency on the part of the villagers has not been systematically related to party and state policies.

Most importantly, the overwhelming majority of the available works on the topic are national case studies. Although the transnational character and implications of the post-World War II processes of collectivization in Eastern Europe are evident, the vast majority of published studies have not transcended the national realm. Few comparative investigations of collectivization or explorations of mutual relations, perceptions, and transfers across national borders have been published to date. Moreover, even those volumes that approached collectivization at the regional level comprise national case studies, and lack systematic comparisons.[58] Among the most notable exceptions, one should mention Karl-Eugen Wädekin's early studies,[59] which are, however, mostly focusing on communist agrarian policies and their impact on the overarching economic and social structures, and recent research by Nigel Swain.[60] The lack of accounts of ideological and policy transfers among the communist states with regard to collectivization is even more anomalous if one approaches the process against the wider background of Sovietization. Studies of collectivization in the GDR, for instance, have provided evidence of a cross-border exchange in agrarian policies between the East German communist elites and political leaders of other "people's republics" even before the leaders of the

56 Graziosi, *The Great Soviet Peasant War*; Viola, *Peasant Rebels*, 44, 130, 180, 233, 235; Viola et al., eds., *War Against the Peasantry.*

57 For an exception, see Fitzpatrick, *Stalin's Peasants.*

58 Francisco, Laird, and Laird, eds., *Agricultural Policies.*

59 Wädekin, *Agrarian Policies.*

60 Swain, *Rural Transition.*

SED officially launched collectivization in 1952.[61] Although the importation of the Soviet model of collectivization and the role of the Soviet advisors in its implementation have been repeatedly stressed in various national contexts, the selective appropriation and adaptation of the Soviet experience(s) in the socialist satellite countries has received only limited attention to date. No systematic attempt has yet been made to assess the multiple interactions among these actors and its impact on collectivization policies at various political levels. In the following section, we review recent approaches to comparative history in an effort to highlight the potential heuristic value of the comparative method, and of related transnational approaches, for the study of collectivization in Eastern Europe.

Historical Comparison, Entanglements and Transfers: New Perspectives

The dominance of the "national paradigm" has been a persistent feature in the writing of history in the modern period. In political history as well as in social history, the nation-state has been routinely employed as the prevalent analytical framework. This feature is valid not only for European states like Germany, Italy, or Romania, which had achieved national unity as late as the second half of the nineteenth century and in the early twentieth century, but also for "stable," early centralized states like Britain and France, where the imperialism of the late nineteenth century reinforced nationalism. The collapse of multinational empires in Central and Eastern Europe after World War I and the failure of Wilsonian universalism in its aftermath reaffirmed, rather than weakened, the nation-state paradigm of history writing. Although, as early the late 1920s, the Belgian Henry Pirenne and historians of the French *Annales* school like Marc Bloch and Lucien Febvre had called for the application of comparative approaches to European history, it was mostly after 1945 that the predominant focus on national history was gradually complemented by regional or continental comparative perspectives.[62]

---

[61] Witkowski, "On the Campaign Trail;" Ross, *Constructing Socialism*; Schöne, *Frühling*, 73–153; Bauerkämper, *Ländliche Gesellschaft*, 157–66. On the Soviet influence, see Scherstjanoi, *SED-Agrarpolitik*, 345–424.

[62] Iggers, *Historiography*, 23–35; Iggers, "Nationalism and Historiography," 15–29; Bahners, "National Unification," 57–68.

Historical comparisons can serve multiple analytical purposes. The basic aim of the comparative method is to identify differences and similarities between two or more analytical units. As such historical comparisons underscore general as well as specific features of the particular case studies under scrutiny, leading to a more comprehensive understanding of the particular objects of analysis and their classification into broader typologies. Apart from the wide scope of applicability and purposes, historical comparisons encourage researchers to take a certain analytical distance from their objects of analysis, thus stimulating methodological reflexivity. Comparative studies also enable historians to call established theoretical paradigms into question, thereby generating new research questions and analytical perspectives.[63]

In the last decades, the comparative method has been increasingly challenged by new transnational approaches. Critics of comparative history assert that the comparative method presupposes a high degree of arbitrary source selectivity. In the process of comparing two or more cases with regard to a specific aspect, issue, or dimension, researchers tend to neatly separate their units of analysis. By ignoring their entanglements, scholars thus run the danger of artificially dissociating the historical phenomena under scrutiny from their immediate contexts (the so-called "Galton's problem").[64] Comparative historiography therefore tends to abstract from particular features in an effort to arrive at historical typologies. In addition, the choice of parameters which prefigure comparative studies is usually influenced by theoretical assumptions or considerations beyond the realm of historical scholarship. In asymmetrical comparisons, for example, historians derive specific analytical categories from their main unit of study and apply them to more minor or secondary analytical objects, thus running the danger of distorting the findings of comparative investigations.

Comparisons do not only require selection and abstraction, but they also depend on defining precise criteria that enable historians to identify and ultimately explain similarities and differences.[65] Moreover, even though comparative history enables historians to overcome the narrow confines of national history, numerous comparative investigations have in fact strengthened the national paradigm by reinforcing the tendency

---

[63] For an overview, see Kaelble, *Der historische Vergleich*; Kocka, "Comparison and Beyond," 40; Kaelble, "Die interdisziplinären Debatten;" Eisenberg, "Kulturtransfer als historischer Process."

[64] See Haupt, "Comparative History," 240; Peters, *Comparative Politics*, 42.

[65] Haupt and Kocka, "Comparative History," 25.

to compare nations rather than other units of comparison. Finally, comparisons do not necessarily transcend historiographical parochialism, as many historians employ the comparative method with the sole aim of confirming the "unique nature" or "exceptional development" of their own countries.[66]

More recently, historians have debated the relationship between comparative history and the history of transfers and entanglements. Numerous new studies are explicitly devoted to the interrelations and mutual influences between two or more units of analysis. The concept of "entanglements" highlights "a relational perspective which foregrounds processes of interaction and intermixture in the entangled histories of uneven modernities."[67] Entanglements denote close relations that draw two or more objects so strongly together that they become inextricably interwoven and can no longer be approached as separate analytical objects. This research perspective underscores the fact that, since in historical reality most units of historical comparison cannot be neatly separated, the world should be better understood as a web of interactions, encounters, and exchanges. As Marc Bloch emphasized as early as the 1930s, historical similarities among neighboring and contemporary societies are often due to mutual influences and multilateral exchanges across boundaries.[68] Consequently a clear-cut separation of neighbouring societies and their discussion in isolation from each other is artificial and may ultimately even distort research findings.

We assert that comparative history and transfer studies are complementary rather than incompatible approaches. On the one hand, far from being obsolete, historical comparisons remain an indispensable method in the historian's toolkit. Efforts to identify and explain similarities and differences among units of research cannot be fully supplanted by the studies of transfers and entanglements between those units. Yet historical comparisons need to be combined with investigations of transfers in order to grasp interrelations among intertwined historical phenomena. Studies of transfers might, in fact, further refine the findings of comparative studies,

---

66 On this issue, see Berger, "Comparative history." For a debate on the link between comparative history and the idea of manifest destiny in American studies, see Grew, "The Comparative Weakness of American History;"

67 Randeria, "Entangled Histories;" Randeria, "Geteilte Geschichte;" Lepenies, ed., *Entangled Histories*.

68 Bloch, "Contribution," 123, 125, 130; see also Haupt, "Historische Komparatistik," 146.

as similarities can be the result of transfers or of common historical challenges.[69] On the other hand, studies of cross-border exchanges and interactions need to employ the comparative method in order to identify differences between transmitting and recipient societies, before, during, and after the processes of transmission. Scholars must thus resist the temptation to blur the difference between the two approaches by trying to artificially "subordinate" historical transfers to comparison and vice versa.

THE COLLECTIVIZATION OF AGRICULTURE AND THE COMPARATIVE STUDY OF COMMUNIST DICTATORSHIPS

To date, scholars have proposed various overarching research frameworks in order to approach the history of communist regimes from a comparative perspective, focusing mostly on the generic concepts of totalitarianism, political religions, Stalinization, and Sovietization. Counter-intuitively, despite its generic and thus comparative nature, the concept of totalitarianism, which dominated Western scholarship on communist regimes during most of the Cold War period, has not stimulated sophisticated comparative analyses.[70] First, the assertion of the almighty, totalitarian character of communist states tended to render the work of historians rather superfluous, since in that scheme of interpretation they could only confirm what was *a priori* inferred. Second, historical research on totalitarianism has focused primarily on the comparison between Stalinism and Nazism, the two rival regimes that seem the closest embodiments of an ideal-type of a totalitarian state.[71] At the same time, proponents of the totalitarian approach have largely refrained from differentiating between

[69] Paulmann, "Internationaler Vergleich," 681. For an inclination to subsume transfers and entanglements under the comparative method, see Haupt and Kocka, eds., "Comparative History," 32.

[70] Seminal studies are Arendt, *Origins of Totalitarianism*; Friedrich and Brzezinski, *Totalitarian Dictatorship*; Gleason, *Totalitarianism*; Söllner et al. eds., *Totalitarismus*; Rabinbach, "Moments of Totalitarianism." For an attempt to overcome the limitations of the totalitarian model of interpreting communist regimes, see Fitzpatrick and Geyer, eds., *Beyond Totalitarianism*.

[71] For recent yet distinct approaches to this topic, see Kershaw and Lewin, "Introduction" in Kershaw and Lewin, eds., *Nazism and Stalinism*, 1–25; Geyer and Fitzpatrick, "Introduction," in Geyer and Fitzpatrick, eds., *Beyond Totalitarianism*, 1–37; and Snyder, *Bloodlands*, esp. 1–20, 379–408.

forms or phases of communist dictatorships, emphasizing instead continuities in repressive practices. Although numerous scholars argued that, from the 1960s to the 1980s, there occurred a transition from totalitarian to authoritarian rule in communist regimes, the concrete similarities and differences between distinct phases in the evolution of communist dictatorships in Eastern Europe have not been systematically pursued. Moreover, recent comparative perspectives tend to widen the scope of research rather than deepen it: for example, novel approaches to "political religions" include both fascist and communist regimes in a grand retrospective view of totalitarian movements and regimes in the twentieth century.[72]

The concepts of "Sovietization," "Stalinization," and "de-Stalinization" have proven more prone to comparative research. However, early works on Sovietization tended to treat Eastern European "satellite" regimes as quasi-colonial replicas or "carbon-copies" of the Soviet model, in view of their complete dependency on Moscow.[73] They also focused preponderantly on the Stalinist period, and emphasized almost exclusively forms of political repression and physical violence that accompanied the forceful imposition of the Soviet model. The assumption of a fully planned, monolithic and all-embracing implementation of a Soviet model "from above" has been, however, gradually modified to allow for a more nuanced understanding of the multi-faceted, open-ended and even multi-directional nature of political transfers in Soviet-dominated Eastern Europe.[74] Several novel comparative insights into the nature and dynamics of Sovietization should be highlighted, in particular. First, while the process of Sovietization of Eastern Europe was, without a doubt, a direct outcome of the military occupation of the region and its forceful

---

72 For a classical perspective, see Voegelin, *Political Religions*. For the more recent debates, see Maier, ed., *Totalitarismus*. For a pioneering work on communism as a political religion, see Lane, *The Rites of Rulers*. For comparative perspectives on fascism and communism as political religions, see mostly articles in the journal *Totalitarian Movements and Political Religions*.

73 For early works on Sovietization, see Raditsa, "The Sovietization of the Satellites;" Senn, "The Sovietization;" Fedynskyj, "Sovietization of an Occupied Area;" Staar, *Poland, 1944–1962*; for collectivization as Sovietization, see J. A., "Sovietization of the Czechoslovak Economy."

74 For a case study of early Sovietization within Soviet Russia, see Borys, *The Sovietization of Ukraine*. For regional perspectives, see Jarausch and Siegrist, eds., *Amerikanisierung*; Lemke, *Sowjetisierung*; Mertelsmann, ed., *The Sovietization of the Baltic States*; Apor, Apor, and Rees, *The Sovietization of Eastern Europe*; Tismăneanu, ed., *Stalinism Revisited*. For an issue-oriented comparative perspective on Sovietization, focusing on education, see Connelly, *Captive University*.

political incorporation into the Soviet camp, it should be stressed that the process did not consist solely of coercive but also of normative or even voluntary forms of political emulation: see in this respect attitudes of "working toward Moscow" in order to gain Soviet political or economic support or cases of self-Sovietization (such as Yugoslavia in Southeastern Europe, Cuba in Latin America, etc.) Second, while the forceful imposition of the Soviet model was directly linked with the Soviet imperial domination and was often accompanied by linguistic or cultural Russification, Sovietization also served, in certain contexts, as a form of nation- and state-building (see the interwar process of state-building in Soviet Central Asia, and the Caucasus, the creation of the Soviet Socialist Republic of Moldova in 1924/1940, and of the GDR in 1955). Third, Sovietization did not simply entail the adoption of an ideologically driven legal-institutional model of socio-political organization resulting in the dominance of the dual party–state structure. It was also conceived as an alternative strategy of economic modernization based mainly on collectivization, socialist industrialization, and urbanization. In the socio-cultural sphere, it entailed the adoption of a peculiar way of life and everyday practices leading to the creation of the "new man." This explains why, in the context of the Cold War, the emulation of the Soviet model was regarded, in certain non-European developing regions, in particular, as a potential strategy of modernization and "catching up" with the West by underdeveloped countries. Fourth, while the power relations between the Soviet Union and Eastern European countries were highly unequal, one should not rule out the existence of practices of unequal negotiation, adaptation to local conditions, or innovation in Marxist-Leninist practices in the satellite countries. From this perspective, new approaches to Sovietization no longer assume an unimpeded and uni-directional transfer of ideas, institutions, and practices from the USSR to its "satellites," but take into account a wide-scale convergence of policies between communist dictatorships, which allowed for a multilateral process of learning, alteration and adaptation of the Soviet model, including also transfers from one satellite to another or from them back to the Soviet Union (see, for example, the impact of the Hungarian New Economic Mechanism on *Perestroika*). Fifth, it should also be stressed, in direct connection to the previous point, that the process of Sovietization was not restricted in time to the period of high Stalinism, but continued, in various forms of manifestation, throughout the communist period. Moreover, although directives emanating from the USSR continued to shape the policies of socialist regimes until 1989, the Soviet influence had receded, to a certain extent, from the late 1950s to the 1980s, a fact which allowed a certain degree of domestic autonomy within the communist camp, leading to increasing diversity. Political diversity was further amplified by the fact that, within the umbrella of *Pax Sovietica*, interstate relations among socialist states were often marked by considerable disagreements over major bilateral issues or policy questions.

In the following, we apply these perspectives on Sovietization to the issue of collectivization in Eastern Europe, pointing out novel ways in which to approach this process from a comparative and transnational perspective. The collectivization of agriculture in Eastern Europe is a comparative topic *par excellence*, since it refers to a complex series of events that occurred in several waves spanning the period 1929 to circa 1965. This 36-year long period was characterized by intensive legal, economic, and political struggles in 12 Eastern European countries for more than 21 continuous years, divided into three collectivization campaigns (1929–1935, 1945–1947, 1948–1965), with two political interplays (1935–1945, and 1953–1956). Collectivization campaigns resulted in the socialization of the agricultural production and its subordination to central economic planning.

Table 1: Land Reforms, the Nationalization of Industry, and Collectivization Campaigns in the USSR and Eastern Europe.

Country	**Land reform**	**Nationalization**		**Collectivization**		
		High	**Mass**	**Begins**	**Halted**	**Ends**
USSR	1917	1917	1917	1929	1931	1935/ 1940
Poland	Sept. 1944	1945[74]	1946–8	1948	1956	n.a
Czechoslovakia	June 1945	1945	1948	1949	1953–5	1960
GDR	Fall 1945	1945–8	1972	1952		1960
Hungary	March 1945	1946–7	1948–9	1948	1953–4 1956/8	1961
Romania	March 1945, 1948–9	1946	1948	1949	1953–5	1962

Country	Land reform	Nationalization High	Mass	Collectivization Begins	Halted	Ends
Yugoslavia	August 1945	1944–5	1946–8	1948	1953	n.a
Bulgaria	March 1946	1944	1947	1945–8	1953–5	1959
Albania	August 1945	1944–5	1946–7	1946–48	1953–7	1967

Source: Chronologies in this volume; Swain and Swain, *Eastern Europe Since 1945*, 120–121.

The first experiment in collectivization took place in the Soviet Union, where the process began in 1929 and ended around 1935. The second wave of collectivization campaigns occurred in the Soviet occupied territories during World War II, namely in the Baltic States (Lithuania, Latvia, Estonia), Moldova, and the eastern territories of Poland in the immediate postwar period (1945–47). The third and most complex wave, from 1948 to the mid-1960s, transformed the agriculture of Soviet satellite countries in Central and Southeastern Europe. Similar campaigns took place outside Eastern Europe, too, in China, Cuba, Southeast Asia, and various African countries.[76]

Investigating these series of campaigns poses considerable challenges to historians, due to their complexities, their multifaceted and entangled nature, and their extensive short- and long-term impact on society and politics. To provide a comprehensive account of the history of the collectivization campaigns in Eastern Europe, one needs to creatively employ the comparative method and supplement it with novel forms of cross-histories. Several forms of comparison are possible, as a function of the

[75] Banks were nationalized in 1948–1949.

[76] For a sample of recent scholarship on some of these cases, see the following works on collectivization and decollectivization in Vietnam: Raymond, "No Responsibility;" "The Insoluble Internal Conflicts." On collectivization in China, see Doak, "China's Road;" Lin, "Collectivization and China's Agricultueal Crisis;" Liu, "Why Did It Go;" Kueh, "Mao and Agriculture;" On the 1958–62 Great Famine in China, see Dikötter, *Mao's Great Famine*; Zhou, *The Great Famine*. For a review of the recent literature on China, see Diamond, "Rural Collectivization."

research aims of the respective investigation and the particular theoretical framework employed in a given historical investigation.

Since collectivization was not pursued simultaneously by communist rulers, a first form of comparison is diachronic, examining the process of collectivization in the Soviet Union in the interwar period, on the one hand, and similar campaigns of collectivization that occurred in post-1945 Eastern Europe, on the other. This form of comparison raises a number of analytical questions: Was there a Soviet model of collectivization? In other words, can one reify the complex history of Soviet collectivization, marked by numerous policy shifts and turns, into a comprehensive pattern? If yes, what were its main legal, economic, and political features? How was this model understood and presented by Soviet elites? How did the communist leaders in East-Central Europe perceive it? Another key diachronic comparative research issue is to test the implementation of the Soviet model in Central and Southeastern Europe. What were the similarities and differences between the interwar process of collectivization in the USSR, the immediate postwar collectivization of annexed territories, and the Soviet-inspired post-1948 campaigns in Eastern Europe? What was the role of the Soviet advisors dispatched to Eastern Europe in this process? How was the Soviet model adapted to local conditions? Were there alternative or even rival models of collectivization developed in Eastern Europe, either building on or rejecting the Soviet pattern?

A second possible form of comparison is synchronic, between the various case studies of collectivization in post-1945 Eastern and Southeastern Europe at intra- or inter-regional levels. Why did the collectivization campaign last for less than 10 years in GDR and Bulgaria but over 15 years in Hungary, Romania, and Albania? Why was the process initiated with vigor but then abandoned in Yugoslavia in 1953 and in Poland in 1956? On which conditions did the success or failure of collectivization campaigns in these countries depend? Can one establish a typology of various forms of communist agricultural transformation? Third, another avenue of comparison concerns the outcome of collectivization and the nature of the socialist agriculture and post-socialist transformation. Was collectivization an economic success or a failure? What was its impact on the economic development of communist countries? What is the relation between forms of collectivization, socialist agricultural systems, and patterns of decollectivization? Moreover, what is the legacy of agricultural transformation under state-socialism in post-communist economies and societies? Finally, in order to understand what was specific to collectivization in Eastern Europe, the scope of the comparative studies of collectivization can be expanded chronologically as well as geographically, to include processes of collectivization in other socialist countries around the world, such as North Korea, China, or Cuba.[77]

These multiple forms of comparison are all equally valid; their employment depends on the analytical aims set by particular researchers, and the overall theoretical framework they adopt. The current volume tests some of these analytical strategies, in an attempt to contribute to the comparative study of patterns of collectivization in Eastern Europe.

AIMS, STRUCTURE, AND THE ORGANIZATION OF THE VOLUME

In order to fill some of the existing lacunae in the academic literature, this volume brings together 15 leading Eastern European, Western European, and North American experts in the history of agriculture and rural societies in communist Eastern Europe, and engages them in a critical rethinking of the history and societal impact of the processes of collectivization. The volume grew out of a research project initiated and coordinated by Arnd Bauerkämper (Freie Universität Berlin) and Constantin Iordachi (Department of History, Central European University, Budapest). Within the framework of the project, two international conferences on collectivization were organized in Berlin and Budapest, funded by the Stiftung zur Aufarbeitung der SED-Diktatur, Berlin, Germany.

Naturally, given the complex and multi-faced nature of the subject-matter under discussion, the volume does not aim to provide an exhaustive analysis of the history of collectivization in Eastern Europe. Instead, the studies included in this collection focus primarily on the interaction between the process of Sovietization on the one hand and forms of national path dependency on the other. They shed light on the timing and main features of collectivization at national and regional levels. While emphasizing the asymmetry of power relations between the USSR and its new "allies," the contributions to the volume conceive the collectivization of agriculture as a complex process integrating exogenous influences and pressures as well as indigenous traditions and actors. In doing so, they call

[77] For comparisons between collectivization and de-collectivization in the USSR/Russian Federation and China, see W. K., "Soviet Agriculture;" Ka-kui, "Agricultural collectivization;" Selden, "Pathways from Collectivization;" Nam, *Contemporary Food Shortage.*

into question established notions of Sovietization as a one-way process of legal-political transfer from the center to the periphery.

The volume is made up of an introduction and 14 research chapters, organized into four parts. These chapters combine in-depth analyses of national case studies with issue-oriented comparative essays, followed by a concluding comparative synthesis. To facilitate the readers' understanding of the complexities of the collectivization campaigns, each chapter focusing on a national case study is followed by a brief chronology of events.

The first part of the volume focuses on the collectivization of agriculture in the USSR in the interwar period and in the Soviet-occupied territories during the war and in the immediate postwar years. In the first chapter, Lynne Viola reviews the history and main features of collectivization in Stalinist Russia. Viola explores the legacy of the Russian agrarian traditions and their impact on shaping the agricultural transformation in the USSR, emphasizing the role of collectivization in the processes of Soviet state-building and social-economic modernization. On this basis, Viola highlights the peculiarities of collectivization in the Soviet Union, and tests its possible characterization as an "experiment" in social engineering, a "model," or a "colonizing venture."

The collectivization of agriculture in Eastern Europe had been preceded by a wave of collectivization in the provinces annexed by the USSR during World War II, namely the Baltic States, Moldova, and Western Ukraine. This first exercise in extending Soviet-style collectivization in Eastern Europe was also important in setting a precedent and a standard to be emulated outside the USSR. In the second chapter, David Feest (University of Göttingen) proposes a comprehensive comparative analysis of the processes of collectivization in Soviet-occupied Estonia, Lithuania, and Latvia. Feest addresses two main analytical questions in relation to the history of collectivization in the Baltic States: Why was agriculture not collectivized immediately after these countries' incorporation into the Soviet Union in July 1940 or after their reoccupation in 1944? How can the rush and pressure to collectivize in 1948–50 be explained? Feest argues that the timing of collectivization in the Baltic States cannot be understood without considering the organizational problems that the Soviet rulers faced in their attempts to subjugate the small republics but also geopolitical considerations. While there were pragmatic political reasons preventing a rush to collectivization in the first years after seizing power, by 1948 domestic conditions as well as international relations had changed so drastically that the Soviets were able to push for forced collectivization. Feest argues that the Soviets approached collectivization in the newly-annexed territories in view of their own experience of the late 1920s and early 1930s, when agriculture was collectivized by mass terror and deportation. He concludes that although collectivization had economically proved to be a failure in the old Soviet republics, it was nevertheless implemented in the Baltic States because it fulfilled important political functions, paving the way for a firmer incorporation of these countries within the Soviet system.

Feest's findings underscore the similarities and differences between the process of collectivization in the Soviet-occupied Baltic States and the other communist countries in Eastern Europe. In the Baltic States, just as in the prewar Soviet Union, Stalin ultimately opted for the mass deportation of *kulaks*, a strategy that was not implemented outside USSR. Aside from this major difference, the collectivization of agriculture in the Baltic States nevertheless shares many common features with other Eastern European countries, relating to the peasants' tradition of private land ownership and attachment to the land, their stubborn if often tacit resistance against collectivization and non-cooperative attitude inside collective farms, and the difficulties faced by authorities in setting up and consolidating viable collectives, especially in the early stage of collectivization.

The second part of the volume deals with processes of collectivization in Central Europe, with chapters devoted to Poland, the GDR, Czechoslovakia, and Hungary. Dariusz Jarosz focuses on the specificity of collectivization in Poland. Why did the process ultimately fail? Why was the collectivization campaign stopped before completion? Jarosz rejects the interpretation that the abandonment of full-scale collectivization in Poland was simply due to the heroic resistance of the peasantry against collectivization, a thesis that would also implicitely but erroneously suggest that peasants were more lenient in joining collective farms in other countries in the region. Instead, Jarosz emphasizes the ambivalent attitude of the Polish elites toward collectivization and their relationship to Moscow, drawing particular attention to Gomułka's views on the issue. Jarosz also underscores the important role played by religion and the Catholic Church in Polish society and politics, and the link between the policies of colonization and national consolidation in the newly-acquired provinces of Silesia and Pomerania.

Jens Schöne focuses on the role of ideology and of asymmetrical entanglements in the collectivization of agriculture in the German Democratic Republic. The chapter starts with a short overview of the process of collectivization, reviewing the basic ideological tenets of Marxist-Leninist agrarian theories and the steps taken by the Socialist Unity Party toward collectivizing agriculture in 1952–53 and 1958–60. It then explores the complex and ultimately asymmetric power structure that defined the relationship between the GDR and the Soviet Union. The main research questions posed by Schöne are: Was there a unique GDR-version of collectivization? How far did the autonomy of the SED actually go? In response to these questions, Schöne concludes that, while the relationship between the GDR and the Soviet Union was one-sided, East German party leaders were nevertheless able, to some extent, to modify the collectivization program in response to endemic local circumstances and demands. Finally, Schöne argues that developments in other socialist countries influenced the ways in which the SED implemented its own agrarian program, as there is evidence that the GDR emulated agrarian policies applied in neighboring states during key moments of transformation.

Jan Rychlík details the general and specific aspects of collectivization in Czechoslovakia, the most urbanized and industrialized country in Eastern Europe. How important was this factor in the collectivization campaign? Rychlík presents a historical overview that takes into account the long tradition of state farms and cooperatives in the Czech and Slovak lands. He also addresses regional differences in the economic development of various regions in Czechoslovakia, such as between Bohemia and Slovakia, and their impact on collectivization at the national level. In his conclusions, Rychlík reviews the main features of the collectivization campaign in Czechoslovakia as compared to other Eastern European countries. He also elaborates on the role of ethnicity in the process of collectivization, in view of the specific experience of the Hungarian-dominated areas in southern Slovakia.

József Ö. Kovács explores the process of collectivization in Hungary, identifying the actors involved in the campaign and delineating its main waves and stages. He accounts for major shifts and turns in the agrarian policy of the Hungarian Communist Party, and assesses the impact of the 1956 revolution on collectivization. Using the methodological tools of social and oral history, and based on a wealth of primary sources—most notably interviews with peasants that provide eyewitness accounts of the agricultural upheaval—Kovács concludes that collectivization profoundly shaped Hungarian society, transforming traditional peasant values and social attitudes.

Part three of the volume concentrates on a cluster of countries from Southeastern Europe, with chapters on Romania, Yugoslavia, Bulgaria, and Albania. Constantin Iordachi and Dorin Dobrincu provide an overview of the collectivization of agriculture in Romania. Based on ample archival resources, Iordachi and Dobrincu advance several hypotheses as to why that process was so long and protracted. Among crucial factors are the size and economic importance of the agricultural sector, the weakness of the Communist Party membership in rural areas, its lack of political experience and legitimacy, and the political rivalry among various leading factions within the party. The case of Romania provides important avenues of regional comparison, mostly with Poland, in view of these countries' size and the importance of the agricultural sector. Why did collectivization succeed in Romania whereas it was abandoned in Poland? Was it because, unlike the Polish communists, the Romanian communist leaders were willing to pay the political price of subjugating the peasantry in order to consolidate their rule and embark on the project of industrialization leading to autarchy? The second possible comparative perspective is between Romania and Albania, two of the least industrialized countries in the Eastern Bloc. In both states, the Stalinist leadership survived attempts at political reforms. Moreover, both Romania and Albania resisted Soviet attempts to institute a division of labor in the communist camp and pushed for forced industrialization. From this perspective, collectivization appears as having a paramount political and economic importance for the evolution of communist regimes in both countries.

Melissa Bokovoy highlights the interactions between officials and peasants in Yugoslavia and the internal and external reasons for abandoning collectivization. Was there a Yugoslav model of collectivization distinct from the Soviet model? Combining approaches pertaining to political and social history, Bokovoy argues that the clashes but also interactions between the central party leadership and peasant culture led to distinguishable "learning curves." She stresses the role of peasants as social actors (as distinct from mere objects of political rule), and the impact of monitoring commissions and agricultural specialists in the collectivization campaign.

In his chapter on Bulgaria, Mihail Gruev analyzes the legacy of the pre-1944 agrarian system, the main stages of collectivization, the policy tools available to the communist party-state in the village, and the long-lasting consequences of the collectivization campaign, which is regarded as "the most radical and all-embracing social cataclysm in modern Bulgarian history." According to Gruev, the Bulgarian campaign can be divided into three subsequent stages: the "first massovization" of 1948–51, which resulted in large-scale peasant resistance, especially in northwestern Bulgaria; a second stage of relative stalemate, between 1951 and 1956, centered on the stabilization of the existing cooperative farm; and the "second massovization," between 1956 and 1959, marked by political and social oppression, which finalized the completion of collectivization in mountainous and peripheral regions as well. Gruev argues that these stages of the collectivization campaign in Bulgaria roughly mirrored the Stalinist campaign of the 1920s and '30s. In his conclusions, Gruev stresses that, similar to Albania, collectivization in Bulgaria was accomplished rather rapidly and that the Bulgarian agrarian model established in 1950s did not undergo substantial changes, as happened, for example, in Hungary.

Örjan Sjöberg explores the history of collectivization in Albania. Apart from its protracted nature, the collectivization of agriculture in this mountainous Balkan state was deeply affected by a wide variety of factors, such as geographic conditions, considerable population growth, the program of large scale population relocation in the villages during the 1950s, and the regime's political emphasis on autarchy. Sjöberg identifies four major waves of collectivization in Albania from the late 1940s to the early 1980s. He relates his chronological overview of collectivization to specific types of collectives, pointing out that Albania largely rejected de-Stalinization. Sjöberg explains the ambivalent impact of Soviet collectivization on agrarian relations in Albania, arguing that the policy of autarchy and the legacy of deeply rooted semi-feudal agrarian traditions explain the specific features of collectivization in Albania.

Part four of the volume focuses on axes of differentiation during these processes of collectivization, incorporating research results of these national case studies. The contributions deal with issues pertaining to social conflicts, ethnic cleavages, or changing relations between center and periphery in the new communist economic and social-political systems in the making. Zsuzsanna Varga (ELTE University, Budapest) provides an issue-oriented analysis of the last stage of collectivization in Hungary between 1957 and 1962. Her contribution concentrates on the changing relationship between the communist regime and rural society after the revolution of 1956. Varga points to the Communist Party's accumulation of expertise and change of policy in collectivization, which she calls a "process of learning." On this basis, she advances an important argument, claiming that the roots of the Hungarian idiosyncratic but highly successful socialist agricultural system can be traced back to the communists' change of agrarian policy after 1956.

Arnd Bauerkämper focuses on collectivization as social practice, in view of the historical narratives and competing memories as sources of agency in the transformation of agriculture in the GDR. His research offers new insights into the multifaceted relationship between memories, remembrance, and party propaganda. On the one hand, the SED leadership managed to successfully utilize collective memories in order to promote their political project of collectivization. On the other hand, narratives of the past that peasants and farmers advanced in the villages ran counter to the official memory politics and thus proved to be a source of resistance to collectivization. In his conclusion, Bauerkämper emphasizes the dynamics of competing versions of history in the forced merger of farmers and peasants, highlighting the impact of different forms of collective memory on the social conflicts accompanying collectivization.

Gregory R. Witkowski investigates the spatial dimension of collectivization, exploring the dynamic interrelationship between the center and periphery in the collectivization campaigns in the GDR, Poland, Bulgaria, and Romania. Witkowski links the process to social dislocation as well as migration from peripheral areas. He argues that the rural population adopted and appropriated central initiatives and directives according to particular needs and the local conditions. The concepts of center and periphery require theoretical underpinning: apart from the geographic divide between the capitals and the hinterlands, they refer to the metaphorical gap between local and national identities and interests. Witkowski also emphasizes the role of local officials as mediators between center and periphery, the impact of regional traditions on the collectivization processes, and the ways in which the dynamics of collectivization repeatedly changed the relationship between centers and peripheries. According to Witkowski, these shifts accentuated the processes of centralization or decentralization.

In the final chapter, Nigel Swain provides one of the most comprehensive comparative treatments of post-1945 collectivization processes in Eastern Europe available to date. Swain's aim is to identify the common features imposed by Stalin's model and its subsequent changes, the path-dependencies that each country faced in its efforts to adopt a socialist strategy of development, and the entanglements between the Soviet Union and the countries that embarked on the collectivization course. The first part of the chapter focuses on a chronology for the implementation of Stalinist collectivization in the region and the changes that took place after Stalin's death. The second part reviews the profiles of the national case studies, pointing to the variants of collectivization adopted in various Eastern European countries and the resulting deviations from the Soviet model of collectivization. The comparison proceeds along certain axes, including postwar land policies and resettlements, the context (political and material) in which collectivization was implemented, the manner in which policies towards *kulaks* were applied, the pace and mode of collectivization, and the final outcome as reflected in the proportion of collectivized land out of total land ownership, the economic evolution of the socialist sector in agriculture, and its relation to the private sector.

Overall, the volume offers a complex comparative overview of collectivization in Eastern Europe, covering the changing legal framework governing property relations during the collectivization campaign, the political dimensions of the decision-making process, the implementation of the principle of "class struggle" in the countryside, and policies towards and impacts on various strata of the peasantry. It also provides new insights into the perceptions and (selective) appropriations of communist agricultural policies among peasants.

The volume opens up new cross-border perspectives on the collectivization of agriculture in Eastern Europe. As emphasized above, this analytical approach requires historical comparisons, but it also necessitates explorations of legal and political transfers and entanglements among Eastern European countries. By assessing the transfer of the Soviet model of collectivization, its practical implementation, and its outcomes, the volume contributes to the reconceptualization of the relationship between path dependency and Sovietization in communist dictatorships in Eastern Europe.[78] We emphasize that collectivization campaigns were shaped by social and ethnic cleavages and conflicts. Exploring the policy of ethnic cleansing and war retribution inside the Socialist bloc, the volume highlights the strong interdependence between class divisions, forced migration, and ethnic minority questions.

While most contributions in this volume utilize historical approaches, the research supersedes disciplinary boundaries in many ways, by combining state-centered interpretations with grassroots perspectives and by integrating various sociological, anthropological, and political science approaches. In doing so, the volume interrelates the structural aspects of the collectivization campaigns with the diverse life experiences of the peasantry. What was the impact of collectivization on everyday lives in rural communities? How was this process perceived by the peasantry? As the contributions in this volume show, perceptions and experiences varied widely, as a function of the particular history of collectivization in a given village and also between males and females as well as between different generations.[79] However, the creation of collective farms through violence and force and the ensuing abolition of private property were largely perceived as a loss of communitarian autonomy and individual independence. As a result, traditional

---

[78] See Schöpflin, "Political Traditions."

[79] See, for instance, Verdery and Humphrey, *Property in Question*; Verdery, *Transylvanian Villagers*; Verdery, *The Vanishing Hectare*; Thelen, "Violence and social discontinuity."

social inequalities and status-based hierarchies lost their foundation, and the deeply entrenched traditional values and established networks of social relations between villagers were severed. At the same time, new opportunities for social mobility within the emerging socialist socioeconomic systems emerged. Peasants were able to ascend the social ladder and achieve leading positions within the collective farms and agro-industrial compounds. Moreover, members of the collective farms benefited from regular working hours, holidays, improvements in housing, social services, and diverse types of financial subventions. Ultimately, however, collectivization should not be assessed by quantifiying its economic costs, achievements and failures, but mostly by taking into account its profound, long-term impact on communist and post-communist societies and politics.

It is our hope that the research provided in this volume will encourage further comparative studies of communist regimes in their economic, social, political, and cultural dimensions. Although this volume focuses specifically on the collectivization of agriculture in Eastern Europe, its approach and research agenda are potentially relevant for similar comparative studies outside Europe, such as in Cuba, Southeast Asia, and various African countries.[80]

BIBLIOGRAPHY

A., J. "Sovietization of the Czechoslovak Economy: The Effects in Agriculture." *The World Today* 9, no. 4 (April 1953): 174–84.

Allen, Robert C. "Imposition et mobilisation du surplus agricole a l'époque stalinienne," *Annales. Histoire, Sciences Sociales* 53, no. 3 (May–June 1998): 569–95.

Apor, Balázs, Péter Apor, and E. A. Rees, eds. *The Sovietization of Eastern Europe: New Perspectives on the Postwar Period.* Washington, DC: New Academia Pub, 2008.

Arendt, Hannah. *The Origins of Totalitarianism.* New York: Harcourt Brace & Company, 1951.

Avram, Cezar. *Procesul de colectivizarea agriculturii în Regiunea Oltenia, 1949–1962* [The process of the collectivization of agriculture in Oltenia Region, 1949–1962]. Ph.D. dissertation, University of Bucharest, 1998.

Bahners, Patrick. "National unification and narrative unity: the case of Ranke's German History" in Stefan Berger, Mark Donovan, and Kevin Passmore, eds. *Writing National Histories: Western Europe since 1800.* London: Bloomsbury Academic, 1999, 57–68.

80 For comparative perspectives, see Kim, "Decision-Making Process;" Klein, "Technologische Innovation."

Balassa, Béla A. "Collectivization in Hungarian Agriculture." *Journal of Farm Economics* 42, no. 1 (Feb. 1960): 35–51.

Barnett, Doak, A. "China's Road to Collectivization." *Journal of Farm Economics* 35, no. 2 (May 1953): 188–202.

Bauerkämper, Arnd. "Vertreibung als Exklusion gesellschaftlicher Führungsgruppen. Die Verdrängung der 'Großbauern' in der SBZ/DDR und die Vernichtung der 'Kulaken' in der UdSSR im Vergleich" in Günther Schulz, ed. *Vertriebene Eliten. Vertreibung und Verfolgung von Führungsschichten im 20. Jahrhundert.* Munich: R. Oldenbourg Verlag, 2001, 125–63.

—. "Collectivization and Memory: Views of the Past and the Transformation of Rural Society in the GDR from 1952 to the Early 1960s." *German Studies Review* 25 (2002): 213–25.

—. *Ländliche Gesellschaft in der kommunistischen Diktatur. Zwangsmodernisierung und Tradition in Brandenburg 1945–1963.* Cologne: Böhlau, 2002.

—. "'Von der Polithistorie zur Sozialgeschichte.' Die Historiographie zur Agrarwirtschaft und ländlichen Gesellschaft in der SBZ/DDR" in Ernst Bruckmüller, et al. eds. *Agrargeschichte schreiben. Traditionen und Innovationen im internationalen Vergleich.* Innsbruck: Studien Verlag, 2004, 63–77.

—. "The Industrialization of Agriculture and its Consequences for the Natural Environment: An Inter-German Comparative Perspective." *Historical Social Research* 29, no. 3 (2004): 124–49.

—. "Europe as Social Practice: Towards an Interactive Approach to Modern European History." *East Central Europe* 36 (2009): 20–36.

—. "Wege zur europäischen Geschichte. Erträge und Perspektiven der vergleichs- und transfergeschichtlichen Forschung," in Agnes Arndt, et al. eds. *Vergleichen, verflechten, verwirren? Europäische Geschichtsschreibung zwischen Theorie und Praxis.* Göttingen: Vandenhoeck & Ruprecht, 2011, 33–60.

Beer, Daniel. "Modernity and Resistance in the Historiography of Stalinism." *Journal of Contemporary History* 40, no. 2 (2005): 363–79.

Berger, Stefan, Mark Donovan, and Kevin Passmore, eds. *Writing National Histories: Western Europe since 1800.* London: Bloomsbury Academic, 1999.

Bezemer, Dirk J. "De-Collectivization in Czech and Slovak Agriculture: An Institutional Explanation." *Journal of Economic Issues* 36, no. 3 (Sep. 2002): 723–45.

Biji, M., ed. *Dezvoltarea economică a României. 1944–1964* [The economic development of Romania, 1944–1964]. Bucharest: Editura Academiei Republicii Populare Române, 1965.

Bloch, Marc. "A Contribution Toward a Comparative History of European Societies" in *Land and Work in Mediaeval Europe: Selected Papers by Marc Bloch.* London: Routledge, 1967, 137–168. Originally published in French: "Pour une histoire comparée des sociétés européennes." *Revue de Synthèse* 46 (1928): 15–50.

Bokovoy, Melissa. *Peasants and Communists. Politics and Ideology in the Yugoslav Countryside, 1941–1953.* Pittsburgh: Pittsburgh University Press, 1998.

Borys, Jurij. *The Sovietization of Ukraine, 1917–1923: The Communist Doctrine and Practice of National Self-Determination.* Edmonton: Canadian Institute of Ukrainian Studies, 1980.

Brainard, Guy Peters. *Comparative Politics: Theory and Methods.* Basingstoke: Palgrave Macmillan, 1998.

Buechler, Hans C. and Judith-Maria Buechler. *Contesting Agriculture: Cooperativism and Privatization in the New Eastern Germany.* Albany: State University of New York Press, 2002.

C., J. "The Peasant in Poland Today: Reactions to Land Reform and Collectivization." *The World Today* 8, no. 2 (Feb. 1952): 71–79.

Calvin, Patricia. "Defining Transnationalism." *Contemporary European History* 14 (2005): 421–39.

Cartwright, Andrew L. *The Return of the Peasant. Land Reform in Post-Communist Romania.* Aldershot: Ashgate, 2001.

Cătănuş, Dan and Octavian Roske, eds. *Colectivizarea agriculturii în România. Dimensiunea politică,* vol. I: *1949–1953* [The collectivization of agriculture in Romania. The political dimension]. Bucharest: Institutul Naţional pentru Studiul Totalitarismului, 2000.

Cernea, Mihail. *Sociologia Cooperativei Agricole* [The Sociology of Agricultural Cooperative]. Bucharest: Editura Academiei Române, 1974.

Connelly, John. *Captive University: The Sovietization of East German, Czech, and Polish Higher Education, 1945–1956.* Chapel Hill: University of North Carolina Press, 2000.

Conquest, Robert, Stephen Cohen and Stephen G. Wheatcroft. "New Demographic Evidence on Excess Collectivization Deaths: Further Comments on Wheatcroft, Rosefielde, Anderson and Silver." *Slavic Review* 45, no. 2 (Summer 1986): 295–99.

Conquest, Robert. *The Harvest of Sorrow: Soviet Collectivization and the Terror-Famine.* Oxford: Oxford University Press, 1986.

Coombe, Rosemary. *The Cultural Life of Intellectual Properties.* Durham: Duke University Press, 1998.

Creed, Gerald. *Domesticating Revolution: From Socialist Reform to Ambivalent Transition in a Bulgarian Village.* University Park, PA: Penn State University Press, 1998.

Damian, Anca, Florin Breazu, and Ion Bălan, eds. *Colectivizarea în Vlaşca, 1949–1950. Documente* [Collectivization in Vlaşca, 1949–1950. Documents]. Bucharest: Vinea, 2002.

Davies, R. W. *The Socialist Offensive: The Collectivization of Soviet Agriculture, 1929–1930.* London: Macmillan, 1980.

—. *Crisis and Progress in the Soviet Economy, 1931–1933.* London: Macmillan, 1996.

Davies, R. W., and Stephen G. Wheatcroft, eds. *The Years of Hunger: Soviet Agriculture, 1931–1933.* New York: Palgrave Macmillan, 2003.

Deletant, Dennis. *România sub regimul communist.* Bucharest: Fundaţia Academia Civică, 1997.

—. *Communist Terror in Romania: Gheorghiu-Dej and the Police State, 1948–1965.* New York: St. Martin's Press, 1999.

Diamond, Norma. "Rural Collectivization and Decollectivization in China—A Review Article." *The Journal of Asian Studies* 44, no. 4 (Aug. 1985): 785–92.

Dikötter, Frank. *Mao's Great Famine: The History of China's Most Devastating Catastrophe, 1958–1962.* New York: Walker & Co, 2010.

Dobeş, Andrea, Gheorghe Mihai Bârlea, and Robert Fürtos, eds. *Colectivizarea în Maramureş. Contribuţii documentare,* vol. I. [Collectivization in Maramureş. Documentary contributions, vol I.]. Bucharest: Fundaţia Academia Civică, 2004.

Dobieszewski, Adolf. *Kolektywizacja Wsi Polskiej, 1948–1956* [Collectivization of the Polish Countryside 1948–1956]. Warsaw: Fundacja im. Kazimierza Kelles-Krauza, 1993.

Dobrincu, Dorin and Constantin Iordachi, eds. *Țărănimea și puterea: Procesul de colectivizare a agriculturii în România, 1949–1962* [The Peasantry and the Power. The Process of Collectivization of Agriculture in Romania (1949–1962)]. Iași: Polirom, 2005.

Dolot, Miron. *Execution by Hunger: The Hidden Holocaust.* New York: W. W. Norton 1985.

Duchhardt, Heinz and Andreas Kunz. "Was heißt und zu welchem Ende betreibt man—europäische Geschichte?" in Heinz Duchhardt et al., eds. *"Europäische Geschichte" als historiographisches Problem.* Mainz: Zabern, 1997, 191–202.

Economakis, Evel. "Soviet Interpretations of Collectivization." *The Slavonic and East European Review* 69, no. 2 (April 1991): 257–81.

Eisenberg, Christiane. "Kulturtransfer als historischer Prozess. Ein Beitrag zur Komparatistik" in Hartmut Kaelble and Jürgen Schriewer, eds. *Vergleich und Transfer. Komparatistik in den Sozial-, Geschichts- und Kulturwissenschaften.* Frankfurt am Main: Campus, 2003, 399–417.

Fallenbuchl, Z. M. "Collectivization and Economic Development." *The Canadian Journal of Economics and Political Science/Revue canadienne d'Economique et de Science politique* 33, no. 1 (Feb. 1967): 1–15.

Fedynskyj, Jurij. "Sovietization of an Occupied Area Through the Medium of the Courts (Northern Bukovina)." *American Slavic and East European Review* 12, no. 1 (Feb. 1953): 44–56.

Feneșan, O. *Dezvoltarea agriculturii în Republica Populară Română.* Bucharest: Editura Agro-Silvică de Stat, 1958.

Fitzpatrick, Sheila. *Stalin's Peasants: Resistance and Survival in the Russian Village After the Collectivization.* Oxford: Oxford University Press, 1994.

—. *Everyday Stalinism. Ordinary Lives in Extraordinary Times: Soviet Russia in the 1930s.* Oxford: Oxford University Press, 1999.

— and Michael Geyer, eds. *Beyond Totalitarianism: Stalinism and Nazism Compared.* Cambridge: Cambridge University Press, 2009.

Francisco, Roland A., Betty A. Laird, and Roy D. Laird, eds. *Agricultural Policies in the USSR and Eastern Europe.* Boulder: Westview Press, 1980.

Frevert, Ute. "Die Europäizität Ostmitteleuropas." *Jahrbuch für Europäische Geschichte* 4 (2003): 189–214.

—. "Europeanizing German History." *German Historical Institute Bulletin* 36 (2005): 9–24.

Friedrich, Carl J. and Zbigniew Brzezinski. *Totalitarian Dictatorship and Autocracy.* New York: Praeger, 1965.

Getty, J. Arch, et al. "Victims of the Soviet Penal System in the Prewar Years: A First Approach on the Basis of Archival Evidence." *American Historical Review* 98 (1993): 1017–49.

Geyer, Michael and Sheila Fitzpatrick, eds. *Beyond Totalitarianism. Stalinism and Nazism Compared.* Cambridge: Cambridge University Press, 2009.

Gilberg, Trond. "The Costly Experiment: Collectivization of Romanian Agriculture" in R. A. Francisco, B. A. Laird, and R. D. Laird, eds. *The Political Economy of Collectivized Agriculture.* New York: Pergamon Press, 1979, 23–62.

Girnius, Kestutis K. "The Collectivization of Lithuanian Agriculture, 1944–1950." *Soviet Studies* 40, no. 3 (1988): 460–78.

Gleason, Abbott. *Totalitarianism: The Inner History of the Cold War.* Oxford: Oxford University Press, 1995.

Graziosi, Andrea. *The Great Soviet Peasant War: Bolsheviks and Peasants, 1917–1933.* Cambridge, Mass.: Harvard University Press, 1996.

Grew, Raymond. "The Comparative Weakness of American History," *Journal of Interdisciplinary History* 16, no. 1 (Summer, 1985): 87–101.

Gribincea, Mihai. *Agricultural Collectivization in Moldavia. Bessarabia During Stalinism 1944–1950.* Boulder: Westview Press, 1996.

Gruev, Michail. *Preorani slogove. Kolektivizaciya i socialna promyana v Bulgarskiya severozapad, 40-te–50-te godini na XX vek* [Re-ploughed Boundaries: Collectivization and Social Change in Bulgarian Northwest, 1940s–1950s]. Sofia: Ciela, 2009.

Halpern, Joel Martin and David A. Kideckel. "Anthropology of Eastern Europe." *Annual Review of Anthropology* 12 (1983): 377–402.

Hartia, S. and M. Dulea. *Constanţa, prima regiune colectivizată* [Constanţa, the first collectivized region]. Bucharest: Editura Politică, 1960.

Haupt, Heinz-Gerhard. "Comparative History" in Neil J. Smelser and Paul B. Baltes, eds. *International Encyclopedia of the Social and Behavioral Sciences,* Vol. 4. Amsterdam: Elseviert, 2001, 2397–403.

—. "Die Geschichte Europas als vergleichende Geschichtsschreibung." *Comparativ* 14 (2004): 85–97.

— and Jürgen Kocka. "Comparative History: Methods, Aims, Problems" in Deborah Cohen and Maura O'Connor, eds. *Comparison and History: Europe in Cross-National Perspective.* New York: Routledge, 2004, 23–39.

—. "Historische Komparatistik in der internationalen Geschichts schreibung" in Gunilla Budde et al., eds. *Transnationale Geschichte. Themen, Tendenzen und Theorien.* Göttingen: Vandenhoeck & Ruprecht, 2006, 137–49.

Heydemann, Günther and Heinrich Oberreuter, eds. *Diktaturen in Deutschland—Vergleichsaspekte. Strukturen, Institutionen und Verhaltensweisen.* Bonn: Bundeszentrale für Politische Bildung, 2003.

Hindus, Maurice Gerschon. *Red Bread: Collectivization in a Russian Village.* Bloomington: Indiana University Press, 1988.

Hollos, Marida. "Ideology and Economics: Cooperative Organizations and Attitudes toward Collectivization in Two Hungarian Communities." *Dialectical Anthropology* 7, no. 2 (Sept. 1982): 165–83.

—, and Bela C. Maday, eds. *New Hungarian Peasants: An East Central European Experience with Collectivization.* New York: Social Science Monographs, 1983.

Hughes, James. *Stalinism in a Russian Province. Collectivization and Dekulakization in Siberia.* New York: St. Martin's Press, 1996.

Hunter, Holland. "Soviet Agriculture with and without Collectivization, 1928–1940." *Slavic Review* 47, no. 2 (Summer 1988): 203–16.

Iancu, Gheorghe, Virgiliu Ţârău, and Ottmar Traşcă, eds. *Colectivizarea agriculturii în România. Aspecte legislative, 1949–1962* [The collectivization of agriculture in Romania. Legislative aspects, 1949–1962]. Cluj-Napoca: Presa Universitară Clujeană, 2000.

Iggers, Georg G. *Historiography in the Twentieth Century: From Scientific Objectivity to the Postmodern Challenge.* Hanover NH: Wesleyan University Press, 1997.

—. "Nationalism and Historiography, 1789–1996: The German example in historical perspective" in Stefan Berger, Mark Donovan, and Kevin Passmore, eds. *Writing National Histories: Western Europe since 1800*. London: Bloomsbery Academic, 1999, 15–29.

Iordachi, Constantin and Balázs Trencsényi. "In Search of a Usable Past: The Question of National Identity in Romanian Studies, 1990–2000." *East European Politics & Society* 17, no. 3 (Summer 2003): 415–54.

Iordachi, Constantin. "Constanţa, the First Collectivized Region: Soviet Geo-Political Interests and National and Regional Factors in the Collectivization of Dobrogea (1949–1962)" in Constantin Iordachi and Dorin Dobrincu, eds. *Transforming Peasants, Property and Power: The Collectivization of Agriculture in Romania, 1949–1962*. Budapest: CEU Press, 2009, 103–40.

— and Dorin Dobrincu, eds. *Transforming Peasants, Property and Power. The Process of Land Collectivization in Romania, 1949–1962*. Budapest: CEU Press, 2009.

Jarausch, Konrad and Hannes Siegrist, eds. *Amerikanisierung und Sowjetisierung in Deutschland 1945–1970*. Frankfurt am Main, New York: Campus Verlag, 1997.

Jarosz, Dariusz. *Polityka władz komunistycznych w Polsce w latach 1948–1956 a chłopi* [The Communist Regime in Poland 1948–1956 and the Peasants]. Warsaw: Wydawnictwo DiG, 1998.

—. "Polish Peasants versus Stalinism: Stalinism in Poland, 1944–1956" in A. Kemp-Welch, ed. *Selected Papers from the Fifth World Congress of Central and East European Studies*. London: Macmillan Press, 1999, 59–77.

—. *Polacy a stalinizm 1948–1956* [The Poles and Stalinism 1948–1956]. Warsaw: Instytut Historii PAN, 2000.

Jaska, E. "The Results of Collectivization of Estonian Agriculture." *Land Economics* 28, no. 3 (Aug. 1952): 212–17.

Jech, Karel. *Soumrak selského stavu* [The Sunset of Peasantry]. Prague: ÚSD AV ČR: 2001.

—. *Kolektivizace a vyhánění sedláků z půdy* [Collectivization and Expulsions of Farmers from their Land]. Prague: Vyšehrad, 2008.

Josifov, Kalin. *Totalitarnoto nasilie v bulgarskoto selo (1944–1951) i posleditzite za Bulgaria* [Totalitarian Pressure in the Bulgarian Village (1944–1951) and the Consequences for Bulgaria]. Sofia: Universitetsko izdatelstvo "Sv. Kliment Ohridski," 2003.

Jowitt, Kenneth. *The Leninist Response to National Dependency*. Berkeley: Institute of International Studies, University of California, 1978.

—. *New World Disorder: The Leninist Extinction*. Berkeley: University of California Press, 1992.

K., W. "Soviet Agriculture as a Model for Asia Countries," *The China Quarterly* 5 (1961): 116–30.

Kaelble, Hartmut. *Der historische Vergleich. Eine Einführung zum 19. und 20. Jahrhundert*. Frankfurt am Main: CampusVerlag, 1999.

—. "Die interdisziplinären Debatten über Vergleich und Transfer" in Hartmut Kaelble and Jürgen Schriewer, eds. *Vergleich und Transfer. Komparatistik in den Sozial-, Geschichts- und Kulturwissenschaften*. Frankfurt am Main: Campus, 2003, 469–93.

Ka-kui, Tse. "Agricultural collectivization and Socialist Construction: The Soviet Union and China." *Dialectical Anthropology* 2, no. 3 (1977): 199–221.

Kasianov, Heorhii. *Danse macabre: holod 1932–1933 rokiv: u polititsi, masovi isvidomosti ta istoriohrafii (1980-ti-pochatok 2000-kh)* [*Danse macabre*: the 1932–1933 Famine in Politics, Mass Consciousness and Historiography (the 1980s – Early 2000s)] Kiev: Nash chas, 2010.

Katona, Paul. "Collectivization of Agriculture in Central Europe." *The World Today* 16, no. 9 (Sep. 1960): 399–410.

Keohane, Robert O. and Joseph S. Nye. "Transnational Relations and World Politics: An Introduction." *International Organization* 25, no. 3 (1971): 329–49.

—. "Transnational Relations and World Politics: A Conclusion." *International Organization* 25, no. 3 (1971): 721–48.

Kershaw, Ian. "Totalitarianism Revisited: Nazism and Stalinism in Comparative Perspective." *Tel Aviver Jahrbuchfür deutsche Geschichte* 23 (1994): 23–40.

— and Moshe Lewin, eds. *Nazism and Stalinism. Dictatorships in Comparison*. Cambridge: Cambridge University Press, 1997.

Kideckel, David A. "The Dialectic of Rural Development: Cooperative Farm Goals and Family Strategies in a Romanian Commune." *Journal of Rural Cooperation* 5, no. 1 (1977): 43–62.

—. *Agricultural Cooperativism and Social Process in a Romanian Commune* Ph.D. Dissertation. Amherst: Anthropology Department, University of Massachusetts, 1979.

—. "The Socialist Transformation of Agriculture in a Romanian Commune, 1945–1962." *American Ethnologist* 9 (May 1982): 320–40.

— and Steven L. Sampson, "Fieldwork in Romania: Political, Practical and Ethical Aspects" in John W. Cole, ed. *Anthropological Research in Romania*. Amherst: University of Massachusetts, 1984, 85–102.

—. *The Solitude of Collectivism: Romanian Villagers to the Revolution and Beyond*. Ithaca: Cornell University Press, 1993.

Kim, Seong-bo. "The Decision-Making Process and Implementation of the North Korean Land Reform" in Kie-Chung Pang and Michael D. Shin, eds. *Landlords, Peasants and Intellectuals in Modern Korea*. Ithaca: Cornell University Press, 2005, 207–40.

Klein, Thoralf. "Technologische Innovation oder soziale Revolution? Chinas Bauern in der Transformation der Agrargesellschaft." *Geschichte und Gesellschaft* 33 (2007): 575–611.

Kligman, Gail. *The Politics of Duplicity: Controlling Reproduction in Ceauşescu's Romania*. Berkeley: University of California Press, 1998.

— and Katherine Verdery. *Peasants under Siege: The Collectivization of Agriculture in Romania, 1949–1962*. Princeton: Princeton University Press, 2011.

Kocka, Jürgen. "Asymmetrical Comparison: The Case of the German Sonderweg." *History and Theory* 38 (1999): 40–50.

—. "Comparison and Beyond." *History and Theory* 42 (2003): 39–44.

—. "Die Grenzen Europas. Ein Essay aus historischer Perspektive" in Gunnar Folke Schuppert et al., eds. *Europawissenschaft*. Baden-Baden: Nomos, 2005, 275–86.

Kõll, Anu Mai. *The Village and the Class War: Anti-Kulak Campaign in Estonia*. Budapest: CEU Press, 2013.

Korboński, Andrzej. *Politics of Socialist Agriculture in Poland: 1945–1960*. New York: Columbia University Press, 1965.

Kueh, Y. Y. "Mao and Agriculture in China's Industrialization: Three Antitheses in a 50-Year Perspective." *The China Quarterly* 187 (Sep. 2006): 700–23.

Labsvirs, Janis. "The Effect of Collectivization on Latvian Agriculture." *Slavic Review* 22, no. 1 (March 1963): 121–25.

—. *The Sovietization of the Baltic States: Collectivization of Latvian Agriculture 1944–1956*. Indianapolis: Taurus, 1988.

Ladejinsky, W. "Collectivization of Agriculture in the Soviet Union I." *Political Science Quarterly* 49, no. 1 (March 1934): 1–43.

—. "Collectivization of Agriculture in the Soviet Union II." *Political Science Quarterly* 49, no. 2 (June 1934): 207–52.

Lampland, Martha. *The Object of Labor: Commodification in Socialist Hungary*. Chicago: University of Chicago Press, 1995.

Langenhan, Dagmar. "'Halte Dich fern von den Kommunisten, die wollen nicht arbeiten!' Kollektivierung der Landwirtschaft und bäuerlicher Eigen-Sinn am Beispiel Niederlausitzer Dörfer (1952 bis Mitte der sechziger Jahre)" in Thomas Lindenberger, ed. *Herrschaft und Eigen-Sinn in der Diktatur. Studien zur Gesellschaftsgeschichte der DDR*. Cologne: Böhlau Verlag, 1999, 119–65.

Last, George. *After the "Socialist Spring:" Collectivization and Economic Transformation in the GDR*. New York and Oxford: Berghahn Books, 2009.

—. "Rural Functionaries and the Transmission of Agricultural Policy: The Case of *Bezirk* Erfurt from the 1960s to the 1970s" in Mary Fulbrook, ed. *Power and Society in the GDR 1961–1979. The "Normalization" of Rule?* New York and Oxford: Berghahn Books, 2009, 76–101.

Lemke, Michael. *Sowjetisierung und Eigenständigkeit in der SBZ/DDR (1945–1953)*. Cologne: Böhlau Verlag, 1999.

Lepenies, Wolf, ed. *Entangled Histories and Negotiated Universals: Centers and Peripheries in a Changing World*. Frankfurt am Main: Campus, 2003.

Lewin, Moshe. "The Immediate Background of Soviet Collectivization." *Soviet Studies* 17, no. 2 (Oct. 1965): 162–97.

—. *Russian Peasants and Soviet Power: A Study of Collectivization*. Evanston, IL: Northwestern University Press, 1968. Original French Edition: *La Paysannerie et le pouvoir soviétique, 1928–1930*. Paris: Mouton, 1966.

—. *The Soviet Century*. London: Verso, 2005.

Lin, Justin Yifu. "Collectivization and China's Agricultural Crisis in 1959–1961." *Journal of Political Economy* 98, no. 6 (Dec. 1990): 1228–52.

Liu, Yu. "Why Did It Go so High? Political Mobilization and Agricultural Collectivization in China." *The China Quarterly* 187 (Sep. 2006): 732–42.

Lindenberger, Thomas, ed. *Herrschaft und Eigen-Sinn in der Diktatur. Studien zur Gesellschaftsgeschichte der DDR*. Cologne: Böhlau Verlag, 1999.

Livi-Bacci, Massimo. "On the Human Costs of Collectivization in the Soviet Union," *Population and Development Review* 19, no. 4 (Dec. 1993): 743–66.

Luciuk, Lubomyr Y., with the assistance of Lisa Grekul, eds. *Holodomor: Reflections on the Great Famine of 1932–1933 in Soviet Ukraine*. Kingston, Ont.: Kashtan Press, 2008.

Maier, Hans, ed. *Totalitarismus und Politische Religionen. Konzepte des Diktaturenvergleichs*. Paderborn, 1986.

Markiewicz, Marcin. *Kolektywizacja wsi w województwie białostockim 1948–1956* [Collectivization of the Countryside in Voivodship of Białystok, 1948–1956]. Białystok: Instytut Pamięci Narodowej, 2010.

Marples, David R., "The *Kulaks* in the Post-War USSR: The West Ukrainian Example." *Soviet Studies* 4 (July 1984): 560–70.

—. "Toward a Thematic Approach to the Collectivization Campaign in the Soviet West (1948–56)." *Canadian Slavonic Papers/Revue Canadienne des Slavistes* 33, nos. 3/4 (September–December 1991): 285–300.

McDonald, Tracy. "A Peasant Rebellion in Stalin's Russia: The Pitelinskii Uprising, Riazan 1930." *Journal of Social History* 35, no. 1 (Autumn 2001): 125–46.

Merl, Stephan. *Bauern unter Stalin. Die Formierung des sowjetischen Kolchos systems 1930–1941.* Berlin: Duncker & Humblot, 1990.

Mertelsmann, Olaf, ed. *The Sovietization of the Baltic States, 1940–1956*. Tartu: KLEIO Ajalookirjanduse Sihtasutus, 2003.

Miernik, Grzegorz. *Opór chłopów wobec kolektywizacji w Województwiekie Kieleckim 1948–1956* [Peasant resistance to collectivization in the Kielce Voivodship 1948–1956]. Kielce: Wydawnictwo TAKT, 1999.

Millar, James R. "Mass Collectivization and the Contribution of Soviet Agriculture to the First Five-Year Plan: A Review Article." *Slavic Review* 33, no. 4 (Dec. 1974): 750–66.

—. "Collectivization and its Consequences: A New Look." *Russian Review* 41, no. 1 (Jan. 1982): 60–67.

Miller, Daniel E. "Collectivization in the 1970s and 1980s in Zamagurie, Slovakia." *Agricultural History* 73, no. 3 (Summer 1999): 281–302.

Mitrany, David. *The Land and the Peasant in Rumania: The War and Agrarian Reform (1917–1921)*. London: Oxford University Press, 1930.

—. *Marx Against the Peasant: A Study in Social Dogmatism*. Chapel Hill: University of North Carolina Press, 1951.

Müller, Dietmar. "Bodeneigentum und Institutionenwandel in Ostmittel- und Südosteuropa 1918–1945–1989." *Zeitschrift für Ostmitteleuropa-Forschung* 61, no. 3 (2012): 332–55.

Mungiu, Alina. *A Tale of Two Villages: Coerced Modernization in the East European Countryside*. Budapest: Central European University Press, 2010.

Nam, Sung-Wook. *Contemporary Food Shortage of North Korea and Reform of Collective Farm*. Munich: Utz, 2006.

Narkiewicz, O. A. "Stalin, War Communism and Collectivization." *Soviet Studies* 18, no. 1 (July 1966): 20–37.

O'Connor, Maura. "Cross-National Travelers: Rethinking Comparisons and Representations" in Deborah Cohen and Maura O'Connor, eds. *Comparison and History: Europe in Cross-National Perspective*. New York: Routledge, 2004, 133–44.

Oláh, Sándor. *Csendes csatatér. Kollektivizálás és túlélési stratégiák a két Homoród mentén (1949–1962)* [Silent battlefield. Collectivization and strategies of survival in the region of the two Homoród rivers]. Miercurea-Ciuc: Pro-Print, 2001.

Olcott, Martha Brill. "The Collectivization Drive in Kazakhstan." *Russian Review* 40, no. 2 (April 1981): 122–42.

Osterhammel, Jürgen. "Transnationale Gesellschaftsgeschichte: Erweiterung oder Alternative?" *Geschichte und Gesellschaft* 27 (2001): 464–79.

Paulmann, Johannes. "Internationaler Vergleich und interkultureller Transfer. Zwei Forschungsansätze zur europäischen Geschichte des 19. und 20. Jahrhunderts." *Historische Zeitschrift* 267 (1998): 649–85.

Pianciola, Niccolò. "The Collectivization Famine in Kazakhstan, 1931–1933." *Harvard Ukrainian Studies* 25, nos. 3/4 (Fall 2001): 237–51.

— and Susan Finnel. "Famine in the Steppe. The Collectivization of Agriculture and the Kazak Herdsmen, 1928–1934." *Cahiers du Monde russe* 45, nos. 1/2, Stratégies impériales: Expansion, colonisation, intégration, conversion (Jan–June 2004): 137–91.

Popescu-Puţuri, Ion et al., eds. *Din lupta PCR pentru închegarea alianţei clasei muncitoare cu ţărănimea muncitoare, în bătălia pentru reforma agrară din 1944–1945*, 3 vols. [From the struggle of the RCP for the establishment of the alliance between the working class and the working peasantry, in the battle for the agrarian reform of 1944–1945]. Bucharest: Editura de Stat, 1955–1960.

Próchniak, Leszek. *Kolektywizacja rolnictwa w regionie łódzkim* [Collectivizations in Voivodship of Łódź]. Łódź: Instytut Pamięci Narodowej, 1993.

Puntscher, Riekmann, Sonja Wodak, and Ruth Wodak. "'Europe for All'—diskursive Konstruktionen europäischer Identitäten" in Monika Mokre et al., eds. *Europas Identitäten. Mythen, Konflikte, Konstruktionen*. Frankfurt am Main: Campus, 2003, 283–303.

Rabinbach, Anson. "Moments of Totalitarianism." *History and Theory* 45 (2006): 72–100.

Raditsa, Bogdan. "The Sovietization of the Satellites." *Annals of the American Academy of Political and Social Science* 271 (Sep. 1950): 122–34.

Randeria, Shalini. "Geteilte Geschichte und verwobene Moderne" in Jörn Rüsen et al., eds. *Zukunftsentwürfe. Ideen für eine Kultur der Veränderung*. Frankfurt am Main: Campus Verlag, 2000, 87–96.

—. "Entangled Histories or Uneven Modernities: Civil Society, Caste Solidarities and Legal Pluralism in Post-Colonial India" in Yehuda Elkana, ed. *Unravelling Ties: From Social Cohesion to New Practices of Connectedness*. Frankfurt am Main: Campus Verlag, 2002, 284–311.

Raymond, Chad. "The Insoluble Internal Conflicts of Agricultural Collectivization in Vietnam." *Crossroads: An Interdisciplinary Journal of Southeast Asian Studies* 15, no. 2 (2001): 41–70.

—. "No Responsibility and No Rice: The Rise and Fall of Agricultural Collectivization in Vietnam." *Agricultural History* 82, no. 1 (Winter 2008): 43–61.

Reese, Roger R. "Red Army Opposition to Forced Collectivization, 1929–1930: The Army Wavers." *Slavic Review* 55, no. 1 (Spring 1996): 24–45.

Rév, István. "The Advantages of Being Atomized. How Hungarian Peasants Coped with Collectivization." *Dissent* (Summer 1987): 335–50.

Robakowski, Kazimierz. *Społeczno-polityczne problemy rozwoju spółdzielczości produkcyjnej w Polsce w latach 1944–1956* [Sociopolitical problems of development of production cooperatives in Poland, 1944–1956]. Poznań: Wyadwnictwo Uniwerystetu Adama Mickiewicza, 1986.

Roberts, Henry. *Rumania: Political Problems of an Agrarian State*. Hamden, Connecticut: Archon Books, 1969.

Robinson, William I. "Beyond Nation-State Paradigms: Globalization, Sociology, and the Challenge of Transnational Studies." *Sociological Forum* 13, no. 4 (Dec. 1998): 561–94.

Rosefielde, Steven. "Excess Collectivization Deaths 1929–1933: New Demographic Evidence." *Slavic Review* 43, no. 1 (Spring 1984): 83–88.

—. "New Demographic Evidence on Collectivization Deaths: A Rejoinder to Stephen Wheatcroft." *Slavic Review* 44, no. 3 (Autumn 1985): 509–16.

Roske, Octavian, ed. *Dosarul colectivizării agriculturii în România, 1949–1962* [The dossier of agricultural collectivization in Romania, 1949–1962]. Bucharest: Parlamentul României, Camera Deputaţilor, 1992.

Ross, Corey. *Constructing Socialism at the Grassroots. The Transformation of East Germany, 1945–65*. Basingstoke: Palgrave Macmillan, 2000.

S., A. V. "Agriculture in Yugoslavia: Problems of Collectivization." *The World Today* 6, no. 11 (Nov. 1950): 469–80.

Sampson, Steven L. and David A. Kideckel. "Anthropologists Going into the Cold: Research in the Age of Mutually Assured Destruction" in Paul Turner and David Pitt Hadley, eds. *The Anthropology of War and Peace*. Hadley, Massachusetts: Bergin and Garvery, 1989, 160–173.

Sanders, Irwin T, and Enno E. Kraehe. *Collectivization of Agriculture in Eastern Europe*. Lexington: University of Kentucky Press, 1958.

Saunier, Pierre-Yves. "Transnational" in Akira Iriye and Pierre-Yves Saunier, eds. *The Palgrave Dictionary of Transnational History*. Houndmills: Pelgrave, 2009, 1047–55.

Scherstjanoi, Elke. *SED-Agrarpolitik unter sowjetischer Kontrolle 1949–1953*. Munich: Oldenbourg, 2007, 345–424.

Schmale, Wolfgang. *Geschichte Europas*. Vienna: Böhlau, 2000.

Schöne, Jens. "Landwirtschaft und ländliche Gesellschaft in der DDR" in Rainer Eppelmann et al., eds. *Bilanz und Perspektiven der DDR-Forschung*. Paderborn: Schöningh, 2003, 254–59.

—. *Frühling auf dem Lande? Die Kollektivierung der DDR-Landwirtschaft*. Berlin: Links Verlag, 2005.

Schöpflin, George. "The Political Traditions of Eastern Europe" in Stephen R. Graubard, ed. *Eastern Europe–Central Europe–Europe*. Boulder: Westview Press, 1991, 59–94.

Selden, Mark. "Pathways from Collectivization: Socialist and Post-Socialist Agrarian Alternatives in Russia and China." *Review* (Fernand Braudel Center) 17, no. 4 (Fall 1994): 423–49.

Senn, Alfred Erich. "The Sovietization of the Baltic States." *Annals of the American Academy of Political and Social Science* 317 (May 1958): 123–29.

Sirc, Ljubo. "Economics of Collectivization." *Soviet Studies* 18, no. 3 (Jan. 1967): 362–70.

Sjoberg, Orjan. *Rural Change and Development in Albania*. Boulder: Westview Press, 1991.

Skultans, Vieda. "The Expropriated Harvest: Narratives of Deportation and Collectivization in North-East Latvia." *History Workshop Journal* 44 (Autumn 1997): 170–88.

Snyder, Timothy. *Bloodlands: Europe Between Hitler and Stalin*. New York: Basic Books, 2010.

Söllner, Alfons, Ralf Walkenhaus, and Karin Wieland, eds. *Totalitarismus. Eine Ideengeschichte des 20. Jahrhunderts*. Berlin: Wiley-VCH Verlag, 1997.

Spulber, Nicolas. "Eastern Europe: The Changes in Agriculture from Land Reforms to Collectivization." *American Slavic and East European Review* 13, no. 3 (Oct. 1954): 389–401.

Staar, Richard F. *Poland, 1944–1962: The Sovietization of a Captive People*. Baton Rouge: Louisiana State University Press, 1962.

Storella, C. J., and A. K. Sokolov. *The Voice of the People: Letters from the Soviet Village, 1918–1932.* Documents translated by C. J. Storella. New Haven: Yale University Press, 2012.

Swain, Nigel. *Collective Farms which Work?* Cambridge: Cambridge University Press, 1985.

—. "Collective Farms which Work" in Teodor Shanin, ed. *Peasants and Peasant Societies: Selected Readings.* New York: Basil Blackwell Inc., 1987, 111–13.

—. "From Kolkhoz to Holding Company: A Hungarian Agricultural Producer Cooperative in Transition." *Journal of Historical Sociology* 13 (2000) 2: 142–71.

—. *The Rural Transition in Post-socialist Central Europe and the Balkans.* Max Planck Institute for Social Anthropology, Working Paper No. 9, 2000.

—. "Decollectivization Politics and Rural Change in Bulgaria, Poland and Former Czechoslovakia." *Social History* 32, no. 1 (February 2007): 1–26.

Swain, Geoffrey and Nigel Swain. *Eastern Europe Since 1945.* New York: St. Martin's Press, 1993.

—. "Deciding to Collectivize Latvian Agriculture." *Europe-Asia Studies* 55, no. 1 (2003): 39–58.

Taagepera, Rein. "Soviet Collectivization of Estonian Agriculture: The Deportation Phase." *Soviet Studies* 32, no. 3 (July 1980): 379–97.

Tănase, Stelian. *Elite şi societate. Guvernarea Gheorghiu-Dej, 1948–1965* [Elites and Society. The Rule of Gheorghiu-Dej, 1948–1965]. Bucharest: Humanitas, 1998.

Tănăsescu, Bogdan. *Colectivizarea, între propagandă şi realitate* [Collectivization, between propaganda and reality]. Bucharest: Globus, 1994.

The Compulsory Collectivization of Independent Farms in the Soviet Zone of Occupation in Germany. Bonn: Federal Ministry for All-German Affairs, 1961.

Thelen, Tatjana. "Violence and social (dis)continuity: comparing collectivization in two East European villages." *Social History* 30 (2005): 25–44.

Tismăneanu, Vladimir. *Fantoma lui Gheorghiu-Dej* [The Phantom of Gheorghiu-Dej]. Bucharest: Univers, 1995.

—. *Stalinism for All Seasons: A Political History of Romanian Communism.* Berkeley: University of California Press, 2003.

—, ed. *Stalinism Revisited: The Establishment of Communist Regimes in East-Central Europe.* Budapest: Central European University Press, 2009.

Tochitch, Desimir. "Collectivization in Yugoslavia." *Journal of Farm Economics* 41, no. 1 (Feb. 1959), 26–42.

Tomasevich, Jozo. "Agriculture in Eastern Europe." *Annals of the American Academy of Political and Social Science* 317, The Satellites in Eastern Europe (May 1958): 44–52.

Tucker, Robert C. *Stalin in Power. The Revolution from above 1928–1941.* New York: W. W. Norton, 1990.

V. M. "Collectivization in Yugoslav Agriculture." *The World Today* 14, no. 2 (Feb. 1958): 80–87.

Verdery, Katherine. *Transylvanian Villagers: Three Centuries of Political, Economic, and Ethnic Change.* Berkeley: University of California Press, 1983.

—. *The Vanishing Hectare: Property and Value in Postsocialist Transylvania.* Ithaca: Cornell University Press, 2003.

— and Caroline Humphrey. *Property in Question: Value Transformation in the Global Economy.* Oxford: Berg Press, 2004.

Vickers, Miranda and James Pettifer. *Albania: from Anarchy to a Balkan Identity*. London: Hurst, 1997.
Viola, Lynne. "Bab'i Bunty and Peasant Women's Protest during Collectivization." *Russian Review* 45 (Jan., 1986) 1: 23–42.
—. "The Campaign to Eliminate the Kulak as a Class, Winter 1929–1930: A Reevaluation of the Legislation." *Slavic Review* 45, no. 3 (1986): 503–24.
—. *The Best Sons of the Fatherland: Workers in the Vanguard of Soviet Collectivization*. Oxford: Oxford University Press, 1987.
—. *Peasant Rebels under Stalin: Collectivization and the Culture of Peasant Resistance*. Oxford: Oxford University Press, 1996.
—. *The Role of the OGPU in Dekulakization, Mass Deportations, and Special Resettlement in 1930*. Pittsburgh: Russian and East European Studies Program, University of Pittsburgh, 2000.
— et al., eds. *The War Against the Peasantry, 1927–1930. The Tragedy of the Soviet Countryside*. New Haven: Yale University Press, 2005.
—, ed. *The Unknown Gulag. The Lost World of Stalin's Special Settlements*. Oxford: Oxford University Press, 2007.
Voegelin, Eric. *The Political Religions*. Lewiston: E. Mellen Press, 1986.
Wädekin, Karl-Eugen. *Agrarian Policies in Communist Europe: A Critical Introduction*. Lanham MD: Rowman & Littlefield Publishers, 1982.
Welskopp, Thomas. "Stolpersteine auf dem Königsweg. Methodenkritische Anmerkungen zum internationalen Vergleich in der Gesellschaftsgeschichte." *Archiv für Sozialgeschichte* 35 (1995): 339–67.
Werner, Michael, and Bénédicte Zimmermann. "Beyond Comparison: *Histoire Croisée* and the Challenge of Reflexivity." *History and Theory* 45 (2006): 30–50.
Wheatcroft, Stephen G. "New Demographic Evidence on Excess Collectivization Deaths: Yet Another Kliukva from Steven Rosefielde?" *Slavic Review* 44, no. 3 (Autumn 1985): 505–8.
Wheatcroft, Stephen G., R. W. Davies, and J. M. Cooper. "Soviet Industrialization Reconsidered: Some Preliminary Conclusions about Economic Development between 1926 and 1941." *The Economic History Review* 39, No. 2 (May 1986): 264–94.
Witkowski, Gregory R. "On the Campaign Trail: State Planning and *Eigen-Sinn* in a Communist Campaign to Transform the East German Countryside." *Central European History* 37 (2004): 400–22.
Yaney, George. "Some Suggestions Regarding the Study of Russian Peasant Society prior to Collectivization." *Russian Review* 44, no. 1 (Jan. 1985): 27–33.
Zhou, Xun. *The Great Famine in China, 1958–1962: A Documentary History*. New Haven: Yale University Press, 2012.

PART I

The Interwar Soviet Model and its Post-1945 Application in the Newly Annexed Territories

Collectivization in the Soviet Union: Specificities and Modalities

LYNNE VIOLA

The collectivization of agriculture in the Soviet Union was an integral component of Stalin's First Five-Year Plan (1928–32). Wholesale collectivization began in the winter of 1929–1930. The pivotal years in collectivization were 1930 and 1931, when the state made its greatest strides to force Soviet peasants into collective farms and to eliminate the so-called *kulaks*.[1] Some 125 million individual peasant households were absorbed into around 200,000 collective farms. Thirty thousand peasants were executed as *kulaks* and close to two million peasants were subject to expropriation and internal exile.[2] Collectivization was implemented mainly by urban cadres with the use of massive violence. By March 1930, Soviet leaders lost control of the process. That month, Stalin published his "Dizziness with Success," which blamed local cadres for the "excesses" of collectivization, rather than admitting any blame on the part of the central

[1] The term *kulak* derives from the Russian word for "fist" and originally described the village usurer or tight-fisted peasant. After the revolution, the term came to describe the (oxymoronic) "capitalist peasant" or village capitalist. Through the 1920s, agronomists and party theorists attempted to pin down a definition of the *kulak*, but with little success, in part because the Russian village remained, on the whole, pre-capitalist. The result of this inability to define the *kulak* was the arbitrary application of the term to all manner of peasantry including wealthier peasants, village elites, and outspoken critics. It is highly ironic that the term was subsequently exported to some Eastern European countries during their collectivization phases.

[2] See Viola, *The Unknown Gulag*, 29–32.

state. Thereafter, collectivization proceeded with far greater central control. The most important grain-growing regions of the Soviet Union completed collectivization by 1932; elsewhere, the process of collectivization developed more slowly, and was completed by the end of the decade.

The collectivization of agriculture served as a major tool of state building in the Soviet Union. Anchored by ideology and shaped in conditions of extreme violence, collectivization was a foundation for Stalin's repressive dictatorship. Yet, while the specificities of circumstance and ideology played an important role in shaping the dynamics of Soviet collectivization, the policy cannot be explained solely by a particularist understanding of Soviet history, a *Sonderweg* (special path) approach to Russia's twentieth century. Soviet collectivization must be placed against the backdrop of the Tsarist past and the modern condition, both of which—and in sometimes contradictory ways—shaped the forms and functions of the "socialist transformation" of the peasantry.

Soviet Collectivization: Contingency, Contour, Meaning

The collectivization of agriculture in the Soviet Union was about more than "taking grain."[3] Collectivization was a tool of state-building—the state in this case referring to the regime as a whole. Stalinist state-building encompassed economic modernization, colonization of the peasantry, geographic expansion, border defense, and political centralization. It was a key component of the Soviet industrial revolution, one with profound political, economic, and social consequences that would play an important role in the bloodshed of 1937 and resound all the way to 1991 and the fall of the Soviet Union. It was also a clash of cultures, a war on peasants aimed at destroying rural culture in its broadest definition and subjugating the countryside to the designs of the centralizing, modernizing state.[4]

Stalin famously described collectivization as a "revolution from above."[5] (He added "with support from below," but that was clearly not the case insofar as it relates to the majority of the rural population.) "Stalin's Revolution" was based on rapid, state-generated industrialization

---

[3] This phrase comes from Lewin, "Taking Grain."

[4] For an elaboration of some of these arguments, see my *Peasant Rebels Under.*

[5] *History of the Communist Party*, 305.

and, in theory, on the forced accumulation of domestic, mainly peasant resources for capital funding in order to create an autarkic, industrialized polity capable of defending itself from "capitalist encirclement."

The decision to proceed in this direction was not an exercise of sheer power or ideological hubris. Instead, it was a choice party leaders made within a context of debate and crisis. In its broadest terms, the policy alternative to revolution from above and to "squeezing" the peasantry was a more gradual and balanced approach to industrialization, based in part on increasing the development of textiles and light industry and on Nikolai Bukharin's politically self-destructive call for the peasantry to "enrich themselves." In theory, these levers of economic development would gradually expand the domestic market, and light industry would eventually rollover its capital resources into heavy industry. This was "socialism at a snail's pace," to borrow another of the less than felicitous phrases associated with the Right Opposition.[6]

Collectivization, in fact, was very much tied into the factional divisions within the Communist Party. Bukharin and the right wing of the party constituted the last real opposition to Stalin. The Right Opposition called for the gradual socialization of the peasantry through the formation of a variety of different types of cooperatives, repeatedly citing as

---

[6] For further information on Bukharin and the "Right Opposition," see Lewin, *Russian Peasants*, and Cohen, *Bukharin*. It should be noted that the two alternatives to economic development had parallels in the prerevolutionary period. The Russian industrial spurt of the 1890s had also been based, in part, on squeezing the peasantry. Through most of the 1880s, Minister of Finance N. K. Bunge had steered the Russian economy cautiously toward balanced growth and the easing of the tax burden on the peasantry. His tenure ended in failure when it became clear that such moderated economic growth could not sustain Russia's increasing need to compete in arms production. Alexander III then appointed I. A. Vyshnegradskii, who, in the late 1880s, coined the phrase, "We must export [grain] though we die," and mercilessly exploited the peasantry in his efforts to shore up government coffers. When reality followed promise in the famine of 1891, Alexander III dismissed Vyshnegradskii and appointed Sergei Witte as his new Minister of Finance. Although Witte's economic policy was far more sound than that of his predecessor, it still relied heavily upon high direct and indirect taxation of the peasantry to fund his program of industrialization. Political considerations—especially defense needs—weighed heavily in the choice of economic policy in those years. Political factors continued to influence the direction of economic decision-making in the Stalin years; state and defense needs ultimately determined which of the alternatives to economic development the state chose at the end of the 1920s. For further discussion, see von Laue, *Sergei Witte*; and von Laue, *Why Lenin*.

a model Lenin's cooperative plan written toward the end of his life. The Right argued that the forced grain procurements of the late 1920s threatened the *smychka*, the union with the peasantry that formed the backbone of the social base of the revolution, and warned that this rupture could lead to civil war between town and countryside. Ultimately, however, they had little support within the party as domestic politics became increasingly radicalized in the face of crisis in the late 1920s. Their formal defeat was proclaimed in tandem with the decision to implement wholesale collectivization at the November 1929 plenum of the Communist Party.[7]

The decision to forge ahead with what became known as the Stalin Revolution was both political and military, and emerged from a series of destabilizing crises that in large part shaped the contours of Stalin's collectivization program. Collectivization grew out of the crisis of the NEP (New Economic Policy), which developed from a combination of structural and economic problems partly of the party's own making, but ideological factors, party infighting, and international threats—though no doubt exaggerated—also played key roles. The war scare of 1927 was an important backdrop for the Stalin Revolution and gave Stalinist leadership levers of political manipulation against the Right Opposition, which in turn set off a chain of reactions that would inevitably lead to a radical solution to the industrialization debate. A series of interrelated crises led to the crisis of the NEP: the war scare, which served as general precipitant; the grain- requisitioning crisis and the resulting upswing in rural violence; urban instability—the dimensions of which are still undefined—sparked by rising food prices, queues, and rationing; and interparty struggles that manifested in a series of well-informed critiques against grain-requisitioning practices. For Stalin and company, the NEP, the peasantry's sabotage of industrialization by way of its seeming withdrawal from the market, and the interplay of domestic and international enemies, all radicalized the atmosphere. And even before the "Great Turn," the secret police (OGPU) launched a series of mass operations, arresting thousands of traders, mer-

[7] It is beyond the scope of this essay to delve into this debate in detail. It should be noted, however, that a fuller exploration of the debate is finally possible with the publication of the five-volume *Kak lomali NEP*, transcripts of the Central Committee plenary of the late 1920s, and the five-volume *Tragediia Sovetskoi derevni*, the first volume of which sheds important new light on the issues at stake. See Iakovlev et al., eds., *Kak lomali NEP*; Danilov, Manning, and Viola, eds., *Tragediia Sovetskoi.*

chants, and NEP men and pursuing a kind of *de facto* dekulakization through the increased use of administrative fines and exile.[8]

In this context, Stalin viewed collectivization as a corollary to industrialization, vital to the military might of the Soviet Union. Yet, collectivization was also about domination, subjugation, and victory over the "internal enemy" (in this case, the *kulak*, the supposed rural capitalist). Stalinist state-building demanded a "tribute" from the peasantry, and the Communist Party explicitly committed itself to a policy of exploitation of the countryside in the interests of state development. It is worth quoting Stalin's speech at the July 1928 Central Committee plenum in order to capture the essence of state policy:

> In capitalist countries, industrialization was usually based not only on internal accumulation but also on the plundering of other countries, the plundering of colonies or vanquished countries, or on substantial loans from abroad. You know that for hundreds of years England used to drain all its colonies, from every continent, and in this way injected additional investments into its industry... Our country differs from the capitalist countries, by the way, in that it cannot and must not engage in the plundering of colonies or in the plundering of other countries in general. Therefore this path is closed to us. But our country doesn't have loans from abroad either. Consequently, this path is closed to us as well. In that case what is left for us? One choice is left: to develop industry, to industrialize the country on the basis of internal accumulation... But where are the sources of this accumulation? As I said, there are two such sources: first, the working class, which creates valuable output and moves industry forward; and second, the peasantry.
>
> The situation in our country with regard to the peasantry in this case is the following: it pays the state not only ordinary taxes, direct and indirect, but it also pays relatively high prices for goods from industry—that is first of all—and it doesn't receive the full value of the prices of agricultural products—that is second of all. This is an additional tax on the peasantry in the interests of developing industry, which serves the whole country, including

[8] For an excellent summary of this period, see Osokina, *Our Daily Bread*, esp. chap. 2–3. For a discussion and documentation of the crisis, see Danilov, Manning, and Viola, eds., *Tragediia Sovetskoi derevni*, vol. 1.

> the peasantry. This is something like a "tribute," something like a surtax, where we are forced to take temporarily in order to sustain and further develop the current rate of industrial growth... This situation, needless to say, is unpleasant. But we would not be Bolsheviks if we papered over this fact and closed our eyes to the fact that, unfortunately, our industry and our country cannot manage without this additional tax on the peasantry.[9]

The "situation," wherein the peasantry was supposed to serve as the primary resource for industrial development, may have been "unpleasant," to quote Stalin, but from his perspective—and that of the majority of the party in the late 1920s—it was absolutely necessary.[10] Given the legacy of the civil war and the general Bolshevik disdain for the peasantry, perhaps it was at some level even seen to be desirable insofar as "class war" and the destruction of the *kulak* allowed for the completion of unfinished business.

Relations between town and countryside were in fact colored by cultural imperialism, long inherent in elite perceptions of the peasantry and sharply accentuated by Marxist-Leninist-Stalinist attitudes toward the village. The peasant served as the Bolshevik "other" through the 1920s and most of the 1930s. Although Marxist-Leninist theory clearly stipulated the existence of class stratification and therefore political differentiation among the peasantry, in practice, Bolshevik and particularly Stalinist practice tended to *kulakize* the entire peasantry, a variation on a theme of E. H. Carr's dictum that "policy determined class" in the Soviet 1920s.[11] And because it was assumed that the peasantry would disappear with the advent of an industrial society, the Bolsheviks could feel justified in quickening the process, careless of the human consequences given the "*kulak* consciousness" of the countryside.

The "colony" status of the countryside both assured and rationalized its plunder, domination, and subjugation by the party-state. Collectivization became a virtual war on the peasantry and its destruction as a culture. The state needed to leap over entire centuries of development in order to establish what it saw as the requisites of industrial modernization. In doing

---

9 Viola, Danilov, Ivnitskii and Kozlov, eds., *War Against the Peasantry*, vol. 1, 98–99. For the Russian original, see Stalin, *Sochineniia*, 157–60.

10 James Millar conclusively disproved the notion that collectivization "paid" for industrialization in the short term. See the classic debate between Millar and Nove, "A Debate on Collectivization."

11 Carr, *Socialism in One Country*, vol. 1, 99. (Volume 5 of *A History of Soviet Russia*).

so, it faced the gargantuan task of transforming a largely traditional peasantry into a serving "colony."

Collectivization represented a clash of cultures in which the centralizing, modernizing state confronted a semi-traditional peasantry determined to preserve and protect its independence, subsistence, and culture (in the broadest meaning of the term). The Soviet Union's "peasant colony" became an object of transformation: when it was not engaged in "class war," it was a "dark mass" to be "tamed" and "enlightened" by the higher forces of the urban working class and Communist Party. Collectivization, therefore, had an inherent civilizing mission, however uncivilized the actual process and results in fact were. The end goal of this "civilizing mission" was to destroy the peasantry as a relatively cohesive and independent cultural entity, and to establish a communist hegemony that would allow the state to dominate and subjugate the countryside. In its effort to establish hegemony over the countryside, the state aimed first at eliminating a series of customary autonomous peasant cultural spaces, traditions, or sites: the *skhod* (or land commune), the mill, tavern, peasant manufacturing, and the market. These measures were an intrinsic part of the Stalinist agricultural revolution and the effort to socialize the peasant rural economy. They were equally vital to the assertion of a communist, urban hegemony over the village. The destruction of churches and persecution of clergy hastened this process, as did the removal of traditional elites (*kulaks* and others), which, in a sense, decapitated the village of traditional sources of authority, legitimacy, and leadership. This process became more urgent as the Communist Party faced a massive peasant rebellion against collectivization and its auxiliary policies. Therefore, the destruction of traditional elites, or in the parlance of the times, the "liquidation of the *kulak* as a class," accompanied collectivization as a composite—indeed *requisite*—measure.[12]

[12] Domination was the other side of subjugation and required a series of substitutions in personnel, governance, and practice. Urban mobilizations and appointments from above (as well as appointments of official *bednota*) were to create a new leadership cadre in the countryside. New institutions, such as the *kolkhoz*, *sovkhoz*, and MTS replaced traditional village institutions, bringing with them their own forms of rule and organization and enabling (in theory) an efficacious siphoning off of grain, produce, taxes, and labor from the countryside. Soviet education, Soviet holidays, and the military indoctrination of peasant soldier recruits provided new legitimizing tools for the new order in the countryside. This is not to say that such practices of domination were always successful. To some extent, the creature that emerged from collectivization was a hybrid—state forms adapted to the reality of rural needs and peasant practices.

Dekulakization was central to collectivization, as it was not only intended to provide capital for the new collective farms through property expropriations, but also to intimidate the majority of the peasantry into joining the new collectives. Following the initial chaos of the first months of collectivization and dekulakization, when local officials used dekulakization to attempt to stem the tide of peasant flight and property destruction, policy on the fate of the *kulak* began to take shape. The *kulaks* would serve state industrialization and "colonization" efforts directly following their forced exile into the Soviet hinterlands. *Kulak* families would be employed in remote, labor-scarce territories with rich natural resources needed for the industrialization campaign. In the process, these families would colonize the northern- and easternmost territories of the Soviet Union, expanding the population (particularly Slavic) into the hinterlands. Plans for the deployment of prison labor had been in place for a number of years (indeed harkened back to Tsarist-era practices); dekulakization "freed up" nearly two million laborers in just under two years. Exiled *kulaks* would build the "other archipelago" of the Stalinist special settlements, a major building block in the formation of the Gulag.[13] Therefore, the forced deportation of the *kulaks* was intrinsic to Soviet industrialization, state building, and the destruction of the traditional village elite.

Military defense provided an important rationalization for collectivization. Just as collectivization was required to provide capital for industrialization, it was also meant to bolster the military strength of the Soviet Union. For Stalin, who feared a "two-front war" and the prospect of Moscow facing both a foreign enemy and a disgruntled peasantry, collectivization would serve as an instrument of political control. Along with ridding the countryside of potential sources of opposition through the dekulakization campaign, the state began what would be the first of a continuing series of purges of border populations, aimed especially at *kulaks* of Polish nationality, as well as at a series of "socially dangerous" categories (former anti-Bolshevik civil war partisans, contrabandists, criminals, etc.). The population was to be replaced by supposed loyalists, sometimes from the ranks of demobilized Red Army soldiers. This policy was meant to strengthen border defense and at the same time to insure the maximum pliability of the rural population.[14]

---

[13] For a discussion of this process, see my *Unknown Gulag*.

[14] Danilov, Manning, and Viola, eds. *Tragediia Sovetskoi derevni*, vol. 2, 167; *RGASPI*, f. 17, op. 162, d. 8, ll. 109–10; *GARF*, f. 9414, op. 1, d. 1944, l. 82. The Ukrainian Council of People's Commissars had already passed a decree on

The crisis-ridden context of collectivization was a central factor in how party leaders shaped the campaign. Ultimately, Stalin's role was pivotal to both the timing and radical nature of the collectivization policy. His trip to Siberia in early 1928, during which he castigated local officials for ignoring the "class struggle" and treating *kulaks* too passively, resulted in the acceleration and radicalization of the state's grain-requisitioning campaigns. Stalin was also personally involved in developing key legislation on collectivization and dekulakization, ensuring that both campaigns would create the "Great Break" he had spoken of in the fall of 1929, a maximally destructive and violent program of forced collectivization.[15] At the same time, neither Stalin's role as a personality nor his power over history should be exaggerated: Stalin was a creature of his times, acting within a political culture that elevated violence and class struggle to a positive virtue and stressed—albeit with convoluted Marxist-Leninist-Stalinist reasoning—the leading role of the vanguard party in moving history forward.

Violence was, in fact, assumed to be a necessary corollary to revolution and an "inevitable" result of the class war with the *kulak*. Violence served as the foundation of the new order in the countryside. The state implemented collectivization as an act of war. Although the actual military was used sparingly in the countryside—largely for political reasons and for fear of sparking insurrection among the army's peasant base—collectivization was implemented using practices which, to a certain extent, replicated military practices.[16] The state deployed the urban party and thousands of industrial workers in a series of mass mobilizations. Plenipotentiary rule was meant to override and overcome what was viewed as the inertia or resistance of a weak local government. Headquarters (*shtaby*) were set up on the *raion* (district) level to conduct operations. And the deportation of the *kulaks* was nothing if not a vast exercise in troop mobilization, substituting families for soldiers but following the basic transport and housing rules for rapid troop deployment. Collectivization would be conducted as a *campaign* in the full meaning of the term. Violence, *proizvol* (arbitrariness), and *peregiby* (excesses) were rampant, reaching staggering proportions and sparking

---

November 13, 1929 to exile "socially dangerous" elements from border zones. See *RGAE*, f. 5675, op. 1, d. 23, ll. 42–41. (sic).

15 See Viola et al., eds., *War Against the Peasantry*, chap. 2 and 4.

16 For information on the use of the army in collectivization, see the original OGPU decree of February 2, 1930 in *GARF*, f. 9414, op. 1, d. 1944, ll. 17–25.

massive village unrest.[17] The violence, moreover, did not stop with the end of the campaign. The *kulak*, an increasingly elusive subject, would continue to haunt the countryside through the 1930s, serving as the main victim in the mass operations (Order 00447) of 1937–38—although, needless to say, the label by then had become meaningless.[18]

Collectivization had a profound impact on Soviet society. Some 12 million peasants moved from village to city during the First Five-Year Plan, and millions continued to leave throughout the 1930s.[19] Draconian requisitioning of agricultural products along with the massive peasant destruction of livestock and animal draught power had left the countryside devastated. An estimated five to seven million peasants died in the 1932–33 famine. The countryside never recovered. Agriculture stagnated, becoming the proverbial "Achilles' heel" of the Soviet economy.

Collectivization was as much an integral part of the Stalinist industrial revolution as it was an exercise in state-building. It was a campaign through which Moscow endeavored to extend a range of controls—political, social, economic, and cultural—over the countryside. The ultimate aim was to create a communist hegemony that would allow for the maximum exploitation of the countryside in the interests of both industrial development and military defense. Collectivization was a massive experiment in social engineering, aimed at the pacification and "socialist transformation" of the countryside. The entire process was considered an essential part of Stalin's "socialism in one country," an effort to develop an autarkic economic system that was both anti-capitalist and indeed anti-western, both imposed by the conditions of "capitalist encirclement" and *desirable* from the point of view of the emerging insular state. The system was born in crisis, thus contributing to its militaristic and increasingly xenophobic form and features. Ideology was never absent, but neither was it defining. Instead, it served most often as a rationale for justifying practices aimed at subjugating and exploiting the peasantry. Collectivization

---

[17] In 1930, there were over 1,000 assassinations and close to 6,000 assaults of rural officials and activists; at the same time, there were over 13,754 peasant "mass disturbances" in the villages (of which 3,712 were exclusively made up of women protesters, with women forming the majority or a significant part of protesters in the other cases). See Viola et al., eds., *Peasant Rebels under Stalin*, 110, 136, 183.

[18] Order 00447 was first published in *Trud*, June 4, 1992, and reproduced in Danilov, Manning, and Viola, eds., *Tragediia Sovetskoi derevni*, vol. 5 (book 1), 330–7.

[19] Fitzpatrick, *Stalin's Peasants*, 80, 96.

was not limited to "taking grain"—indeed, in that respect, collectivization was something less than an unmitigated success.

BEYOND PARTICULARISM I: THE WEIGHT OF THE PAST

Soviet collectivization cannot be understood apart from the immediate contingencies of the late 1920s and the desiderata of the Stalinist dictatorship. At the same time, Soviet collectivization can be situated within a broader context that takes into account the legacies of the past.

The Soviet experiment was, of course, never entirely free of the weight of the past. Despite revolution and Soviet claims of a new world, the *Soviet* cannot be understood in the absence of the legacy and logic of the Russian empire. Needless to say, this is anything but a new proposition; instead it has a venerable historiographical tradition in the West.[20] But at a time when ideology is often at the forefront of interpretation, it is worth revisiting this line of argumentation.

The Russian Revolution of 1917 did not alter the basic themes of Russian history. The imperatives of geography, soil, and climate; the role of the state and state centralization; expansion; modernization; and the primacy of defense needs did not disappear in the wake of the revolution. To a certain extent, one could argue that these themes were if anything accentuated as central power disintegrated with the revolution and was reconstituted through the various political and economic processes associated with Stalinism.

The most important series of continuities between patterns of pre-revolutionary and post-revolutionary historical development revolve around the role and nature of the state. It is a truism of Russian history that the state served as the primary agent of historical change. Its defining role is explained in various ways: by reference to the sheer size of the country; the dependent nature of social estates, especially the gentry; a largely absent middle class and weak civil society; and agrarian backwardness and rural under-government, to offer a far from exhaustive list. Whichever one of these factors one highlights, it is incontestable that the Russian state

---

[20] See for example Berdyaev, *Origins of Russian Communism*; von Laue, *Sergei Witte*; von Laue, *Why Lenin*; Pipes, *Russia Under*; Timasheff, *The Great Retreat*; Tucker, "Stalinism as Revolution;" Tucker, *Stalin in Power*.

was an "interventionist" state long before the advent of any recognizable "modernity."

The history of the Muscovite and Imperial periods is to a great extent a history of state building from the center out. The Muscovite princes "gathered" the lands (and the revenues); the Imperial monarchs extended their reign in an almost continuous process of geographical expansion and (often failed) administrative reform. State building emanated from the center; power was centrifugal. Military defense needs and the absence of natural geographical boundaries pushed state building and expansion. New territories were populated by a continuous process of migration and Russian colonization—voluntary and involuntary—and the borders were shored up by the strategic placement of service gentry (through gifts of land) in new areas.

In the Imperial period, state building and expansion often came as a response to the West—England, France, Austria, and Prussia. The problem of Russian "backwardness" was both real and constructed. It was real in terms of Russia's comparative position to the West: Russia was continually playing "catch up" with its economy in order to compete militarily with Western powers. And it was constructed both in the sense that the West may not have been the best yardstick with which to measure Russian development and in the sense that the problem of backwardness became a key element of Russian economic development and was a central issue for the intelligentsia when considering Russia's future.

The state was the main benefactor of economic modernization, especially during periods of its greatest momentum. Economic development worked in tandem with state building: as the economy progressed, the state strengthened its powers. Economic development also reinforced expansion, as the state sought outlets to ports and access to natural resources; colonization then followed in the wake of expansion.

Russian state building, economic modernization, and expansion depended upon the maximum centralization of resources. The Russian state was an extraction state, heavily dependent upon the land and villages for its revenues—all the more so given the slow and weak development of urbanization and the largely agrarian basis of the country. Both state and towns developed in an exploitative symbiosis with the countryside. Taxes, grain, labor, and cannon fodder were siphoned from the countryside as the Russian state developed. The countryside found itself in a subordinate economic position.

As cultural Westernization came to accompany economic Westernization, the countryside fell behind. This resulted in a cultural and economic gulf between town and countryside. In the course of the nineteenth and early twentieth centuries, the peasantry came to be viewed as "other" (whether in terms of its "Russianness" or maintenance of tradition, its primitivism, its inherent "revolutionary instincts," or its supposed destiny to disappear with the advent of a modern industrial society). Both the state and the radical intelligentsia increasingly attempted to "civilize" the peasantry. For the state, however, the peasantry remained first and foremost an economic resource for development.

At the risk of extreme simplification, Russian history is reduced to its bare essentials in order to demonstrate the *Russian* features of collectivization. Soviet collectivization can be seen as an extreme extension of state building and economic extraction from the pre-revolutionary Russian tradition. As with expansionist Imperial Russia, collectivization was state initiated and state centered, and hampered by weak provincial and rural government. It was a radical exercise in state building—an attempt to extend political, economic, and cultural controls over the countryside. It was a massive exercise in resource extraction. It was also an agent of expansion and colonization, as the social and political detritus of collectivization was siphoned off in dekulakization to "colonize" the north and east, which helped to create an army of forced labor. And a variant of the same cultural imperialism which reduced the peasantry to "other" in Tsarist Russia continued to animate Bolshevik policy as the peasantry came to be viewed as an enemy class, destined to disappear with industrialization, in need of "proletarian" tutelage *(shefstvo)* and deemed "socially dangerous" by the secret police. Finally, military defense—both real and exaggerated—played a key role in the entire assembly of policies that together constituted the Stalin Revolution.

This is not, however, an argument for a *Russian* particularist understanding of Soviet collectivization. The fiercely autarkic nature of Soviet development under Stalin, its anti-Western and anti-capitalist nature, surely set it apart from the Russian tradition. The scale of the policy and the repression necessary to implement the campaign also distinguished it from earlier Tsarist practices. And, of course, Marxist-Leninist-Stalinist ideology cast cultural imperialism in an entirely new light.

Nevertheless, it is important to keep in mind the burden of the past when attempting to contextualize collectivization. This is important not because collectivization was perceived as a kind of "second serfdom," as many peasants concluded, or that collectivization "looked like" serfdom, as some Western historians argue. History by analogy explains little. Rather, collectivization, in part, addressed a series of chronic structural problems of Russian state and economic development and thus, in radical, indeed tragic form, served as a continuation of key themes of Russian history.

BEYOND PARTICULARISM II: THE "MODERN" STATE

It is clear, then, that the "Soviet experiment" did not operate in a vacuum. Neither sheer power and ideological hubris nor the heritage of the past entirely explains collectivization. It is therefore useful to expand our scope of analysis yet again by revisiting another line of explanation, one that extends outward, seeking to understand Soviet developments within the context of comparative history.

This emphasis, like that on the Russian past, has a long tradition in Western historiography. Its earliest manifestation was in the totalitarian and modernization models.[21] The totalitarian model, of course, arose as an attempt to understand the political similarities among communist and fascist states in the prewar period. It would later be not only the foundation of Western Sovietology, but also of Western Cold War ideology and U.S. foreign policy.[22] The modernization model, though clearly biased toward the West, focused on industrialization and agrarian revolution as fundamental aspects of modernization, inserting the Soviet Union into a larger comparative framework of development.

Comparative approaches to Soviet history in the 1990s—while still largely Western-centered—are far more expansive. The so-called "modernity school" focuses on the role of the interventionist state in the twentieth century and its attempts to "mold" its polity and society according to rational, "Enlightenment" ideals.[23] A less Western-oriented variant, derived from James Scott's work on "high modernism," places Russia within the greater context of largely failed experiments in *creating* the modern state.[24]

---

[21] On the history of the concept of totalitarianism, see Gleason, *Totalitarianism*; for a classic example of the genre, see Friedrich and Brzezinski, *Totalitarian Dictatorship*. The most sophisticated and intellectually challenging account remains, of course, Arendt, *Origins of Totalitarianism*. For a summary description of the modernization approach, see Black, *Dynamics of Modernization*.

[22] See for example Spiro and Barber, "Counter-Ideological Uses."

[23] The founding work is Kotkin, *Magnetic Mountain*.

[24] Scott, *Seeing Like a State*.

The modernity school is associated with a group of scholars at Columbia University in the 1990s. Their focus is largely on the state, which they view as the child of Enlightenment, bent primarily toward molding and sculpting society according to the supposedly rational goal of bettering it. Policies as diverse as social welfare, universal education, dekulakization, and the Nazi "final solution" can all be subsumed into this paradigm. Adherents distinguish between the liberal and illiberal state, as well as examining a series of particularist factors (primarily ideology) in order to understand the scale of repression that occurred in the Soviet Union and Nazi Germany, but still see a common denominator in the activity and ethos of the modern state.

Following the work of Zygmunt Bauman on the Holocaust, the shorthand of the "gardening state" has served as a useful metaphor for the activities of the modern state and especially the modern "illiberal" state.[25] Both the Soviet Union and Nazi Germany were avid, omnipotent, "gardeners," weeding out what they viewed as alien, enemy, or subhuman, and cultivating (or attempting to cultivate) "a new man." This kind of gardening then can be used to explain the forced population movements, police activities, and repression of the Stalin years, including the "liquidation of the *kulak* as a class."

While the modernity school bases its comparisons primarily within a Western context, Scott expands the field of comparison beyond the Western world. His focus is broader in that his analysis combines both the state's intentions (the focus of the modernity school) and the reality on the ground—that is, how state policies were put into practice, received, and altered. Scott uses the term "high modernism" to describe the state's desire for "mastery of nature (including human nature), and, above all, the rational design of social order commensurate with the scientific understanding of natural laws."[26] In the context of Soviet history, high modernism, in combination with an authoritarian state, transformed modern techniques in social engineering into societal disaster. In the case of collectivization, the policy was implemented from the outside on subjects in a void of local knowledge and within the context of an illiberal polity.[27]

Modernity and high modernism are closely related concepts and provide useful tools for understanding the Stalin Revolution and collec-

[25] Bauman, *Modernity and the Holocaust*. For an application of this concept to Soviet history, see Weiner, "Nature, Nurture."

[26] Scott, *Seeing Like a State*, 4.

[27] See for example my article, "The Aesthetic of Stalinist."

tivization. It may be that Scott provides the more useful comparison for understanding the Soviet state given that he does not limit his focus on the West. The Soviet Union, after all, remained a largely agrarian nation with a large, traditional peasantry and a relatively small industrial proletariat circa 1930. The modernity school fails to account for the underlying impact of long-standing socioeconomic structures, thereby missing the key points of why the Soviet state loomed so large over society and why there was such a huge disjuncture between the vision of the state and its reality on the ground, one of the most persistent and often ignored themes of Soviet (Russian) history.[28]

An equally salient point of comparison comes from pre-twentieth century state building. Here I am thinking of the process Eugene Weber described for centralizing France, which he labeled "internal colonization," a process whereby the state gradually extended its control—political, economic, and cultural—to the disparate regions and peasantries of France.[29] This model is particularly useful in understanding the state-building aspects of Soviet collectivization. The process and some of the underlying motivations were comparable; the results were, of course, profoundly different.

A broader comparative perspective offers new ways of understanding collectivization. Collectivization was nothing if not a massive exercise in social engineering aimed at developing, overnight, a new society and economic polity. Undesirable elements—the "weeds," to use the Bauman metaphor—were mercilessly removed. An attempt to create "a new man"—the *kolkhoznik*—followed, through education, new labor patterns, and military indoctrination. All of this was inspired by the strategy of centralized, supposedly "rational planning," which aimed at "human betterment." The disastrous results of the policy arose in part because of the brutal nature of the endeavor and its ideology, and in part because of the absence of real local knowledge and real planning, and ineffective local administration.

It is important to note, however, a series of important factors that may not fit so seamlessly within the modernity approach. First of all, the Soviet Union, unlike Nazi Germany for example, cannot be characterized as a "strong state." State administration was, on paper, highly centralized; at the same time, rural government was weak, inefficient, and inadequately staffed. As a consequence, policy in the countryside was not implemented

---

[28] Viola, "The Aesthetic of Stalinist."

[29] Weber, *Peasants Into Frenchmen*.

through routine channels of government. Instead, policy erupted in a "campaign style," through plenipotentiary rule and local sweeps ordered by Moscow, after which inertia reigned. The state was neither strong nor effective in achieving the results of its vision. The history of collectivization and dekulakization is a tale of tragedy based on the complete disconnect of an already unrealistic central policy and the inability to implement it. As a result, Moscow governed largely by means of repression. Repression became not a manifestation of state strength, but an expression of state weakness, a substitute for routine mechanisms of governance.[30]

The state, moreover, did not act in a vacuum. Instead, it confronted a fairly strong and complex society made up of cohesive communities with local traditions and identities, able and willing to at least attempt to defend themselves.[31] The "garden" was not so malleable, it turned out, and this too affected the outcome. The general economic context was also quite different. It is debatable whether Russia, with its huge peasantry and agrarian economy, could ever be categorized as "modern," in comparison to the West's version of modern. And here arises the greatest disjuncture of all: the intentions and visions of the center may have been "modern," but the capacity for realizing them, the process, and the end results were not.

Nonetheless, a broader, comparative perspective may illuminate important features of collectivization by placing them within the context of the modern state as well as within the context of Weber's idea of "internal colonization." Comparative perspectives may help us to understand the state building and social engineering dimensions of collectivization, as well as the immense scale of this endeavor. At the same time, a more comparative perspective highlights the particularities of the Russian twentieth century, which in turn may help us in understanding the tragedy of the Soviet experiment.

Conclusion

It is clear our understanding of collectivization goes beyond "taking grain." Historical contingency was central to the process; at the same time, the peculiarities of Russian historical structures and developments, carried

---

30 See Viola, "The Aesthetic of Stalinist Planning," and Viola, "The Campaign to Eliminate the Kulak."

31 See Viola, ed., *Contending with Stalinism*—especially articles by Viola, Rossman, and Northrop.

over into the Soviet period, also affected the collectivization campaign's possibilities and outcomes. A comparative perspective allows us to judge what the Soviet state may have shared in common with other twentieth-century states—whether in terms of modernity, high modernism, or the more traditional modernization—thereby providing insight into certain aspects of state development, intervention, and ethos.

Although this broader understanding may diminish or better balance the ideological components of collectivization, it does not cancel them out. The Soviet Union's position as a revolutionary, self-proclaimed socialist state among hostile capitalist nations—the always vaunted capitalist encirclement of communist propaganda—continually served to radicalize domestic politics and increase the urgency of industrialization. The very idea of capitalist encirclement, compounded by a bloody civil war that had torn the nation apart, helped create an atmosphere of omnipresent enemies.

Soviet collectivization also very prominently featured ideological motives and goals. It was cast as a "class war." The collective farm was meant to enshrine collectivist values and practices that would lead to the "socialist transformation" of the peasantry. Yet it is clear to historians of collectivization that the actual reality of socialist transformation was very different. In fact it may have been ideology in praxis, or perhaps more accurately, ideology-in-the-becoming, one of the more quintessentially Soviet features of collectivization. Evgeny Dobrenko has written that "Russia entered Europe in the 20th century just as it had according to Custine—through discourse and representation, passed off this time as 'socialism.'"[32] This formulation is apt because it grounds the Soviet experience in both the modern (the European in reality) and the Russian (of Potemkinism and "discourse and representation") in an entirely novel way. It points to the failure and unreality of Soviet ideology rather than to its realization. It points to the fact that ideology was whatever discourse dictated, and inevitably and always located in the future, in the same way that socialist realism was idealized as socialism-in-the-becoming.

Collective farming remained very much a socialist agriculture-in-the-becoming. There was nothing particularly socialist in Stalin's collectivization practices or in their implementation—unless one's definition of socialism is equated with violence. What emerged from collectivization was a hybrid of peasant practices and collectivist forms, which resulted by the end of the Soviet period in the emergence of the ultimate in no-risk

---

32 Dobrenko, "Socialism as Will," 707.

peasant farming and social welfare. Insofar as collectivized agriculture is concerned, ideology remained strong in the party rhetoric but in practice the actual forms themselves stayed firmly planted in rural Russia.

The Soviet Union was a hybrid modern state—and its hybrid status is key to understanding the role of Soviet ideology. Moored in the pre-twentieth-century socioeconomic world of an agrarian economy and a largely traditional peasantry, the Soviet Union set out to leap over what had been a century of organic European social and economic development achieved through far more gradual state engineering. If the intentions were modern and the discourse Soviet, the path to modernity, and the raw materials the Soviet state sought to shape it, was distinctly un-modern. And it is here that we see most clearly how the past and the modern were fused together to form this hybrid Soviet state.

It may also be worth questioning the very meaning of the term "modern" in reference to Soviet modernization. Stalin's attempt to build "socialism in one country" represented an alternative to Western models of development, a path to modernity never meant to replicate the results of Western capitalism. In its purist form, Stalin's "socialism in one country" was a radical reaction against those models. Soviet modernization was anti-market, anti-trade, and indeed anti-Western to the extent that "Western" could be equated with "capitalist." The system that would thus emerge would be an autarkic economic system, both imposed by the conditions of "capitalist encirclement" and *desirable* from the point of view of the emerging insular state.

The Soviet superpower was a leviathan in bast shoes. It was, most importantly, an *extraction state* dependent upon a high level of centralization and force to administer its vast territories and continue to "squeeze" its population for economic development and military power. Ultimately, popular support for collectivization remained too weak to sustain its growth; after the death of Stalin, the Soviet state was largely unable and unwilling to resort to the violence this required. When the Soviet Union came to an end in 1991, its fall was more nearly a result of the hollowness behind its facade than the failure of its ideology or the victory of Ronald Reagan's United States. In this light, Soviet collectivization was a forced, but ideologically rationalized stopgap measure designed to transform the Soviet Union in minimal time from an agrarian nation into a modern industrial state.

Soviet collectivization became a model for collectivization in Eastern Europe and elsewhere. There are multiple ironies in this development, not the least being that Soviet collectivization occurred in part under conditions historically specific to the Soviet Union. It was, moreover, an improvised endeavor, and a partially botched experiment without precedent, a less-than-perfect model other communist states would come to emulate. Furthermore, the shape of collective farming in the USSR changed over time as the system evolved and, in particular, after the death of Stalin in 1953. The model, therefore, was both dogmatic (in its ideological articulation) and pliable (in its dynamic evolution), enabling a variety of forms of collectivization elsewhere.

At the same time, collectivization in the Soviet Union and in Eastern Europe shared features that are useful in gauging its meaning and impact across borders. First, the decision to collectivize (or not) was based primarily on *political* rather than economic considerations. It was tied into internal Communist Party factional struggles as well as, in the case of Eastern Europe, the Soviet Union's determining influence. As a result, the application of collectivization policies and the evolution of collective farming were often improvisational and provisional, not to mention less than satisfactory from an economic point of view. Second, collectivization followed war—World War I and the Russian Civil War in the case of the Soviet Union, and World War II in the case of Eastern Europe—occurring against a backdrop of heightened fears of international conflict, as well as within the context of domestic political and social strains. These processes varied from country to country, and determined what course, or what level of violence, party leaders chose in implementing their collectivization campaigns—which in turn affected the success of their endeavors. Third, the state played a central role in the collectivization process, often in the face of weak local administration. The center-periphery divisions, which sometimes evolved a result, often impeded collectivization efforts and weakened the ability of the state to exact its full demands upon the countryside. Fourth, collectivization generally had limited support among the peasants or farmers, neither of which acted according to Marxist-Leninist assumptions regarding class and class struggle. As a consequence, there were varying levels of peasant and farmer resistance to collectivization. In many cases peasants exhibited similar forms of resistance across communist states (e.g., women's revolts, provocational rumors, destruction of livestock, flight). Consequently, collectivization had a similar impact on demographics throughout the socialist region, resulting in a large and continuing out-migration from village to town. Finally, ideology was always used as the prime rationale for collectivization across Eastern Europe, although, at the same time, it served more as a gloss than a fixed determinant of practice, leaving the impression of a kind of Potemkin socialism, as it had in the Soviet Union.

Still, the historical-contextual specificities of Soviet collectivization complicate our understanding of its function as a "model" for collectivization in Eastern Europe. As a purely theoretical construct, as an ideological "blueprint," Soviet collectivization certainly was exploited as a model for agricultural socialization. But Soviet collectivization arose in historically specific circumstances, shaped by persistent structural constraints and, in part, by practices of modernity. In this sense, collectivization in Eastern Europe was in large part a foreign import. What had been a ruinous policy within the Soviet Union, based on violence and incorrect assumptions about the peasantry, was transported into a variety of very different social and economic settings within Eastern Europe. The Soviet Union's experiment in collective agriculture was reified and superimposed upon Eastern Europe's agrarian order. What began as experiment ended as socialist parody.

BIBLIOGRAPHY

ARCHIVAL SOURCES:

OGPU decree of February 2, 1930 in *GARF* (Gosudarstvennyi arkhiv Rossiiskoi Federatsii, State Archive of the Russian Federation), f. 9414, op. 1, d. 1944, ll. 17–25.

RGAE (Rossiiskii gosudarstvennyi arkhiv ekonomiki, Russian State Archive of the Economy), f. 5675, op. 1, d. 23, ll. 42–41. (sic).

RGASPI, f. 17, op. 162, d. 8, ll. 109–110; *GARF*, f. 9414, op. 1, d. 1944, l. 82.

BOOKS AND ARTICLES:

History of the Communist Party of the Soviet Union (Bolsheviks). Short Course, 1939 ed. San Francisco: Proletarian Publishers, 1976.

Arendt, Hannah. *The Origins of Totalitarianism.* New York: Harvest, 1973.

Bauman, Zygmunt. *Modernity and the Holocaust.* Ithaca, NY: Cornell University Press, 1989.

Berdyaev, Nicolas. *The Origins of Russian Communism.* Ann Arbor: University of Michigan Press, 1972.

Black, Cyril E. *The Dynamics of Modernization: A Study in Comparative History.* New York: Harper, 1967.

Carr, Edward Hallet. *Socialism in One Country, 1924–1926.* London: MacMillan, 1958, vol. 1. (Volume 5 of *A History of Soviet Russia*).

Cohen, Stephen F. *Bukharin and the Bolshevik Revolution: A Political Biography, 1888–1938.* rev. ed. Oxford: Oxford University Press, 1980.

Danilov, Viktor Petrovich, Roberta T. Manning and Lynne Viola, eds. *Tragediia Sovetskoi derevni: Kollektivizatsiia i raskulachinanie. Dokumenty i materialy v 5-kh tomakh. 1927–1939* [The Tragedy of the Soviet Countryside: Collectiviza-

tion and Dekulakization. Documents and Materials in 5 volumes. 1927–1939]. Moscow: ROSSPEN, 1999–2006, vol. 1: Mai 1927–noiabr' 1929.
Dobrenko, Evgeny. "Socialism as Will and Representation or What Legacy Are We Rejecting?" *Kritika* 5, no. 4 (2004): 675–708.
Fitzpatrick, Sheila. *Stalin's Peasants: Resistance and Survival in the Russian Village after Collectivization.* Oxford: Oxford University Press, 1994.
Friedrich, Carl J. and Zbigniew K. Brzezinski. *Totalitarian Dictatorship and Autocracy.* New York: Praeger, 1965.
Gleason, Abbott. *Totalitarianism: The Inner History of the Cold War.* Oxford: Oxford University Press, 1995.
Iakovlev, Aleksandr N. et al., eds. *Kak lomali NEP: Stenogrammy plenumov TsK VKP (b) 1928–1929 gg. v 5-kh tomakh.* [How NEP was Broken: The Stenograms of the Plena of the Central Committee of the All-Union Communist Party (Bolsheviks) 1928–1929, 5 vols.] Moscow: Materik, 2000.
Kotkin, Stephen. *Magnetic Mountain: Stalinism as a Civilization.* Berkeley: University of California Press, 1995.
von Laue, Theodore H. *Sergei Witte and the Industrialization of Russia.* New York: Atheneum, 1974.
———. *Why Lenin? Why Stalin? Why Gorbachev? The Rise and Fall of the Soviet System.* 3rd ed. New York: Harper Collins, 1993.
Lewin, Moshe. *Russian Peasants and Soviet Power: A Study of Collectivization*, trans. by Irene Nove. New York: W. W. Norton, 1975.
———. "'Taking Grain:' Soviet Policies of Agricultural Procurements before the War" in Moshe Lewin, ed. *The Making of the Soviet Union.* New York: Pantheon, 1985, 142–77.
Millar, James R. and Alec Nove. "A Debate on Collectivization: Was Stalin Really Necessary?" *Problems of Communism* 25 (July–Aug. 1976): 49–62.
Osokina, Elena. *Our Daily Bread: Socialist Distribution and the Art of Survival in Stalin's Russia, 1927–1941*, trans. by Kate Transchel. Armonk, NY: M. E. Sharpe, 2001.
Pipes, Richard. *Russia Under the Old Regime.* New York: Scribner, 1974.
Scott, James C. *Seeing Like a State: How Certain Schemes to Improve the Human Condition Have Failed.* New Haven: Yale University Press, 1998.
Spiro, Herbert J. and Benjamin R. Barber. "Counter-Ideological Uses of 'Totalitarianism.'" *Politics and Society* 1, no. 1 (1970): 3–21.
Stalin, Iosif. *Sochineniia* [Works], 13 vols. Moscow: Gospolitizdat, 1946–52.
Timasheff, Nicholas S. *The Great Retreat: The Growth and Decline of Communism in Russia.* New York: Dutton, 1946.
Tucker, Robert C. "Stalinism as Revolution From Above" in Robert C. Tucker, ed. *Stalinism: Essays in Historical Interpretation.* New York: W. W. Norton, 1977, 77–108.
———. *Stalin in Power: The Revolution from Above, 1928–1941.* New York: W. W. Norton, 1990.
Viola, Lynne. "The Campaign to Eliminate the Kulak as a Class, Winter 1929–1930: A Reevaluation of the Legislation." *Slavic Review* 45, no. 3 (1986): 503–24.
———. *Peasant Rebels Under Stalin: Collectivization and the Culture of Peasant Resistance.* Oxford: Oxford University Press, 1996.

———, ed. *Contending with Stalinism: Soviet Power and Popular Resistance in the 1930s*. Ithaca, NY: Cornell University Press, 2002.
———. "The Aesthetic of Stalinist Planning and the World of the Special Villages." *Kritika* 4, no. 1 (2003): 101–28.
———. Viktor Petrovich Danilov, Nikolai Alekseevich Ivnitskii, and Denis Kozlov, eds. *The War Against the Peasantry, 1927–1930*, trans. by Steven Shabad. New Haven: Yale University Press, 2005.
———. *The Unknown Gulag: The Lost World of Stalin's Special Settlements*. Oxford: Oxford University Press, 2007.
Weber, Eugen. *Peasants Into Frenchmen: The Modernization of Rural France, 1870–1914*. Palo Alto: Stanford University Press, 1976.
Weiner, Amir. "Nature, Nurture, and Memory in a Socialist Utopia: Delineating the Soviet Socio-Ethnic Body in the Age of Socialism." *American Historical Review* 104, no. 4 (1999): 1114–55.

THE COLLECTIVIZATION CAMPAIGN IN THE SOVIET UNION: CHRONOLOGY OF EVENTS

Lynne Viola

1917

March 12: Tsarist regime overthrown in Petrograd.

March 15: Formation of the Provisional Government. Abdication of Tsar Nicholas II.

November 7: Overthrow of the Provisional Government. Second Congress of Soviets opens in Petrograd, with a Bolshevik majority.

November 8: Promulgation of decree nationalizing land.

1918–1920

Russian Civil War and Policy of War Communism, the latter featuring forced grain requisitions and attempts to stir up class war in the villages under the communist "food dictatorship." Some mainly localized attempts at collectivization. Famine ensues in 1921.

1918

June 11: Committees of Poor Peasants *(kombedy)* created.

November 2: Extraordinary tax of ten billion rubles levied on propertied classes of city and countryside.

November 8: Dissolution of Committees of Poor Peasants in most parts of country.

November 21: Nationalization of internal trade.

1920

November 29: Decree nationalizing small industries, including handicrafts.

1921–1929

Era of the New Economic Policy featuring a mixed economy, peasant communal land tenure, and regulated free market in agricultural products.

1921

Famine following Russian Civil War.

March 8–16: Tenth Congress of the Communist Party and the introduction of the New Economic Policy.

March 16: Forced grain requisitions of the Civil War era replaced by tax-in-kind, later by a money tax. Regulated trade in grain and other agricultural produce legalized.

July 7: Decree denationalizing small industries.

1922

October 30: Land Code, legalizing all forms of land tenure.

1923–1924

Scissors' Crisis: price imbalance between industrial and agricultural goods.

1926

Grain harvest reaches prewar levels. Amount of grain marketed does not, setting up the political-economic context of the grain procurement crisis which begins in 1927–1928 and continues into collectivization.

1927

December 2–19: Collectivization placed on the agenda of the Fifteenth Congress of the Communist Party.

1928

January: Introduction of "extraordinary measures" in grain requisitioning (forced grain procurements by mobilized urban grain requisitioning brigades); slow death of NEP and mixed economy from this point to collectivization.

January 4: OGPU directive to arrest private traders.

July 9: Stalin calls for a "tribute" from the peasantry to pay for industrialization.

1929

January: Acceleration of repressive measures against "*kulak* terrorists."

April–May: First Five-Year Plan ratified. At 16th party conference, plans project 9.6% of peasant population to be collectivized by 1932–33 economic year, and 13.6% by 1933–34 economic year.

May–June: Ural-Siberian method sanctioned and extended through the Soviet Union (general village assemblies to endorse grain requisition plans and assign quotas to villagers on the basis of a "class approach").

July 11: SNK SSSR decree calling for expansion of camp system and transfer of all prisoners serving more than 3 years to OGPU for work in colonization and economic exploitation of far north and east (thus setting up conditions for *kulak* influx into camps from 1930).

October 3: Politburo authorizes OGPU and Commissariat of Justice to deal "decisively and quickly" with *kulak* terrorists.

November 7: Stalin publishes "The Year of the Great Turn: On the Occasion of the Twelfth Anniversary of the October Revolution," claiming that the peasantry was flocking to the collective farms "by whole villages, volosts, districts."

November 10–17: November Plenum ratifies transition to wholesale collectivization. Official defeat of the Right Opposition.

December 5, *1929 to January 5, 1930*: Politburo commission, chaired by Iakovlev, on collectivization.

December 27: Stalin announces that the *kulak* will be "liquidated as a class" at the Conference of Marxist Agronomists.

1930

January 5: Communist Party Central Committee decree "On the Pace of Collectivization and State Assistance to Collective-Farm Construction," officially launches wholesale collectivization. Plans projected for wholesale collectivization of major grain regions by fall 1930 or spring 1931 at the latest, other grain regions by fall 1931 or spring 1932 at the latest. Silence regarding collectivization of the remaining countries, thereby allowing for the "initiative of the vanguard."

January 15–30: Politburo creates commission, chaired by Molotov, for the elaboration of measures against the *kulaks*.

January 30: Politburo decree "On Measures for the Liquidation of *Kulak* Farms in Districts of Wholesale Collectivization." "Kulakdom" divided into three categories: category one—heads of households to be arrested, followed by execution or internment in labor camps, while their families to be subject to forced deportation following expropriation; category two—entire family to be subject to forced deportation, following expropriation; category three—left in place for time being and expropriated of "means of production," but not allowed to join collective farms (eventually dekulakized in next round or taxed out of existence).

January–April: Arrests of first-category *kulaks* (approx. 30,000 heads of households will be executed in 1930 and 1931). Category originally estimated at 60,000, but plan overfilled to result in closer to 280,000 arrests between January and October 1930.

February 2: OGPU directive 44/21 "On Measures for the Liquidation of the *Kulak* as a Class."

February: Deportations of families of first-category *kulaks* and second-category *kulak* families begin, to be settled in "special settlements" in remote territories rich in resources where *kulaks* will serve as labor force. Their first task will be to actually build the special settlements (in 1930, as many as 337,563 families subject to some form of dekulakization, with or without exile. Of these, approximately 115,231 families deported to Siberia, Urals, and Northern European Russia).

March 1: Model Charter of Agricultural Collectives *(arteli)*. 57.2% of all households collectivized by this time (with far higher rates in individual regions—e.g.,

Central Black Earth, 83.3%; Moscow Region, 74.2%; North Caucasus, 79.4%; Lower Volga, 70.1%; Middle Volga, 60.3%).

March 2: Stalin publishes "Dizzy with Success: Concerning Questions of the Collective-Farm Movement." Temporary retreat from collectivization as peasants quit the collective farms in droves, with overall percentages of households collectivized falling from 57.2% on March 1 to 38.6% in April, 28% in May, and downward to a low of 21.5% in September before the next collectivization wave begins.

March 9 to August 13: Tolmachev commission under Sovnarkom RSFSR to coordinate special resettlement issues established (first train loads of *kulak* families already en route).

April 1: Shmidt commission under Sovnarkom USSR to coordinate special resettlement issues established.

April 5: Bergavinov commission to review incorrect exiles in North begins work.

July 30: Decree liquidating peasant land societies in districts of wholesale collectivization.

Summer: Families of special settlers begin trek into hinterlands. Male heads of households had been sent into hinterland immediately, while families lingered in towns and cities of exile until further transport possible.

September 1: Deadline for construction of special settlements (not met).

Fall: Wholesale collectivization renewed. New, smaller waves of dekulakization ensue.

1931

February 20: Politburo resolution "On *Kulaks*" plans to exile 200,000 to 300,000 *kulak* families to Kazakhstan (in view of lack of preparation to receive the exiles in Kazakhstan, numbers reduced in May 1931. Still, in the end, some 265,795 families are exiled to Urals, Siberia, Kazakhstan, and Northern European Russia in 1931).

March 11: Politburo establishes the Andreev Commission on *kulak* exile and resettlement.

May: Andreev Commission calls for OGPU takeover all special settler affairs. Second large wave of dekulakization begins and extends through the summer.

May 17: SNK RSFSR decree on organizing Red Army collective farms (subsequently used to repopulate famine area collective farms and to settle the border).

July 1: Official OGPU takeover of all special resettlement affairs.

July 3: Central Executive Committee decree indicating *kulaks* can be rehabilitated after five years (prior to this, there was no fixed sentence or term. Very few will actually be rehabilitated after the five-year mark).

July 20: Politburo declares that period of mass exile of *kulaks* "basically" over (but not really).

October 25: OGPU issues "Temporary statutes on the rights and responsibilities of special settlers and on the administrative functions and rights of the settlement administration."

By the end of 1931, roughly 60% of all peasant households are collectivized in the Soviet Union.

1932–1933

Famine.

1932

Wholesale collectivization of basic grain regions completed by end of 1932.

May 6: Decree permitting collective farm trade in bread (after fulfillment of government requisitions).

June 25: Decree on Revolutionary Legality.

August 7: Decree on Guarding the Property of Government Enterprises, Collective Farms and Cooperatives and on Strengthening Socialist Property (punishments of 5–10 years incarceration in concentration camps for theft).

December 6: Ukrainian state/party decree on blacklisting villages guilty of sabotaging grain requisitioning (blacklisting then led to deportations of entire villages—339,327 new exiles join *kulak* special settlers in 1932 and 1933).

Late 1932: Introduction of internal passports, but not to peasantry.

1933

January 27: OGPU order to organize *politotdely* (under OGPU leadership) in Machine-Tractor Stations.

May 8: Central Committee and Sovnarkom (again) end period of mass exiles (again).

1934

May 27: USSR Central Executive Committee decree on rehabilitation of *kulaks* after 5 years.

1935

January 25: Central Executive Committee decree on rehabilitation ruling that rehabilitation does not give *kulaks* right to leave place of exile.

February 17: New Model Collective Farm Charter, guaranteeing collective farmers their garden plots *(uchastki)*.

1937

February: Reiteration that *kulaks* cannot leave places of exile upon rehabilitation. NKVD address suggesting that special settlers still hostile to Soviet power be sent to labor camps for 3–5 years.

April 13: Politburo Resolution on sabotage in the Commissariat of Agriculture and the Commissariat of State Farms.

July 30: NKVD Operational Order 00447 "On the operation to repress former *kulaks*, criminals, and other anti-soviet elements" (approximately 46,000 *kulak* special settlers arrested; total, all categories of population, of est. 1,575,259 people arrested, of which 1,344,923 convicted; 681,692 executions; most arrests are through mass operations rather than elite purge, meaning large, but inestimable percent of peasants). Size of labor camp population will reach 2 million by early 1939, with population of "special (or labor) settlements" roughly equal.

1938

December 19: Decree on extending collectivization to 30,000 *khutory* (peasant farms with fields and house enclosed) in Belorussia.

1939

May 27: Decree extending collectivization to all *khutory* in Soviet Union in 1939 and 1940 (affecting 447,400 farms).

1942

April 11: State Defense Committee directive authorizing draft of special settlers.
October 22: NKVD directive on emancipation of families of special settler soldiers.

1945

January: Decree granting all deported *kulaks* the rights of Soviet citizens, but denying them the right to leave their place of exile.

1946–1947

Postwar famine.

1946

September 19: "On measures to liquidate breaches of the [collective farm] statutes" (all lands acquired by private persons or institutions to be returned to collective farms).

1948

Tax-in-kind introduced on individual households and on production from private plots.
February 21: Directive to exile from Ukraine collective farmers not fulfilling *truoden* (labor day) requirements or leading anti-soviet, parasitical life. Decree extended to rest of Soviet Union in June.
May 21: Decision to collectivize agriculture in Baltic States, Bessarabia, Western Ukraine, and Belorussia begins.

1950

Khrushchev carries out campaign to amalgamate small collective farms (number of collective farms reduced in this year from 236,900 to 123,700, and soon after to 93,000).

1954

August 13: Central Committee directive lifting all restrictions on *kulaks* exiled in 1929–33, including the right to leave their place of exile.

1954–1956

Virgin Lands campaign; transformation of some collective farms into state farms. Further amalgamations of collective farms: number of collective farms fall from approximately 125,000 at end of 1950s, to 69,100 in 1958, to 36,000 by 1980 (resulting in increasingly large sizes for collective farms).

1958
Abolition of Machine-Tractor Stations.

1976–80
Peasants receive internal passports.

1990s
1990–91: Series of decrees fully "exonerating" the *kulaks* (and other victims of Soviet repression).
February 1990: Government resolution, "Fundamentals of Land Legislation" (Soviet citizens permitted to lease but not own land for peasant farming).
December 1991: Government resolution, "On the Procedure for the Reorganization of Collective and State Farms" (reform to privatize collective farms, transforming them into "joint-stock" companies, agricultural cooperatives, limited liability companies, etc).
October 1993: Government resolution, "Regulation of Land Relations and the Development of Agrarian Reform in Russia" (state and collective farm members given certificate of landownership for share of farm's land; allows for ability, with restrictions, to purchase agricultural land).

The Collectivization of Agriculture in the Baltic Soviet Republics, 1944–1953

DAVID FEEST

Two analytical questions are of outstanding importance concerning the history of the collectivization in the Baltic Soviet Republics: first, why was agriculture not collectivized immediately after the incorporation of the Baltic states into the Soviet Union in July 1940 or after the region's reconquest in 1944; second, why was collectivization implemented so hastily in 1948–1949. These issues cannot be understood without taking into account the organizational problems Soviet leaders faced when attempting to assert dominance over the three Baltic republics, as well as the ideological vocabulary that was used to conceptualize these problems and the ways in which this vocabulary corresponded to existing interpretations of the Soviet experience in the late 1920s and early 1930s, when agriculture had been collectivized in the old republics by means of mass terror and deportation. Overall, the answer to both these questions can shed additional light on the question of why Soviet authorities considered it necessary to collectivize at all.

The similarities of the collectivization politics in the three Baltic countries are striking. While collectivization was hardly an issue in the immediate postwar years, a radical land reform program was implemented by the Soviet authorities in order to gain local support as well as to punish real or alleged enemies of the new regime. Meanwhile, existing agricultural cooperatives were taken over by the new regime. It was not until May 1947 when leaders in Moscow demanded more concrete plans for establishing collective farms from the party leaders of the three republics, yet the results were meager. Moreover, against all ideologically founded expectations, it was primarily mainly the richer farmers who, overburdened by high taxes, joined the collective farms *(kolkhozes)*, while the small holders preferred to retain their private plots. Large-scale collectivization was achieved in March 1949 when party leaders carried out mass deportations of wealthier peasants, the so-called *kulaks*, and other "enemies of the people." These hastily established collective farms were even less productive than the small private farmsteads had been after the Soviet land reform.

In this paper, I will argue that although collectivization had been an economic failure in the old Soviet republics, it nevertheless provided important political as well as symbolic functions that outweighed the economic failures. While in the early years after the change in regime there were reasons for party leaders to pursue a slower course, by 1948 the internal as well as external situation had fundamentally changed, after which Moscow adopted a hard-lined approach which resembled the violent collectivization drive in the prewar Soviet Union, crushing aspirations by local communists of instituting a "third way."

Collectivization in the Baltic republics has until now mainly been treated from a single country perspective. Although before 1991 Soviet historians were bound by strict ideological limits, some nevertheless provided useful and reliable facts from the archives.[1] At the same time, Western historians (mainly of Baltic descent) compensated for the lack of direct access to archival sources by offering critical examinations from previously published material.[2] The opening of the archives after the Soviet Union collapsed has provided a wealth of new research opportunities. The historical commissions in the Baltics have put much effort into making new findings from the archives available to a broader public.[3] Other scholars have used the archives to research case studies on collectivization and the role of the respective Communist Party organizations.[4] These new research trends have allowed for an interpretative survey of Soviet policy in all three republics, stressing the similarities in the organization of and reac-

---

1 In this chapter I especially draw on the following unpublished candidate theses: Ruusmann, *Põllumajanduse taastamine ja kollektiviseerimine*, 1967; and Kivimaa, *EKP tegevus vabariigi*.

2 On Lithuania: Zunde, *Die Landwirtschaft*; Biggart, "Collectivization of Agriculture;" Girnius, "Collectivization of Lithuanian Agriculture." On Latvia: Labsvirs, *Sovietization of the Baltic States*. On Estonia: Taagepera, "Soviet Collectivization;" Taagepera, "Soviet Collectivization: The Deportation Phase." On the three Baltic Republics: Taagepera and Misiunas, *The Baltic States*.

3 On collectivization, see especially Bleiere, "Repressions against Farmer;" Paavle, "Sovietization of Agriculture."

4 Swain, "Deciding to Collectivize;" Feest, *Zwangskollektivierung im Baltikum*.

tions to the Russian case. However, the lack of detailed local case studies in a common Western language or in Russian makes it hard to carve out the *differences* between these country cases. An exception is the recent work of historian Anu Mai Kõll, who has provided a micro-study of kulakization that abandons the top-down, political-historical approach and instead emphasizes the modes of participation by local executive committees.[5] More comparative studies of this kind might clarify key differences in what were broadly similar collectivization projects in the Baltic republics.

Understanding Collectivization

Studying collectivization in the Baltic republics might help to clarify the underlying impetus and drive for collectivization by Soviet leaders. The exclusively economic interpretation, which is based on the notion that collectivized agriculture was necessary for industrialization, has been largely refuted by the groundbreaking work of Viktor P. Danilov.[6] The economic argument makes even less sense in the Baltic contexts, as party elites were aware of the devastating effects that collectivization could cause in the short term. Therefore, the party's reasons for pursuing collectivization likely fell beyond immediate economic considerations. In her contribution to this collection, Lynne Viola has drawn attention to the general concepts behind the Soviet approach to governing. Utopian visions of creating a new human being and images of a "gardening state" were not specific to the Soviet Union but were general traits of the "high modernism" that was especially fervent in the totalitarian states of the twentieth century.[7] Among these, however, the Bolshevik state stands out for its relative weakness, especially in the countryside where the Communist Party had few supporters and constantly struggled for control. Fueled by Marxist slogans and an ideological militancy, Bolshevik elites increasingly perceived the countryside as a battlefield where the struggle between good and evil was

5 Kõll, *Village and the Class War*.

6 Danilov, *Sozdanie material'no-technicheskich*. In his book on the Soviet restructuring of the Estonian economy, Olaf Mertelsmann generally argues in favor of the economic explanation for collectivization but refers to works on the economic theory of the 1920s rather than to the postwar situation in the Baltic republics. See Mertelsmann: *Der stalinistische*.

7 See the contribution by Lynne Viola in this volume.

taking place.[8] Party leaders did not search for root *causes to problems* but rather looked for *culprits*, assigning the lack of progress in building the utopian state to an "omnipresent conspiracy."[9] Thus, collectivization was always more than a struggle for grain—it was a struggle with the enemies of socialism.

That is not to say that Bolshevik leaders were not interested in the economic benefits of collectivized agriculture. But political power and economic success cannot be fully disentangled. A deeply rooted distrust of the private market made even an inferior state-controlled economy seem preferable to any kind of private market. "Agricultural goods that disappeared into the private economy," according to Paul Gregory, "were regarded as lost, even if the private market eventually sold these goods in urban markets."[10] The Stalinist "revolution from above" finally put an end to the "anarchy of the market" and disposed of the alleged enemies of the socialist state at the price of a dramatic decline in production.

The process of enforcing Bolshevik power in the countryside and distinguishing between loyalists and "enemies" was always accompanied by ideological conceptualizations aimed at justifying the actions taken by the party. Yet it is often difficult to discern how far the analytical functions of these theoretical concepts actually went, or whether they merely served as justifications for actions rooted in raw power politics. These difficulties of interpretation become especially pronounced when considering the postwar years of the Baltic republics compared to the Soviet experience of the 1920s and 1930s. In the latter context, terms like the "Lenin cooperative plan," the "class struggle in the countryside" or the "voluntary path to collectivization" did not describe how collectivization had eventually been accomplished in the old Soviet republics, but rather, how the Bolsheviks used this rhetoric to promote its programs. The use of these terms in the postwar years was ambiguous, however. For those who took this rhetoric more literally, they rallied for a return to certain ideals (and illusions) of the 1920s and for possible alternatives to development. For others, these terms were just a code for the way collectivization had *historically* been enforced—with violence and terror. This ambiguity was increased by the

[8] This has been stressed especially by Baberowski, "Stalinismus 'von oben.'"

[9] The term is borrowed from Gábor Tamás Rittersporn's article, which, however, does not focus on the rural areas. Rittersporn, "The Omnipresent Conspiracy." For the significance of conspiracy theory see also: Getty, "Afraid of their Shadows."

[10] Gregory, *Political Economy of Stalinism*, 31.

fact that during the early postwar years, the pragmatism that had been introduced during the war years seemed to allow for the possibility of a less brutal approach to agricultural policy. This was even more so in the Baltic Soviet Republics, which had only come under Soviet rule shortly before the war and the agricultural market was still mainly private. Therefore, the Baltic states serve as an example of the complex dynamics between politics, economics, and ideology in the Soviet system.

Why the Delay?

Although leaders in Moscow would hardly have permitted the newly acquired republics to retain economic systems incompatible with those of the old republics, collectivization was not on the agenda shortly before and directly after World War II. Nor was it even discussed as a long-term plan. In Latvia, the first discussions of collectivization were dispelled by the government as "mere rumor." In Estonia, the first Secretary of the Communist Party Nikolai Karotamm also claimed that rumors about collectivization were inventions of enemies or based on harmful prejudices. Collectivization was strictly off-limits.[11]

One of the reasons for this was a "*policy of reassurance*" which had been central to Soviet politics since the occupation and subsequent incorporation of the Baltic republics into the Soviet Union in 1940. The formation of governments comprised of left-wing intellectuals rather than of communists had driven home the message to locals and foreign observers alike that the Baltic republics were to receive special treatment from Moscow. Mass deportations on June 14, 1941, however, crushed these illusions about the character of Stalinist rule.[12] Still, when the Red Army reentered the Baltic republics after two years of German occupation, from 1942 to 1944, public propaganda resumed in the same vein. Considerations of foreign policy and the relation of the Baltics to its former allies were most prominent, as was the case when the Baltic Soviet Republics were represented by their own foreign ministers in the Paris Peace Conference from July to October 1946. Such symbolic maneuvers had a strong influence on how local party

---

[11] Labsvirs, *Sovietization*, 13; Kivimaa, *EKP tegevus vabariigi*, 25.

[12] According to the most reliable summary, approximately 10,800 people were deported from Estonia, 15,400 from Latvia, and 18,000 from Lithuania: Brandes, Sundhaussen and Troebst, *Lexikon der Vertreibungen*, 226, 390. For the lower numbers put forth by the MGB, *Narody stran Baltii*, 268.

elites envisioned future developments of their respective republics. While in the mid-1940s Latvian communists still envisaged a "third way," some of their Estonian counterparts were even discussing a "Mongolian way" which would transform Estonia into a semi-independent republic.[13] Hence, rapid collectivization fit neither into Moscow's *policy of reassurance* nor into the long-term plans of some local communists.

Apart from the effort to regain public confidence that was still shaken by the mass terror of 1941, there were also economic considerations involved in Moscow's more careful approach to collectivization. After the war's destruction, small-scale subsistence farming was actually an important source of economic recovery for the Baltic states. What the economic historian Wlodzimierz Brus has stated concerning Hungary and Poland is applicable to the Baltic Soviet Republics as well: That small farmsteads were able to start production "with an intensity of labor undeterred by cost calculations in money terms."[14] Under these circumstances it would have been unwise to jeopardize the minimal economic standards (and the food supply of the Red Army) with forced collectivization, which as experience had shown, would have at least initially led to a breakdown of agricultural production.[15]

In addition, the Communist Party's power base in the countryside was too weak to enforce collectivization in the early postwar years. While the guerilla fighters who attacked Soviet institutions and functionaries were never a threat to the Soviet power as a whole—even in Lithuania where armed resistance was the strongest—these armed resisters nevertheless posed a serious menace to the functioning of rural institutions.[16] The line between resistance fighters, people hiding in the woods to evade conscription and terror, and rural inhabitants was never clearly drawn. The woods, where locals and armed fighters convened, proved impenetrable for the occupation force. In Estonia, the so-called "forest brethren" were even said to have maintained ties in Tallinn's KGB headquarters.[17] Con-

---

[13] Swain, "Deciding to Collectivize," 40; Kuuli and Lõbu, eds., *Eesti 1939–1940*, 71.

[14] Brus, "Postwar Reconstruction," 595.

[15] For Lithuania this argument has been brought forward by Girnius, "Collectivization of Lithuanian," 464. For Latvia, Geoffrey Swain has explained the decision not to collectivize amongst others with "the chaos on the countryside and the need to meet delivery targets." Swain, "Deciding to Collectivize," 40.

[16] For an overview of anti-Soviet resistance, see the collection: Anušauskas, ed., *Anti-Soviet Resistance*.

[17] Laar, "Armed Resistance Movement," 220.

sequently, defeating these armed resistance forces and building up a power base in the countryside were the most urgent aims of Soviet leaders, and collectivization was seen as the means to achieving these goals.

This strategy was no doubt approved at the highest levels of the Communist Party. While Andrei Zhdanov, Andrei Vyshinskii and Vladimir Dekanozov had been responsible for the "policy of reassurance" in early 1940, after the war, Communist Party Central Committees Bureaus were implemented in each of the Baltic capitals to supervise local politics.[18] As Geoffrey Swain has shown for Latvia, until 1947 these bureaus were the central authorities in charge of communicating a more careful approach to collectivization.[19]

Redistributing Land and Taking Over the Cooperatives

The peculiarities of the land reforms in the Baltic states

Winning local support and identifying alleged enemies in the countryside was the initial concern of the Soviet leadership. When occupying the Baltic Soviet Republics in 1940, the first means of achieving this aim was a large-scale land reform, expropriating farmsteads of more than 30 hectares and redistributing to rural inhabitants with little or no land. This reform continued after 1944, although under quite different circumstances and far more aggressively. In the postwar setting, 30 hectares were only taken as an upper margin, while the size of farmsteads could be reduced to as little as 20 hectares. In comparison to other postwar land reform programs in Eastern Europe and in the Soviet Occupation Zone in Germany, these standards were extremely rigid.[20] If the prewar reform of 1940 could

---

18 In Estonia the Communist Party's Central Committees Bureau was headed by Nikolai Shatalin and later by Georgii Perov, in Latvia by Vassilii Ryazanov, and in Lithuania by Mikhail Suslov.

19 Swain, "Deciding to Collectivize," 40.

20 While for instance in the Soviet Occupation Zone of Germany or in Poland only those farms where agricultural land exceeded 100 hectares were partially dispossessed. In Hungary the maximum amount was fixed to 114 hectares for peasants and 57 hectares for non-peasants, in Czechoslovakia to 250 hectares and to 50 hectares respectively from 1947 on. For East Germany see Bauerkämper, "Die Bodenreform," 10. For Poland see Brus, "History," 592. For Hungary see Hollos and Maday, "Introduction," 3. For Czechoslovakia see Brus, "History," 591.

hardly claim to have drawn a line between poor and well-to-do peasants as two clearly distinguishable groups, the postwar reform met this claim even less. Land reform hit right into the center of the village. Fixing 30 hectares as a legal maximum can only be explained with by the fact that in the interwar years the Baltic republics had already carried out radical land reforms, during which the old estates had practically been abolished.[21] Therefore, new (and in many ways ahistorical) land-based divisions were needed in order to help create the ideologically envisioned class struggle. In addition, land reform also served as an instrument for punishing peasants suspected of collaborating with the German occupation force without ever drawing a clear line between the different grounds for expropriation. The disintegration of the rural social structure proved to be a difficult task.[22] Not only was the solidarity within the village stronger than expected, but it also followed historical patterns the Bolsheviks failed to understand. Also, the beneficiaries of the redistribution—the landless peasants and small dwellers—often were overburdened by the task of setting up new or enlarging old farmsteads, while agricultural inventory was lacking, state support was minimal, and the periods during which new farmers were exempted from the delivery quotas were very short. Furthermore, since the long-term presence of Soviet power in this region seemed doubtful to many farmers, a great number of peasants had little interest in the support offered by the state and expropriated land often went uncultivated.[23]

This problem went beyond the agricultural sector. By failing to establish some form of patronage in the countryside, the Soviet authorities' attempts to rearrange the rural population according to their ideology had failed. Where they had expected poor peasants to enthusiastically embrace what the Soviet power offered and rich peasants to be unwilling to relinquish their privileges, party leaders in fact encountered a complex mix of traditional customs and inter-village solidarity that could not be explained

---

[21] Pistohlkors, "Estland, Lettland."

[22] See for instance: Feest, "Terror und Gewalt."

[23] While in Lithuania about 45 percent of the dispossessed land remained in the state land fonds, in Latvia it was 42 percent and in Estonia 33 percent. Zunde, "Landwirschaft," 6. Labsvirs, *Sovietization*, 29. Feest, *Zwangskollektivierung im Baltikum*, 517. On peasants refusing to take additional land see also Zubkova, *Pribaltika*, 170. Girnius, who claims that holding back the land was a conscious policy decision, underestimates the objective reasons on the side of the peasants as well as the need for the Soviet power to build up a supporting class in the countryside: see Girnius, "Collectivization of Lithuanian," 468.

in socioeconomic terms. If in Soviet Russia, the division between poor (*bednyaki*), middle (*srednyaki*) and rich peasants (*kulak*) had taken a long time to develop in the consciousness of the peasants, for the Baltic peasant who had experienced a thorough redistribution twenty years before, these divisions made even less sense.[24] Therefore, in spite of existing tensions within the villages especially since the late 1930s, cultivating a "class war" in the Baltic countryside proved to be a failure. There was no grassroots pressure in favor of Soviet solutions that would eliminate the village elites so as to bring about mass collectivization. Even the party's own supporters in the countryside had trouble reconciling ideology and reality and often did not carry out what was expected of them.[25]

The intermediaries: local party workers

Party officials on the municipal level were never trusted by the members of the new power apparatus. Since for these cadres, knowledge of the local languages and circumstances was essential, the party leadership faced the dilemma of having to entrust the implementation of agricultural politics to people whose knowledge of Soviet ideology and the everyday practices of Stalinist rule were often rather weak. In order to ensure a minimum degree of loyalty, admission to the local party organizations during the early postwar years was almost exclusively restricted to Estonians, Latvians, and Lithuanians who had grown up in one of the old Soviet republics or who had been recently demobilized from the Red Army. As a rule, party workers in the villages were young and had joined the Communist Party during the war. In the Lithuanian case, these members were "notoriously uneducated and inexperienced men, suspicious of most of their charges,"[26] which applied to the other two republics as well. Since there was hardly any time to train them for their jobs, they were mostly unprepared for their tasks.[27]

---

[24] Sheila Fitzpatrick has pointed out how these terms were quite ambiguous in the Russian context as well and how they were used in a historical as well as an ideological sense: Fitzpatrick, *Stalin's Peasants*, 30–3, 249–52. On group identities in Estonia see Kõll, *Peasants on the World Market*. Generally on the three Baltic republics see Pistohlkors, "Estland, Lettland."

[25] For the Estonian case see Feest, "Dealing With..." 93–108.

[26] Girnius, "Collectivization of Lithuanian," 461.

[27] Feest, *Zwangskollektivierung im Baltikum*, 98–101.

Local party workers found different ways to deal with the near impossibility of implementing land reform in the prescribed way. While some willingly accepted the central leadership's conspiracy rhetoric offered to explain the shortcomings, others followed a more cooperative approach that enabled them to fulfill plan quotas. Central party leaders often complained that rural party workers were too closely involved with local society. Drinking with the *kulaks* was not only common, but also became shorthand for the unwanted fraternization between party members and wealthier village elites. Due to weak central control, corruption was widespread and local party workers sometimes levied special taxes, extorted goods and vodka from the population, or misused their positions in other ways, as noted regularly in the reports of the Central Committee.[28]

Faced with these difficulties, the party leadership resorted to patterns of explanation and practices that had been developed in the late twenties and now were claimed to be valid in any given context. Rather than addressing root causes for this kind of unruliness, party leaders replaced local cadres.[29] In Lithuania, in 1945–46 the number of members removed from party and government positions was almost equal to the number appointed, while in Estonia between 1945–47, 50 to 65 percent of the party organizers and 100 percent of the chairmen of the Executive Committees in the communes were replaced.[30] The situation became more stable after 1947, but was never fully stabilized.

Taking over the cooperatives—first steps towards collectivization?

Party leaders adopted another strategy for exercising its power by taking over existing cooperative organizations. A network of cooperatives—mainly consumer, milk, animal husbandry, and credit cooperatives—had existed in the region since the late 19th century. The communist regime immediately began to reorganize and centralize these cooperatives, especially after 1947. They also purged the cooperatives' leadership and

[28] Feest, *Zwangskollektivierung im Baltikum*; Bleiere, "Repressions against Farmers," 242–53.

[29] On cleanings in the regional party organizations in the old republics from 1927–1929, see Merl, *Die Anfänge*, 98–100.

[30] Girnius, "Collectivization of Lithuanian," 462, generally about cadre policy: pp. 461–64; Feest, *Zwangskollektivierung im Baltikum*, 152.

richer peasants, while broadening the base of permanent members.[31] At the same time, the Communist Party not only ensured its power base in the countryside, but also conformed to the ideological demands that had grown out of the politics and theoretical discussions of the 1920s. Collectivization in the old Soviet republics, according to the official interpretation, had been the goal of the "cooperative plan" designed by Lenin. This plan—which was in fact a rather loosely knit collection of ideas drawn from a number of Lenin's writings—was based on the notion that within a socialist society, marketing, credit, and consumer cooperatives that were actually capitalist organizations would slowly evolve into the "higher form" of socialist production cooperatives, or Soviet *kolkhozes*, in which labor and production would become socialized.[32] Although cooperatives in fact hardly ever grew into *kolkhozes* in the prewar Soviet Union, Lenin's "cooperative plan" remained a part of the party's ideological inventory even after the war.

Was there at least a potential chance that, in contrast to the Russian example, the restructuring of cooperatives could become a first step in the path to collectivization in the Baltic republics? At first sight, the growing number of cooperatives with a broader range of tasks and the dominance of machine cooperatives seems to indicate a move in this direction.[33] A closer look, however, shows that cooperatives hardly ever extended to forms of collective production. First, Baltic peasants traditionally perceived these cooperatives as organizations aimed at facilitating rather than replacing private farming. Second, the dispersed location of the farmsteads made it difficult to encourage collective farming of cooperative land while maintaining the single farm system. This became especially noticeable after 1947, when the party implemented a rushed decision to introduce the so-called "communal farms" from the cooperatives.[34] The Estonian case clearly shows that the peasants, who were absorbed by the work on their on private farmsteads, were neither able nor willing to till the fields of cooperatives, which had been stripped of their voluntary character almost entirely. Eventually, hired hands had to fill the gaps and a few coopera-

---

31 For Lithuania: Efremenko, "Razvite form," 169–71. For Estonia: Ruusman, "O roli."

32 Lenin, "O kooperatsii," 369–77; Lenin, "O prodovol'stvennom," 205–45; Lenin, "Zasedanie I s-ezda," 24–5.

33 Ruusman, "O roli," 175.

34 A few farms of that type had existed before.

tives even gave the land to private use.[35] Even from a Soviet point of view, these "communal farms" could only function through a "reinforcement of market relations."[36] Prospects were limited for these farms to grow into Soviet-type *kolkhozes*. Even in Soviet Lithuania, where 352 cooperatives were eventually turned into *kolkhozes* by November 1949, it seems doubtful that this had evolved naturally.[37] In the Baltic republics, as in Russia, *kolkhozes* did not evolve out of cooperatives but rather replaced them by decrees from above.[38]

The Collectivization Drive, 1947–49

Acceleration of collectivization in 1947

Since neither pressure from below nor a systematic development of the cooperatives can be credited with having prepared the groundwork for the collectivization drive that started in 1947, scholars have looked to politics for explanations. However, political explanations are hardly straightforward. Geoffrey Swain has described the main role of Moscow's representatives in the Baltic Soviet Republics—the Communist Party Central Committees Bureaus—as causing confusion "first by delaying and then by accelerating the collectivization drive."[39]

Changes inside and outside the Soviet Union led to the adoption of a new strategy in 1947. First, the deterioration of relations with the Western allies and the beginning of the Cold War convinced leaders in Moscow that a separate treatment for the western republics of the USSR was no longer feasible. Second, as early as August 1946, reports about the *kolkhozes* in the old Soviet republics had shown that the pragmatic relaxation of *kolkhoz* discipline during the war years had led to a massive loss of control. Collective farms were dissolving, while corruption and withdrawal

35 Feest, *Zwangskollektivierung im Baltikum*, 304.

36 Matin and Bronštein, *Eesti NSV põllumajanduse*, 66.

37 That is the claim of Efremenko, "Razvite," 173.

38 Biggart, "Collectivization of Agriculture," 60. This point has been made in Soviet historiography as well. For all the western republics, see Kabanov, "NSV Liidu läänerajoonide," 39. For Soviet Estonia Ants Ruusmann already made clear in 1971 that cases in which "simple cooperatives step-by-step grew into *Toz*es and *artel*'s did not exist in Estonia." Ruusman, "O roli," 176.

39 Swain, "Deciding to Collectivize," 40.

from collective work and theft were increasing.[40] Again, party leadership did not seek explanations for this situation in the rigid and unproductive *kolkhoz*-system that forced the peasants to violate the rules in order to survive. It rather blamed these problems on peasants' unruly and hostile attitude.[41] Already by September 1946, a "Council for Kolkhoz Matters" had been founded within the Central Committee of the CPSU in order to more closely see to it that ideological standards were obeyed.[42]

The drought in parts of central Russia, the Ukraine, and Moldavia in 1946 was a key factor to events in the Baltic Soviet Republics, where decent harvests in 1946 left surpluses for sale, and peasants profited from the rising demand. Refugees from the famine-stricken countries traveled to the newly acquired republics to buy grain and potatoes. This situation underlined the importance of the Baltic agriculture, and also fueled fears of the "anarchy of the market," evident in the uncontrolled mass movement and trade. For the Baltic peasants, the so-called "sack people" from the old republics portrayed the misery of the *kolkhoz* system. Their appearance alone conveyed a picture that Soviet propaganda was unable to combat. In popular rhymes, Lithuanians denounced these migrants as "Stalin's beggars," while Estonian peasants ridiculed the "sack people" with the song: "I am poor Russian *kolkhoznik*."[43] For leaders in Moscow, these migrants evoked memories of the black market and speculation in the late 1920s, which in striking similarity had been called "sack carrying."[44] Especially in the tense situation due to the drought, the existence of these "sack people" indicated the need for reliable access to grain.[45] Again, considerations went beyond economic reasoning. In the light of the problems of implementing the land reform, the lack of control over the local cadres, and falling productivity due to corruption in the state and party apparatus, the question of whether the party organizations in the republics were able and willing to transmit central power down to the village level became imperative.

---

40 Popov, *Krest'janstvo*, 24; Channon, "Stalin and the Peasantry."

41 Zubkova, *Russia after the War*, 66.

42 Dunmore, *Soviet Politics*, 80.

43 Girnius, "Collectivization of Lithuanian," 473; Feest, *Zwangskollektivierung im Baltikum*, 368.

44 Merl, *Anfänge*, 63–66.

45 Elena Zubkova stresses the change of atmosphere between Moscow and the Baltic Communist Parties due to the drought. Zubkova, *Pribaltika*, 172–73.

Preparing for collectivization

Moscow's shift in attitude did not result in an immediate change of agricultural politics in the Baltic republics. Plans for collectivization were developed cautiously. Concrete programs for collectivization were first discussed in April 1947 at a meeting with the first secretaries of the Estonian, Lithuanian, and Latvian Communist Parties—Nikolai Karotamm, Antanas Sniecus and Jānis Kalberžins—in the Central Committee Orgburo of the Communist Party of the Soviet Union (CPSU). In May 1947, the Central Committee issued the decree: "On founding *kolkhozes* in the Lithuanian, Latvian and Estonian SSR,"[46] which established a special commission for collectivization in the Baltic Soviet Republics, headed by Andrei Zhdanov. The Communist Party Central Committees were requested to submit proposals for establishing *kolkhozes* by that June.[47] Apart from these steps, little else happened with regard to implementing collectivization.[48] The party's strategy of secrecy left peasants ignorant of the coming changes. Secrecy was extended to the lower echelons of the Communist Party itself. Therefore, while the collectivization drive in the winter of 1929 to 1930 had basically been accomplished without the village Soviets, consisting mainly of local people,[49] in the case of the Baltic republics, neither were the Soviets nor the party organizations on the municipal level fully informed about the decision to start collectivization, and they were excluded from the founding of collective farms in 1947.[50] Accordingly, in the lower strata of the Communist Party the notion that collectivization could be implemented in the remote future—or not at all—was widespread.[51] Some party workers even thought that the dispersed settlement typical for the Baltic region made collectivization impossible.[52]

Apart from the initial ignorance with regard to the collectivization plans, local party workers had other reasons for their reluctance toward the campaign. They knew well how unpopular collective farms were among

46 *Reshenia partii*, 427–33.
47 *Reshenia partii*, 428.
48 Swain, "Deciding to Collectivize," 45; Feest, *Zwangskollektivierung im Baltikum*, 351; Girnius, "Collectivization of Lithuanian," 468.
49 Merl, *Anfänge*, 97.
50 Kivimaa, *EKP tegevus vabariigi*, 163.
51 Biggart, "Collectivization of Agriculture," 57; Swain, "Deciding to Collectivize," 45.
52 Bleiere, "Repressions against Farmers," 246.

the peasants, many of whom had heard firsthand about Soviet *kolkhozes* from the "sack people." Also, the Russian experience of the 1920s and 1930s was well known in the Estonian, Latvian, and Lithuanian villages. Before the first *kolkhoz* in Soviet Estonia was founded on the island Saaremaa in summer 1947, local peasants wrote a letter of protest to a representative of the Supreme Soviet of the ESSR, in which it openly referred to the bad experience of the old Soviet republics, pointing out how long it would take to make up for the losses it would cause: "Most of us know the problems of collectivization in the USSR—this takes years, sometimes decades."[53] In the current situation, they concluded, a collective farm would deliver less products to the state than the private peasants did.[54]

The party leadership in the republics was in agreement with these arguments, which may also have reflected their own fear of destroying what was left of agricultural production with a quickly implemented collectivization drive. In a report to the Central Committee of the CPSU, the first secretary of the Communist Party of Estonia, Nikolai Karotamm, made it clear that "speeding up the pace of collectivization could cause damages to animal husbandry in the republic during a short period of time and it would take years to liquidate them."[55] Here, the reference to the mass slaughtering of livestock in Russia as a reaction to the collectivization drive is quite obvious. The plans that the Baltic party leaders hastily compiled and presented to the Central Committee Orgbureau were modest and moreover differed substantially from the *kolkhozes* actually founded later that year. At the end of 1947, some 17 model *kolkhozes* were founded in Lithuania, 30 in Latvia, and 5 in Estonia.[56] Since much effort was put into meeting the demand of the collectivization decree that their members should be small dwellers, the first *kolkhozes* had to be heavily subsidized by the state.

By the summer of 1948, the atmosphere had begun to change. While within the Moscow party, leader Andrei Zhdanov, who obviously had been one of the architects of the "politics of reassurance" in the Baltic republics, began to lose power. Internationally, the Cold War was in full gear.[57]

---

53 Feest, *Zwangskollektivierung im Baltikum*, 356.

54 Feest, *Zwangskollektivierung im Baltikum*, 356.

55 Tõnurist, "Traagiliste sündmuste," 34.

56 Biggart, "Collectivization of Agriculture," 63; Bleiere, "Repressions against Farmers," 247; Feest, *Zwangskollektivierung im Baltikum*, 362.

57 Biggart points out the significance of the "Truman Doctrine" from March 1947 as well as the interest by the Polish and Czech communist parties in receiving Marshall Aid in June 1947. Biggart, "Collectivization of Agriculture," 61.

At the same time, relations with Yugoslavian leader Josip Broz Tito were steadily worsening, demonstrating that even close allies could turn away from the CPSU.[58] As for the Baltic Soviet Republics, growing tax arrears may well have nurtured fears that they too were following an independent path.[59]

Tensions were also mounting between party leaders within the Baltic Soviet Republics. Friction between local party workers and officials coming from the old Soviet republics became more apparent, as it was clear land reform had failed to provide reliable allies in the countryside. While national rhetoric played an important role in an attempt by the party to create loyal local elites, Moscow was suspicious of nationalism for its potential to diminish its centralized power.[60] As the party apparatus in all three republics was soon dominated by Russian non-locals, negative views on national prerogatives were gaining significance.[61] This opened up a new space for strategizing. Immigrant party workers—as well as strategically minded locals—who were dissatisfied with the Latvian, Estonian, or Lithuanian leadership increasingly "nationalized" the existing problems by blaming the situation on the unwillingness of local cadres to crack down on their co-nationals. They found an open ear in Moscow. It is telling that while in Estonia the party leadership accused rural party workers of feeling "too much at home" in their villages, the party leaders of the republics themselves were later reproached for a "family-like approach" to the problem of the well-to-do farmers, the *kulaks*.[62] While trying to install

---

[58] Hahn, *Postwar Soviet Politics*, 98.

[59] On the tax arrears see Swain, "Deciding to Collectivize," 54.

[60] While in Russia national sentiments could never play this part, in Ukraine by 1930 "local nationalism" was seen as an inhibiting factor. See for example: Hildermeier, *Geschichte der Sowjetunion*, 477. On changes in the attitude towards nationality in the Estonian case, see Feest, "Neo-Korenizacija."

[61] In Lithuania, by January 1, 1948 only 18.5% of the Lithuanian Communist Party were Lithuanians, while in the other Baltic communist parties the contingent of the majority national group was 43.5% in Estonia and 53% (in 1949) in Latvia respectively. For Lithuania see Girnius, "Collectivization of Lithuanian," 463; for Estonia: *Kommunisticheskaia partiia*, 108–9. For Latvia: Taagepera and Misiunas, *The Baltic States*, 80. In all cases, those party members who were Lithuanian, Latvian, and Estonian by origin, but often did not speak the respective language since they had grown up in one of the old Soviet Republics, were counted in. This explains the considerably lower number of Lithuanians in comparison to that of Estonians and Latvians in the respective party organizations, since only few Lithuanians resided in the Soviet Union before the war.

[62] Feest, *Zwangskollektivierung im Baltikum*, 155–56, 344.

their own networks, the Soviet leadership increasingly suspected local networks as an obstacle.

In these power struggles, collectivization became an important issue, insofar as it was seen as a measure of who was running the country. Likewise, attitudes toward collectivization functioned as the crucial signifier of loyalty toward party leaders in Moscow. It was likely alarming news for the centralized leadership that even the village communists were reluctant to join the collective farms.[63] Again, the striking similarity to the old experience eventually may have contributed to the decision to solve the problem via similar means. Also in Russia the attitude of most village communists toward collectivization had been very reserved. Their justifications bore resemblance to their Baltic counterparts: In many cases they blamed their wives for refusing to consent to collectivizing family farms.[64] In both cases, the local milieu—including the closest family connections—proved to be a large hurdle to the Bolshevist's claim for power. To these obstacles, in the Baltic states one has to add national issues and tensions, which increasingly preoccupied the Moscow leadership.

In winter 1947, in an attempt to exert control at the village level, the same local party workers who had been denied information about the collectivization plans only a couple of months earlier were now accused of not having done enough for establishing *kolkhozes*. At the same time, leading politicians in the field of agriculture were criticized and dismissed in all three Baltic Soviet Republics for failing to address the class struggle in the countryside.[65] Collectivization became a key part of the party's agenda. However, the effects of the collectivization campaign were not long lasting. Although at the beginning of 1948 a sharp increase in the number of collective farms was carried out, as soon as the pressure was taken off lower party organs, the speed of collectivization decreased. Many party workers understood the collectivization drive of 1947–1948 as just another short-

---

63 Feest, *Zwangskollektivierung im Baltikum*, 371–74. Taagepera, "Soviet Collectivization," 273.

64 For Russia: Merl, *Anfänge*, 104. For Estonia: Feest, *Zwangskollektivierung im Baltikum*, 372. In the Baltic Soviet Republics female resistance to collectivization never reached the level of the Russian *bab'i bunty* that had forced the Soviet power to back down on issues of the right to own a private cow and a private plot. After the war, such riots were especially widespread in Poland. See Viola, "Bab'i Bunty;" Merl, *Anfänge*, 148–53; Merl, *Bauern unter Stalin*, 257–58; Jarosz, "Polish Peasants," 63.

65 Bleiere, "Repressions against Farmers," 246; Feest, *Zwangskollektivierung im Baltikum*, 218–20.

lived campaign. In some cases, rumors spread that the lack of productivity meant collectivization would be abandoned for good.[66]

Tax policy and its results

If in 1948 there was a single factor which motivated people to join *kolkhozes*, it was the rise in taxes.[67] Regarding this policy, the Soviet power resorted to an instrument that had been developed in 1929–1930, implementing the so-called "extraordinary measures" of grain procurement. In the Baltic Soviet Republics, taxes on agricultural goods had been quite moderate up to the summer of 1947, with delivery obligations making up most of the peasants' dues. From 1947, agricultural taxing became the most important instrument in coercing peasants to join collectives. Although the tax rates themselves stayed the same, the amount of allegedly produced goods, on the basis of which the required payments were calculated, was raised drastically. When in addition to this, the fiscal reform introduced in December 1947 destroyed the savings of the peasants, prices on the free market fell by as much as 50 percent, making farming in general increasingly unfeasible.[68] Because payments were indirectly dependent on the means of production only small farms with little land and inventory could survive. As a result, during 1947 and 1948, most larger farms were destroyed.[69]

Sharp changes in taxation had far-reaching results. One of these was the phenomenon of "self-liquidation:" many farmers gave up farming and used what they had left to pay their outstanding taxes and left for the cities in order to find industrial work. This caused a sharp decline in agricultural production.[70] Wealthier peasants reacted by establishing or joining collective farms. The paradoxical result was that the small landholders clung to their private farms, while richer peasants established collective farms.

---

[66] Kivimaa, *EKP tegevus vabariigi*, 215; Swain, "Deciding to Collectivize," 45.

[67] Taagepera, "Soviet Collectivization."

[68] Ruusmann, *Põllumajanduse taastamine*, 359; Bleiere, "Repressions against Farmers," 42. On the monetary reform in general see Zubkova, *Russia after the War*, 51–55; Chlevnjuk, "Die sowjetische," 1033–34.

[69] Bleiere, "Repressions against Farmers," 245; Feest, *Zwangskollektivierung im Baltikum*, 380–81; Girnius, "Collectivization of Lithuanian," 469.

[70] Bleiere, "Repressions against Farmers," 245.

Thus, *kolkhozes* were run by farmers who did not represent the ideological foundations of collectivization.

The new tax also singled out a special kind of richer peasants, the *kulak*. This category had until then been used exclusively as a propagandistic term in the Baltic Soviet Republics. In many cases, local variants were used, like for instance the term "grey baron" in Estonia, to signify a peasant who had become wealthy in the decades after the emancipation of the serfs in the Baltic provinces. Now, however, *kulak* became a technical term applied to a whole strata of villagers who were seen as class enemies and criminals. According to tax law in the Baltic republics, basically any person proved to be a *kulak* who had used hired labor on a regular basis, received income from machines or from mills, or had rented out land during the German occupation or after the war.[71] As had been the case in the Soviet Union, strict definitions were not as important as the conviction that a peasant had in some way exploited his fellow peasants in the recent past.[72] Also, as Anu Mai Kõll has shown, the question of how an individual had behaved during the German occupation figured into decisions by local executive committees as a kind of "*hidden criteria*" that had even more weight in the postwar kulakization than allegedly exploitative conduct.[73] Still, in the outcome the parallel to the Russian case is striking. While in some villages the number of "kulakized" farms were exorbitantly high, other party activists claimed that there were no *kulaks* at all in their villages.[74] Many others used the opportunity to settle old bills. Being on such a list had grave consequences for the respective peasant. A certain percentage was added to the alleged income of *kulak* farms, quickly raising them into a higher tax bracket. A further rise of this percentage in 1948 made it clear that all *kulak* households would be ruined within a short time. Thousands of peasants appealed this label and in many cases their names were taken off the lists. Again in a parallel to the Soviet Union, the party leadership unleashed local campaigns, causing bitterness and friction within the populace, in order to later condemn the excesses.[75]

---

71 Tõnurist, ed. *Eesti NSV põllumajanduse*, 231. See also Bleiere, "Repressions against Farmers," 244.

72 For Russia see Fitzpatrick, *Stalin's Peasants*, 32. For an example in the Baltic republics see Feest, *Zwangskollektivierung im Baltikum*, 393.

73 Kõll, *The Village*, especially chapter 4. See also: Kõll, "Tender Wolves."

74 Viola, *Peasant Rebels*, 88–9.

75 On Latvia: Bleiere, "Repressions against Farmers."

At the end of 1947 in Estonia 5.3 percent of the farmsteads were declared *kulak*, in Latvia 4.1 and 2.4 in Lithuania.[76] These farmers were ruined economically and labeled as class enemies. Many *kulaks* were tried for sabotage when eventually failing to pay the exorbitant taxes demanded of them.[77] With few exceptions, the peasants on the *kulak* lists were to be among the victims of mass deportations in March 1949.

The question of violence

Despite the destructive results of the tax policy, some historians have interpreted the years up to 1949 as offering opportunities other than the path toward mass violence, which was eventually taken. These scholars claim that collectivization up to that point was achievable only by economical pressure. The term "mini NEP" has been used to refer to the coexistence of state and private forms of production in which collective forms would only be formed on the basis of "voluntary collectivization" and "self liquidation" on a purely economic basis.[78] Whether there was the possibility to reach collectivization by purely economic pressure remains a key question. Considering collectivization in its broader context makes it difficult to endorse this viability of the economic argument, however. Indeed, not only the collectivization decree but also numerous subsequent decisions by the local party stressed the absolutely voluntary character of collectivization. But as discussed above, these terms could be used in both a literal as well as historical sense and in 1948 the "historical" interpretation clearly gained ground for reasons discussed above. The images of class enemies that had been propagated by the party served as a central explanation for everything that was going wrong in the country, justification for the fight against anti-Soviet partisans and brutal actions against anyone suspected of being against the new system. Postwar land reform served as a useful tool in such political struggles.

[76] Feest, *Zwangskollektivierung im Baltikum*, 170, 400; Bleiere, "Repressions against Farmers," 244; Biggart, "Collectivization of Agriculture," 63.

[77] This procedure led to irritations even among local party workers, who wondered whether it was correct to try people by criminal or civil means "if everybody knows well that the accused has no possessions to pay the tasks and, if one had it, the accused has no rights to sell them in order to settle accounts with the state." See Pelkaus, ed., *Policy of Occupation*, 349–50.

[78] Labsvirs, *Sovietization*, 83. See also Swain, "Deciding to Collectivize," 41.

The "class struggle in the countryside" was not only accepted as an explanation for shortcomings, but was demanded. Although the term *kulak* had not been defined in the Baltics before 1947, it had always played a central role in those scenarios. There was no prospect of living peaceably with the *kulaks*, as they were portrayed as being stubborn and inflexible. Consequently, when the term was finally defined, the label often remained impossible to escape. Therefore *kulaks* were not to be admitted as workers in the factories or in the *kolkhozes*. Ideologically, no loophole was left to peacefully integrate the *kulaks* into the society.[79]

Although the use of violence to force regular peasants to join the *kolkhoz* was not in the interest of the party leadership, it seems inevitable that the party reverted to this path. Elena Zubkova concludes from her extensive study of party documents that between 1947 and 1948 Moscow did not approve of the pressure exerted on peasants in order to make them join the *kolkhozes*.[80] But as the party leadership became increasingly dissatisfied with the slow pace of collectivization especially in the end of 1948, local party workers were likely to be blamed for weak propaganda work or even criminalized if they did not resort to more aggressive and even violent means of pressuring the peasantry. In this way, violence was in fact a necessary consequence of what was demanded from the party workers.

Mass Collectivization

Terror

The destruction of the more productive farmsteads through higher taxes led to a sharp decline in agricultural productivity and an alarming downfall in animal husbandry, which again fit in well with slogans about the wreckage committed by *kulaks* and "enemies of the people." The mass deportations that were carried out in 1948 and 1949 show how collectivization and class warfare had become inextricably bound. In Lithuania, in the spring of 1948 an estimated 43,300 people were deported to Siberia (Operation Vesna). Less than a year later, in March 1949, nearly 100,000

---

[79] This has been noted concerning Russian *kulaks*: "In order to survive, the former *kulaks* were virtually obliged to violate Soviet law in some way, and that is what they regularly did." Fitzpatrick, *Stalin's Peasants*, 241.

[80] Zubkova, *Pribaltika*, 179.

people were deported from Lithuania, Latvia, and Estonia, collectively.[81] The composition of the victims clearly shows that collectivization was not the only aim of these operations. The deportations were directed broadly against rural society. Apart from the families of alleged *kulaks*, families of already deported so-called "enemies of the people"—that is partisans and other "nationalists"[82]—were among the victims, which is why the majority of all people deported in March 1949 were women and children. In Lithuania, in 1951 another 19,000 *kulaks* were deported.[83]

As a result of this program, collectivization was implemented at a rapid speed, although differing in the three republics. Fearing deportation, peasants quickly agreed to give up their independent farms. By May of 1949, 66 percent of all farms in Estonia and 71.6 percent in Latvia had joined collective farms, whereas in Lithuania even by the end of 1949 only 60 percent of the farms were collectivized and further deportations were carried out. By the end of 1950, however, 90 percent of the Lithuanian farms were collectivized.[84]

Putting kolkhozes to work: the socialist agriculture

The establishment of *kolkhozes* after 1948 was not successful: many existed only on paper, and peasants often still worked their old plots using their old equipment. In Estonia, some farmers were unaware they were actually members of a collective farm.[85]

In addition, the newly appointed *kolkhoz* chairmen often selected potential members and excluded others. The first secretary of the Estonian Communist Party, Nikolai Karotamm, complained in May 1949 that "small dwellers, receivers of new land, elder small dwellers, families of middle farmers who were in debt and other poor farmsteads" had not

---

[81] In Operation Priboi, 33,496 people from Lithuania, 41,445 from Latvia, and 20,660 from Estonia were deported.

[82] These differences are often ignored, for instance when Geoffrey Swain claims that in the whole "50,000 alleged Latvian *kulaks*" were deported. Swain, "Deciding to Collectivize," 39.

[83] Zubkova, *Pribaltika*, 181.

[84] Feest, *Zwangskollektivierung im Baltikum*, 443; Bleiere, "Repressions against Farmers," 247; Girnius, "Collectivization of Lithuanian," 461, 471–72.

[85] Girnius, "Collectivization of Lithuanian," 461; Zubkova, *Pribaltika*, 184–85; Feest, *Zwangskollektivierung im Baltikum*.

been admitted into some *kolkhozes*.[86] Some chairmen had made arbitrary factors, like knowledge of the Estonian language, a precondition for membership.[87] Many chairmen did not object to admitting peasants who had been labeled as *kulaks*. In Latvia, over 200 *kulaks* were expelled from collective farms in 1948. In other cases the local party organs and district administration bent the rules in order to keep experienced agricultural workers. Some party workers simply had different ideas about the danger these alleged former village exploiters posed. In Estonia, a local party organizer justified his lenient attitude toward four *kulaks* in the local collective farm with the statement that he had "not noticed any infiltration work on their side," while another was reprimanded for thinking it possible to "re-educate the *kulaks*."[88] The chairman of a Latvian collective farm made it clear that he saw no problem with *kulaks* working in collective farms as rank-and-file members.[89] One Lithuanian collective farm went even further: In an official decision it specified the exact amount of aid they would supply the partisans in the forests with.[90]

The party leadership's fear over personal connections and customs surviving in the collective farms even after collectivization is not without precedent. In the 1930s, the hunt for *kulaks* in the *kolkhozes* had gone on for years.[91] After the war, the endurance of old traditions within the collective farms was conceived as a problem throughout the Soviet Union. In May 1950, the newly appointed head of the Central Committee Department for Agriculture, Nikita Khrushchev, noted that in spite of a strict policy since 1946, many *kolkhozes* still displayed semi-private traits.[92] Therefore, they were amalgamated into larger units. The amalgamation campaign was also carried out in the Baltic Soviet Republics, although here the collective farms had hardly been established. The changes were dramatic. In Lithuania, the 6,500 *kolkhozes* that had existed before the amalgamation were reduced to 1,809 by 1954; in Estonia, 2,213 *kolkhozes* in 1950 were amalgamated to 937 in 1952.[93] Taking the effects into con-

86 Feest, *Zwangskollektivierung im Baltikum*, 449.
87 Kivimaa, *EKP tegevus vabariigi*, 280.
88 Feest, *Zwangskollektivierung im Baltikum*, 456.
89 Swain, "Deciding to Collectivize," 53.
90 Zubkova, *Pribaltika*, 185.
91 See especially Fitzpatrick, *Stalin's Peasants*, 238–46.
92 Boterbloem, *Life and Death*, 142.
93 Biggart, "Collectivization of Agriculture," 66; Feest, *Zwangskollektivierung im Baltikum*, 461.

sideration, this campaign has been called a "second collectivization."[94] Lacking the necessary equipment, large-scale farms that now stood in the place of small *kolkhozes* could not function. They also could no longer profit from the advantages of small-scale farming. The amalgamation eliminated the sense of responsibility that peasants had maintained in terms of identifying with their animals and equipment in the small *kolkhozes*. The "petite-bourgeois" producer that Khrushchev had aimed to eliminate was replaced by the new collectivized Soviet man, characterized by his own peculiar strategies of survival and adaptation.

Conclusion

While in the late 1920s and early 1930s collectivization was pursued by a trial-and-error politics and increasingly violent policies, after the war, strategic methods and their ideological justification were established. Some historians, such as Latvian scholar Daina Bleiere, have therefore claimed that collectivization in the Baltic republics was "applied in a systematic and consistent manner and yielded tangible results in spite of the generally negative attitude toward the occupation regime in rural areas."[95] In this article I have argued that the postwar politics in Estonia, Latvia, and Lithuania were more complex. Although, at the political-ideological level, the collectivization of agriculture was declared a priority, neither the timetable of collectivization nor its mode of implementation was clear in the postwar years. Moreover, strategic consideration in foreign and domestic affairs made it initially more feasible to choose a slower course of collectivization.

Due to this ambiguity, which was strengthened by a certain amount of secrecy, local communists often developed their own understanding of how agricultural politics should be pursued. However, hopes for an alternative path to socialist agriculture were doomed from the outset, and the more the party's attempts to exercise control failed, the more the elites drew on the interpretations and methods of the 1920s and 1930s. Since the notion of systemic errors was a taboo, enemy activity was the only acceptable explanation for collectivization's shortcomings. Internal tensions within the region as well as international tensions of the Cold War further complicated this picture.

---

[94] Boterbloem, *Life and Death*, 142.

[95] Bleiere, "Repressions against Farmers," 242.

Connecting collectivization to the destruction of enemies became a central concern for party elites in Moscow once they believed they had lost control of the countryside. This conception was vindicated by local party members, who willingly espoused the slogans of defeating "bourgeois nationalism" and *kulaks* whenever peasants rebelled or even to punish their personal rivals.

Violence was a prerequisite of collectivization. Just as the Baltic peasants would only join collective farms by brutal force, the Bolsheviks were unable to conceive collectivization in any other terms than class war, in which *kulaks* were the enemy, even after most of the farms had been collectivized. While collectivization in the Baltics may not have been planned to match the Russian example from the outset, certain patterns inherited from the prewar collectivization did determine the way problems were interpreted and dealt with.

In contrast to the Russian example, collectivization in the Baltic republics was also closely tied to the question of national solidarity. When in March 1950 the Estonian state and party leadership were cleansed, the party adopted a lenient attitude not only toward *kulaks* but also toward "bourgeois nationalism," which was often regarded as a sentiment of the former. These purges showed, however, that leaders in Moscow were determined to control the situation. It served as a warning to Latvia and Lithuania as well.[96]

BIBLIOGRAPHY

BOOKS AND ARTICLES:

Anušauskas, Arvydas, ed. *The Anti-Soviet Resistance in the Baltic States.* Vilnius: Du Ka, 1999.

Arch Getty, J. "Afraid of their Shadows: The Bolshevik Recourse to Terror, 1932–1938" in Manfred Hildermeier, ed. *Stalinismus vor dem Zweiten Weltkrieg. Neue Wege der Forschung.* München: Oldenbourg, 1998, 169–92.

Baberowski, Jörg. "Stalinismus 'von oben.' Kulakendeportationen in der Sowjetunion 1929–1933." *Jahrbücher für GeschichteOsteuropas* 46 (1998): 572–95.

Bauerkämper, Arnd. "Die Bodenreform in der Sowjetischen Besatzungszone in vergleichender und beziehungsgeschichtlicher Perspektive. Einleitung" in Arnd Bauerkämper, ed. *"Junkerland in Bauernhand?" Durchführung, Auswirkungen und Stellenwert der Bodenreform in der Sowjetischen Besatzungszone.* Stuttgart: F. Steiner, 1996.

---

96 On the "Estonian Affair" see Zubkova: "Fenomen," 89–101; and Feest, *Zwangskollektivierung im Baltikum*, 427–39.

Biggart, John. "The Collectivization of Agriculture in Soviet Lithuania." *East European Quarterly* 9 (1975/76): 53–75.

Bleiere, Daina. "Repressions against Farmers in Latvia in 1944–1953" in Valters Nollendorfs et al, eds. *The Hidden and Forbidden History of Latvia under Soviet and Nazi Occupations 1940–1991*. Riga: Latvijas vēstures institūta apgāds, 2005, 242–53.

Boterbloem, Kees. *Life and Death under Stalin: Kalinin Province, 1945–1953*. Montreal: McGill-Queens University Press, 1999.

Brandes, Detlef, Holm Sundhaussen, and Stefan Troebst. *Lexikon der Vertreibungen. Deportation, Zwangsaussiedlung und ethnische Säuberung im Europa des 20. Jahrhunderts*. Vienna: Böhlau, 2010.

Brus, Włodzimierz, "Postwar Reconstruction and Socio-Economic Transformation" in Michael Charles Kaser and Edward Albert Radice, eds. *The Economic History of Eastern Europe 1919–1975, Volume 2*. Oxford: Clarendon, 1986, 564–641.

Channon, John. "Stalin and the Peasantry: Reassessing the Postwar Years, 1945–53" in John Channon, ed. *Politics, Society and Stalinism in the USSR*. New York: St. Martin's Press, 1998, 185–209.

Chlevnjuk [Khlevniuk], Oleg. "Die sowjetische Wirtschaftspolitik im Spätstalinismus und die Affäre 'Gosplan.'" *Osteuropa* 9 (2000): 1031–47.

Danilov, Viktor P. *Sozdanie material'no-technicheskich predposylok kollektivizatsii sel'skogo khozjajstva v SSSR*. Moscow: Akad. Izd. AN SSSR, 1957.

Dunmore, Timothy. *Soviet Politics 1945–53*. New York: St. Martin's Press, 1984.

Efremenko, A. P. "Razvite form sel'skochoziaistvennoi kooperatsii v Litovskoi SSR" in *Problemy agrarnoi istorii sovetskogo obshchestva*. Moscow: Izd. Nauka, 1971.

Feest, David. "Terror und Gewalt auf dem estnischen Dorf." *Osteuropa* 6 (2000): 656–71.

—. "Dealing With the Unruly Reality: Rural Party Workers in Estonia, 1944–1950" in *Padomju okupācijas režīms baltijā 1944–1959. Gadā: Politika un tās sekas. Starptautiskās konferences materiāli 2002. gada 13.–14. jūmijs, Rīga* [The Soviet Occupation Regime in the Baltic States 1944–1959: Policies and their Consequences. Materials of an International Conference, June 13–14, 2002]. Riga: Latvijas vēstures institūta apgāds, 2003, 93–108.

—. "'Neo-Korenizacija' in den baltischen Sowjetrepubliken? Die Kommunistische Partei Estlands nach dem zweiten Weltkrieg." *Zeitschrift für Geschichtswissenschaften* 54 (2006): 263–80.

—. *Zwangskollektivierung im Baltikum. Die Sowjetisierung des Estnischen Dorfes 1944–1953*. Vienna: Böhlau, 2007.

Fitzpatrick, Sheila. *Stalin's Peasants. Resistance and Survival in the Russian Village After Collectivization*. Oxford: Oxford University Press, 1994.

Girnius, Kestutis K. "The Collectivization of Lithuanian Agriculture, 1944–1950." *Soviet Studies* 40 (1988): 460–78.

Gregory, Paul R. *The Political Economy of Stalinism. Evidence from the Soviet Secret Archives*. Cambridge: Cambridge University Press, 2004.

Hahn, Werner G. *Postwar Soviet Politics. The Fall of Zhdanov and the Defeat of Moderation, 1946–53*. Ithaca: Cornell University Press, 1982.

Hildermeier, Manfred. *Geschichte der Sowjetunion 1917–1991. Entstehung und Niedergang des ersten sozialistischen Staates*. Munich: Beck, 1998.

Hollos, Marida and Bela C. Maday, ed. *New Hungarian Peasants: An East Central European Experience with Collectivization.* New York: Columbia University Press, 1983.

Jarosz, Dariusz. "Polish Peasants versus Stalinism" in A. Kemp-Welch, ed. *Stalinism in Poland, 1944–1956. Selected Papers from the Fifth World Congress of Central and East European Studies, Warszawa, 1995.* New York: St. Martin's Press, 1999, 59–77.

Kabanov, V. "NSV Liidu läänerajoonide põllumajanduse tähtsamaid probleeme" [The most important problems of agriculture in the western regions of the USSR], in Edgar Tõnurist, ed., *Sotsialistliku põllumajanduse areng Nõukogude Eestis* [The development of socialist agriculture in Soviet Estonia]. Tallinn: Valgus, 1976, 32–46.

Kivimaa, Ervin. *EKP tegevus vabariigi põllumajanduse kollektiviseerimisel aastail 1944–1950. Dissertatsioon ajalooteaduste kandidaadi kraadi taotlemiseks* [The activity of the CPE collectivizing the agriculture of the republic. Dissertation to obtain the degree of a candidate]. Unpublished manuscript in the filial of the Estonian State Archives, 1970.

Kõll, Anu Mai. *Peasants on the World Market: Agricultural Experience of Independent Estonia 1919–1939.* Stockholm: Almqvist & Wiksell International, 1994.

—. "Tender Wolves. Identification and Persecution of Kulaks in Viljandimaa, 1940–1949" in Olaf Mertelsmann, ed. *The Sovietization of the Baltic States.* Tartu: Kleio Ajalookirjanduse Sihtasutus, 2003, 127–49.

—. *The Village and the Class War, Anti-Kulak Campaign in Estonia.* Budapest, New York: Central European University Press, 2013.

Kommunisticheskaia partiia Ėstonii v tsifrakh, 1920–1980: Sbornik statisticheskikh dannykh. Tallinn: Eesti Raamat, 1983.

Kuuli, Olaf and Erni Lõbu, eds. *Eesti 1939–1940. Fakte, probleeme, meenutusi* [Estonia 1939–1940. Facts, problems, remarks]. Tallinn: Eesti Raamat, 1989.

Labsvirs, Janis. *The Sovietization of the Baltic States: Collectivization of Latvian Agriculture 1944–1956.* Indianapolis: Taurus, 1988.

Lenin, Vladimir Ilich. "Zasedanie I s-ezda sel'skokhozyaistvennykh rabochikh petrogradskoi gubernii 13 Marta 1919," *Polnoe sobranie sochinenii,* vol. 38. Moscow: Gosudarstvennoe Izdat. Politicheskoi Literatury, 1963, 22–30.

—. "O prodovol'stvennom naloge (Znachenie novoi politiki i ee usloviia)," *Polnoe sobranie sochinenii,* vol. 43. Moscow: Gosudarstvennoe Izdat. Politicheskoi Literatury, 1963, 205–45.

—. "O kooperatsii," *Polnoe sobranie sochinenii,* vol. 45. Moscow: Gosudarstvennoe Izdat. Politicheskoi Literatury, 1964, 369–77.

Matin, Valentin and Mikhail Bronštein. *Eesti NSV põllumajanduse kollektiviseerimine ning selle sotsiaalsed ja majanduslikud tulemused* [The collectivization of agriculture in the Estonian SSR and its social and economical consequences]. Tallinn: Eesti Riiklik Kirjastus, 1959.

Merl, Stephan. *Die Anfänge der Kollektivierung in der Sowjetunion.* Wiesbaden: O. Harrassowitz, 1985.

———. *Bauern unter Stalin. Die Formierung des sowjetischen Kolchossystems 1930–1941.* Berlin: Duncker & Humblot, 1990, 257–58.

Mertelsmann, Olaf. *Der stalinistische Umbau in Estland. Von der Markt- zur Kommandowirtschaft.* Hamburg: Kovac, 2006.

Narody stran Baltii v usloviakh stalinizma, 1940-e–1950-e gg. Pod red. N. Bugaia. Stuttgart: Ibidem-Verlag, 2005.

Paavle, Indrek. “Sovietization of Agriculture” in *Estonia since 1944. Reports of the Estonian International Commission for the Investigation of Crimes Against Humanity.* Tallinn: Estonian International Commission for the Investigation of Crimes against Humanity 2011, 37–78.

Pelkaus, E., ed. *Policy of Occupation Powers in Latvia, 1939–1991: A Collection of Documents.* Latvijas Valsts arhīvs. Nordik, 1999.

Pistohlkors, Gert von. “Estland, Lettland und Litauen 1920–1940” in Wolfram Fischer, ed. *Handbuch der europäischen Wirtschafts- und Sozialgeschichte. 6: Europäische Wirtschafts- und Sozialgeschichte vom Ersten Weltkrieg bis zur Gegenwart, Vol. 6.* Stuttgart: Klett-Cotta, 1987, 729–68.

Popov, V. P. *Krest'janstvo i gosudarstvo (1945–1953).* Paris: YMCA-Press, 1992.

Resheniia partii i pravitel'stva po khozjajstvennym voprosam v piati tomakh, vol. 3, 1941–1952. Moscow: Gosudarstvennoe Izdat. Politicheskoi Literatury, 1968.

Rittersporn, Gábor Tamás. “The Omnipresent Conspiracy. On Soviet Imagery of Politics and Social Relations” in Nick Lampert and Gábor Tamás Rittersporn, eds. *Stalinism: Its Nature and Aftermath. Essays in Honour of Moshe Lewin.* London: M. E. Sharpe, 1992, 101–20.

Ruusman [Ruusmann], Ants. “O roli sel'skokhoziaistvennoi kooperatsii pri vosstanovlenii i sotsialisticheskom pereustroistve sel'skogo khoziaistva v Ėstonskoi SSR (1945–1950gg.)” in *Problemy agrarnoi istorii sovetskogo obshchestva.* Moscow: Izd. Nauka, 1971, 173–76.

———. *Põllumajanduse taastamine ja kollektiviseerimine Eesti NSV-s aastail 1944–1950. Dissertatsioon ajalooteaduste kandidaadi teadusliku kraadi taotlemiseks* [The restoration and collectivization of agriculture in the Estonian SSR 1944–1950. Dissertation to obtain the degree of a candidate]. Unpublished manuscript in the Estonian Academy of Science, Tallinn, 1967.

Swain, Geoffrey. “Deciding to Collectivize Latvian Agriculture.” *Europe-Asia-Studies* 55 (2003): 39–58.

Taagepera, Rein. “Soviet Collectivization of Estonian Agriculture: The Taxation Phase.” *Journal of Baltic Studies* 10 (1979): 263–82.

———. “Soviet Collectivization of Estonian Agriculture: The Deportation Phase.” *Soviet Studies* 32 (1980): 379–97.

——— and Romuald J. Misiunas. *The Baltic States: Years of Dependence, 1940–1990. Expanded updated edition.* Berkeley: University of California Press, 1993.

Tõnurist, Edgar, ed. *Eesti NSV põllumajanduse kollektiviseerimine: Dokumentide ja materialide kogumik* [The collectivization of agriculture of the Estonian SSR. Collection of documents and materials]. Tallinn: Eesti Raamat, 1978.

———. “Traagiliste sündmuste aasta” [The year of the tragical events] in *Ausalt & avameelselt EKP Keskkomitee VIII pleenumist, Karotammest ja Käbinist, hinge harimatusest. Koost. E. Hion* [Frankly and openly about the eighth plenum of the CPE Central Comitee, Karotamm, Käbin and the lacking nobleness of heart. Compiled by E. Hion]. Tallinn: Perioodika, 1989, 31–62.

Viola, Lynne. “*Bab'i Bunty* and Peasant Women's Protest during Collectivization.” *The Russian Review* 45, no. 1 (January 1986): 23–42.

———. *Peasant Rebels under Stalin: Collectivization and the Culture of Peasant Resistance.* Oxford: Oxford University Press, 1996.

Zubkova, Elena. "Fenomen 'mestnogo natsionalisma': 'Estonskoe delo' 1949–1952 godov v kontekste sovetizatsii baltii." *Otechestvennaya Istoria* 3 (2001): 89–101.
———. *Pribaltika i Kreml'. 1940–1953*. Moscow: ROSSPEN, 2008.
———. *Russia after the War: Hopes, Illusions, and Disappointment, 1945–1957*. London: Sharpe, 1998.
Zunde, Pranas. *Die Landwirtschaft Sowjetlitauens*. Marburg/Lahn: Herder-Institut, 1962.

THE COLLECTIVIZATION OF THE AGRICULTURE IN THE BALTIC REPUBLICS, 1944–1951: CHRONOLOGY OF EVENTS

1944

September 17: Supreme Council of the Estonian SSR: "Law concerning the restitution of the land taken away by the German occupants to the Estonian peasants."

September 30: Supreme Council of the Lithuanian SSR: "Law concerning the liquidation of the consequences of the German occupation in agriculture."

A land reform is initiated in all Baltic Soviet Republics, scaling down all farmsteads with more than 30 hectare land to down to 20 hectares and punishing those accused of collaboration with the Germans or of cooperation with the anti-Soviet partisans reducing their farmsteads down to 5–7 hectares and taking most of their animals and equipment.

1945

March 18: Fourth Five-Year Plan of the USSR.

September: Founding of the first *kolkhoz* in the Latvian Soviet Republic.

September 19: Founding of a "Council for *kolkhoz* matters" in the Central Committee of the CPSU, headed by Andrei Andreev.

October 15: Decree of "On the Work of the CC of the Lithuanian CP" by the Central Committee criticizes the insufficient *kolkhoz* propaganda in the Lithuanian Soviet Republic. The decree is discussed in all three Baltic Soviet Republics.

1947

February 4: Vyacheslav Molotov informs the Bureau of the Communist Party of Latvia that collective farms are now top of the agenda.

February 10: The Paris Peace Traty is signed.

February 21–26: February plenum of the CPSU. Speech by Andreev focusing on Soviet agriculture. No reference to the collectivization in the Baltic republics.

April 16: The First Secretaries of the Baltic Soviet Republics, Nikolai Kartotamm, Antanas Sniecus, and Jānis Kalberžins are called to the Orgburo of the CPSU in order to discuss collectivization.

May: First postwar *kolkhoz* founded in the Lithuanian Soviet Republic.

May 21: Decree of the Communist Party of the Soviet Union titled "On founding *kolkhozes* in the Lithuanian, Latvian and Estonian SSR."

May 31: Detailed Plans by the First Secretaries of the Baltic Communist Parties on the course of collectivization. These plans have no practical consequences whatsoever for the actual collectivization drive.

August 23: First *kolkhoz* in the Estonian SSR.

August 30: Decree 654 by the Council of Ministers of the USSR on "The taxation of farmsteads in the Estonian SSR." Dramatic rise in taxes. From now on taxation is one of the main means of collectivization. The decree also gives a definition of who was to be counted as *kulak*.

December 12: Resolution of the Central Committee of the Communist Party of Latvia and the Council of Minister of the LSSR with sufficient conditions for the classification of a farm as *kulak*.

1948

January: A control brigade from Moscow visits Estonia and reports on "Deficiencies in the work of the party organizations." They are accused of slowing down the process of socialist reconstruction.

January 26: Decision by the bureau of the Central Committee of the Estonian Communist Party: "On the work of the party organizations in founding *kolkhozes*," declaring that the time is ripe for speeding up the collectivization drive.

May: Collectivization is accelerated in all three Baltic Soviet Republics without showing notable results.

May 22: Mass deportations in the Lithuanian Soviet Republic (Operation Vesna). 43,300 people are deported.

July: Dramatic rise in agricultural taxation, ultimately destroying any further prospects for the *kulak* farmsteads.

October 15–16: Twenty-Fourth Plenum of the Communist Party of Estonia. Collectivization speed and size of *kolkhozes* are criticized. First Secretary Nikolai Karotamm blames "*kulaks* and band-dogs of Anglo-American imperialism" in his speech.

October 21: First secretary of the Estonian Communist party, Nikolai Karotamm, and chairman of the Council of Ministers of the ESSR, Arnold Veimer, propose to deport the *kulaks* to the industrial northeastern regions of Estonia.

October 26–27: Twelfth Plenum of the Communist Party of Latvia. First Secretary of the Communist Party of Latvia states that Latvia's collective farms would soon be numbered not in hundreds but in thousands.

December 1–2: Fifth Congress of the Communist Party of Estonia: First secretary Nikolai Karotamm and Minister for Agriculture Alexander Puusepp are criticized for slow speed of collectivization. The congress decides on the course of mass collectivization.

1949

January 1: Decree on the founding of *kolkhozes* in the Estonian SSR. Contains the decision about the deportation of *kulaks* and "enemies of the people."

January 9: Party congress of the Latvian SSR, deciding on mass collectivization.

January 28: Resolution No. 390–138 by the Council of Ministers of the USSR, fixing the categories of people to be deported.

March 17: Resolution by the USSR Council of Ministers "On measures of help for the agriculture of the Lithuanian, Latvian and Estonian SSRs."

March 21: Resolution No. 282 by the Council of Ministers of the Latvian SSR: "On the deportation of *kulak* families out of the Latvian SSR."

March 21: Mass deportations in the Lithuanian, Latvian, and Estonian Soviet Republics (Operation Priboi). 33,496 people from Lithuania, 41,445 people from Latvia, and 20,660 people from Estonia are deported.

Dramatically rising collectivization rates.

1950

March 21–26: Eighth Plenum of the CP of Estonia. Cleansing of the party leadership.

March 30: Decree of the CPSU about the amalgamation of small *kolkhozes*.

1951

October 3: Mass deportations in Lithuania (Operation Osen). 19,000 people are deported.

PART II

Land Collectivization in Central Europe

The Collectivization of Agriculture in Poland: Causes of Defeat

DARIUSZ JAROSZ

The collectivization of Polish rural society and agriculture from 1948 to 1956 has recently been the subject of numerous new studies based on sources inaccessible to researchers before 1989.[1] Thanks to these studies, we now have a better understanding of how agricultural policy was shaped and modified during the period in which production cooperatives—called *spółdzielnie produkcyjne*, the Polish counterparts of Soviet collective farms—were organized. Although our understanding of peasant attitudes and resistance toward collectivization policies has advanced considerably, the attitudes and behavior of those who were subject to direct collectivization pressure still warrants additional research.

Compared to similar collectivization programs implemented across East European states during communism, in Poland the process was an exceptionally slow, limited, and overall, a failed policy. In 1955, production cooperatives covered just over 2.1 million hectares, or 11 percent of the overall land and just 6 percent of peasant farms—the lowest percentage among any country in the Soviet-dominated Socialist camp.[2] Although Polish peasants, like those in Yugoslavia and Czechoslovakia, were able to choose from four

---

[1] On the collectivization of Polish agriculture see Dobieszewski, *Kolektywizacja*; Robakowski, *Społeczno-polityczne*; Kaliński, "Forsowna kolektywizacja;" Korboński, *Politics of Socialist*; Jarosz, *Polityka władz*.

[2] Compare this with the aggressive policy of collectivization under Stalin in the 1930s and the percentage of collectivization in Soviet Russia—close to 97 percent. See Suny, *The Soviet Experiment*.

types of collectives, a majority resisted joining collective farms. As in Yugoslavia under Tito, in Poland collectivization was halted in 1956 by the new regime under Władysław Gomułka. As a result, the second stage of collectivization, which had been in progress from the late 1950s and which had brought about the liquidation of private farms across Eastern Europe, was never pursued in Poland. Why was that the case? What was it that defeated collectivization á la polonaise? This article addresses these questions.

Transformation of Rural Society and Agriculture, 1944–48

When collectivization began in 1948, fundamental socioeconomic changes had already occurred in the Polish countryside. These changes would have a strong impact on how the policy unfolded over the next eight years. The most important factor in this regard was land reform, which in Poland had occurred in two stages: the first followed a September 1944 law mandating estates larger than 50 hectares or 100 hectares be parceled.[3] The second was the policy of agricultural resettlement of Poland's "Regained Lands"—those territories annexed by Poland according to the postwar resolutions at Yalta and Potsdam (see Map 1).[4] The average size of the newly

[3] The actual threshold depended on the region, as following: 100 hectares in total (irrespective of proportion of arable land in that total) in the voivodships of Poznań, Silesia (from 1950—Katowice), and Pomerania (from 1950—Bydgoszcz). A limit of 50 ha of arable land existed in other areas that before September 1, 1939, had belonged to the Second Republic of interwar Poland.

[4] The term Regained Lands was used after 1945 to describe the post-German lands incorporated by Poland as a result of the Potsdam Conference. After 1950, due to the new administrative division of Poland—in which 22 smaller administrative units were set up with 17 voivodships (provinces), and 5 voivodship cities—the Regained Lands were referred to as western and northern lands. The Regained Lands were then divided up between 11 voivodships of which six (Koszalińskie, Opolskie, Szczecińskie, Wrocławskie, the City of Wrocław and Zielonogórskie) consisted exclusively of Regained Lands, and two—Gdańskie and Olsztyńskie—were a mixture of old and new lands. Their total area was 102,800 km^2, which constituted 32.9% of the current size of Poland. The population of the Regained Lands was distinguished by numerous social-demographic characteristics. In 1950, the settlers constituted 78% of the population. In 1948, there were 16.3 marriages per 1000 inhabitants in these eight voivodships (the overall national indicator was 13.3), in 1950 there were 40.8 births per 1000 inhabitants (in the old lands—that is, in the remaining part of the country—it was 30.7). See Rybiński, *Ziemie Zachodnie*, 35–43.

formed farms was 6.9 hectares, and the average size of an additional lot was 1.9 ha.

As a result of these policies, approximately one million peasants received land. Out of this number, about 380,000 plots were allotted from the parceled lands in the old parts of Poland, and 620,000 plots were created from the Regained Lands.[5] In total, about six million hectares of land was redistributed, over four million of which derived from the Regained Lands. By 1949, a third of all Polish farms had been created as a result of these two stages of agricultural reform.[6]

Map 1: Poland During the Collectivization Period.
(after the administrative division of 1950)

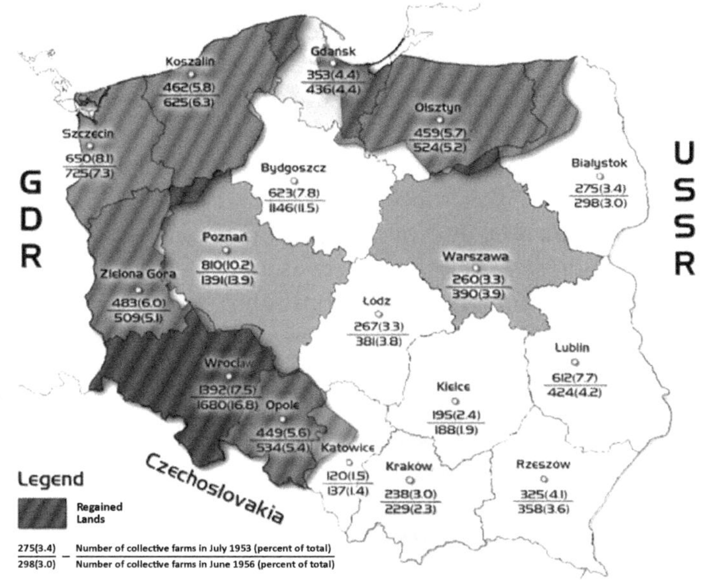

Source: Author's map and own calculations.

---

[5] Parcelization was carried out in that part of postwar Poland (that is, the old lands) that before September 1, 1939, had belonged to the Second Republic (interwar Poland). However, because the land was too scarce in relation to the number of those qualifying for this redistribution, many were supplied farmsteads in the post-German territories.

[6] Słabek, "Reformarolna," 60–61.

Communist Party leaders pursued agrarian reform in order to broaden their social base in the countryside, which had a profound impact on the political sympathies of the rural population.[7] The Communist Polish Workers' Party (*Polska Partia Robotnicza*—PPR), and to some degree the Polish Socialist Party (*Polska Partia Socjalistyczna*—PPS), which was increasingly controlled by the communists, were popular among newly landed peasants, who were former farm laborers that had benefited from the recent land redistribution. Social conflicts were closely linked to political differences. To that point, the most important criterion dividing rural Poland, both in economic and in social terms, was the size of the farms, which determined the level of affluence and prestige of its owner. After the agrarian reform, the new standard for a farmer's status was the farm's origin: if the land was acquired as a result of the postwar parcelization, this was perceived as worse than if it were inherited—passed down from one generation to the next. This distinction, so rooted in rural culture, was underscored by the prevailing political divisions in the countryside. The most important political gulf separated PPR members (or *dworusy*, a disdainful term for the farm laborers) from farmers who had inherited their land from their families and joined the ranks of the anti-communist Polish Peasant Party (*Polskie Stronnictwo Ludowe*—PSL). The emergence of that party was hailed in the countryside for its opposition to the Polish Workers' Party, whose supporters were branded as pro-collectivization atheists and criticized for promoting non-Polish rule. The destruction of the PSL (1947–49) and the forced merging of the communist PPR and PPS (in December 1948) into the Polish United Workers' Party (*Polska Zjednoczona Partia Robotnicza*, PZPR), coupled with a push for collectivization, were followed by a general decline in peasant enrollment in political organizations, as they were generally reluctant to participate in organizations controlled by the communists.[8]

Agricultural reform created an additional social divide in the countryside. For the rural population, the land titles acquired as a result of postwar reform were inferior to what had been inherited. This had a significant affect on societal differences and attitudes towards collectivization. Hereditary farmers were more inclined to resist the program than

---

[7] Some of the party leaders, excluding Władysław Gomułka, unofficially declared that land reform was a transitional phase and that the ultimate aim was collectivization.

[8] Dobieszewski, "Nomenklatura partyjna," 192–94; Ocena działalności organizacyjnej.

the *dworusy*, who were familiar with working collectively on prewar farms. Moreover, hereditary farmers suffered from a lack of essential equipment in their newly created farms and adapted to the role of peasant-farmers with much more difficulty.

An even more complicated set of societal conflicts emerged in the Regained Lands. By the end of 1945, more than 1.63 million people were resettled in the Regained Lands, an estimated 400,000 from territories annexed by Poland from the Soviet Union. By 1948, the Polish population in these territories had grown to 5.5 million, which included over 935,000 indigenous inhabitants, almost 2.5 million settlers from "former lands" (central and eastern lands of new Poland, which belonged to prewar Poland), 1.3 million settlers from the USSR and 0,5 million children under four years of age (no-division into groups of origin). The Regained Lands were inhabited also by the small numbers of Belorussians, Jews, Lithuanians, Czechs, Roma, Greeks, Germans and 150,000 Ukrainians, who had been forcibly resettled from southeastern Poland as a part of Operation Vistula in 1947.[9]

The groups maintained their distinct cultures: relations between them were strained, often influenced by prejudices and hardened stereotypes. When the Polish settlers arrived, their contact with the population of former Eastern Prussia was hostile and bitter. For many new settlers, the locals were simply "Germans"—since they spoke German and were Lutherans in Masuria—which reinforced the stereotype of the indigenous evangelical German.

As a result, Germans were the stripped of their farms. Following the agricultural reform decree of September 1944—which granted the state the right to claim ownership of property belonging to German citizens or to Poles who supported the enemy—land owned by citizens of the German Reich and Polish citizens of German nationality, along with estates that had been abandoned, were taken over by the state for redistribution. On this basis, the resettlement action plan encompassed local populations with German citizenship, who in the course of the war had been forcibly evacuated or voluntarily resettled inside Germany. Subsequent laws did not take into account the difficult situation of the local population evacuated to Germany or moved temporarily. When they returned, they often found settlers already living on their properties. They had to undergo complicated, protracted and ineffective vetting procedures to

[9] Dziurzyński, *Osadnictwo rolne*, 353–57.

establish their Polish identity in order to recover their land.[10] According to estimates, these disputes affected 24,000 farms altogether. Some 20 percent of the land remained in indigenous hands in 1945. Peasant complaints, found in Polish archives and sent to various resettlement authorities, reveal how painful and emotionally charged conflicts over these farms became.

Many of the shortcomings in the resettlement plan were a matter of timing. The Communist Party viewed the resettlement policy as an important strategy of political legitimization, which it needed not only in the international arena but also at home. As a result, the party stressed the rapid execution of the resettlement action plans, which was reflected in the various solutions proposed. Leaders abandoned the plan to exchange farms from the old lands for those in the new lands once they realized what strong emotional attachments peasants had to their home villages. This, as a rule, put settlers from central Poland (voivodships of Warszawa, Kielce, Łódź, and Kraków—see Map 1) in a privileged position in relation to other peasant groupings in the western lands, which was not effective in stabilizing conflicts stemming from the resettlement program. Peasants frequently treated their second farms in the western lands as a source of supply for their first ones in the old lands, stripping the second farms of valuable assets and then abandoning them. There was a feeling that land reform was temporary and the majority of settlers, uprooted and culturally distinct, suspected that Polish control over these lands would not be permanent. This had a strong impact on the way the farms were managed, making it far easier to find newly landed peasants joining production cooperatives than the hereditary farmers of central Poland.

Securing full ownership titles of new farms was a complicated process and often subject to political fluctuation. It required the district (*poviat*) agrarian settlement commission to issue a ruling on the farm's borders and its price, and a ruling on the execution of the title deeds that transferred the ownership of the land to the settler. Tensions and conflicts were intensified by administrative chaos, coupled with incompetence and corruption by local authorities.[11] As elections of January 1947 approached, the process of ownership transfer was accelerated as the communists sought to ensure broader social support. The new course initiated in July 1948, which aimed at setting up *kolkhoz,* was irreconcilable with the consolida-

[10] Dziurzyński, *Osadnictwo rolne*, 112–13.

[11] Jarosz, "Chłopi na Ziemiach;" Osękowski, *Społeczeństwo Polski.*

tion of peasant private property, so that the communists abandoned the process of ownership transfer described above.

The Organization of Polish Collectivization

Poland's collectivization policy was devised by Communist leaders during party meetings in 1948 and 1949. The initial decision to begin organizing farm cooperatives was made at the plenary sessions of the Central Committee of the PPR during July and September 1948. At the end of 1948 and the beginning of 1949, party leaders drafted resolutions on the program's basic tenants and created three types of collective ventures: Soil Cultivation Associations (Type I); Agricultural Manufacturing Cooperatives (Type II); and Agricultural Cooperative Groupings (Type III). Each was distinguished by their degree of socialization in fieldwork, land, agricultural machinery, equipment, and livestock. Following the Soviet model of collectivization, those peasants who were often called "*bogacze wiejscy*" ("rich rural men"), and later *kulaks,* were not allowed to join any type of cooperatives. As in Soviet Russia, *kulaks* were the main focus of communist propaganda in the countryside, but unlike the USSR, *kulaks* in Poland were not deported. Instead, the prime weapon against the *kulaks* was economic: they were subjected to extraordinarily high tax rates and mandatory supply quotas for milk, potatoes, livestock, and cereals. They were also denied use of the state's newly built network of machine centers, which mechanized agriculture and helped meet the needs of nascent collectives.[12]

In December 1950, the Politburo created a new type of cooperative, Agricultural Cooperative Association (Type Ib), after concluding that Types II and III were not suitable for peasants. This new collective did not require that horses be put in a common pool, and did not require members to work a minimum of days, two policies of Type II and III farms that peasants resisted. Instead, members were encouraged to contribute their livestock, agricultural equipment and farm buildings to the collective farms. Income was to be distributed not only by the amount of

[12] In December 1950, the Politburo of the Central Committee of the United Polish Workers' Party (CC of the PZPR) decided to nominate Political Divisions of POM (*Państwowe Ośrodki Maszynowe*—State Machinery Centers). Their task was to exercise constant control over the existing collectives and promote the formation of new ones. *Kulaks* were denied access to POM machines.

labor contributed but also by the amount of equipment and stock that was volunteered.[13]

As in the other East European countries, the post-Stalin thaw had a strong influence on collectivization programs in Poland. This became obvious in 1956, when leaders began to rethink Poland's agrarian policy.[14] In response to the growing problem in agricultural supply, party leaders decided to increase aid to individual peasant farmers and began to discuss the collapse of *kulak* farms. The resolutions of the seventh plenary session of the Central Committee of the PZPR (July 18–28, 1956) upheld the thesis that "production cooperatives should play a leading role in the struggle to increase agricultural production in the 5-year plan." A significant new departure in the collectivization model to date was the assertion that "the bigger and good cooperatives deserve to be given the opportunity to purchase tractors"—that is, ones they would own, which until then they were unable to do. Moreover, the possibility of admitting "loyal *kulaks*" as members of consolidated production cooperatives was recognized, on condition that they enjoyed the confidence of their village.[15] Thus, two significant elements of the Soviet collectivization model were never pursued in Poland.

The most radical change to Poland's collectivization policy occurred at the October 1956 plenary session of the Central Committee, when Władysław Gomułka, a former victim of Stalinism, was named First Secretary (October 19–21, 1956). In acceptance his speech, Gomułka criticized the achievements of production cooperatives, but claimed they were necessary because collectives constituted a higher form of socialist production. Despite the reservations of the communist leadership, Gomułka managed to push through his reform policies, after which, collectives were organized voluntarily and as self-governing units, with the right to purchase machines for agricultural production and ancillary work plants. With Gomułka's reforms, a vast majority of cooperatives collapsed in a short period of time (from October to December 1956), which helped trigger the Polish October of 1956.

Beyond political factors, the failure of collectivization can be linked to factors related to Polish rural society. Peasants resisted collectivization not only because of its Soviet lineage, but also because for most Polish peasants, a *kolkhoz* was an institution alien to traditional rural culture. Much

---

[13] *Statuty spółdzielni.*

[14] For a comparative perspective, see the other contributions to this volume.

[15] *Uchwały Komitetu Centralnego*, 145, 196–97.

information on this topic can be found in the surviving PZPR archives and those of the pro-Communist United Peasant Party, as well as in the archives of the Poland's Ministry of Public Safety, which controlled all "anti-state activities." These archives contain records of meetings held by party activists and political police who were sent to the countryside to win peasants over to the idea of cooperatives. At these meetings, peasants frequently complained that being stripped of their hereditary land was destroying institutions that made surviving peasant life possible, and that new solutions were never provided. For instance, peasants asked representatives what material safeguards would be given to older people in lieu of the traditional institution of children and family providing a basic existence for the older generation in exchange for their inheritance. Likewise, a young married couple asked: "When a young couple marries and wants to build a house, who will give them the land and small-holding when their parents haven't got it?"[16]

Discussions over certain statutes showed that peasants were unrepentant in their criticism of the state's new policy of remunerating members of collective farms. Comparing the collective to private-owned farms, they wondered whether the lack of personal interest in economic results would make members apathetic and anticipated there would be reluctance and a lackadaisical approach to work. They also expressed fears of an unequal distribution and pay, as well as of theft of common property.[17] Peasants who were not in debt anticipated that they would have to service the debts of other members upon joining teams.[18]

Numerous statements by peasants demonstrated that they saw production cooperatives as a denial of their traditional way of life. For instance, peasants insisted that they would rather suffer hunger and punishment than work in a *kolkhoz*.[19] Sometimes Polish peasants articulated their aversion to collectivization in terms of a defense of their freedom and independence. The worry was that "if no one rules, I'll get up when I feel like it, I'll work when I want to, I'll decide for myself."[20] Cooperatives, in their opinion, were to change their daily rhythm of life so radically that they questioned their identity as a social group. Finally, as many peasants expressed, the organizational solutions risked the bureaucratization of pro-

---

16 Protokół z zebrania organizacyjnego.
17 "Relacja Artura Starewicza," 111.
18 Notatki Stefana Żmijki.
19 Ocena przeniesienia uchwały.
20 Sprawozdanie inspektorów.

duction cooperatives, or that the "cooperative will be governed by bureaucrats, while we will be doing all the work."[21]

Another feature of the collectivization policy, which sparked peasant's aversion to communist authorities was the ideologically motivated struggle with the *kulaks*. The slogans of class war were taken into the countryside with the aim of disrupting the existing social structure. The struggle between various groups of peasants, divided up in accordance with Stalinist ideology, challenged the prevailing hierarchies of prestige and social status and the power relations in the rural government structures, as well as the rationality of farm management. The Communist Party policy of reconstructing the rural class system took no account of the existing social relations. The policy of liquidating larger farms did not result in class warfare as the authorities imagined, but rather, sympathy with the big farmers and defense against state repression. The implementation of agricultural policy made rural society more prone to close ranks against the rulers than against the *kulaks*. Hence numerous reports analyzing the situation in various villages ended with assertions that "there is no dividing line between the poor and the *kulaks* ... The *kulaks*, who are not isolated by the rest of the community, set the tone in the countryside."[22] However, this did not mean to suggest there were no anti-*kulak* attitudes. One can find evidence of this in archival material about anonymous denunciations of "the rural rich," in order to replace them in political offices.[23] Peasant resistance to collectivization in Poland was in large measure a criticism of the Soviet model for agrarian reform that was applied elsewhere in Eastern Europe.[24] That criticism would become the anchor of resistance to the policies of collectivization.

The Implementation of Collectivization Policy: Peasants and Apparatchiks

Pressure to set up cooperatives was reflected in how communist leadership planned its collectivization program—and in this respect, Polish collectivization methods showed a great deal of congruence with the Soviet

21 Sprawozdanie z odprawy powiatowego aktywu.
22 Sprawozdanie z wyjazdu ekipy agitacyjnej.
23 Raport WUBP w Rzeszowie.
24 See Constantin Iordachi and Dorin Dobrincu's contribution to this volume.

model.[25] Because these planning methods had far-reaching consequences on collectivization, they deserve special treatment.

In September 1948, Politburo member Hilary Minc decided that production cooperatives could comprise 1 percent of all farms in Poland in a year's time.[26] At a conference of the Provincial Committees of the PZPR in March 1949, Roman Zambrowski, a member of the Central Committee's Politburo responsible for agricultural affairs, discussed the party's decision to set up 200 farm groupings of various types following the 1950 harvest.[27] Although such plans for collectivization were officially condemned as favoring "distortions," (a euphemism used by the communists to describe illegal and brutal methods applied in order to force peasants to join the collective farms) by the end of 1950, the party did not abandon the program.[28] Decision-makers in Warsaw had decisive influence on what methods were adopted by the local authorities: plans were formulated in Warsaw and imposed across the provinces, counties, and municipalities.

In August 1950, the PZPR County Executive Committee (*Egzekutywa Komitetu Powiatowego*—KP) in Drawsko (voivodship of Koszalin—see Map 1) decided the county's farms could be collectivized by the end of July 1951. Party leaders then decided to move the deadline earlier to March 1951 and later even earlier, to December 20, 1950. Later, the secretary of the KP of the PZPR in Drawsko received a telephone call from the head of the Agricultural Division of the KW of the PZPR in Koszalin, who told him to speed up the process or a county in the province of Posnania would be the first to collectivize. As a consequence, leaders passed a resolution to cut the deadline even shorter, and in the course of the first three weeks of 1950 there was a cooperative in each municipality.[29] A similar race for collectivization occurred in many other regions of Poland. This dynamic tempted authorities to adopt no-nonsense, shortcut methods to set up group farms. But the program was met with such resistance among peasants that local leaders, under pressure from Moscow, resorted to forcing farmers to joining cooperatives. This dynamic can now be reconstructed using party documents that describe the subsequent "distortions" in cooperative-forming initiatives.

---

[25] Fitzpatrick, *Stalin's Peasants*, 55.

[26] Jarosz, *Polityka władz*, 20.

[27] Stenogram odprawy sekretarzy.

[28] Stenogram narady krajowego aktywu.

[29] Protokół nr 17 z posiedzenia.

But peasant resistance was becoming hardened, and breaking the rural population was becoming increasingly difficult. The rural population was not waging a war against state authorities, but rather against the local government. Contradictory directives from Warsaw, meanwhile, pressured local representatives: they were held accountable for the number of cooperatives planned, despite the fact that collectives were to be set up voluntarily according to official documents and public pronouncements. But growing peasant resistance meant voluntary collectivization was increasingly unlikely. Meanwhile, the Central Committee of the PZPR passed resolutions against the so-called "distortions," imposing penalties on those guilty of breaking the law. These penalties were enforced in the county of Gryfice in 1951 and in the province of Lublin in 1953.[30] At the same time, law-abiding officials were accused of excessive liberalism in the face of the "class enemy" and were also punished. Between 1948 and 1953, most local *apparatchiks* succumbed to party attitudes and applied pressure to rural farmers to collectivize, and only a minority adopted the peasant perspective and worked to sabotage collectivization efforts. In the later period, from late 1953 to October 1956, as the liberalization of the Stalinist agrarian policy progressed, officials gradually became less repressive toward peasant resistance. The political thaw and increase in state aid for individual farms meant that it was more difficult to put pressure on farmers to join production cooperatives.

Thus, when analyzing what factors contributed to the failure of collectivization in Poland, it should be noted that while peasant resistance was critical, so too were attitudes of local *apparatchiks*. Because of the program's unpopularity, many officials were convinced that setting up collective farms at any price could cast a shadow over their careers. In addition, many officials had roots in the countryside and were sympathetic to rural perspectives against collectivization. The Polish example, therefore, is similar to the collectivization experience in other communist states. With new research on collectivization in Yugoslavia and the German Democratic Republic, respectively, Melissa Bokovoy and Gregory Witkowski both argue that the adversarial relationship between communist elites and

[30] For instance, in May 1951, the Politburo published a declaration on the Gryfice case emphasizing the need for restraint in dealing with the peasants. The PZPR county committee in Gryfice (Voivodship of Szczecin) used force in conducting illegal searches for grain among individual peasants. See Kura, *Aparat bezpieczeństwa*, 226–39.

the rural population has been oversimplified.[31] They show that local *apparatchiks* frequently were more apt to side with the peasant point of view than those of party superiors.

This does not mean that some local officials did not behave brutally. The surviving source materials on "distortions" contain descriptions of various methods applied in order to force peasants to join the collective farms.[32] As in other countries in this region, administrative sanctions were especially irksome. The pretexts for their imposition were infringements such as failure to install adequate fire prevention in one's attic or an untidy yard. Often peasants were offered the opportunity to avoid the fine if they joined a collective farm. The same type of statements were often made by tax collectors and commissions set up on an ad-hoc basis. They exploited financial arrears in order to put pressure on peasants resisting the collective economy.[33]

Another widespread tactic the party used to pressure peasants to join collectives was to penalize farmers for not meeting quotas. Between 1952 and 1955, close to 600,000 farmers were penalized for failing to meet the quotas, and from 1953–1955, more 10,000 people received sentences in court. The climax of these repressive measures was 1953, when more than 254,000 peasants were punished with jail sentences or fines.[34]

The application of repression was especially painful when the consent of some peasants was already secured in a given village to set up a collective farm, taking the form of an intentionally malicious exchange of land between the new cooperative members and the farmers continuing on an individual basis. Such an exchange was to lead to the creation of one or several cooperative estates in villages. They were given arable land that due to the size as well as the quality of the soil lent itself to mechanical cultivation and the achievement of good production results.

The decree issued on August 16, 1949 allowed for such an exchange if it was demanded by the rationality of the agricultural system.[35] Article 2 in the decree stated that everyone would receive land of the same value in

---

[31] See Witkowski and Bokovoy's contributions on the GDR and Yugoslavia, respectively, in this volume.

[32] These materials are letters of complaints sent by peasants to state and party autorities preserved in Central Archives of Modern Act (Archiwum Akt Nowych).

[33] Sytuacja w spółdzielniach województwa łódzkiego.

[34] Jarosz, *Polityka władz*, 1998, 228–37.

[35] Dziennik Ustaw, no 48, poz. 36.

exchange for the farmland that they had left. Where that was impossible, lands of higher or lower value were allocated by way of exchange with the proviso that the difference in value would not exceed 5 percent of the value of the land owned to date. In the face of numerous abuses, the minister of agriculture issued a disposition on August 16, 1951, on the exchange of land for production cooperatives.[36] It established the general principle that as little change in boundaries as possible should happen and that it should not impede productivity. Both the mentioned decree and the disposition did not eliminate incidents of peasants being forced to exchange their land. The system was subject to corruption and irregularities.

A large proportion of peasant complaints as well as reports by party functionaries and state officials of the regional inspectorates contained descriptions of inequitable exchanges of lands. From 1950 to 1951, numerous county committees violated peasant rights of ownership by giving them land with inferior quality of soil and even fallow land. Frequently there were cases of peasants given small plots or land several kilometers away from buildings and without access by road. In some cases, peasants lost not only their conveniently located plots but also outbuildings or wells.[37]

When administrative pressure failed, the final weapon was state terror. The Citizens' Militia (*Milicja Obywatelska*), the Security Office (*Urząd Bezpieczeństwa*) officers, and occasionally units of the Polish army, carried out these repressive measures, often with brute force. A frequent practice was to detain peasants who opposed joining cooperatives at Citizens' Militia stations and Security Office buildings without formally logging the incarceration,[38] keeping some farmers for days without food and water. Another common method was to detain mothers with large families, knowing their absence would punish an entire household as well as keep her from her land.

Although measures used by the state against Polish peasants were frequently marked by cruelty, it is difficult to quantify the scale of such repression. In the contemporaneous system there were no direct legal sanctions to enforce collectivization (court cases could only penalize peasants for failing to meet quotas but not for not joining production cooperatives). Overall, Polish collectivization did not claim as many lives as the collectivization program did in Stalinist Russia in the late 1930s.

---

36 Dziennik Urzędowy Ministerstwa Rolnictwa, no 14, poz. 89.
37 Zestawienie meldunków z terenu, 1951.
38 Wyciągi ze sprawozdań słuchaczy, 1950.

PEASANT REACTIONS TO COLLECTIVIZATION

The party's adoption of a strict collectivization program took Polish peasants by surprise. Villagers were afraid of the process because it changed their daily routines in ways that were difficult to predict. By 1948 and 1949, peasants had begun to express their fears of cooperative farms: many rural inhabitants could not imagine what the economy would be like; others believed that they would be evicted from the land and that state-owned farms would be set up.[39]

Villagers associated collectivization with three institutions: Soviet collective farms, state-owned estates, and rural cooperatives. For instance, a majority of peasants in the Białystok province had experienced firsthand Soviet collectivization, which took place between 1939 and 1941. Others had heard of the program from Red Army soldiers stationed in their villages.[40] Stories spread quickly of unproductive and poorly managed farms, and peasants believed that collective farms were worse than a fire, because at least one could rebuild after a fire.[41] Other statements show that some villagers were aware of the brutality of Soviet collectivization, especially those peasants in the Regained Lands, for whom cooperative farms were part of their personal wartime experience.

Poorly managed state-owned farms also reinforced peasant attitudes against cooperatives. Nearly all recorded peasant opinions pointed to their abysmal performance. In numerous statements, peasants reported that state farms were a bad example for the nascent production cooperatives and that peasants are better at cultivating their own soil.[42] The activities of rural supply and distribution cooperatives also did little to enhance the reputation of production cooperatives. Peasants accused local organizers of mismanaging agricultural centers, and criticized boards staffed by incompetent officials chosen by the government. Leaders of cooperatives were also held responsible for inefficiently distributing goods to shops.[43] Peasants complained that if municipal trading cooperatives did not work, then other forms of production cooperatives

[39] "Relacja Artura Starewicza," 1993, 112.
[40] As stated in a special Central Committee report of March 1949.
[41] Sprawozdanie z wyjazdu, 1949.
[42] Sprawozdanie z konferencji, 1949.
[43] Protokół ze zjazdu powiatowego, 1949.

would also fail.[44] It is worth noting that in later years peasants would refer to the example of already existing cooperatives as evidence of the inferiority of collective farms.

Rumors provide lucrative insight into peasant attitudes toward collectivization. As discussed by Arnd Bauerkämper and Sheila Fitzpatrick in this volume, rumors were particularly popular among peasants in the Soviet Union. In Poland, peasants also spread stories about the perils of collectivization, which they saw as a lethal threat to their way of life.[45] According to one, children would be taken away from collective farmers and brought up in state nurseries. In Szczecińskie province in October 1950, peasants said the harvests from the cooperative fields would be requisitioned by the government and what was left would not be available to them until spring. According to a more macabre rumor circulated in Syców county in 1949, older and infirmed peasants would be turned into soap once the cooperative was established.[46]

When collectivization began in 1948, it triggered a wave of shock and fear among Polish peasants, much like it did in Soviet Russia and in other Eastern European countries. Party supporters noticed frequent work interruptions, and in autumn 1948, there was a standstill in sowing and reductions in livestock. They also noted a decline in the purchasing of agricultural tools and artificial fertilizers, as well as in the construction of farm outbuildings.

In response to their fear of collectivization, along with rumors of the imminent outbreak of World War III, peasants rushed to buy essential commodities.[47] In Łódzkie province, many farmers harvested unripe wheat, fearing others might thresh their crops before they did. Alcohol consumption increased, as did pilgrimages to the shrine of the Holy Madonna at Częstochowa.[48] By 1948 and 1949, peasants were already demonstrating passive resistance to the program by refusing to attend party meetings—a typical form of passive resistance, absenteeism became perhaps the most common reaction to collectivization in rural Poland.[49]

---

[44] Sprawozdanie z narady aktywu, 1949.

[45] Fitzpatrick, *Stalin's Peasants*, 47.

[46] Jarosz, "Pogłoski jako wyraz," 39–55.

[47] The important role of this reaction is also highlighted by Melissa Bokovoy in relation to Yugoslavia. See her contribution to this volume.

[48] "Relacja Artura Starewicza," 106–16.

[49] Protokół rejonowej konferencji, 1950.

There were also instances of hostility toward organizers of production cooperatives. Out of 650 incidents of political terror documented in 1950, the vast majority occurred in the countryside.[50] Reports by party activists and the Polish political police *(Urząd Bezpieczeństwa)* suggest that organizers of cooperatives in various villages received anonymous threats and were the subject of aggression. Likewise, it was common for members of collective farms to be stigmatized by the community, or deliberately ignored in public places.[51] Their wells were poisoned, their dogs killed, and powdered glass was mixed into their haystacks.

Establishing a production cooperative often triggered strong opposition among peasants. Land that had belonged to families for generations was surveyed and then divided up, often resulting in mass protest. Between 1949 and 1950, mass demonstrations against cooperatives gradually increased. In 1949, the Ministry of Public Safety (MBP) registered three demonstrations in the country. In the first half of 1950 there were 14, and in the second half of 1950, there were 76.[52] These protests occurred even when those not joining cooperatives were offered better quality and more conveniently located land.

Another activity that caused numerous conflicts and violent reactions by peasants was the plowing of new farmland by tractors. Tractors had become not only the symbol of an externally imposed and painful process of collectivization but it also signified a change in traditional farming methods.[53] Between 1949 and 1953, the number of people who protested against plowing ranged from several hundred and several thousand. In some instances, these activities were organized with participants from neighboring villages.[54]

What were the social differences in attitudes and behavior toward collectivization? Records from villages where cooperatives were set up show that collective farms were for the most part founded by those who most benefited from postwar land reform and by settlers who were the prewar proletariat and the rural poor. These farmers had been pulled out of their misery by land redistribution and in particular by postwar resettlement. The tendency of farmers to join cooperatives, therefore, was determined by farm origin. The basic dividing line ran between heredi-

---

[50] Spuścizna M. Moczara, t.24.
[51] Sprawozdanie z województwa poznańskiego.
[52] Spuścizna M. Moczara, t.24.
[53] AZHRL, Nowe Pamiętniki Chłopów (NPCh), Pamiętnik nr 793.
[54] Sprawozdanie Franciszka Dziekana, 1950.

tary and the newly made farmers, the latter of whom had frequently been pulled out of their natural local environments. They found it more difficult to support the anti-collective movement. Three quarters of all the farms that belonged to this category were concentrated in northwestern Poland, where the largest numbers of production cooperatives were concentrated. The farms existing in that region, including those absorbed by cooperatives, were larger than in other regions of the country. Most villages where cooperatives were established had land that had been parceled out and settled in 1945–46. Even in mixed communities composed of beneficiaries of land reform and hereditary farmers, collectivization was promoted and supported by the beneficiaries of land reform almost exclusively.[55]

Smallholder farmers were generally loath to join cooperatives, and non-agricultural work became a more favorable solution. In most family farmsteads, there was a surplus of free hands, and at the same time, the demand for industrial labor in Poland's new command economy was enormous. As across all Eastern Europe and the USSR, industrial production was given preference over agriculture. Despite this, collectivization resulted in the gradual rise in the peasant-worker population (peasant farmers engaged in some non-farm work)—an estimated 15 percent of farmsteads by 1950 drew income from non-farming sources, which independent farmers preferred over joining collective farms.

The limited attractiveness and prospects of production cooperatives for Polish peasants created a wave of migration into urban centers. More than 1.8 million people moved to cities between 1951 to 1955—most were young Poles for whom the city was a considerable social advance that promised higher living standards.[56] As a result, the rural population dropped from 63.8 to 55.1 percent between 1948 and 1956, and the urban population increased from 36.2 to 44.9 percent.[57]

As indicated above, there is a clear correlation between hereditary farmers and the level of resistance against collectivization, a trend reinforced by cultural differences in certain regions. Historical conditions caused particularly strong opposition against collective farms in villages inhabited by the petty nobility as in Mazowsze and Podlasie counties (voivodships of Warszawa and Białystok). Their adherence to noble tra-

---

55 Dobieszewski, *Kolektywizacja*, 63; Jarosz, *Polityka władz*, 116; Próchniak, *Kolektywizacja w regionie łódzkim*, 243.

56 Gawryszewski, *Przestrzenna ruchliwość*, 7.

57 *Rocznik Statystyczny*, 13.

ditions and particularly strong attachment to hereditary property ensured that collectivization would not progress quickly in these counties.[58]

It is also worth noting that resistance to collectivization is closely correlated with the strength of anti-communist resistance. A number of anti-communist armed groups were able to operate in central and eastern Poland with support of local communities, where collectivization was least successful, as compared to northern and western Poland, where there were more communist supporters and less resistance to cooperative farms.[59] Likewise, the disparity in attitudes toward production cooperatives can be linked to nationality: in the eastern provinces, Belarusian peasants were more likely to join collective farms than Polish farmers, due to their historical-cultural roots. Research shows that Belarusian peasants adapted more readily to production cooperatives since working on Soviet collective farms—following the Soviet annexation of eastern Poland on September 17, 1939—was not a new experience. Most were acutely aware of the costs of resistance, and they also knew how to navigate the system to derive maximum benefits for themselves. Moreover, they did not perceive new communist authorities in Poland as a hostile and alien force, but rather as an opportunity for social advancement. For these reasons, anti-system postures among Belarusian peasants, in comparison to Polish nationals, were muted.

Scholars have stressed the pivotal role of the Catholic Church in reinforcing opposition to communism and to collectivization in particular.[60] A majority of Belarusian peasants were Orthodox Catholic, a church less apt to spread anti-collectivization dogma among their parishioners than their Roman Catholic counterparts.[61] In Stalinist Poland, authorities attempted to force the Roman Catholic Church to support collectivization, even issuing a proviso in April 1950 obliging the Episcopate to tell clergy not to resist the cooperative movement, claiming the program was rooted in the assumption that human nature strives for voluntary social solidarity, which has the common good as its goal.[62] An important achievement of state authorities in its struggle to control the Catholic Church was its creation of "patriot-priests," who aimed to implement

[58] Biernacka, *Wsie drobnoszlacheckie*, 192; Bartkowski, *Tradycja i polityka*, 365; Markiewicz, *Kolektywizacja*, 374–379.

[59] *Atlas polskiego podziemia*, 2007, LX–LXI.

[60] See for example: Ekiert, "Rebellious Poles;" Chrypinski, "Catholic Church."

[61] Urban, "Wybrane problemy," 32.

[62] "Porozumienie zawarte."

the party's programs and agenda.[63] However, this effort had little impact on collectivization. Political police reports indicate instead that Catholic clergy consolidated resistance to collectivization in their parishes.

Numerous sources underscore the notion that religion played a significant role in reinforcing anti-collectivization attitudes in rural Poland.[64] Religious activities were a central element of protests that erupted when collectivization efforts began. Women prayed in the fields, sang religious hymns and carried icons.[65] In one village, in the district of Kozienice in central Poland, wooden crosses shrouded in black crepe were stuck in the ground outside the homes of peasants who planned to participate in a trip to a production cooperative in western Poland.[66] In the community of Laszki, in the district of Jarosław, pro-cooperative agitators were greeted by a big cross, indicating that they were people bringing evil to the whole village. There are also reports of a brisk trade in religious paraphernalia—typically pictures of patron saints who could protect the community against the calamity of collectivization.[67] This was accompanied the so-called Lublin miracle. On July 3, 1949, a nun kneeling in front of a picture of the Blessed Virgin in Lublin Cathedral noticed a drop, the size of a berry, under her left eye. News about the "weeping" icon spread across the city and the whole country. Lublin became the destination of pilgrimages from the farthest corners of the land. Numerous other "local miracles" spread in rural areas between 1949 and 1950.[68] Especially popular among villagers were prophecies, in the form of letters from God, with warnings against joining production cooperatives.[69]

Surprisingly, anti-collectivization attitudes were shared to a large degree by those who should have supported most ardently: members of the PZPR, and its subordinate, the Peasant Party (ZSL). In August 1950, an estimated 56 percent of local Communist Party members and 42 percent of Peasant Party members belonged to collectives. From 1951–55, just one out of three Communist Party members and one out every six Peasant Party members belonged to cooperatives.[70] Although the compo-

---

[63] Żaryn, "Księża-patrioci."
[64] *Sprawozdanie WUBP w Krakowie*, 1949.
[65] *Sytuacja polityczno-organizacyjna*, 1954.
[66] Informacja nr 397/1063, July 3, 1952.
[67] *Sprawozdanie miesięczne*, 1950.
[68] Sołtysiak, "Cud w Lublinie."
[69] *Meldunki z terenu*, 1950.
[70] Jarosz, *Polityka władz*, 433–434; Próchniak, *Kolektywizacja w regionie łódzkim*, 133.

sition of collectives was clearly more "red" than the rest of the village and peasant group as a whole,[71] the degree of political loyalty did not meet the expectations of decision-makers in Warsaw. Communist Party and Peasant Party members supported the communist agricultural policy in the previous period, when peasants received land. Communists could thus count on a relative popularity in the countryside. After 1948, however, when the communist agriculture policy was associated in peasants' minds with the confiscation of land, they largely turned against it.

Records also show that women played a central role in resistance to collectivization. As they had in Soviet Russia, Polish women interrupted cooperative meetings, heckled pro-collectivization leaders and supporters, sang religious hymns and recited prayers, stomped their feet when supporters spoke, and walked out of meetings in protest.[72] They also showed spirited resistance against teams of surveyors marking out the new collective farms. In the village of Bałtów, in the province of Kielce, a group of some 60 women and 30 men armed with staves attempted to disrupt surveying.[73] Similar incidents were also recorded in villages near Łódź and Lublin. In Rzeczyca Ziemiańska, in the district of Kraśnik, a group of women, along with some men, obstructed the surveyors by singing traditional Polish patriotic and religious hymns.[74] In May 1953, a group of women prevented surveyors from marking out the cooperative estate in the village of Grzmiąca, also in the Kielce region. The PZPR secretary of the Municipal Committee and members of the newly formed collective were beaten up by women.[75]

Women played a significant and often decisive role in protesting newly formed collectives, often laying down in front of tractors when old estates were plowed.[76] In one of the largest protests, which occurred on September 16, 1950 in the village of Śmiłowice in the voivodship of Kraków,

---

71 The number of peasants enrolled in PZPR was 194,528 in 1949 (14 percent), and 175,569 in 1956 (12.7 percent). In cooperative farms, the number of PZPR peasant members was 3,529 in 1949, 45,030 in 1953, and 58,033 in 1956. See Dobieszewski, *Nomenklatura*, 191–94.

72 Informacja nr 570/2080, July 27, 1953; Próchniak, *Kolektywizacja w regionie łódzkim*, 46, 51.

73 This incident occurred on March 17, 1950. See *Biuletyny dzienne Ministerstwa*, 589.

74 Biuletyn Gabinetu MBP, March 9, 1950; Ocena Egzekutywy KW PZPR, August 19, 1950.

75 Biuletyn Gabinetu MBP, May 9, 1953.

76 Biuletyn Informacyjny nr 2/186, February 1, 1950.

an estimated 200 women laid down in front of the tractor pretending to faint, and when local activists dragged the women away from the tractor, the women threw dirt at them and insulted them, and then beat the tractor driver and a party secretary with staves and hoes.[77]

This reaction and the role women played in resisting collectivization is similar to those reported by Lynne Viola in her analysis of the *bab'i bunty* in the Soviet Union in the 1930s.[78] Polish women protesting collectivization carried infants in their arms, just like those in the Soviet Union—a calculated strategy based on the assumption that the arrestment of or violent repression against mothers was, from a moral point of view, more discomforting for the authorities than arresting men. Such a strategy in depoliticising the conflict proved to be effective. Party leaders often branded women who participated in such protests as uneducated and classless (*baba*).

THE EFFECTS OF THE COLLECTIVIZATION POLICY

The production cooperative movement developed unevenly. To be sure, the number of production cooperatives was steadily growing: there were 2,199 at the end of 1950; 3,056 in 1951; 4,890 in 1952; 8,060 in 1953; 9,178 in 1954; and 9,790 in 1955. But crises in the development of the movement led to temporary dips in these figures, mostly in the period from May 1951 to July 1952 and from August to December in 1953. From 1949 to 1956, the number of families belonging to cooperatives gradually increased, and in June 1956 it reached 216,000, representing roughly six percent of the overall number of peasant families. At the time, they accounted for 11.2 percent of peasant-owned agricultural land. The size of production cooperatives in Poland were gradually shrinking. In 1952, their average size was over 220 hectares, while they comprised just over 180 hectares in 1955. The average collective farm was made up of not many more than 20 families.

As they did in other communist countries, Polish farmers developed methods of subverting collectivization, such as passing on their livestock to fictitious owners, refusing to hand over all of their land, or finding ways to

[77] Sprawozdanie Franciszka Dziekana, 1950.

[78] *Bab'i bunty* were women's riots characterized by female hysteria, irrational behaviour, unorganized protest, rage, and violence. See Viola, *Peasant Rebels under Stalin*, 181–204.

cultivate on private plots.[79] Cooperative farmers—especially women and the younger generation—were generally reluctant to participate in team farming.

Problems with collectivization were exacerbated by corruption and conflicts over dividing up the cooperatives. Cooperative boards and their leaders were accused of forming cliques and nepotism. Occasionally, these boards were controlled in whole or in part by members of a single family. In extreme cases, cooperatives consisted of the chairman's family, who treated the land as the family's own private property and maneuvered to prevent others from gaining access to the plot.[80] This put them in a more favorable position: they had easier access to the ever-scarce goods and to state loans that helped support collective farms. Hence, not long after the program began, cooperatives had already become synonymous with theft.

Adding to their unpopularity, cooperative farms were far less productive than private agricultural ventures. The party's push to mechanize work in the field was failing. The most productive farmers were those who did not rely on the state's POM services, and those who kept their horses. Meanwhile, cooperatives were mounting considerable debt, as less productive collective farms were unable to pay back state loans, and therefore, farmers were not benefiting from their membership, which made joining increasingly less attractive.

The Breakup of Production Cooperatives after 1956

Although new sources have enabled researchers to explain why collectivization failed during its initial years, a much more difficult task for scholars is to answer why Poland evaded collectivization after Gomułka came into power in 1956. While continued public aversion to production cooperatives, as noted in opinion surveys from the 1960s to '80s, was certainly important, other factors were equally significant.

Key among these was Stalin's death, after which state leaders across the socialist camp faced a "crisis of legitimacy," and in response, aimed to readjust policy directives from Moscow in order to quell growing domestic social unrest. Following Khrushchev's "cult of personality" speech in February 1956, Poland and Hungary were rocked by popular

[79] Uchwała Biura Organizacyjnego KC PZPR.

[80] Odpis listu zbiorowego, July 1954.

protest, as the masses called for greater political, economic, and cultural freedom. That June, mass protests and strikes by Polish workers in Poznań led to violent clashes with authorities, during which at least 49 protestors were killed and hundreds arrested, triggering the anti-communist Polish Uprising.[81]

De-Stalinization had initiated, too, a shift in hard-lined attitudes among party leaders. Even so, Gomułka's initial critique of the Stalinist model of collectivization, expressed in his acceptance speech in 1956, did not necessarily indicate that he had altogether abandoned the communist idea of a socialized agrarian economy.[82] Mindful of the experiences in the Stalinist period, he did to pursue a soft-line policy and abandoned more repressive approaches to collectivizing rural Poland.

Following Gomułka's 1956 speech, the Central Committee faced an avalanche of collective farm dissolutions. Of the 10,203 collectives registered on September 30, 1956 only 1,534 still existed by December 31, 1956.[83] Most collectives were dissolved in western Poland, the region with the greatest concentration of cooperative farms. Those in the central and southeastern provinces, where support for the communist party was highest, were more enduring. Another noticeable feature was the rapid collapse of the cooperatives regarded as "model farms," particularly in cases where the scale of state interference in support of their activities had been greatest.[84] On more than one occasion, these farms dissolved violently.[85] Beyond the political thaw, the rush to dissolve collective farms was also driven by the fear that favorable political conditions might not last long.

During late socialism, Polish leader Edward Gierek was equally reluctant to pursue a hard-lined policy of collectivization, fearing it might spark a mass uprising. When he came to power in 1970, Gierek continued Gomułka's soft policy toward collective farms and even offered economic incentives to peasants who handed over their land voluntarily to the state in exchange for pensions. In the 1980s, the idea of enforcing a strict agricultural policy would have meant opening a new field of conflict with a society that was already in a very rebellious mood. In order to appease the population, and to strike an accord between the state and the masses,

---

[81] Jastrząb, "Rozstrzelano moje serce w Poznaniu," 227; Machcewicz, *Rebellious Satellite*, 87–124; Ekiert, *Political Crises*, 216–19.

[82] VIII plenum KC PZPR, 1956, 21–46.

[83] Dobieszewski, *Kolektywizacja*, 122.

[84] Jarosz, *Polityka władz*, 152.

[85] Jarosz, *Polityka władz*, 150–54.

the Polish Peoples' Republic Constitution, drafted in July 1983, carried a provision protecting private family farms as an enduring institution of the country's economic system.

COLLECTIVIZATION *Á LA POLONAISE*: CAUSES OF DEFEAT

Although scholars have long attributed the defeat of collectivization exclusively to peasant resistance, we now know the situation was far more complex. Peasant resistance was indeed the driving force behind the program's failure: it clashed with rural, social, national, and religious traditions in ways the party had not predicted. Yet new records reveal that resistance among rural farmers was hardly universal, and that peasant attitudes toward collectivization were dictated by prewar conditions, cultural backgrounds, religion, gender, and age. The beneficiaries of postwar land reform tended to favor the program, whereas hereditary farmers often rejected it. Likewise, the party's emphasis on industrial production fueled a mass migration to urban centers, which accelerated the decline in agricultural production in rural Poland.

Hence, mismanagement by party leaders also contributed the program's failure. Initially, leaders in Warsaw adapted the Soviet model of collectivization, hoping to persuade peasants to join cooperatives by promoting socialist ideology and anti-*kulak* dogma. When that failed, they resorted to state terror and repression. Local party *apparatchiks*, who were held responsible for these actions, became ever more cautious and increasingly less zealous in implementing Stalinist agrarian policy. Gradually, leaders adopted a soft-line stance toward the program, often changing policies in response to public resistance.

Despite new research in this area, scholars are still unable to answer whether the Polish leaders were motivated by a genuine commitment to the idea of collectivization, or whether they were merely towing the socialist line and following orders from the Kremlin.[86] One conclusion is certain: The failure to collectivize was an argument Moscow would wheel out when dealing with leaders in Warsaw and confronting the issue of the

86 The latter is suggested in the memoirs of the former party leader Kazimierz Barcikowski, who was from December 1970 the secretary of the CC of the PZPR responsible for agriculture and later became the minister of agriculture. See Barcikowski, *U szczytów władzy*, 34–35, 70.

frequent "political crises" in Poland.[87] The question remains as to why Moscow tolerated Poland's failure to collectivize.

BIBLIOGRAPHY

ARCHIVAL SOURCES:

Biuletyn Gabinetu Ministra Bezpieczeństwa Publicznego, March 9, 1950 (Former Cenralne Archiwum Ministerstwa Spraw Wewnętrznych – CANSW, Ministerstwo Bezpieczeństwa Publicznego – MBP 1199/37).

Biuletyn Gabinetu MBP, May 9, 1953 (Former CAMSWiA, MBP1199/42).

Biuletyn Informacyjny nr 2/186, February 1, 1950 (Archiwum Akt Nowych – ANN, Komitet Centralny Polskiej Zjednoczonej Partii Robotniczej – KC PZPR, 237/VII-113)

Informacja nr 397/1063, July 3, 1952 (AAN, KC PZPR, 237/VII-142, k. 6).

Informacja nr 570/2080, July 27, 1953 (AAN, KC PZPR, 237/VII-144, k. 81).

Meldunki z terenu nr 73/339, April 7, 1950 r. (AAN, KC PZPR, 237/VII-121).

Notatki Stefana Żmijki, Warszawa, October 26, 1949 (Archiwum Zakładu Historii Ruchu Ludowego – AZHRL, Stronnictwo Ludowe – SL 92, k. 88).

Ocena działalności organizacyjnej ogniw ruchu ludowego, September 1, 1950 (AZHRL, Naczelny Komitet Zjednoczonego Stronnictwa Ludowego–NK ZSL, Wydział Organizacyjny), 308, k. 89–90.

Ocena Egzekutywy KW PZPR w Lublinie przebiegu realizacji planu budowy spółdzielni produkcyjnych, August 19, 1950 (AAN, KC PZPR, 237/XII-193, k. 158).

Ocena przeniesienia uchwały Biura Politycznego i omawianie statutów na POP w gromadach, Łódź, April 14, 1951 r. (AAN, KC PZPR, 237/XII-97, k. 49).

Odpis listu zbiorowego członków spółdzielni produkcyjnej w Wielogłowach do Polskiego Radia, July 1954 (Archiwum Ministerstwa Rolnictwa i Gospodarki Żywnościowej - AMRiGŻ), 14/3.

Pamiętnik nr 793.(AZHRL, Nowe Pamiętniki Chłopów).

Protokół nr 17 z posiedzenia Sekretariatu KC PZPR, August 31, 1951 r. (AAN, KC PZPR, 1647, k. 30–33).

Protokół rejonowej konferencji roboczej aktywu ZSL w Będzinie, November 23, 1950 (AZHRL, NK ZSL, II/34, k. 7).

Protokół ze zjazdu powiatowego PSL w Lublinie, November 6, 1949 (AZHRL, PSL 19, k. 5).

Protokół z zebrania organizacyjnego członków PZPR, ZSL, ZSCh, bezpartyjnych, January 25, 1950 r. w gromadzie Wielka Wieś woj. Kraków (AAN, KC PZPR, 237/XII-92, k. 54).

Raport WUBP w Rzeszowie na temat sytuacji na wsi, February 1949 (Former CAMSW, 17/IX/14, t. 3a).

[87] Barcikowski, *U szczytów władzy*, 73; Rolicki, *Edward Gierek*, 85.

Sprawozdanie Franciszka Dziekana z podróży służbowej na teren powiatu miechowskiego, November 10–12, 1950 r. (AAN, KC PZPR, 237/XII-55).
Sprawozdanie inspektorów, AMRiGŻ, Ministerstwo Rolnictwa (MR), 14/11.
Sprawozdanie miesięczne z prac spółdzielni produkcyjnych powiatu Jarosław za miesiąc styczeń, February 13, 1950 r. (AAN, KC PZPR, 237/XII-113, k. 21).
Sprawozdanie WUBP w Krakowie za kwiecień 1949, (Former CAMSWiA, MBP 17/IX/7, t. 2).
Sprawozdanie z konferencji w sprawie spółdzielczości produkcyjnej, Warsaw, March 18–19, 1949 (AZHRL, Polskie Stronnictwo Ludowe (PSL), 216, k. 9).
Sprawozdanie z narady aktywu powiatowego PZPR w powiecie Gostynin (woj. poznańskie), October 30, 1949 (AAN, KC PZPR, 237/VII-237, k. 196).
Sprawozdanie z odprawy powiatowego aktywu partyjnego spółdzielni produkcyjnych, Wrocław, October 30, 1949 r. (AAN, KC PZPR, 237/VII-79, k. 78).
Sprawozdanie z województwa poznańskiego (AZHRL, NK ZSL, II/306, k. 61–64).
Sprawozdanie z wyjazdu ekipy agitacyjnej do gromady Kraszew w ramach łączności wojska ze wsią, January 13, 1952 (Centralne Archiwum Wojskowe (CAW), Zarząd Polityczny Centralnych Instytucji Ministerstwa Obrony Narodowej (ZPCIMON), IV.502.3.162, k. 22).
Sprawozdanie z wyjazdu na teren województwa białostockiego, March 26–28, 1949 (AAN, KC PZPR, 237/VII-172, k. 37).
Spuścizna Mieczysława Moczara (AAN, t. 24).
Stenogram narady krajowego aktywu spółdzielni produkcyjnych z załącznikami, January 22, 1953 r. (AAN, KC PZPR,237/XII-132, k. 21).
Stenogram odprawy sekretarzy wojewódzkich, March 10, 1949 r. (AAN, KC PZPR, 237/V-17, k. 1–61).
Sytuacja polityczno-organizacyjna po III kwartale [1953 r.], (AAN, KC PZPR, 237/VII-3832).
Sytuacja w spółdzielniach województwa łódzkiego w latach 1955–1956 (AAN, Rada Spółdzielczości Produkcyjnej–RSP, 1/43, k. 36–37).
Uchwała Biura Organizacyjnego KC PZPR o wykonaniu uchwały Biura Politycznego w sprawie spółdzielczości produkcyjnej (AAN, KC PZPR, 237/V-3, k. 68).
Wyciągi ze sprawozdań słuchaczy Centralnej Szkoły Partyjnej z pobytu na spółdzielniach produkcyjnych w woj. rzeszowskim, Warsaw, March 31, 1950 (AAN, KC PZPR, 237/XII-63, k. 16–19).
Zestawienie meldunków z terenu nr 9/16, May 1–15, 1951 (AAN, KC PZPR, 237/VII-137, k. 243).

PUBLISHED DOCUMENTS:

Biuletyny dzienne Ministerstwa Bezpieczeństwa Publicznego 1949–1950 [Daily newsletters, the Ministry of Public Security], edited by Łukasz Kamiński. Warsaw: Instytut Pamięci Narodowej, 2004.
"Porozumienie zawarte między przedstawicielami Rządu Rzeczypospolitej Polskiej i Episkopatu Polski" [The agreement concluded between the representatives of the Government of the Republic of Poland and the Polish Episcopate] in Peter Raina, ed., *Kośćół katolicki a państwo w świetle dokumentów 1945–1989, t.1., Lata 1945–1959* [Catholic Church and State in the Light of Documents 1945–1980, vol. I, 1945–1959]. Poznań: W Drodze, 1994, 232–35.

"Relacja Artura Starewicza o nastrojach społecznych na wsi w sierpniu 1948 r." [Artur Starewicz's report on public opinion's state of mind in the coutryside in August 1948], *Dzieje Najnowsze* no. 2 (1993): 111.

Statuty spółdzielni produkcyjnych [Statutes of production cooperatives]. Warsaw: Związek Samopomocy Chłopskiej, 1951.

Uchwały Komitetu Centralnego Polskiej Zjednoczonej Partii Robotniczej od II do III Zjazdu [Resolutions of Central Committee of the Polish United Workers' Party from Second to Third Congress]. Warsaw: Książka i Wiedza, 1959.

VIII plenum KC PZPR [The Eighth Plenum of the CC of the PZPR], *Nowe Drogi* 10 (1956): 21–46.

BOOKS AND ARTICLES:

Atlas polskiego podziemia niepodległościowego 1944–1956 [An atlas of the Polish Underground Independence Movement]. Warsaw-Lublin: Instytut Pamięci Narodowej, 2007.

Barcikowski, Kazimierz. *U szczytów władzy* [In the highest authority]. Warsaw: Wydawnictwo Projekt, 1998.

Bartkowski, Jerzy. *Tradycja i polityka. Wpływ tradycji kulturowych polskich regionów na współczesne zachowania społeczne i polityczne* [Traditions and politics. The influence of Polish cultural tradition on modern social and political behavior]. Warszawa: Wydawnictwo akademickie 'Żak,' 2003.

Biernacka, Maria. *Wsie drobnoszlacheckie na Mazowszu i Podlasiu. Tradycje historyczne a współczesne przemiany* [The petty noble villages in Mazovia and Podlasie. The historic traditions and contemporary changes]. Wrocław-Warszawa-Kraków: Ossolineum, 1966.

Chrypinski, Vincent. "The Catholic Church in Poland" in Pedro Ramet, ed. *Catholicism and Politics in Communist Societies.* Durham: Duke University Press, 1990, 119–41.

Dobieszewski, Adolf. *Kolektywizacja wsi polskiej 1948–1956* [Collectivization of the Polish countryside 1948–1956]. Warsaw: Fundacja im. Kazimierza Kelles-Krauza, 1993.

———. "'Nomenklatura' partyjna—spojrzenie od wewnątrz (1948–1989)" [The Party "Nomenklatura"—A look from the inside] in Przemysław Wójcik, ed. *Elity władzy w Polsce a struktura społeczna w latach 1956–1981* [Power elites in Poland and social structure]. Warsaw: Akademia Nauk społecznych, 1994, 192–94.

Dziurzyński, Patrycy. *Osadnictwo rolne na Ziemiach Odzyskanych* [The agricultural settlement in the Regained Lands]. Warsaw: Ludowa Spółdzielnia Wydawnicza, 1983.

Ekiert, Grzegorz. *Political Crises and Their Aftermath in East Central Europe.* Princeton: Princeton University Press, 1996.

———. "Rebellious Poles: Political crises and Popular Protest under state Socialism 1945–1989." *East European Politics and Societies* 11, no. 2 (May 1997): 299–338.

Fitzpatrick, Sheila. *Stalin's Peasants. Resistance and Survival in the Russian Village After Collectivization.* Oxford: Oxford University Press, 1994.

Gawryszewski, Andrzej. *Przestrzenna ruchliwość ludności Polski. Bibliografia (lata 1896–1990)* [Spatial mobility of population of Poland. Bibliography (Years 1896–1990)]. Warsaw: Continuo, 1997.

Jarosz, Dariusz. "Pogłoski jako wyraz świadomości potocznej chłopów w Polsce w latach 1949–1956" [Hearsay as an expression of peasants' consciousness in Poland, 1949–1956], *Dzieje Najnowsze* no. 3 (1993): 39–55.

———. "Chłopi na Ziemiach im Obiecywanych" [Peasants in their Promised Lands and Territories]. *Regiony*, no. 1–3 (1998): 108–37.

———. *Polityka władz komunistycznych w Polsce w latach 1948–1956 a chłopi* [The Communist regime in Poland 1948–1956 and the peasants]. Warsaw: Wydawnictwo DiG, 1998.

Jastrząb, Łukasz. *"Rozstrzelano moje serce w Poznaniu." Poznański Czerwiec 1956 r. – straty osobowe i ich analiza* ["My heart has been shot in Poznań." Personal losses and its analysis]. Poznań: Comandor, 2006.

Kaliński, Janusz. "Forsowna kolektywizacja rolnictwa (1948–1956)" [Forced Collectivization of Agriculture, 1944–1956]. *Kwartalnik Historyczny* 1 (1984): 111–36.

Korboński, Andrzej. *Politics of Socialist Agriculture in Poland: 1945–1960.* New York: Columbia University Press, 1965.

Kura, Antoni. *Aparat bezpieczeństwa i wymiar sprawiedliwości wobec kolektywizacji wsi polskiej 1948–1956* [The security apparatus and the justice system to the collectivization of the Polish countryside, 1948–1956]. Warsaw: Instytut Pamięci Narodowej.

Machcewicz, Paweł. *Rebellious Satellite: Poland 1956.* Palo Alto: Stanford University Press, 2009.

Markiewicz, Marcin. *Kolektywizacja wsi w województwie białostockim 1948–1956* [Collectivization of the countryside in Voivodship of Białystok, 1948–1956]. Białystok: Instytut Pamięci Narodowej, 2010.

Osękowski, Czesław. *Społeczeństwo Polski zachodniej i północnej w latach 1945–1956. Procesy integracji i dezintegracji* [Society of Western and Northern Poland 1945–1956. Processes of integration and disintegration]. Zielona Góra: Wyższa Szkoła Pedagogiczna im. Tadeusza Kotarbińskiego, 1994.

Próchniak, Leszek. *Kolektywizacja w regionie łódzkim* [Collectivizations in Voivodship of Łódź]. Łódź: Instytut Pamięci Narodowej, 2003.

Robakowski, Kazimierz. *Społeczno-polityczne problemy rozwoju spółdzielczości produkcyjnej w Polsce w latach 1944–1956* [Sociopolitical problems of development of production cooperatives in Poland, 1944–1956]. Poznań: Wyadwnictwo Uniwerystetu Adama Mickiewicza, 1986.

Rolicki, Janusz. *Edward Gierek: przerwana dekada* [Edward Gierek: The interrupted decade]. Warsaw: Polska Oficyna Wydawnicza "BGW," 1990.

Rybiński, Rajmund. *Ziemie Zachodnie i Północne w półwieczu 1945–1995* [The Western and Northern Territories 1945–1995]. Toruń: Wydawnictwo Adam Marszałek, 1997.

Słabek, Henryk. "Reforma rolna" [Land reform] in Janusz Kaliński and Zbigniew Landau, eds. *Gospodarka Polski Ludowej 1944–1955* [Economy of People's Poland]. Warsaw: Książka i Wiedza, 1986.

Sołtysiak, Grzegorz. "Cud w Lublinie" [Miracle in Lublin]. *Karta* 9 (1992): 121–36.

Suny, Ronald Grigor. *The Soviet Experiment: Russia, the USSR, and the Successor States*. Oxford: Oxford University Press, 1998.
Urban, Kazimierz. "Wybrane problemy życia kościelnego i polityki wyznaniowej państwa" [Selected Problem of Church Life and Religion Politics]. *Białostocczyzna* no. 4 (1991): 32.
Viola, Lynn. *Peasant Rebels under Stalin: Collectivization and the Culture of Peasant Resistance*. Oxford: Oxford University Press, 1996.
Żaryn, Jan. "'Księża-patrioci'—geneza powstawania formacji duchownych katolickich" [Priests-patriots. The genesis and formation of Catholic clergymen]. *Polska 1944/5–1989. Studia i materiały* 1 (1995): 123–50.

LAND COLLECTIVIZATION IN POLAND. CHRONOLOGY OF EVENTS, 1944–1956

1944

July: Polish Committee of National Liberation (*Polski Komitet Wyzwolenia Narodowego*, PKWN), forms in Moscow on the basis of the Polish Workers' Party, and is installed in Lublin. Edward Osóbka-Morawski, PPS leader is appointed PKWN chairman; Andrzej Witos as a head of the PKWN Department of Agriculture.

July 22: Manifesto proclaimed the imminence of land reform in liberated provinces. The state would seize all estates owned by the Germans and traitors, regardless of size, and all other estates over 50 or 100 hectares, depending on the province. The land would be expropriated without compensation but the owners would receive some sort of payment. The expropriated land is to be assembled into a Land Fund and, except for estates to be used for special purposes, is to be distributed at a minimum payment to farm laborers, small tenants, farmers, and medium peasants burdened with large families. The average size of the newly created farms is to be 5 hectares and the existing small farms were to be increased to that size.

August: Compulsory deliveries (*świadczenia rzeczowe*) of agricultural products are introduced.

September 6: Land reform decree. All land owned by the state, the Germans, the "traitors and collaborators" regardless of size, and Polish estates over 50 hectares of agricultural land or 100 hectares of total area (except in the three western provinces of Poznan, Pomerania, and Silesia, where 100 hectares of total area was the limit) are to be taken over immediately by the state without indemnity. The future of estates belonging to religious bodies is to be determined by the *Sejm*. Land Offices are to prepare a list of seized estates, and the communal land reform commissions are to prepare lists of farm laborers and peasants with farms below 5 hectares.

October 9: PPR decision to reduce the authority of the Land Offices in favor of direct participation of the workers and peasants. Included among the land recipients are middle peasants burdened with large families. Conception of land reform as a social revolution.

October 10: Witos's resignation from PKWN. Department of Agriculture is placed directly under the chairman of PKWN. PKWN appoints special land reform commissioners for every province and county, responsible for allocating expropriated estates to the villages, which were to elect special land distribution commissioners. Between 1944 and 1949, as a result of the expropriation of land and settlement on the post-German lands allocated to Poland, over 6,070,100 hectares of land is granted to peasants. About 814,000 new farms are established and 254,400 are agriculturally upgraded. The average size of the newly formed farms is 6.9 hectares, and the average size of an additional lot is 1.9 ha.

December 31: Provisional Government is formed on the basis of Polish Committee of National Liberation (PKWN). E. Osóbka-Morawski becomes Prime Minister.

1945

February 4–11: Yalta Conference. Stalin, Roosevelt, and Churchill confirm Polish–Russian frontier along the river Bug and agree that Poland's loss of territory in the east should be compensated by some former German territory in the west. The "Big Three" insist that representatives of the parties supporting the Polish government in London should join members of the Soviet-sponsored administration in Warsaw (the Provisional Government). This government was to prepare free democratic elections.

June 21: At a conference in Moscow, the Provisional Government of National Unity is finally formed. A majority of ministerial posts go to communists. Only six ministers are non-communist politicians. E. Osóbka-Morawski becomes Prime Minister; Stanisław Mikołajczyk, former Prime Minister of the Government-in-Exile in London (to October 1944), becomes his deputy and minister of agriculture and land reform.

July 17–August 2: Potsdam Conference. Allied leaders fix the Polish western frontier on the river Oder and western Neisse and insist that future general elections in Poland be free and unfettered. New Poland extends over 311.7 thousand square kilometers and is inhabited by 23.9 million people, which means a 30% decline in comparison to the prewar population.

July 8: Polish Peasant Party (*Polskie Stronnictwo Ludowe*—PSL), headed by Stanisław Mikołajczyk is created.

1946

January 3: Decree on nationalization of all factories employing fifty or more workers, as well as smaller ones of particular importance.

June 6: Authorities announce that compulsory deliveries will be abolished.

June 30: A national referendum, as a "rehearsal" before the parliamentary election takes place (three questions: 1. Are you for abolishment of the Senate? 2. Are you for continuity of reform? 3. Are you for the Polish western frontier along the Oder–Neisse line?). Communist recommends that people answer "three times yes," whereas Mikołajczyk recommends no to the first question. The polls were falsified (in reality 27 percent had voted "three times yes," while at least 33 percent had voted "three times no").

1947

January 19: Falsified election in Poland. Political police (*bezpieka*) nullified the list of PSL candidates in one-fifth of all constituencies. Over 30 district party boards are suspended, 149 candidates for Members of Parliament are arrested, with tens of thousands of local activists taken into preventive custody. The pro-Soviet block of communists gain an overwhelming majority of seats, whereas Mikołajczyk's party took a mere 10%.

February: Boleslaw Bierut elected as a new President of Poland. New government headed by Józef Cyrankiewicz, leader of the Polish Socialist Party (*Polska Partia Socjalistyyczna*) and fully dependent on PPR.

May: The "battle over Trade" is announced (campaign privileged state-run manufacturing and tightened control over other sectors).

October: Mikołajczyk's escape from Poland to Great Britain. Fall of peasant political opposition.

1948

July 6–7: Plenary of Central Committee of PPR. Hilary Minc announces the beginning of cooperative movement (collectivization) in agriculture.

August 31–September 3: Plenary of CC PPR. A short communiqué issued by the PPR Central Committee announces that Gomułka resigned as secretary-general of the party. Resolution accuses Gomułka of, among other things, the desire to postpone the struggle against the capitalist elements in the countryside, that is, of opposition to collectivization. Minc announces that the maximum target for 1949 was collectivization of 1 percent of all individual farms in Poland.

October 22: Cardinal Hlond dies. He is replaced by the bishop of Lublin, Stefan Wyszyński.

December: A new communist party: the Polish United Workers' Party is established (*Polska Zjednoczona Partia Robotnicza*—PZPR).

1949

February: The Central Planning Office is replaced by the State Economic Planning Commission (*Państwowa Komisja Planowania Gospodarczego*—PKPG), modeled on the Soviet Gosplan.

June: Formation of the Central Trade Union Council (*Centralna Rada Związków Zawodowych*—CRZZ) dominated by communists.

November 9: Soviet commander, Konstanty Rokossowski, the head of the Northern Armed Forces Group, which is stationed in Poland, is appointed minister of defense and Polish marshal. Sovietization of Polish army.

December 31: About 240 cooperatives and 30 State Machine Centers (*Państwowe Ośrodki Maszynowe*—POM) are established.

1950

July: The "Law on the Six-Year Plan of Economic Development and construction of the foundations of Socialism" proposes an increase of 158 percent in industrial output over the level achieved in 1949 and a 50-percent increase in agricultural production, including a 68-percent increase in livestock.

December: Political Departments of State Machine Centers are established.

December 31: 2,199 of agricultural cooperatives are registered.

1951

May: Politburo publishes declaration on Gryfice case emphasizing the need for restraint in dealing with the peasants. The PZPR county committee in Gryfice (Szczecin province) uses force in conducting illegal searches for grain among individual peasants.

July 23: The regulations for mandatory milk quotas are published. Failure to deliver quotas results in harsh punishment (arrest or up to three years imprisonment).

August 2: Gomułka and his wife are detained and held in a specially prepared villa in Miedzeszyn, on the outskirts of Warsaw.

October 8: The regulations for mandatory potatoes quotas are published.

December 31: 3,056 of agricultural cooperatives are registered.

1952

February 15: Compulsory deliveries on livestock are imposed. Failure to deliver quotas results in harsh punishment (arrest or up to three years imprisonment).

April 24: The regulations for mandatory milk quotas are published.

July 22: Polish Parliament (*Sejm*) passes the constitution of the Polish People's Republic, which is personally approved by Stalin.

December 31: 4,478 of agricultural cooperatives are registered.

1953

September 24: Authorities decide to "isolate" the head of the Polish Catholic Church, Cardinal Stefan Wyszyński.

December 31: 7,772 of agricultural cooperatives are registered.

1954

March 10–17: The Second Congress of the PZPR. Reorientation of economic policy and reducing growth targets for the remainder of the six-year plan. Party intends to provide more aid for the private sector by fixing the amount of compulsory deliveries at the 1953 level, issuing larger credits and bigger supplies of farm implements, increasing the supplies of consumer goods in the rural trade network.

December 31: 9,322 of agricultural cooperatives are registered.

1955

December 31: 9,790 of agricultural cooperatives are registered.

1956

June 28: Revolt of the workers and citizens of Poznań ("Black Thursday") is suppressed by army divisions. About seventy civilians die in fighting, together with eight soldiers, policemen, and security officials; several hundred people are injured.

July: Seventh plenary session of the Central Committee of the Polish United Workers' Party: acquisition of tractors by cooperatives is allowed and *kulaks* are permitted to join them.

September 30: 10,203 of agricultural cooperatives are registered encompassing about 190,000 families and 2,100,000 ha, which is 11 percent of the overall area of peasant-owned land.

October 19–21: Eighth Plenum of the Central Committee. Gomułka's return to the leadership. Resolution stresses that weak cooperative farms "which discredit cooperative farming" must be dissolved; all attempts by the state to keep them going were condemned as harmful. The idea of collectivization is "necessary because it liquidates all form of exploitation of man by man, imbues peasants with feeling of deep social community and community of work, eases the difficult and toilsome work of the peasants through the use of machines, and opens up the possibilities of achieving big crops." Condemnation of the illegal practices which are applied to destroy *kulak* farms.

October 26: Cardinal Wyszyński is released from detention.

December 31: 1,954 of agricultural cooperatives are registered.

1957

January: Directives concerning Farm Policy are approved jointly by PZPR and United Peasant Party (ZSL). Liquidation of the Rural Machine Stations and the transfer of the machines held by them to agricultural circles.

March: 83 percent of all cooperatives, containing 86 percent of the members and 87 percent of the area, are dissolved.

1970–1980

Various conceptions for socializing agriculture are tried (mainly acquiring peasant land in exchange for pensions and passing them on to State Agricultural Farms).

t1983

July: Statement assuring the permanence of private family farms in the economic system of the country is included in the constitution of the Polish People's Republic.

Ideology and Asymmetrical Entanglements: Collectivization in the German Democratic Republic

JENS SCHÖNE

March 15, 1953 was just like any other Sunday in the German Democratic Republic: the *Free Farmer*, a weekly paper distributed by the Society of Farmers (VdgB) was published. But this particular issue focused entirely on one extraordinary event: the death of Josef Wissarionowitsch Stalin. The head of the Ministry and Secretary of the Central Committee of the Communist Party and the absolute leader of the Soviet Union had died only ten days before, at the age of 74. A tangible sadness—obligatory for some, sincere for others—had settled over the entire Eastern Bloc. This heartfelt emotion was expressed in a telegram sent by the Central Committee of the VdgB, a German communist mass organization: "The death of Josef Wissarionowitsch Stalin, the architect of communism, the creator of collective industry, the leader of the Soviet people and contributor towards the general forward evolution of man, fills us with great sorrow... Stalin's teaching has laid the path for the working farmers of the German Democratic Republic, teaching us how to create socialism within our local communities, enriching both our land and our lives."[1] The content of the telegram exactly reflects the orientation of the agrarian politics of the GDR during the 1950s. It also foretells the direction the future would take. After the end of the Second World War and the beginning of the Soviet occupation, the entire agrarian structure east of the River Elbe underwent purposeful, ideologically driven reform. A fundamental

---

1 *Der Freie Bauer*, 2.

part of this land reform, which took place approximately between 1945 and 1948, was the reorganization of the traditional German small-town elite and its institutional support structure. Beginning in 1952, collectivization was slowly implemented. This process was neither continuous nor unchallenged, yet it was organized along very clear, goal-oriented guidelines. The fact that most other Soviet-dependent states followed similar guidelines proves that the GDR had little choice but to move away from a system of private land ownership and to accept the full collectivization of its agriculture.

This chapter examines the collectivization of the agriculture in the GDR as an example of the broader Soviet-modeled program implemented across the Communist Bloc. The goal is not to record all facets of the development of the "comrade's collective," nor is it to outline the entire history of the process in the GDR. Instead, the chapter will focus on central problems crucial to the collectivization process, and in the context of the socialist-constructivist politics in agricultural circles at the time: namely, ideological foundation of the collectivization process versus the changing goals of policy makers in East Berlin and in Moscow. The chapter is framed by several underlying questions: Did ideology play a deciding role in the collectivization process or were practicalities on the ground the primary factors of influence? Did the Soviet model, which was certainly followed, evolve into a unique GDR-version of collectivization during the process? How far did the independence of the Socialist Unity Party of Germany (SED) actually go—in other words, did the SED assert its independence despite the clear hegemony of the Communist Party of the Soviet Union (KPdSU)?

In order to address these questions, it is necessary to first provide a short overview of the course of collectivization in the GDR, including the basic ideology of Marxist-Leninist agrarian theories, which will illustrate the practical influences that "comradeship" had in the GDR at the time. In addition, the chapter will examine the deciding steps taken by the SED toward collectivizing the agriculture in 1952–1953 and 1958–1960. It will also explore what room for maneuvering—if indeed there was any—the SED had at its disposal, specifically in terms of the party's capacity to gain power with respect to the Soviets. In order to answer these questions in depth, it will become necessary at points to draw comparisons to the collectivization experiences in other countries under Soviet hegemony. Indeed, developments in these countries influenced the ways in which the GDR implemented its own agrarian reforms. The GDR did not follow these examples stringently, but rather adopted neighboring policies only in moments of imminent transformation. The relationship between the GDR and the Soviet Union was more one-sided. Moscow stood as the undeniable center point of influence and opinions for the party leaders in the GDR. This does not in any way imply, however, that SED leaders strictly followed the Soviet collectivization model. Rather, party leaders—like those in other Soviet-controlled states—modified the program in response to endemic circumstances and demands. The chapter explores in depth the complex and ultimately asymmetric power structure that defined the relationship between the GDR and the Soviet Union.

From Private Enterprise to Collective Farms: Stages of Collectivization

Collectivization in the GDR occurred in three chronological stages: the initial push toward collectivization (1952–1953), followed by a period of relaxation (1953–1958), and then an aggressive push toward collectivization when all privately held farmland was officially dissolved (1958–1960 for the administrative organization of the country, see Map 1 below).

The state formally announced the collectivization program and its goals in July 1952. Prior to this declaration, rumors of pending "land reforms" had been widely circulating in many villages, but the party frequently and vehemently denied these speculations. The *New Germany*, the daily newspaper of the SED, had for example publicly announced at the end of April 1952 that "talk of a potential collectivization movement is a huge falsehood. No good farmer should take part in these discussions; our democracy provides a stable agrarian system, we provide our farmers with security, we support them, we protect them from harm and recognize their economic rights and power."[2] But when the SED met at its party conference in Berlin that July, the party's rhetoric had completely changed. Leaders announced a program for the "building of socialism" in the countryside and along with it, the creation of a system of agricultural production controlled by the party-state—and thus closely tied to central industrial planning. As long-time communist leader and general secretary of the party Walter Ulbricht explained: "Various working farmers from diverse areas within the GDR have independently decided that they would like to form collective farms and have sent an appeal to the Second Party Confer-

[2] *Neues Deutschland*, 5.

ence to grant their request ... The Central Committee of our party has carefully studied the letters of the farmers from these collectives and feels that this completely uninfluenced choice reflects a huge step forward."[3]

Two points from Ulbricht's speech stand out: first, that collectivization was legitimized by the party's claim that it had the support of local farmers; and second, that the system was based on and facilitated by freedom of choice to join collectives. Before the conference, however, the party had initiated more than a dozen missions to several villages; following their visits, these villages quickly adopted collective farms. Only after the fall of the GDR and the opening of relevant archives did the facts behind the foundation of these collectives become known: in every case, these "exemplary models" were orchestrated by SED-cadres, who had themselves been acting on official orders from Berlin. Every farmer who cooperated with the orders of the agricultural ministry—whether knowingly or not—had a part to play in the party's "official" legitimization of collective farming. Yet the "spontaneous" desire expressed by these farmers petitioning to form agricultural collectives was in fact a falsehood orchestrated by the SED. That is not to say that the first agricultural collective farming projects (LPG) were not formed freely—only that the idea did not originate with the farmers themselves. Until 1952, mergers of farms had been dissolved by central decrees, as the SED leaders had previously wanted to avoid any impression they were planning wide-reaching "Sovietization" by supporting collectivization programs. However, an opportunity soon presented itself: joint work on the farms had begun to balance out single producers within certain agricultural sectors. This applied above all to the small farms that had originated in 1945 during the so-called "land reforms" and had either poor business plans, outdated or failing equipment, or that had survived economically only through large state handouts. Many owners of such failing businesses, the so-called "new farmers," were eager to embrace the impulses of the SED leadership, thereby ensuring their own economic survival. This applied more so to business owners in the agricultural sector, whose business models were either too similar to the collective farming approach or because they simply were not approved by the SED.[4]

Despite some measure of willingness among rural populations, serious challenges to the collectivization campaign quickly emerged. Beside the

---

[3] *Protokoll der Verhandlungen der II Parteikonferenz*, 109–10.

[4] See for example Schier, *Alltagsleben*, 112–28.

institutional, material, and logistical issues that plagued the process, the number of farmers willing to undertake collectivization was markedly small. For economically successful farms, such a step made little sense even despite the extensive incentives issued by the party to LPG members. In theory, the number of farming collectives should have risen to 1,815 by the end of 1952, but the majority of these farms remained insolvent and could only continue to function with state support. The merger of several independent farms had in most cases clearly shown itself to be unsuccessful. Nevertheless, collectivization was politically so desired that these problems failed to deter the course of agrarian reform. Political, not economic considerations remained the decisive factor.

In what was a purely political decision, the SED thus required that successful farmers join the LPG. The move was rejected by these farmers because collective membership failed to guarantee economic gains, and at the same time, went against traditional rural customs and systems. For this reason, party leaders were determined to change the basic character of the collectivization process. At the Tenth Plenum of the Central Committee of the SED in November 1952, party leaders decided on a new strategy they hoped would boost support for the program by imbibing the collectivization program with a new sense of ideological urgency. Power struggles within the SED, and a certain euphoria inspired by ideas of "forward-movement," culminated into a policy to instigate a "forced class war" that would stigmatize those who opposed the planned political course. Along with the massive administrative, legislative, and logistical structures that had been developed to support the collectivization movement, the party was now committed to punishing, harassing, and jailing those who dissented against the program. This policy was formally announced in the party's February 1953 decree, "Order to Ensure the Safety of Agricultural Production for Betterment of the People," which established punitive measures for those who did not comply with the GDR's collectivization program. Countless other decrees were released at the same time, many aimed at reducing the viability of private business ownership: mandatory delivery requirements were increased, taxes were raised, and access to raw materials, credit, and technology was severely curtailed. These new laws applied mostly to so-called "big farmers," or owners of agricultural businesses larger than 20 hectares of arable land. When these farmers failed to fulfill the increased quotas, their land was reappropriated by the state for use in farming cooperatives. In less than six weeks, the government confiscated more than 6,500 farms, amounting to more than 200,000 hectares of usable land, which was immediately organized into collective farms. These measures did not occur without consequences: though the number of LPG lands had increased, these new collectives of the so-called "socialist sector" managed to till only about 5 percent of its usable land in the first month. SED leadership was thus clearly favoring political and ideological measures over the interests of the majority, demonstrated by the fact that these politically-driven acts managed to negatively affect the majority of citizens living in rural areas. The confiscation of the big farmers' land subsequently led to a sharp drop in agricultural production, lowered living standards among East Germans, and finally, led to a large increase in the number of refugees fleeing the GDR. In the first quarter of 1952, just before the beginning of the collectivization push, 455 farmers left the GDR with their families; the following year that number increased by more than 1,000 percent to 5,681. The need for agricultural workers thus grew exponentially more urgent, which in turn only exacerbated the problem.

The efforts of collectivization were more than just an isolated political act; cooperative farming represented only one facet of an SED program to bring about the "socialization" of the entire society. For this reason, the SED continued to campaign relentlessly for their program despite the fact that they were going against the wishes of the majority of the population.[5] After Stalin's death, however, the GDR leadership was forced by its "big brother" in Moscow to rethink its previous policy. The SED leadership therefore formally ended the collectivization drive and "forced class conflict" in May 1953. Although the SED did not abandon its goal—the creation of a socialist state—the process was pursued in a more moderate fashion. SED leaders had hesitated to change course but ultimately had no choice: in June 1953, the party announced the "New Course" which disbanded LPGs into private farms. The new decree, however, failed to address any consequences among party and state leadership who had led the country into the economic crisis. This mistake had severe consequences, as demonstrated by the popular "People's Protest" on June 17, 1953. The protest-building process had begun several days earlier in smaller villages throughout the GDR. Since June 12, most of rural GDR had been rocked by upheaval, though protests were stronger in some regions. The party's retreat from collectivization was seen as an end to the SED dictatorship that would ultimately lead to the reunification of the two German states. When the unrest reached urban centers several days later, and with the regime on the verge of collapse, Soviet

[5] See Werkentin, "Der totale soziale Krieg."

troops arrived to crush the protests, bringing the first phase of collectivization to an end.

The campaign for full collectivization took a backseat during the next five years, as the SED worked to reconsolidate its power in the wake of mass unrest following Stalin's death. However, the party remained committed to the ideology of collectivization and its desire to fulfill it. Low productivity of collective farms remained an issue of concern. Although party officials frequently touted the rate of growth as being higher in cooperative ventures than in private firms, the actual production capacity of private farmers remained markedly higher. In 1957, the LPG yield listed grain production at approximately a third less than that of private farms, with production of potatoes and pork at an abysmal 50 percent lower in LPGs than in the private farming sector.[6] This resulted in two major problems: first, it undermined the policy of the central state, and second, it called into question the party's rhetoric that only large LPG farms would successfully survive the future. At the 19th Soviet Communist Party meeting in 1956, KPdSU leaders decided to begin a careful de-Stalinization of the entire Eastern Bloc, which promoted an intense debate among leaders in the GDR over whether to continue collectivization. Some of this debate centered on whether to allow the private industrial sector to continue on its course in order to give newer cooperatives time to recover and stabilize. This would allow cooperatives to secure their footing within the agricultural sector, after which they would develop into the only means of production.[7] Alternatives to collectivization were popular among villagers, but these options undermined Marxist-Leninist teachings.

As mass riots shook communist dictatorships in Hungary and Poland, the orthodox part of SED leadership seized decisive control: any attempts to move away from the current socialist model were branded as a "counter revolutionary" and brutally suppressed. The party turned its focus to the full collectivization of all private lands. The party's unshakeable belief in the superiority of large-scale production, and its decisive move away from any possibilities of reform, meant that SED leadership would have to aggressively promote their agricultural policies. Since the national uprising of 1953, the number of LPG members had risen to 230,000, comprising 1.7 million hectares of tilled land, according to official statistics. Although these numbers still represent only a fraction of the total

[6] Bundesarchiv (hereafter BArch), DK 1, 5913, 1958.
[7] See for further information Scholz, *Bauernopfer der deutschen.*

agricultural endeavors within the GDR, the SED leadership clearly felt confident enough to pursue their far-reaching collectivization plan. At the 5th Party meeting of the SED in July 1957, party leaders announced that in order "to demonstrate the clear autonomy of the German Democratic Republic, we, the socialist leadership, have designed a plan to reach the pro-head production level of food stuffs (as well as the most important consumer goods) produced by West Germany and to eventually surpass that number."[8] A central condition of this plan, made widely known by officials, was the rapid collectivization of all farmland. This plan was fundamentally grounded in socialist ideologies of modernization, despite the fact that the state lacked necessary requirements for its implementation. The basic problem soon became apparent: private farmers continued to be more productive than collectives.

Another crisis soon emerged which the SED had failed to predict: following the party's anti-big farmer efforts aggressively implemented since 1948, "middle farmers"[9] had become more productive and prosperous. This stood in direct opposition to the agrarian concepts of the SED, as middle farmers should have been effectively neutralized during the "class conflict," thus ensuring higher production for LPGs. In contrast to party expectations, "middle farmers" succeeded in bypassing the collectivization process. At the end of 1959, after leaders analyzed the issue, the SED could reach only one conclusion: middle farmers not only avoided conforming to LPG standards, they sought to build their farms up to the level previously held by the big farmers. According to party leaders, the success of middle farmers "stood now in danger of being—whether directly or indirectly—opposition to the socialist goals of the party and state."[10]

This situation manifested more clearly during an even more pressing crisis of lowered production levels. Other factors soon became evident which forced SED leadership to deal with the crisis: the tempo of the collectivization process was too slow and a bad harvest had already been predicted for that year (1959). Furthermore, the Soviet Union was not ready to provide necessary financial support to overcome the GDR's crisis. In addition, the GDR increasingly lagged behind West Germany with regard to industrial competence and competitiveness.[11] At the beginning of 1960, SED leadership made a decisive and politically consequential move

[8] *Protokoll der Verhandlungen des V. Parteitages*, 1959, 70.
[9] Those with arable land of between ten and twenty hectares.
[10] Stiftung Archiv (SAPMO-BArch), DY 30/IV 2/7/370, 95.
[11] Bauerkämper, *Ländliche Gesellschaft*, 181–91.

by finally ceasing its previous attempts to think about alternatives to the orthodox socialist agricultural model, opting instead to follow regional examples of "successful" cooperatives based on the Soviet collectivization program. At the time, more than half of all arable land in the GDR was still held privately. Over 400,000 farmers and their families, who overwhelmingly still possessed healthy businesses, were suddenly bombarded by a campaign aimed to convince them of the efficacy of the socialist agricultural system. Despite the official rhetoric, the party's "convincing process" was based on coercion. The laws on which the farmers could have previously relied to protect their businesses were abolished, along with the system that had been implemented to deal with complaints made by farmers. Propagandists flooded rural areas: anyone who resisted was threatened with jail or exile, mandatory delivery requirements of goods were arbitrarily raised, and the sale of agricultural products by individual farmers was banned. Approximately 100,000 members of the "Operation Brigade" overwhelmed villages as part of the Socialist Spring, working tirelessly toward a single goal: the installation of a closed network of cooperative producers in the GDR. This three-month-long campaign, which came from central SED leadership in East Berlin, produced immediate results and brought private farming to an end in the GDR. Of more than 800,000 private farms that existed in the GDR in the early 1950s, only about 20,000 remained.[12] Although Soviet influence had had a decisive and direct affect on early collectivization efforts in the GDR, this was no longer the case.

Structurally, collectivization had thus been achieved. The SED had spent eight years establishing a socialist agriculture, always claiming that the only viable agricultural structure was that of the LPG. Yet the opposite was true: production yields and food supplies in the early 1960s had dropped dramatically, and the number of inhabitants leaving the GDR in favor of West Germany continued to increase. Despite what leaders claimed, the GDR was on the brink of an economic breakdown that was undermining the regime's political power. In response, leaders began to build the Berlin Wall to stem the tide of East German migration and bring to a close any hope of reunification with West Germany.

Although collectivization and its negative consequences had much to do with the coercive measures used by the SED leadership, the use of these extreme intimidation measures did stabilize the socialist agri-

---

[12] For further details see Schöne, *Frühling auf dem Lande*, 198–220.

cultural system. In the 1960s and 1970s, however, conservative reforms were implemented to target specialized, large-scale structural agricultural industry. The SED claimed its policies were founded on the basis of "scientific and technical research," but in reality the program was narrowly tailored to fit the centrally planned economic structure and party initiatives. Meanwhile, the horizontal rather than vertical organization of the agricultural industry on the one hand, and the consequential separation of animal from plant production on the other, led to a number of negative results to which the party responded with a series of "apologetic" reforms during the 1980s. Nevertheless, the basic collectivization model continued unquestioned. The flawed system was finally dismantled after the Peaceful Revolution of 1989–1990. Only then did the agricultural system of the GDR finally fall apart, leaving behind large and obsolete farming structures.[13]

The Essential Root: Agriculture in Communist Ideology

The following section will illuminate the basic determinants of the Marxist-Leninist ideology with reference to collectivization. In addition, this text will analyze the different interpretations of the theory as represented through the SED's collectivization program, as well as those in other Eastern Bloc countries.

The post-Lenin communist movement was designed as an international movement, with its homeland in the Soviet Union and its fundamental ideology based on Marxist-Leninist teachings. Communism was understood as something that could be applied in many places in the world, an ideology that in practice allowed for certain flexibility. The political protagonists in Moscow were well aware of this fact. In 1938 Stalin wrote a "Short Introduction to the History of the KPdSU (B)," clearly directed at communist leaders in other countries. He wrote:

> Marxism, it was asserted, required flexibility of theory and practice. What was suitable for one historical situation was not automatically applicable to another. The application of Marxism-Leninism's teachings had to be adjusted from generation to generation. Organization, slogans, class struggle and interna-

[13] Schöne, *Das sozialistische Dorf*, 142–67.

> tional relations had to be adapted to circumstances that would require constant reconsideration by the communist leadership. Although the October Revolution in Russia was to be regarded as the greatest event in the liberation of humankind, the supreme goal had yet to be attained: the spreading of the communist order to the entire world.[14]

Here, Stalin explicitly describes a model for a universal society but leaves room, at least theoretically, for different approaches in achieving the final goal. He saw this flexibility as a strategy that would ensure the durability of communism. This ideological flexibility also applied to the realm of agricultural planning and to rural society as a whole. Post-World War II, this flexibility was made known to all leaders of other communist parties as they sought to claim power in their respective regions. Overwhelmingly, these leaders had been schooled in Marxist-Leninist teaching in Moscow, and they had a very clear understanding of the coming difficulties. It was understood that the Soviet model should be used, but space was strategically left for open "experimentation." The widely circulated historical theory that communism was a monolithic system implemented in a fashion adhering exactly to a strict, step-by-step ideology does not prove itself to be the case. After the Second World War, Stalin made clear that he would bring the socialist model to the entire Eastern Bloc; the regimes of these regions, strategically placed into power by Moscow, were well aware of this fact. Moscow's dominance remained unquestioned: "It was important for Stalin to keep party-lines static, specifically his interpretation of the model of Marxist-Leninist teaching—i.e. he had a clear idea of what had to be accomplished by a certain point and which elements of the teachings had to be implemented."[15] The use of Marxist-Leninist teachings defined the work of Soviet-dependent regimes, as in the GDR where party leaders believed policies "are only comprehensible when we view them in reference to ideological demands and goals."[16]

Since the 1920s, the Communist Party in Germany had wholeheartedly embraced "Stalinization." For this reason, the party remained unpopular in rural areas—a fact not so surprising when considering that the movement and its basic theories were centered on urban and indus-

[14] Service, *Comrades*, 180.
[15] Gellately, *Lenin, Stalin und Hitler*, 790.
[16] Judt, *Geschichte Europas*, 145.

trial workers. That the Communist Party could not count on support from farmers had far-reaching consequences. The effects of industrialization—which caused widespread poverty among a greater portion of the population—sparked a discussion within the Communist Party over methods of dealing with these negative consequences. After having convinced the industrial sector of the superiority of communist ideology, urban intellectuals attempted to convince the rural population as well. Land in rural areas was largely in the hands of a few owners; the idea of overarching land reform seemed appealing to farm workers and small farmers who hoped it would resolve the inequities of the existing system. Communist leaders believed collective farming would not only stop people from leaving the countryside for better employment prospects in urban centers, but cooperatives would also neutralize the growing social problems caused by industrialization.

This idea itself was tied explicitly to Karl Marx and Friedrich Engels, who argued in the *Manifesto of the Communist Party* (1848) for the "expropriation of free-holders' [land]." They stated:

> Of course, in the beginning, this cannot be effected except by means of despotic inroads on the rights of property, and on the conditions of bourgeois production; by means of measures, therefore, which appear economically insufficient and untenable, but which, in the course of the movement, outstrip themselves, necessitate further inroads upon the old social order, and are unavoidable as a means of entirely revolutionizing the mode of production.[17]

These words form the basis of the central creed of agricultural politics within the communist agenda. In the Soviet Union and in other Soviet-dependent states, the *Manifesto of the Communist Party* was a reference for communist movement and for different methods that could be taken to achieve communist ends, as well as an inherent belief that foolish choices made by any given communist movement would right themselves through the enthusiasm present in each movement. Thus, the primacy of communist politics and ideologies had to do with a certain assumption of exaggerated voluntarism from its followers. From this—as was evident in the GDR, where the political protagonists had little knowledge of agriculture

---

[17] Marx and Engels, *Manifesto*, 45.

itself—grew many problems and issues. Nonetheless, Marx and Engels had predicted certain problems regarding rural workers, describing different theories of alliances between supporters and opponents within the working class. They argued that because of economic struggles, small farmers would be eager to give up their private farms and join the proletariat struggle, and big farmers who wished to maximize their profits through the use of wage workers would have to be overthrown. Regarding middle farmers, Marx and Engels could only provide a vague answer. According to their ideology, these farmers could develop in either direction: their hybrid position—i.e. their potential alliance with either small or large farmers—put them in an untrustworthy category. Because they owned their means of production, middle farmers were at once both exploiter and exploited. After the "proletariat revolution," these farmers were next in line to be neutralized in the war for the working class. Engels clearly spelled out the final goal of agricultural reform: "the common possession of the means of production is thus set forth here as the sole principal goal to be striven for."[18] At its core, this meant nothing less than the complete overhaul of privately held farms, independent of their size. Comradeship and communally held production were defining aspects of communist ideology, a principle Engels believed would guide farmers towards the socialist cause. Collectivization is thus the most important goal of communist agricultural politics, because it rests at the foundation of communist agricultural theory.

In the beginning of the 20th century, Lenin implemented Marx and Engels' theory into practice in the USSR. In retrospect, it is clear that Lenin modified the original model established by Marx and Engels. First, he expressed much more urgently than either Marx or Engels the need for different kinds of interlocking ties between workers and farmers. Villagers could only improve their lives by supporting the working class—the main interest of the party—while the workers needed the security inherent in a coalition with agricultural producers. Crucially, however, Lenin's hope for greater communication between these two factions did not apply to all farmers. In his second step—based on the writings of Marx and Engels—he separated the farmers into different groups that he predicted would be their future after the "proletariat revolution." While farm workers and small farmers were intrinsically similar and therefore tied to the working class—and would thus experience strong support from the revolution-

[18] Engels, "The Peasant Question in France and Germany."

aries—middle and big farmers would be economically and politically "neutralized:" the revolutionary proletariat had to be ready "to deal this stratum a most determined, ruthless and smashing blow at the very first signs of resistance."[19] Only after such a decisive move would other land workers be able to accept the new state.

The first step to be taken by communist agricultural leaders would therefore be the total consolidation of privately held land: the fate of the big farmers would be either exile or internment. The reappropriated land would then be broken up into pieces and divided among small and middle farmers, thus supporting the urban and rural "proletariat" and removing the presence of big farmers in the villages. Lenin anticipated that this aggressive step against the most prosperous big farmers would cause a temporary sink in production. Again, here lay the fundamental assumption that low production rates would reverse over time and lead to wide-reaching economic gains. The goal of this agricultural project was not just to increase yields for the benefit of the population but also to bring agriculture and the rural social structure in line with Marxist-Leninist theory. But because of the farmer's attachment to private property ownership, as well as ownership over the means of production, Lenin, too, was deeply suspicious of rural farmers. For this reason, he developed this third thesis: that the workers, and above all the party, would have to act as teachers, educating the inhabitants of rural areas. This would only be possible, however, when the "class conflict ... came to the villages" in order to finally convince the "uncultured" or the "landed society" to join the communist cause. The transfer of industrial workers to the villages was seen as inevitable, in order to facilitate the necessary transformation, even if pressure and violence were necessary. The "force of example" would have to be used, but should it fail, there should be no hesitation or regret in a "ruthless struggle" to bring about the desired results of the revolution.[20] Lenin's unmistakable message had wide-reaching effects on the development of agriculture in all Soviet-dependent states. Farmers were envisaged as an uneducated, faceless mass, allowing the working class and the avant-garde to form an allegiance—an alliance that would ultimately bring about revolution. This final claim formed the basis of communist agrarian policy and undeniably, the foundation for its practical implementation in the Eastern Bloc.

---

[19] Lenin, "Preliminary Draft Thesis," 158.

[20] Lenin, "Preliminary Draft Thesis," 157.

Map 1: The administrative organization of the GDR.

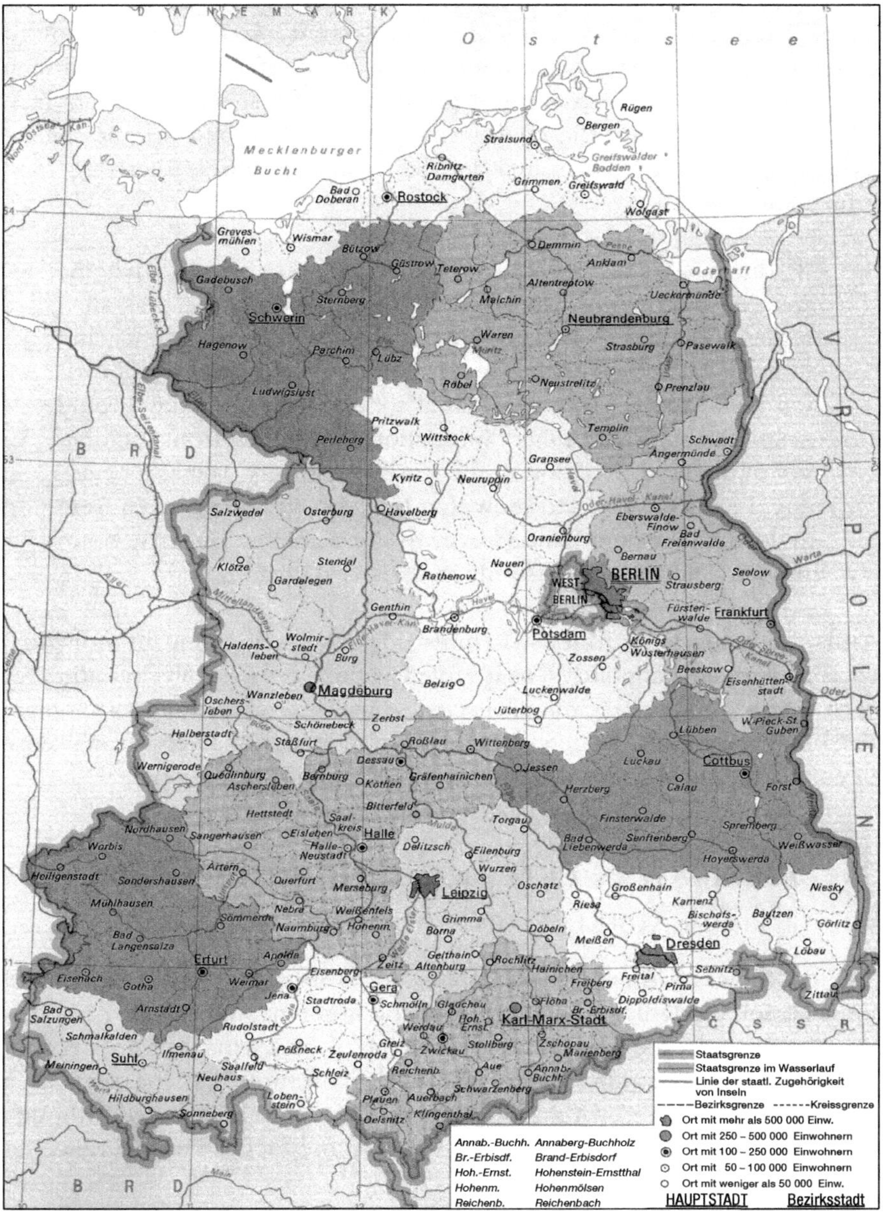

Source: *Atlas für die 4. und 5. Klasse,* Gotha: VEB Hermann Haade Geographisch–kartographische Anstalt 1989, 4.

After Lenin, all agricultural reform efforts by communist leaders attempted the step-by-step transformation of privately held to collective lands. After the establishment of a "proletariat" state, which involved ousting large landowners and neutralizing middle farmers, the next step was to restructure all agricultural production into socialist cooperatives. Lenin also formulated the relationship between land reform and collectivization, though he did so in a decidedly vague manner. Regarding land reform, he noted:

> We did not want to impose on the peasants the idea that the equal division of the land was useless, an idea which was alien to them. Far better, we thought, if, by their own experience and suffering, the peasants themselves come to realize that equal division is nonsense. Only then could we ask them how they would escape the ruin and kulak domination that follow from the division of the land. Division of the land was all very well as a beginning. Its purpose was to show that the land was being taken from the landowners and handed over to the peasants. But that is not enough. The solution lies only in socialized farming.[21]

Thus, the underlying Marxist-Leninist ideology is apparent in individual socialist agricultural policies, as it served as the basis for collectivization for each new communist regime. Collectivization became the long-term objective of all communist leaders: "Stalin was acting out of a very strong sense of conviction, as were his followers in the Eastern Bloc nations."[22]

Ideology in Practice: Collectivization, Flexibility, and Moving Targets

The collectivization process was preceded by necessary land reforms and a succession of ideologically motivated and implemented events. Although there are some exceptions, the three steps of first, land reform, second, "increased class conflict," and last, collectivization, can be seen as a process that occurred within all states under Soviet hegemony. Although historical research occasionally differs on the details in reference to

[21] Lenin, "Speech at a Meeting," 175.
[22] Baberowski, *Der Rote Terror*, 155.

the Soviet Zone of Occupation (SBZ), it is undeniable that this process was initiated by the Soviets. As in every other country dominated by the Soviets, the German communists' power was merely symbolic.[23] This was also the case during the collectivization process, although this would gradually change over the years. Therefore, the political leadership of the GDR undertook significant efforts to study and apply the experiences of the collectivization processes in other communist countries. Hungary was frequently used as a model, especially when leaders were considering basic changes to GDR's collectivization policy. In 1952, party leaders visited Budapest on a diplomatic mission to study the collectivization process in Hungary, looking at specific administrative examples, the practical success of Hungary's efforts, and the consideration of the Marxist-Leninist theory. These reports were used in East Berlin by leaders of the SED, as well as in pamphlets distributed to each rural region. Handwritten reports show that both Walter Ulbricht and the agricultural GDR ministry were issued this information.[24]

The Hungarian model remained a reference point for GDR policy makers throughout the process. From 1953–57, during the period of relaxed collectivization, few records of contacts between Hungary and the GDR exist. This changed dramatically, however, after the SED leadership embarked on an aggressive pursuit of full collectivization of the agricultural structure. Beginning in 1958, East German and Hungarian party and state ministries began bilateral talks, exchanging information on problems as well as ideas on how to resolve roadblocks to collectivization and restore the program. Unlike at the beginning of the decade, when its first attempt at collectivization failed, the GDR quickly surpassed Hungary in its collectivization rate. At the same time, the SED expressed a desire to learn from the Hungarian experience with the possibility of enhancing its own model. This information exchange process "allowed these two partners to realize that the problems occurring in reference to land collectivization frequently overlapped." These overlapping issues are, of course, unsurprising, as the development of the agricultural programs in each country was based on the same Marxist-Leninist ideology and Soviet collectivization experiences. Despite regional differences, this fact inevitably generated similar effects. Debates between representatives of each country thus

---

[23] See Leonhard, *Die Revolution*, 503–13.

[24] Stalin also emphasized the importance of the Hungarian experience. See Scherstjanoi, *SED-Agrarpolitik*, 356.

dealt more with practical concerns, and the ideology itself was infrequently discussed. Notable here is the use of the previously mentioned flexibility of the model, although neither country ever questioned the basic truth of the ideology itself. The Hungarian embassy sent the following message to the GDR in January of 1961: "a report will be sent in shortly regarding the completion of the socialization of the rural industry of the UVR."[25] In fact, the arrival of this report took over a year, but the assumption that full collectivization would eventually occur was never questioned. In light of the similar experiences of Eastern Bloc nations, SED leaders made efforts to keep tabs on the land reform developments in neighboring countries as well as to continue talks with the Hungarians. Although this information was used only minimally, a record of the experiences of other communist states was nonetheless helpful for party leaders. In August of 1952, an analysis of the ongoing collectivization process clearly stated that "besides the experiences of the Soviet Union, the development of collectivization in other countries has given us useful examples." The relevant ministry within the GDR thus attempted to document the collectivization process in a number of communist states, including but not limited to Poland, Czechoslovakia, Romania, and Bulgaria. The collected materials allowed the party and leaders in East Berlin to keep informed of new developments and to use the information for new policy directions.

Despite this, SED leaders deviated distinctly from the opinions of economic specialists. Core arguments for collectivization were based entirely on ideological guidelines rather than economic reality. The developments in Poland, for example, in their view clearly rested on the central tenet of Lenin's "class conflict" model: "the politics of restriction and elimination of serfdom is the only political direction that will save the small and middle farmers from the ruining exploitation of wealthy villagers."[26] But Poland at the time had a small- and middle-sized farming structure, meaning that the rhetoric of "serfdom" and "village wealthy" found little relevance to their political problems, outside of a desire to implement the "class conflict." This example further explicates the priorities of Communist Party leaders. Up to this point, their relationship could have been described as symmetric and balanced, but it quickly became clear that the Soviet Union was playing the crucial role in influencing leaders in the GDR. Despite some "flexibility" within the dynamic of the

[25] BArch, DE 1, 21900, 223, 209.
[26] BArch, DK 1, 5687, 30, 33.

relationship between the two countries, the political leaders of the GDR viewed Moscow as the unquestioned hegemonic center of power. This fact also applied to the realm of agrarian policy. In reference to collectivization, a widely read publication released in the mid-1970s made no reference to questioning the basic ideology:

> Lenin's cooperation plan lies as a central, scientifically derived tenet of the plan to reform agrarian policy and guides the political course of the party in reference to farming during the period of transition from capitalism to a socialist structure. This cooperation plan contains within it support for objective laws that will guide society's development, outlining the fact that the basic interests of the working class and the working farmers coincide, and describe the construction of a socialist society [...] Lenin's meaningful plan, first tested by revolutionary, defining transfer from capitalism to socialism in the USSR, is of international significance. Communist and working parties the world over study Lenin's plan for the socialist reconfiguration of the agricultural system [...] Lenin's direction for the farming transfer to socialism represents a compass and goal for all communist and working parties.[27]

This most accurately describes the political direction and actions of the GDR in the 1950s. Lenin's theory, with its grounding in Marxist thinking, clearly acted as a "compass" for the actions of the German communists in their attempt to implement Socialist collectivization. But "Sovietization" had consequences, in terms of mass social unrest and migration. Still, Stalin's model was not wholly embraced or implemented. Again, the expressed "flexibility" of the process led to each region's own individual means, especially during de-Stalinization. As long as fundamental elements of the ideology were not called into question, regional variations were tolerated and regional contributions accepted.[28] Party leaders in the Soviet Union never defined the precise meaning of the Cooperation Plan, nor did they outline which parts were particularly important to Lenin.

[27] *Lenin und die KPdSV*, 1974, 12, 20.

[28] See Service, *Comrades*, 293–303; Feest, *Zwangskollektivierung im Baltikum*, 477–84.

Nevertheless, challenges did occasionally emerge throughout the process of collectivization. After the Second World War, the aggressive transformation of regions under Soviet hegemony did little to relieve the unstable position of the agricultural sector. Land reforms more often resulted in such overwhelming failure they frequently revealed a central conflict between ideology and reality. Furthermore, the competition between the two governmental systems of both German states led to a pressing political imperative to continuously raise production yields. But on the other hand, the fragmentation of the arable land and the insufficient support of the new farmers generated increasing dissatisfaction that the party failed to cope with. The continuous decline in production was not surprising, as it had been predicted by Marx, Engels, and Lenin. Despite these factors, the party moved to its next step: the complete structural overhaul of the agriculture structure.

The crucial catalyst for the first step came from Moscow. After the spring of 1952 and the failure of the Stalin Note—a document that indicated Soviet expansion further west would not be possible—the then Soviet-dependent states were put on a fast track toward socialism. This was consistent with the desires of the previously installed communist regimes in these regions, but only now had it become possible for these regimes to take decisive steps forward. This applied particularly to the GDR, which had until this time held back because of Stalin's hopes to reunite the two Germanies under a communist regime, a desire Stalin had abandoned. SED leaders were summoned to Moscow where they received concrete direction on their new policy course. On April 7, 1952, Stalin gave SED leaders approval to initiate the transformation of the GDR based on Marxist-Leninist dogma. Any internal political opposition soon dissipated, as what had previously existed only as an ideology could now be implemented in reality.

Until the "People's Protest" of June 1953, the leadership of the GDR made almost no policy decisions without Soviet approval. The Soviet Military Administration (SMAD) and later the Soviet Control Commission (SKK)—the organs of hegemony and power in the GDR—were firmly in control. These two institutions implemented fundamental decisions made by Moscow in every region of the GDR. An agrarian department was created under the leadership of A. D. Stupow, who had previously worked in Bulgaria covering the topic of collectivization. Every East German, from the SED leadership to the local functionaries, was aware of Russian control—and the Soviet collectivization model was thus ensured as the central ideological goal of the GDR. At the second SED Party Conference in July 1952, the SKK emphasized that collectivization should be of greater immediate concern within SED party efforts: "Party organs occupy themselves above all on the propagandizing of the decisions made at the party conference and in most cases, do not work on building a mass movement to fulfill these decisions. The result of this is clear at the village level, where we have problems implementing collectivization."[29]

Not only were Soviet collectivization experiences of the last thirty years implicitly referenced here, but so too was the party's focus on the "avant-garde" as being at the center of the societal transformation—a phenomenon Lenin had also frequently desired. The influence of the Soviet side remained complex and multi-layered and frequently required a certain level of anticipated obedience from the SED. The guidelines under which the SKK operated were frequently changed—it requested technology experts from the Soviet Union and farmer delegations were sent to "Lenin's land" in order to study the positive effects of collectivization. But in the villages of the GDR, the influence of the Soviet Union was viewed critically; the radical measures frequently did not resonate positively among the farmers. A German "new farmer," referencing the Soviet delivery of new technology, wrote in August 1952: "These large harvesting machines seem to have only been delivered in order to compel us farmers to enthusiastically support production cooperates, despite the fact that these machines are only appropriate for farming on a very large scale."[30] And in an even more clearly stated condemnation: "In the Soviet Union, the farmers are compelled to embrace the collectivization model. We are being compelled to embrace production cooperatives. Those who hang themselves have it better than we do."[31]

In the hurried rush to the class conflict, such complaints went unheard. In close agreement with the Soviet occupiers, the SED massively increased the tempo of the collectivization process and tried as much as possible to fit reality with the needs of ideology. Thus, entrance into the LPG had to be "voluntary," despite the fact that until 1952–53, this had not been a requirement in the majority of cases. The fight against large farmers was thus accelerated—following Lenin's call to deliver a "decisive blow" against the most prosperous producers. For this reason, and despite abysmal production rates, political leaders made no effort to change course.

---

29 Scherstjanoi, *SED-Agrarpolitik*, Dokument 15, 49.

30 Brandenburgisches Landeshauptarchiv (BLHA), Rep. 401, 59, 14.

31 BArch, DK 1, 5893, 47.

Party leadership in East Berlin relied on Stalin without question and believed that any problems would be solved by growing enthusiasm for the movement, thus eventually bringing about a positive turn. Notwithstanding these expectations, however, it became clear that a wholesale, uninterrupted implementation of Stalin's model would not be successful. Although there were clear deviations from the Soviet example—illustrated by the establishment of three LPG-types instead of just one—the noted flexibility was not sufficient to carry out collectivization in the GDR without opposition. The complete societal reconstruction undertaken by the SED only led to increased problems, which reached a peak in early summer of 1953 that nearly led to the regime's collapse. For a significant period of time, the SKK had warned against the massive negative results of a forced socialist course. But the order to intensify collectivization had, after all, come from Moscow. After Stalin's death, the failures of collectivization were increasingly acknowledged. Following this, SED leadership was ordered to end forced collectivization and close down any LPGs formed on involuntarily terms. Those SED politicians who refused to accept this order were summoned to Moscow in June 1953 to pledge their allegiance to the new policies proscribed by the Soviet Union. This action firmly outlined the fundamental power dynamic between East Berlin and Moscow. Soon this was made evident to the world when Soviet troops invaded East Germany and brought an end to the "People's Protest," effectively securing the dictatorship of the SED.[32]

The first collectivization push was controlled entirely by the Soviets: from beginning to end, the SED followed orders from Moscow. Moreover, the hegemonic power of the USSR played a significant role in the practical creation of the LPGs, which was significantly shaped by both the Soviet experience and ideology of collectivization. It is thus correct to define the collectivization of the GDR in the beginning of the 1950s as a "Sovietization" process; but it is important to make clear that the collectivization model was not enforced by the USSR but rather enthusiastically embraced by East German leadership. The SED never questioned the Soviet model and made great efforts to adopt it as much as possible in anticipation of the coming transfer of power. That the GDR only ever hesitated once—at the point when the collectivization process was recalled by the Soviet Union in 1953—shows the clear determination of the SED to remain true to the original, ideologically bound model. Certainly, GDR leaders' deci-

[32] For further details see Schöne, *Frühling auf dem Lande*, 134–53.

sions were curtailed and orders from Moscow acted as the deciding factor. But this changed during the following years, as the SED leadership exhibited some freedom, which it used enthusiastically to continue supporting the model. This party's zealous desire to follow the given model is evident in the second collectivization phase, which began in 1958.

At the Fifth Party Meeting of the SED in July 1958, party leaders decided on the next collectivization phase. But this time, unlike in 1952, there was no initial pressure from Moscow; instead, it was the competition between the two German states that provided the impetus for this newly forced collectivization push. Again, SED leadership believed without a doubt that the GDR would only survive the contest between the two systems if it embraced the socialist production method and brought it to full fruition. As the GDR had received the nod of formal independence from the Soviet Union, the SED now sought to use this newly acquired freedom. The reform of agrarian policy—and collectivization—came into political focus. Again, the results that the rulers had hoped for failed to meet expectations. Although the number of LPG members grew, the requirements and conditions of collectivization quickly led to greater dissatisfaction, which became evident as early as 1959, when the production yields remained low and acceptance of cooperatives in villages limited. These negative circumstances occurred not only in the GDR—at the end of the decade and within the entire Eastern Bloc, problems with agricultural policy continued to plague socialist regimes. At the same time, the Soviet Union remained unwilling to provide financial support to the ailing agricultural sector, despite numerous requests from satellite states. This situation again demonstrated that regions conquered by the Soviets after the Second World War continued to depend on its hegemonic power, despite the Soviet emphasis on its so-called "modest politics." In November 1959, Nikita Khrushchev invited "faithful comrade Ulbricht" to Moscow to discuss the problems of agricultural reforms. Although Khrushchev stressed he was not set on any particular policy and that each country had "its own unique qualities that can be clearly seen in the development of agricultural policy," this was not the true nature of the power relations between the Soviets and the GDR. Khrushchev stated that it was the duty of both country's agricultural ministries to issue reports so Moscow could "stay informed of developments in agricultural policy in both our countries."[33] Khrushchev did not invite any discussion on the

[33] SAPMO-BArch, DY 30/3727, 1959, 42.

growing problems of the dependent regime, nor did he provide any solutions for these issues.

This putative invitation, premised on the GDR's increasingly evident industrial problems and meant to enforce the SED's compliance with collectivization, had an incalculable catalyzing effect on the GDR. At that moment, it became clear that certain central tenets of the ideology failed to adhere to reality. The unwillingness of the middle farmers to embrace collective farming especially stood out, and brought the idea of a socialist large industrial sector into question. Lenin had written that middle farmers joining the socialist cause was an unquestionable condition for villages to the embrace the communist stance. As this acceptance by the middle farmers grew increasingly unlikely, party leaders saw the necessity that an alternative—class warfare and "the ruthless battle against the bourgeoisie"[34]—would have to be used to subdue the remaining private farmers.

Leaders in East Berlin began to assess the experiences of the Soviet Union and other "brother states." This led to a renewed sense that the party's course, which aligned itself to ideologically determined goals, should continue unbroken despite supposed independence from Soviet hegemony. According to an example from Poland: "The experiences of the brother states, especially that of the USSR, shows that a quick and steady rise in the growth of agricultural production is guaranteed through collective industrial forms ... For this reason, our party sees the realization of this plan of increase in agricultural production as inseparable from the socialist transformation of the villages."[35] The expectation of a quick and measured increase in production did not describe reality in Poland, where collectivization had recently been decreased, but it nevertheless demonstrates how expectations of the communist worldview remained sacrosanct.

The reports and analyses addressed at the Moscow meeting made it clear that the agricultural policy of the GDR was not satisfactory. Production yields remained far below what had been projected. In the middle of January 1960, Ulbricht informed Khrushchev of this situation. And following this, important changes in the agricultural policy were outlined a month later by Ulbricht at the meeting in Moscow: "To our mind, an increase in yield agriculture and in animal butchery is only possible through new reforms of the agricultural system." Although Ulbricht did not explicitly say full collectivization would occur in the GDR, that con-

[34] Lenin, "Preliminary Draft Thesis," 157.

[35] SAPMO-BArch, DY 30/J IV 2/2/J, 647.

clusion was implied. The measures included the future building of oversized animal stalls, the specialization of animal and plant production, the establishment of new machine technology, the full mechanization of farm equipment, and other measures to support conditions of a socialist large-scale production plan. Ulbricht went on the offensive, offering an agricultural reform plan because he was in fact campaigning for support for the coming transformation. He hoped for direct aid from Moscow for the GDR, where the collectivization push had been a drag on public funds.[36]

For the GDR, the end of individual farming now came quickly. In the Socialist Spring of 1960 almost all individual producers were forced to join the LPG. In fact, this was just the consequent application of Lenin's cooperation plan. During the first phase of collectivization (1952–53), Moscow had massively and directly influenced the process. The Socialist Spring in contrast was realized by the SED alone. Its leaders now knew in theory and in practice what to do. But "what was done in name of industrial rationality was in fact not rational, but instead ideological."[37]

Conclusion

Several days after the end of collectivization, Walter Ulbricht stood before the Central Committee of the SED and concluded: "One could say that we, in the German Democratic Republic, have successfully fulfilled Lenin's plan for cooperative agriculture. Lenin's plan acted as a guideline for us and we have benefited much from the many experiences that the working farmers in the Soviet Union have gathered and allowed us to use."[38] Ulbricht expressed what he saw as the fundamental point of reference for the party-state, the same ideology that had transformed the agricultural policy in the GDR: the Soviet Marxist-Leninist model. Ulbricht's statements also revealed how, despite the failures of collectivization, the party remained true to the rhetoric of its success, and refused to publicly admit the program's evident, widely known failures. Marx, Engels, and Lenin had all written that negative economic decisions could be reversed by inspiring enthusiastic support for communist ideology that in reality did not occur. For these and other reasons, the relationship between SED leaders and

[36] SAPMO-BArch, DY 30/IV 2/7/188, 1960, 83.
[37] Hildermeier, *Geschichte der Sowjetunion*, 391–92.
[38] SAPMO-BArch, DY 30/IV 2/1/231, 1960, 7.

Moscow remained in constant flux. Both sides were aware of the inherent flexibility of the collectivization process but both were also aware that this flexibility had limits. Although SED leadership examined the experiences of other socialist states in search for ways to recreate its own policy, the party never deviated much from the fundamental elements of the original Stalinist model. It can thus come as no surprise that finally, with a few exceptions, no independent model for collectivization and production cooperatives ever emerged in the GDR. Stalin's model was much more formative and influential than a number of alternatives suggested by the SED elite. For East German communists, independence from the Soviets was not a prime goal, and therefore, the party remained dominated by Moscow, which was always recognized by the SED as the legitimate center of power.

Indeed, how the Soviets would react to dissention in other states was a question that plagued communist leaders throughout the Soviet era. The regime had shown it would respond to uprisings with military force, as was the case in Hungary in 1956 and in Czechoslovakia in 1968. These actions sent a strong signal to other communist states that pushing the limits of the Soviet system too far would come with dire consequences. Though deviations from the given model did occur—in the realm of collectivization, Poland stands as a clear example—no regime dared to question the foundations of the asymmetrical relationship of dependent regimes to the Soviet hegemony.

The process of collectivization in the GDR and across the Eastern Bloc was historically predetermined following the invasion of Soviet troops during World War II. Once Moscow's power was established, the question was not if, but rather, when and how collectivization would occur in various satellite states. While this power dynamic was an important factor, the SED pursued collectivization and other socialist policies out of genuine ideological conviction and zealous support for the Marxist-Leninist and Soviet project—and also, it could be argued—to satisfy Soviets leaders.

BIBLIOGRAPHY

ARCHIVAL SOURCES

Bundesarchiv [Federal Archives] (BArch):

Fond DK 1, Files: 5913 (Analyse des Ministeriums für Land- und Forstwirtschaft über die Entwicklung der LPG 1958); 21900/209, 223; 5687/30, 33 (Schreiben des Instituts für Wirtschaftsforschung vom 6.8.1952); 5893, 47 (Information der Redaktion *Freier Bauer* an das Ministerium für Land- und Forstwirtschaft vom 17.12.1952).

Brandenburgisches Landeshauptarchiv [Federal Archive of Brandenburg] (BLHA): Rep. 401, 59, 14 (Bericht über die Meinung der Bauern zur Arbeit sowjetischer Mähdrescher im Kreis Ruppin und Gransee vom 8.8.1952).

Stiftung Archiv der Parteien und Massenorganisationen im Bundesarchiv [Foundation Archive of Political Parties and Mass Organizations in Federal Archives] (SAPMO-BArch):

Fond: DY 30/IV 2/7/370, file: 95 (Einschätzung des erreichten Entwicklungsstandes der sozialistischen Ungestaltung der Landwirtschaft, der Festigung der LPG und VEG sowie der Entwicklung der Produktion).

Fond: DY 30/3727 (Büro Ulbricht), files: 42–51, cit. 42 (Schreiben Chruschtschows an Ulbricht vom 14.11.1959).

Fond: DY 30/J IV 2/2/J, file: 647 (Einschätzung der Lage in der Landwirtschaft der Volksrepublik Polen, o. D.).

Fond: DY 30/IV 2/7/188, files: 77–99, cit. 83 (Diskussionsbeitrag des Genossen Walter Ulbricht auf der Landwirtschaftskonferenz am 2. und 3.2.1960 in Moskau).

Fond: DY 30/IV 2/1/231, file: 7 (Referat auf der 8. Tagung des Zentralkomitees, 30.3. bis 2.4.1960).

BOOKS AND ARTICLES

Baberowski, Jörg. *Der Rote Terror. Die Geschichte des Stalinismus.* München: Deutsche Verlagsanstalt, 2003.

Bauerkämper, Arnd. *Ländliche Gesellschaft in der kommunistischen Diktatur. Zwangsmodernisierung und Tradition in Brandenburg 1945–1963.* Vienna: Böhlau, 2002.

Der Freie Bauer. Wochenzeitung der Vereinigung der gegenseitigen Bauernhilfe (BHG), issue: 15.3.1953.

Engels, Friedrich. "The Peasant Question in France and Germany." *Marxists Internet Archive*, October 31, 1993. http://www.marxists.org/archive/marx/works/download/Engles_The_Peasant_Question_in_France_and_Germany.pdf (accessed on February 18, 2013).

Feest, David. *Zwangskollektivierung im Baltikum. Die Sowjetisierung des estnischen Dorfes.* Vienna: Böhlau, 2007.

Gellately, Robert. *Lenin, Stalin und Hitler. Drei Diktatoren, die Europa in den Abgrund führten.* Bergisch Gladbach: Gustav Lübbe Verlag, 2009.

Hildermeier, Manfred. *Geschichte der Sowjetunion. Entstehung und Niedergang des ersten sozialistischen Staates.* Munich: Verlag C. H. Beck, 1998.

Judt, Tony. *Geschichte Europas von 1945 bis zur Gegenwart.* Vienna: Hanser, 2006.

Lenin, V. I. "Speech at a Meeting of Delegates from the Poor Peasants' Committees of Central Gubernias, November 8, 1918," in *V. I. Lenin. Collected Works, vol. 28* (July 1918–March 1919). Moscow: Progress Publishers, 1974, 171–78.

Lenin, V. I. "Preliminary Draft Theses on the Agrarian Question for the Second Congress of the Communist International," *V. I. Lenin. Collected Works, vol. 31* (April–December 1920). Moscow: Progress Publishers, 1974, 152–64.

Leonhard, Wolfgang. *Die Revolution entlässt ihre Kinder.* Köln: Kiepenheuer & Witsch, 2005.

Marx, Karl and Friedrich Engels. *Manifesto of the Communist Party.* Chicago: Charles H. Kerr & Company, 1906.

Neues Deutschland, Zentralorgan der Sozialistischen Einheitspartei Deutschlands, issue: 24.4.1952.

Protokoll der Verhandlungen der II. Parteikonferenz der Sozialistischen Einheitspartei Deutschlands, 9. bis 12. Juli 1952 in der Werner Seelenbinder-Halle zu Berlin. Berlin: Dietz Verlag, 1952.

Protokoll der Verhandlungen des V. Parteitages der SED, 10. bis 16.7.1958 in der Werner-Seelenbinder-Halle in Berlin. Berlin: Dietz Verlag, 1959.

Scherstjanoi, Elke. *SED-Agrarpolitik unter sowjetischer Kontrolle 1949–1953.* Munich: R. Oldenbourg Verlag, 2007.

Schier, Barbara. *Alltagsleben im "sozialistischen Dorf." Merxleben und seine LPG im Spannungsfeld der SED-Agrarpolitik 1945–1990.* Münster: Waxmann Verlag GmbH, 2001.

Scholz, Michael F. *Bauernopfer der deutschen Frage. Der Kommunist Kurt Vieweg im Dschungel der Geheimdienste.* Berlin: Aufbau Taschenbuch Verlag, 1997.

Schöne, Jens. *Das sozialistische Dorf. Bodenreform und Kollektivierung in der Sowjetzone und DDR.* Leipzig: Evangelische Verlagsanstalt, 2008.

———. *Frühling auf dem Lande? Die Kollektivierung der DDR-Landwirtschaft.* Berlin: Ch. Links, 2007.

Service, Robert. *Comrades: A History of World Communism.* Cambridge: Harvard University Press, 2007.

W. I. Lenin und die KPdSU über die sozialistische Umgestaltung der Landwirtschaft. Berlin: Dietz Verlag, 1974.

Werkentin, Falco. "Der totale soziale Krieg. Auswirkungen der 2. Parteikonferenz im Juli 1952" in *Jahrbuch für Historische Kommunismusforschung.* Berlin: Aufbau Verlag, 2002.

THE COLLECTIVIZATION OF AGRICULTURE IN THE GERMAN DEMOCRATIC REPUBLIC: CHRONOLOGY OF EVENTS, 1945–1960

Arnd Bauerkämper, Jens Schöne, Greg Witkowski

1945

June 11: The German Communist Party (KPD) calls for a comprehensive land reform.

July 14: Under the leadership of the KPD, a broader block of "anti-fascist democratic parties" is built to govern German territory.

July 17–August 2: The victorious Allies (UK, USA, USSR) meet at the Potsdam Conference to determine the specifics of their administration of occupied Germany.

September 2: The KPD announces the beginning of the land reforms. All private farms with more than 100 hectares of land are to be seized and redistributed. In addition, all land held by "Nazis and war criminals" is to be seized. This land is to be redistributed by locally formed committees to land-poor peasants and those without land.

September 3–10: The provinces in the Soviet Occupation Zone (*Sowjetische Besatzungszone*, SBZ), starting with Saxony, passes laws calling for land reform.

1946

March 17: The first conference of the newly formed Peasant Mutual Aid Society (*Vereinigung der gegenseitigen Bauernhilfe*, VdgB) is held. Communist leaders hope that this mass organization will help them bring their policies to the countryside.

April 21–22: The KPD, with Soviet support, forces the fusion of their party with the Social Democratic Party (SPD) in the SBZ. The new party, which is led primarily by leaders of the former KPD, is called the Socialist Unity Party of Germany (SED). Walter Ulbricht becomes the leader of the SED.

1947

June 14: The first governing body for the SBZ, the German Economic Commission (*Deutsche Wirtschaftskommission*, DWK), is formed.

September 9: The Soviet Military Government (*Sowjetische Militäradministration in Deutschland*, SMAD) issues Order Nr 209, which seeks to build houses and farming structures throughout the countryside. This ambitious plan is never realized, as the resources prove unavailable in the postwar period.

December 20: The Chairman of the Christian Democratic Union (CDU) in the SBZ is forced out of his position by the SMAD because of his opposition to the land reforms.

1948

Throughout the year, the SED builds a socialist agricultural system modeled on the USSR. State Farms (*Volkseigene Güter*, VEG) are built from seized and deserted land. Machine Loaning Station (MAS) aids in the planting and harvesting of crops by providing machinery for state and private farms. At the same time, the SED increases its attacks on traditional agricultural institutions and on "class enemies" in the countryside. The so-called large peasants (*Großbauern*), who had more than 20 ha of land, are targeted.

March 9: The DWK becomes responsible for the economy. It controls the economy through central planning.

April 29: The SMAD calls for the end of the Land Reform by June 1. Yields continue to drop as land is abandoned.

April 29: The German Democratic Peasants' Party (DBD) is founded. It was one of the Bloc parties Soviet administrators hope will integrate peasants into the communist system. The party supported SED agrarian policies.

June 29–30: The SED declares its first Two-year Plan for the development of the economy. The plan emphasizes industrial development.

1949

January 25–28: The SED holds its First Party Conference. Party members complain that development in the countryside is advancing too slowly.

March 16–17: In order to gain control over the traditional farming cooperatives, the SED sets up an umbrella association.

October 7: The German Democratic Republic is founded.

October 12: The official head of government, Minister President Otto Grotewohl, declares that rationing of food (other than meat and fats) would end in 1950.

1950

August 17: The first Five-year Plan, which the SED prepared, is approved by the East German Parliament (*Volkskammer*). Agricultural production is to exceed prewar levels.

September 29: The Council for Mutual Economic Assistance (COMECON) admits the GDR as a member.

November 20: Traditional farming cooperatives are forced into the VdgB, an SED-aligned mass organization.

1951

March 15–17: The Fifth Meeting of the SED Central Committee seeks to further secure the relationship of the SED with peasants. Thereafter, small and middle peasants receive additional support while large peasants (*Großbauern*) are further targeted.

October 8: Rationing ends for all food other than meat, fats, and sugar.

1952

March 10: Josef Stalin offers the western powers the possibility of unifying Germany under numerous preconditions. The western rejection of this so-called "Stalin Note" leads to greater emphasis on the development of the GDR as a separate state.

April 1–7: A delegation of top East German officials visits Moscow and meets with Stalin and other Soviet leaders. Stalin declares that collective farms (*Landwirtschaftliche Produktionsgenossenschaften*, LPG) should be used to transform the countryside.

June 3: The Politburo of the SED decides to implement agricultural collectivization.

June 4: SED regional functionaries are told to promote the formation of collective farms but to keep party participation as hidden as possible.

June 6: With encouragement from these functionaries, peasants in a number of villages speak to representatives from the Ministry for Agriculture and call for the formation and support of collective farms.

June 8: The GDR's first LPG is founded in the Thuringian village of Merxleben. This collective farm receives massive economic and political support from its formation.

July 1: The Politburo delineates model statutes for three different types of LPG. The three differ in terms of the amount of private property one could continue to own and the rewards for giving property to a collective. These models are used throughout the GDR as new collectives formed.

July 9–12: The SED's Second Party Conference calls for the "Construction of Socialism" in the GDR. For agriculture, this means an increase in collectivization.

July 24: The government (*Ministerrat*, Council of Ministers) declares that collectives and their members will receive benefits from the government including lower taxes, greater access to fertilizer, etc.

October 5–15: The Soviet Communist Party holds its XIX Party Meeting. Thereafter, the SED strengthens its efforts to collectivize agriculture and to fight "class enemies" in the countryside.

October 20: The Soviet Control Commission warns that forced collectivization is creating a food crisis.

November 20–22: The 10th Meeting of the SED Central Committee calls for the redoubling of efforts to collectivize and defeat class enemies, especially the large peasants (*Großbauern*).

December 5–6: The First LPG Conference, organized by the SED, supports the collectivization process.

December 19: The Council of Ministers calls for MAS to be called Machine Tractor Stations (*Maschinen-Traktoren-Stationen*, MTS), aligning it more closely with the name used throughout Eastern Europe.

1953

February: There are almost 2,800 LPG founded throughout the GDR. The SED leadership still seeks greater collectivization.

February 19: On the basis of an order designed ostensibly to secure agricultural production, more than 6,000 private farms are seized in six weeks. Most of the land is given to collective farms.

March 5: Death of Stalin.

April 23: The Secretariat of the SED issues the decree "Industry Workers to the Countryside." The campaign is designed to advance collectivization and increase the presence of SED members in the countryside.

May 21: Under direction from Moscow, SED leaders tell regional functionaries that the continued formation of collective farms is "not practical" at that time. The fear is that collectivization would have an adverse effect on yields.

June 1–5: SED leaders travel to Moscow. Soviet leaders tell them to stop forcing the "construction of socialism."

June 9: As a result of the discussions in Moscow, SED leaders call for a "New Course." For agriculture, this means a temporary end to the process of collectivization.

June 11: The government calls for a "New Course." Many East Germans see this as an admission that previous policies, and even the regime itself, had failed. In the days that follow, the public discourse in villages change as many see the regime near its end.

June 16–17: East Germans rise up against the regime. While there had been unrest in some villages, the protest in Berlin causes demonstrations in many towns and villages throughout the GDR. In over 400 communities with populations under 2,000, villagers protest in one form or another. As a result, collectivization is not continued and some members seek to leave their LPG.

July 24–25: The 15th Meeting of the SED Central Committee takes place. At this meeting, Walter Ulbricht reasserts his authority. The committee declares that collectivization remains the central goal in the countryside.

August 25: A technical school for collective farms is opened in Meißen.

1954

January 22–23: The 17th Meeting to the SED Central Committee takes place. The transformation of the countryside is the central concern. The committee calls for the SED to take a more active role in the process.

November 12–14: The 21st Meeting of the SED Central Committee is held. The committee orders the strengthening of MTS and of the delegation of workers to the countryside.

December 11–12: The Third LPG Conference declares that for the first time, large peasants could join an LPG under specific circumstances.

1955

September 20: The Soviet Union grants the GDR "full sovereignty."

November: Politburo member Kurt Vieweg, an agrarian specialist, calls for an alternative agrarian program in which both collective and private farms would remain in existence. He is sharply criticized by "orthodox communists" for this plan and eventually imprisoned.

1956

March 24–30: At the Third Party Conference, the SED clearly orients its policy again toward collectivization and the implementation of socialist production methods.

June 16: An LPG Advisory Board is formed by the Council of Ministers to support the collectivization process.

July 17–29: The 28th Meeting of the SED Central Committee calls for "all resources" to be dedicated to the formation and support of LPG.

1957

January 30–February 1: The 30th Meeting of the SED Central Committee calls for the greater integration of peasants in the socialist system. It rejects all proposals to reform the agricultural system.

February 26–28: The Fifth LPG Conference also rejects reform proposals. It seeks the intensification of "socialist competition."

July 23–26: Critics within the SED are given long prison sentences for their positions. This decision begins a cleansing of the party and signals a clear rejection of any reform.

November 14–16: Leaders of the SED take part in a meeting in Moscow in which "revisionism" is declared the primary danger to the buildup of socialism.

1958

May 29: Rationing ends in the GDR for all foods.

July 10–16: The Fifth Party Meeting of the SED declares that the GDR will surpass the living standard of the FRG within three years. The agricultural sector of the GDR is not prepared to reach this goal.

1959

February 20–22: The Sixth LPG Conference calls for the full completion of collectivization.

June 3: For the first time, a "Law on Agricultural Collectives" is passed.

September 16–17: The DBD declares all of its members should join an LPG.

October 1: Following the Soviet model, the GDR passes a Seven-year Plan. The previous Five-year Plan remains unfulfilled.

November 14: Khrushchev invites Ulbricht to a meeting in February 1960 in Moscow in order to discuss agricultural issues.

December 12: The county Eilenburg, near Leipzig, declares it has been completely collectivized. It becomes a model for forced collectivization in other counties in the following months of the so-called "Socialists Spring."

1960

January 13: Ulbricht writes to Khrushchev that decisive efforts are being made in the agricultural transformation of the GDR.

January 14–15: The SED district leadership in Rostock declares that it will be the first to fully collectivize its district. It is likely that this announcement was coordinated with SED leaders in Berlin but no records of this exchange have been found.

January 15: The SED Central Committee writes a letter to local party officials throughout the GDR calling for the early completion of the plan and the strengthening of agricultural collectives.

January 27: At a meeting with SED District leaders, Ulbricht pronounces collectivization the most important goal. Forced collectivization follows throughout the GDR.

February 2–3: In a meeting in Moscow, Ulbricht declares that because of the completion of collectivization the GDR will no longer suffer from food shortages.

February/March: Forced collectivization occurs throughout the GDR. About 100,000 people participate in brigades that seek to convince peasants to join an LPG. Violence is often employed. In the end, about 400,000 peasants and about half the agricultural land in the GDR are forced into collectives.

March 5: Rostock declares that all private farms have been collectivized. It is the first district to do so.

April 14: Karl-Marx-Stadt becomes the last district to declare that all private farms have been collectivized. It would take years before the LPG follows SED directives and farms their land collectively.

April 15: The GDR government officially ends collectivization. Although it is still theoretically possible for an LPG member to leave the collective farm, land will no longer be returned to anyone who left.

There is a dramatic increase in the number of peasants who emigrate to the FRG during and after forced collectivization. This number reaches crisis levels and is only stopped by the construction of the Berlin Wall starting on August 13, 1961. This wave of emigration as well as the results of forced collectivization leads to a dramatic fall in agricultural yields. In the years after the building of the wall, the economy is finally stabilized. In the countryside, collectives begin to farm according to their directives.

Collectivization in Czechoslovakia in Comparative Perspective, 1949–1960

JAN RYCHLÍK

This paper provides a general survey of the agricultural collectivization process in Czechoslovakia, which took place between 1949 and 1960, in an attempt to explain the political and economical motivation behind this campaign and its long-term consequences. To fully understand the process of collectivization, it is necessary to explain property relations and economic realities in Czechoslovak agriculture prior to the beginning of the collectivization campaign, and to also explore the evolution of the socialist collectivized agriculture after 1960. Chronologically, the long and arduous process of collectivization can be divided into five main stages: the preparatory period (1945–1949); the first wave of collectivization (1949–1953); the interim period (1953–1956); the second wave of collectivization (1956–1960); and the third and final wave of collectivization (1975–1980).

The collectivization of agriculture as a research topic emerged in the Czechoslovak historiography during the communist period (1948–1989). The few works published during this period describe the socialist collectivization in accord with the official political ideology, as a necessary step in the process of socialization of the economy. Despite its ideological bias, some of these works provided useful statistical data.[1] However, the first

[1] See Cambel, ed., *Kapitoly z dejín*; Cambel, Skrip, and Vanko, *Roľnícka politika*; Cambel, Kratochvílová, and Pešek, *Dějiny kolektivizace*; and Cambel, *Päťdesiate roky*. This latter book had generally been written before November 1989. Cambel expanded some parts and probably omitted others. The details and statistics included in this book are particularly useful.

unbiased works on collectivization were published only after the fall of the communist regime in Czechoslovakia in November 1989. The Central State Archive (now the National Archive) also published several editions of relevant documents.[2] Some young historians focused on the second, successful phase of collectivization in Czechoslovakia.[3] There was also an attempt to explain the process collectivization within the framework of the development of Czechoslovak economy.[4]

Recent Czech and Slovak historiographies[5] agree, by and large, that collectivization was not part of the initial program of the Czechoslovak Communist Party (*Komunistická strana Československa*—KSČ) after World War II and that the idea of collectivization was adopted in 1948 at Soviet instigation and in accordance with the Soviet model. Czech and Slovak scholars have also proven that the development of agriculture in Czechoslovakia was similar to that of other so-called socialist countries, except for Poland and Yugoslavia, where collectivization was abandoned. Yet, the recent historiographical wave has not yet provided an explanation to an intriguing issue: Why was socialist agriculture in Czechoslovakia an economic failure in the first wave but generally successful in the second wave? This present paper tries to explain this evolution.

Czechoslovak Agriculture after World War II

When analyzing agriculture in Czechoslovakia, one must differentiate between the Czech lands in the west and Slovakia in the east. In the Czech lands, agriculture was highly developed. The farmers were relatively well educated: illiteracy was almost completely eliminated by the end of the 19th century, and many peasants attended special winter schools where they learned about modern methods of farming. The level of mechanization was also at a relatively high level. Most of the Czech farms were small or middle-sized, comprising up to 20 hectares of land. This was partly a result of the post-World War I land reform. According to the law 215/1919 of April 16, 1919,[6] in principle all arable land over 150 hectares, or all

[2] See Juněcová ed. *Zemědělské družstevnictví*, 1995; Pšeničková, ed. *Zemědělské družstevnictví.*

[3] See Březina, ed. *Závěrečná fáze.*

[4] See Průcha, ed. *Hospodářské a sociální dějiny*, 351-72.

[5] See Jech, *Soumrak*; Jech, *Kolektivizace a vyhánění*; Hlavová, *Kulak.*

[6] *Sbírka zákonů a nařízení státu československého.*

lands over 250 hectares belonging to one and the same owner, were expropriated with compensation. In practice, several exceptions to this principle were made which enabled former owners to claim back part of the land exceeding 150 hectares. The distribution of the expropriated land was regulated by the law 81/1920 of January 30, 1920. Every Czechoslovak citizen working in agriculture could apply for land, being obliged, in exchange, to pay the price of the plot received, usually within twenty years. The new owner could obtain a state loan covering up to 90% of the price or a loan from a private bank with a fixed interest rate which was guaranteed by the state. In the process of expropriation, however, there were many exceptions and part of the land was temporally returned to the former owners under special conditions: the owner could not sell or transfer the returned land without the consent of the State Land Office. There was also a special tax imposed on the temporarily returned land.

In Slovakia the situation was different. While in the fertile south there were prosperous big and middle-sized farms, the farms in the other parts of the country were typically small, unprofitable, and did not intend to produce for the market. The agricultural reform in Slovakia was far less effective because local peasants did not have enough money to pay even the basic amount required by the State Land Office. The number of middle-sized farms increased, however. At the same time, there remained many large estates of sometimes more than 500 hectares.

The results of the first land reform in Czechoslovakia (1919–1938) were the following: the state took over a total of 1,312,721 hectares of arable land, out of which only 868,601 hectares were redistributed. The reform aimed at destroying the big estates, which often were in the possession of German and Hungarian nobles, and establishing strong peasant family farms. The project was never completed and at the end of the interwar period there were still many large estates over 150 hectares. Overall, however, the number of middle-sized farms increased, while the significance of large estates decreased. In 1930, there were 178,923 farms between 2–5 hectares, 93,263 farms between 5–10 hectares, 61,300 farms between 10–20 hectares, and 21,259 farms between 20–50 hectares.[7]

In October 1938, as a result of the international conference in Munich (September 30, 1938), Germany occupied the border territory of Czechoslovakia (the Sudetenland). One month later, on November 2 1938, the southern parts of Slovakia and Ruthenia were annexed to Hungary. On

[7] Rychlík, "Pozemková reforma."

March 14, 1939 Slovakia proclaimed its independence under the protection of Germany and the following day the rest of Bohemia and Moravia was attached to Germany as a so-called protectorate. German and Hungarian authorities partly revised the Czechoslovak land reform, claiming that it harmed the German and Hungarian owners. In Slovakia, a new land reform was carried out: law 46/1940 of February 22, 1940[8] allowed the State Land Office to take over for compensation and to redistribute the land of Jewish owners and foreigners. The Slovak land reform, which was in fact part of the anti-Jewish measures (the Aryanization of Jewish property) did not have much effect and did not increase the number of middle-sized farms because there were relatively few Jewish land owners: only 22,441 hectares was distributed to Slovak small-holders. Also, part of the confiscated land (16,311 hectares) was not redistributed to private owners but remained in the possession of the state.[9] On the whole, the social structure of the Czech and Slovak countryside remained generally unchanged during the war: small and middle farms were dominant but larger estates also existed.

From a "Specific Path to Socialism" to Collectivization

In post-World War II Czechoslovakia, communism was not imposed by Soviet military forces like in most of other East European countries. The Czechoslovak Communist Party was a legal party with a strong representation in the Czechoslovak National Assembly (Parliament) even before 1939. Communists seized power in Czechoslovakia by a *coup d'etat* carried out on February 25, 1948. Before that, however, in the "transition period" of 1945–48 known as the Third Czechoslovak Republic, the country was governed by a coalition of six parties.[10] In July 1945, the government decided to implement a new land reform targeting mostly the property

[8] *Slovenský zákonník.*

[9] Rychlík, "Príprava a priebeh."

[10] These parties were: the Communist Party of Czechoslovakia (KSČ), The Communist Party of Slovakia (KSS), Slovak Democratic Party (DS), Czechoslovak Social Democratic Party (ČSSD), Czechoslovak People's Party (ČSL), and Czechoslovak National Socialist Party (ČSNS). The Communist Party of Slovakia was formally independent but in fact followed orders from Prague, so in reality communists were represented in government twice. The "Czechoslovak parties" were in fact the Czechs Parties. They had no organizations in Slovakia and the DS had no organizations in the Czech lands.

of German and Hungarian owners. In the Czech lands, according to the decree 12/1945 of June 21, 1945,[11] the estates of the German and Hungarian owners and the land of "the traitors of the nation" were confiscated without compensation. During the years 1945–46, these lands were redistributed to Czech and Slovak smallholders and landless people for a symbolic price. In Slovakia, a similar decree (4/1945) was adopted on February 27, 1945 and later twice amended.[12] The new land reform was a serious political issue. The strongest political party of the interwar period, the Republican Party of Farmers and Smallholders (*Republikánská strana zemědělského a malorolnického lidu*), also known as the "Agrarian Party," was banned in 1945. All political parties tried to capture the votes of the former agrarians. As the land reform in the Czech lands was completely controlled by the communists, they proved to be the most successful in the countryside during the last free elections of 1946. In Slovakia, on the other hand, the regional Ministry of Agriculture was in the hand of the Democratic Party (DP), which in fact was the old Slovak wing of the Agrarian Party, acting under a new name. Because the land reform in Slovakia was carried out by the Democratic Party, the party won the elections in that region.

On July 11, 1947, law 142/1947 was passed "on the revision of the first, 1919 land reform." The new law affected the land that was returned to the original owners under various exceptions. In extraordinary situations, it was possible to "revise" the land exceeding 50 hectares belonging to one owner. In the Czech lands at the beginning of 1948—in the early days of the communist coup—there was already little demand for land among the peasants because most of their needs had already been met. Due to the expulsion of more than two million of Germans from the Sudetenland in 1946–47, the availability of land in mountainous border areas highly exceeded demand. According to materials from the Czechoslovak Ministry of Agriculture, stored now in the National Archives in Prague (*Národní archív České republiky*—NA ČR), in the Czech borderlands there was a lot abandoned land that nobody wanted.

Before, and even shortly after the coup, the communists talked about a "specific road to socialism" that was to be based on the private sector in agriculture, similar to the model later introduced in Poland and Yugoslavia. It is difficult to say to what extent they were committed to this policy. On February 29, 1948, four days after the coup, Klement Gottwald, the

[11] *Sbírka zákonů Československé republiky.*

[12] *Sbierka nariadení*, decrees 4/1945, 104/1945, and 64/1946.

communist Prime Minister and Chairman of the Communist Party emphasized in a speech delivered to the delegates of the members of Congress of Peasant Committees:[13] "If somebody comes to your village and says that there will be *kolkhozes* in Czechoslovakia, be aware that he is an agent provocateur of the reaction. Expel such persons from the village! There will be no collectivization in Czechoslovakia, we shall go on our own way."[14]

Regardless of the fact that there was now enough available land, the communists announced that they would provide a new redistribution of arable land according to the principle that "the land belongs to those who are working on it." On March 21, 1948, law 44/1948 was passed by the National Assembly (the Parliament). The law amended law 142/1947: according to the new version even land under 50 hectares was affected by the "revision" if the owner did not work on it. Simultaneously, law 46/1948 "On the New Land reform" was passed. The law allowed private property of land only up to 50 hectares. Only a person physically working in agriculture could possess land. All the estates over 50 hectares, including the property of the Catholic Church, were expropriated. The new land reform was a typical example of an action carried out strictly for political needs. The actual goal lying behind the reform was to destroy larger estates once and for all.[15] This is what really happened—and caused big problems in food supplies. Still, according the original idea of the Czechoslovak communists, family farms were to remain the basic form of agriculture in Czechoslovakia, while the state farms were to be the second segment of agriculture. This was the result of the fact that the state now possessed much land. In fact, state farms had been set up in Czechoslovakia already in the interwar period as a result of the confiscation of the former imperial estates by the state after the fall of the Austrian-Hungarian monarchy in 1918. But after the war, the size of the state land increased due to the lands confiscated from the members of the German minority, as not all the lands (especially in the mountainous border areas) were dis-

[13] Peasant Committees were the local committees that carried out land reform in the villages.

[14] *Rudé právo*, March 1, 1948.

[15] On land reform after World War II see Rychlík: "Pozemková reforma v českých;" Rychlík, "Pozemková reforma." Between 1945 and May 1, 1951, 4,143,149 hectares were expropriated, of which were 2,135,798 hectares of agricultural land and the rest mainly forests. The smallholders obtained 59.8% of the agricultural land, and the rest remained in the possession of the state. See Rychlík, "Pozemková reforma v českých," 18–19.

tributed to the peasants. In fact, some of the allocated lands were soon abandoned because the peasants coming from the lowlands could not adapt to the severe conditions in the mountains. Also, the lands confiscated according to the new law 46/1948 remained mostly in the possession of the state. Thus, the Czech and Slovak communists probably did not consider it necessary to establish "collective farms" in 1948 but planned to extend gradually the number of state farms. Klement Gottwald, who was elected president in May 1948,[16] talked in the summer of 1948 about the need for "continuation of the pre-February policy toward peasants," e. g. the support of small and middle-sized family farms and state farms.[17]

In the former Archive of the Central Committee of the Communist Party, not a single internal party document dated before 1948 addresses the issue of collectivization,[18] nor do any such documents exist from the period shortly after the February coup. Arguably, the Czechoslovak communists really believed that collectivization would not take place. The criticism from the Yugoslav leadership and the split between Tito and Stalin probably forced the Czechoslovak communists to abandon their dreams of going their own way in socialism. It should be noted, that—regardless of the fact that collectivization in Yugoslavia was taking place at the same time—the resolution of Cominform on June 26, 1948 criticized Yugoslav communists for the "wrong" policy in agriculture, more specifically for not fighting against capitalist elements in the village. In reaction to the resolution, Rudolf Slánský, the Secretary General of the KSČ (later sentenced to death in a show trial that took place on November 20–27, 1952), announced in a meeting of the Presidium of the Central Committee on June 28, 1948, that "it will be necessary to revise our policy [in agriculture]."[19]

Collectivization was first discussed in a closed session of the Presidium of the Central Committee of the Communist Party on October 14, 1948.[20] Shortly before that, Gottwald visited Stalin in Crimea; it is possible that the change in his attitude toward collectivization was connected with this visit.[21] On the plenary session of the Central Committee

[16] Gottwald replaced Edvard Beneš on June 14, 1948, who was president from 1935–38 and 1940–48 (during World War II in exile) and had resigned on June 2.

[17] Gottwald, *Spisy*, 112–13.

[18] This archive is now part of The National Archive of the Czech Republic (NA ČR) in Prague where it forms a special fund (f. ÚV KSČ).

[19] NA ČR, sv. 2, a. j. 125, PÚV KSČ 28. 6. 1948.

[20] NA ČR, sv. 4, a. j. 138, PÚV KSČ 14. 10. 1948.

[21] Cambel, *Päťdesiate roky*, 52.

on November 17–18, 1948, Gottwald publicly announced the program of a "gradual liquidation of capitalist elements in agriculture."[22] The decision of the Central Committee on November 18, 1948 was the basis for the law on "Single Collective Cooperatives" of February 23, 1949, which we will discuss further. Finally, on May 25, 1949, Gottwald approved the program of collectivization in his speech to the delegates of the Ninth Congress of the Communist Party. Regardless of what had been said only a year ago, he proclaimed that "the village must be won for socialism" and that the Communist Party policy had to be based on the support of small and middle-sized peasants, while big farmers, representing the capitalist element, were to be isolated.[23] The Congress adopted the resolution that "it is necessary to realize the Law on JZD [...] Cooperatives can provide the peasants with the advantages of progressive farming, earlier a privilege of big farmers."[24] By the decision of the Ninth Communist Party Congress (May 25–29, 1949), the way to collectivization was definitively paved. Originally, Gottwald expected that the collectivization program would take place gradually after the necessary preparatory measures. Massive collectivization was to start only after two or three years, e.g. in 1951 or 1952. Gottwald probably even believed that peasants would join cooperatives voluntarily. However, all these expectations proved to be illusions.

Land collectivization in Czechoslovakia displayed certain specific features. First, there was a lot of state land available, due to the preceding land reforms. Thus, the state had enough land to form collective farms but there was not enough labor to till the soil. Second, there had been a tradition of cooperatives in Czechoslovakia, mainly in the Czech lands. The leadership of the cooperatives was usually in the hands of rich peasants (*sedláci*), but middle peasants and smallholders benefited from cooperatives as well. Cooperatives existed not only in the Czech lands, but also in Slovakia. The cooperatives owned mills, stores, dairies, a chain of shops selling food, bakeries, and slaughterhouses. There were also cooperative banks. After the first agricultural reform of 1919, some cooperative farms had been established by the agricultural workers and smallholders—

[22] NA ČR, sv. 5–6, a. j. 20, ÚV KSČ 17–18. 11. 1948. See also: Hlavová: *Kulak*, 22.

[23] NA ČR, f. ÚV KSČ, 00/09, sv. 2, a. j. 55. Protokol IX. řádného sjezdu Komunistické strany Československa 25–29. 5. 1949 [Shorthand Report of the Ninth Congress of the Communist Party of Czechoslovakia]. The shorthand was published in limited copies for internal use only.

[24] NA ČR, f. ÚV KSČ, 00/09, sv. 2, a. j. 55, 102.

members of the Social Democratic Party. In Slovakia, the tradition of the so-called cooperatives for collective pasture (*pasienkové komposesoráty*) was strong. In the spring, these cooperatives took from the peasants their livestock (mainly sheep), which were then pastured together until the autumn when the animals were returned to the owners. Milk and mainly the popular special sheep cheese (*bryndza*) were distributed to the peasants according to the number of their sheep.

The communists in Czechoslovakia tried to use these traditions in the process of collectivization.[25] The law 69/1949 of February 23, 1949 and decree 75 of the Ministry of Agriculture from March 17, 1949, which constituted the legal base of collectivization, demanded all existing cooperatives in agriculture be transformed into the "Single Collective Cooperatives" (*Jednotné zemědělské družstvo* or JZD in Czech and *Jednotné roľnícke družstvo* or JRD in Slovak).

THE CZECHOSLOVAK VERSION OF THE COLLECTIVE FARM: THE JZD/JRD

Technically, the JZDs were operated by representatives elected by all members at general assemblies. In practice, however, the general assembly was simply provided with a single list of candidates for the ruling board (*představenstvo*), including the candidate for the chairman (*předseda*). This list was only formally approved by the general assembly. The leadership was basically imposed "from above" by the district state and political authorities, for example by the district secretary of the Communist Party. The ruling board in fact was not responsible to the general assembly or to the members of the JZD, but to the state and party authorities. They also decided what the JZD was to produce, for what price, and defined marketing of the production of these enterprises. All JZDs, as well as state farms, which composed almost one third of the arable land, were subordinated to the state economic plans.

From the legal point of view, there was one main difference between the Soviet *kolkhoz* and Czechoslovak JZD. In the USSR all land belonged to the state. It meant that legally the land used by *kolkhoz* was state land. In Czechoslovakia, on the other hand, the private property of the land was not abolished. The land remained formally in the possession of the peas-

[25] Cambel, *Päťdesiate roky*, 86.

ants who "voluntarily" joined a JZD. Likewise, in principle, buildings, newly purchased machinery etc. did not belong to the state but to the JZD as a corporative body. On the other hand, JZDs were also very often using state property for free. There were several ways this could happen. When a peasant member of a JZD died, the land was inherited by his or her children as any other property. The heirs could not sell or lease the land. If they were not members of the JZD, they were forced to allow the cooperative to use it for free. The heirs often rejected the land which was of no use to them. If there was no other heir, it fell to the state, which usually gave the land for use to the nearest JZD. Also, land that was confiscated from peasants as part of their punishment for various "crimes" (usually for not delivering the compulsory deliveries or not paying taxes) fell to the state and subsequently handed over to JZD.

Small private property in agriculture was tolerated in Czechoslovakia. Unlike Soviet collectivization, peasants were not forced to join a JZD with all their animals. Small animals like hens, chickens, and geese remained private property. By contrast, large animals like cows, horses, and sheep were usually taken by the JZD, but peasants were sometimes allowed to keep one cow or goat for their own needs. The peasants also kept the so-called *záhumenek* (house plots) of up to half a hectare for their personal needs. Until June 1, 1953, the system of rationing that had been introduced during World War II persisted in Czechoslovakia. All people working in agriculture were excluded from the rationing food system, however. House plots were thus important for their own survival.

Four types of JZDs were established and classified according to the level of collective work and common property. In the first type, peasants were only working in fields that belonged to particular owners, whereas the animals, machinery and also the harvest remained the property of particular farmers. In the ("highest") type IV JZD, all peasants worked together, their machinery and animals were in common cooperative stables and the harvest belonged to the JZD. In 1951, the communist leadership increased pressure to transform the "lower" variants of JZD into the "highest" type. Technically, compensations were to be paid both for animals and for machinery[26] taken from the peasants. However, in most of the cases no compensation was granted. The machinery often was confiscated as a punishment

[26] The legal ground for the confiscation of machinery was article 6 of law 27/1949 of February 2, 1949 "On mechanization of agriculture," which empowered the state authorities to take over the private agricultural machines for compensation if the owner did not use the machines effectively.

for "sabotage." If compensation was paid, it did not cover in any way the real price of the respective animal or machine. In addition, the government introduced a currency reform on May 30, 1953. It was based on the "class principle." For the "capitalist elements" and all citizens who were excluded from the rationing system, the exchange rate was 50 old crowns for 1 new crown. This in fact meant annulment of all compensations—and also savings. As a result of the currency reform, the farmers were left with no money. They were unable to buy the stock necessary for further farming.

THE SOCIAL COMPOSITION OF THE RULING BOARDS OF JZDS AND ITS CONSEQUENCES

While in the "old" cooperatives the leadership was in the hands of big farmers (*sedláci*), the ruling boards of the JZD were chosen according to the "class principle." The Communist Party after the Ninth Congress ordered their members in the countryside to help to establish the JZDs. Indeed, the first JZDs were often composed simply of communists living in a particular village. Thus, the members of the boards were mainly smallholders or even landless people. They had no experience with farming big plots. Some of them had not brought any property into the JZD that they had joined. They did not have any authority among middle and big peasants.

Middle-sized peasants resisted collectivization most strongly. When their opposition proved futile, they preferred to leave the villages. The attitudes and conduct of most big farmers were more ambivalent. Most of them offered some resistance, but it should be noted that in some places the well-to-do farmers tried to establish a JZD and rule it. This practice was soon prohibited by the communist party apparatus, which considered it a violation of the "class policy in the village."

Since 1951, the term *kulak* was used in party propaganda. This word did not exist in Czech or in the Slovak language. Initially, party propaganda used the word *vesnický boháč* (*dedinský boháč*, in Slovak), i.e. the "rich villagers." The word *kulak* was borrowed from Russian, and it became obvious that the campaign against *kulaks* was modeled on Soviet patterns. It should be noted that no law or at least Communist Party directive ever explained who was a *kulak*, nor was the communist leadership able to offer any definition.[27] The decision of who was a *kulak* generally

[27] Jech, *Soumrak*, 61.

depended on the decision of local party and state authorities, which usually took into consideration local conditions. This meant that in some poorer areas, where there were few rich farmers, even middle peasants with no machinery and less than 15 hectares were considered to be *kulaks*.[28]

The question of whether rich peasants should be allowed to join the JZDs was never definitively settled during the whole period of collectivization. The law 69/1949 on JZDs had no provisions against *kulaks*. They could be members of the JZD/JRD and initially they were even welcomed as good specialists. As it has been already said above, in some villages, rich peasants even decided to form JZDs in order to protect their property, and to use poorer members of the cooperative as working power. But by 1950 and mainly in 1951 the situation changed. From 1952, the *kulaks* were not allowed to join collective farms, even if they wished to do so. This was a result of the overriding ideological approach: The *kulaks* were considered to be "class enemies," who were suspected of joining collective farms in order to subvert and destroy them. *Kulaks* who were already members of JZDs were in most cases expelled. Only very rarely were exceptions granted and a *kulak* could remain, but only as a rank and file member. Under no circumstances could *kulaks* become members of the ruling boards. As *kulaks* could not be members of JZDs and, simultaneously, it was practically impossible for them to continue private farming, their only option was to work as agricultural workers in the state farms or to leave the village and find a job elsewhere. But, as we shall see further, in the late fifties and sixties former *kulaks* were accepted as JZDs members again.

Generally speaking, the forced collectivization of 1951–53 led to economic disruption and chaos in agriculture. In 1953, JZD and state farms cultivated about 45 percent[29] of all arable land, but almost all of them were economically unstable. JZDs and state farms were unable to produce enough agricultural products for the market. The situation worsened especially in the production of grain. As a result, rationing of bread, abolished in 1948, had to be re-introduced in 1951.

---

[28] Hlavová, *Kulak*, 18–19.

[29] In the Czech lands, JZDs worked 30.5 percent of all land (32.5 percent of arable land), in Slovakia 31.2 percent (33.9 percent of arable land). The state farms worked on 11 percent of all Czech lands and 6.5 percent in Slovakia. The average size of a collective farm was 261 hectares in the Czech lands, and 554 hectares in Slovakia. The average size of the state farm was much bigger: 3,280 hectares in the Czech Lands and 4,333 hectares in Slovakia. See Průcha, "Vývoj zemědělské," 41.

STAGES AND METHODS OF COLLECTIVIZATION

The intensive propaganda for collective farming started in 1949 and culminated in 1951–53. From 1951, this propaganda was accompanied by a very aggressive campaign against the *kulaks*. In 1954–55 the campaign was temporally stopped, but it was renewed in 1956 and mainly in 1957–58.

From the start, the process of collectivization in Czechoslovakia was based on economic pressure, which was after 1950 combined with open violence. Theoretically, the farmers were to be "convinced" by the propaganda campaign of the Communist Party that it was in their interest to join the collective farms. However, the party leadership themselves had serious doubts about the success of such a campaign if it was not supported by force. From the beginning, the communists did not want to allow private peasants to continue their activity and maintain their property. Thus, they had no program for private farmers. The communists knew very well that the middle and rich farmers would not join collective farms voluntarily.

Most commonly, economic pressure was applied, like mandatory deliveries of agricultural products. This system had been introduced by the German occupation authorities during the Second World War according to a pattern which had existed already during the First World War. Each farmer was allowed to keep only a part of his production that was needed for families and farming. All products that exceeded this limit were bought by the state at fixed prices. The system of compulsory deliveries was not only kept in force after 1945 but pursued even more strongly. In addition, the system of forced deliveries was extended to Slovakia, where many products had not been rationed during the war.[30] Introduction of compulsory deliveries in Slovakia meant that state purchases of agricultural products from peasants increased twice and in some commodities even three times.[31] Decrees 6/1949 and 7/1949 of January 25, 1949 based the compulsory deliveries on "voluntarily contracts" between the local administration (*Místní národní výbor*—Local National Committee) and particular peasants, and were renewed every year. However, if the peasant refused to sign such a contract, the level of his compulsory deliveries was established

[30] As mentioned above, Slovakia was in 1939–45 a formally independent state under German protection. Up to August 1944, it was not occupied by the German army. The standard of living was much higher there than in Bohemia-Moravia, which was occupied in 1939.

[31] Cambel, *Päťdesiate roky*, 73.

arbitrarily by the district administrative office (*Okresní národní výbor*—District National Committee). The level of deliveries depended on the quality and quantity of land in the possession of the particular peasant. Theoretically, peasants were allowed to sell production exceeding the contracted quotas on the free market. However, in practice the farmers rarely had such exceeding production. By 1950, the prescribed deliveries for rich and middle farmers were increased for the following year. The quotas could hardly been met. Sometimes the peasants were asked to produce agricultural commodities that they never produced before or could hardly be produced in the particular region due to the local climatic conditions.[32] The process of deliveries was directed mainly against the rich farmers, who were stigmatized as *kulaks* in 1951 for the first time. Very often *kulaks* were former smallholders who obtained their property from the land reform in 1945 or 1947. The official propaganda claimed that there was not enough food in stores in the cities because *kulaks* did not meet the quotas and sold their products on the black markets. Increased quotas were extended to all arable land in 1952–53. The Central Committee of the Communist Party on its session of December 16–17, 1953, decided that the quotas were to be increased by 10 percent for *kulaks*. In 1955, when pressure slightly eased, this increase was repealed.[33]

The forced deliveries of agricultural products often meant the economic collapse and liquidation of farms. The income obtained from the state usually did not cover even the costs of production. In the first years, farmers tried to increase their production, which would allow them to sell part of it (after fulfilling the quotas) in the free market. Soon they realized, however, that the increase in their production usually meant also an increase of compulsory deliveries in the contract for the next year and that there was no way out. It should be emphasized that the rationing system existed in Czechoslovakia until June 1953. Commodities necessary for agriculture, like machinery and petrol, were not freely available and private farmers usually were not able to buy them except on the black market. Deliveries also forced farmers to focus more on producing food for their families. All people working in agriculture were also excluded from the system of food rationing. While before the war, farmers usually did not bake bread at home but bought it at the shop, they had to do so in the late 1940s and early 1950s.

---

32 Cambel, *Päťdesiate roky*, 72.
33 NA ČR, f. ÚV KSČ, 02, 1953.

On July 12, 1950 the new penal code (law 86/1950) and also the new administrative penal code (law 88/1950) were approved, both becoming active from August 1, 1950. According to Article 88 of the administrative penal code, "non-fulfillment of contracts" (*nedodávky*) was now a criminal offence punishable by heavy fines (100,000 crowns) or up to three months in jail. Fines or jail sentences were carried out not by courts but by administrative authorities. Big farmers were also often accused of intentional non-fulfillment, which the courts classified as attempts of sabotage according to Articles 84 and 85 of the penal code. In such cases, the courts usually sentenced the peasant to jail for several months or even years and ordered the confiscation of his property. Another method of "liquidating" the *kulaks* was excessive taxation. The land and income taxes of 1948 (*pozemková daň, daň z příjmu obyvatelstva*) were based on the size and quality of the landed property, and they were progressive.[34] On December 11, 1952, new tax laws introduced even steeper progressive taxation.[35] In addition, according to Article 7 of the Land Tax Law, the district administrative authorities were allowed to increase the land tax for particular taxpayers (e.g. peasants) by 30 percent. More land and more production now meant not only higher compulsory deliveries but also higher land and income taxes. In many cases, the farmers and peasants were not able to pay the taxes that had been imposed on them. Subsequently, they were tried for tax evasion, sentenced, and their property was confiscated.

Confiscation enhanced the size of the state land, which could not be passed to new holders. In Czechoslovakia, the principle of the so-called "forced lease" had been introduced by law in 1947. The "forced lease" meant that the local administration could order a farmer to lease the abandoned or confiscated land and to work on it.[36] It should be noted that the size of abandoned lands increased in the early 1950s, because many farmers had left their villages and moved to the cities in order to work in factories. Despite the existing laws, the remaining farmers refused to cultivate the confiscated or abandoned land, for they feared increases in compulsory deliveries and taxes. Even members of state farms and JZDs refused to take up the land for the same reason. The final result of the

[34] Law 48/1948 of March 20, 1948 (Income Tax Law) and law 49/1948 of March 21, 1948 (Land Tax Law).

[35] Law 77/1952 and 78/1952.

[36] Article 3 of law 55/1947 of April 1, 1947.

"class war in the Czechoslovak village" was a growth of abandoned land and a decline in agricultural production.

The compulsory deliveries were a nightmare to the peasants. Some farmers tried to escape the increased quotas by dividing their land amongst members of their family. This also enabled farmers to be transferred from the category of *kulak* to the category of the "middle peasant." The authorities soon stopped this practice, however. They used the law of 1947, which did not allow the division of farms comprising less than 20 hectares and another law of 1948 according to which the transfer of land to another person required the consent of the district administrative office.[37] By the end of 1951, it had become clear that there were only two options for the peasants in a long-term perspective: either to join "voluntarily" a JZD or leave the village and find a job in industry. As industrialization, mainly in Slovakia, needed many workers, this perspective was attractive to farmers and peasants.

Economic pressure was combined with non-economic repression. In October 1951, the leadership of the Communist Party decided to expel the "most dangerous *kulaks*" from their villages and to settle them with their families in uninhabited border areas.[38] On October 22, 1951, a secret meeting of the ministers of national security (Ladislav Kopřiva), justice (Štefan Reis), and the interior (Václav Nosek) took place. Their decision was incorporated into the Secret Order of the Minister of National Security that was published only for internal use on the same day.[39] The families of the farmers whose property had been confiscated or who had been sentenced for non-fulfillment ("sabotage") were to be "resettled" in other parts of the country. Between September 1, 1951 and August 23, 1953, when the action was stopped, a total of 1,421 peasant families were "resettled," mainly to the border areas where they were forced to work as agricultural workers in state farms.[40]

---

[37] Law 139/1947 of July 3, 1947 and law 45/1948 of March 21, 1948.

[38] Jech, *Soumrak*, 119.

[39] For the text of the secret order, see Jech, *Soumrak*, apendix, doc. 3. (unpaginated); see also Hlavová, *Kulak*, doc. 5, 102–5.

[40] Archiv Národního zemědělského muzea [Archives of the National Museum of Agriculture], Prague, collection of materials on collectivization; these materials were displayed at the exhibition on collectivization in 2008.

The Policy of the "New Course" in Collectivization

On March 5, 1953, Stalin died. Gottwald died just two weeks later, on March 14, 1953. After Gottwald's death, Antonín Zápotocký, former trade union boss and prime minister in 1948–53, was elected the new president of the republic by the National Assembly. Unlike Gottwald, Zápotocký was not simultaneously the chairman of the KSČ. This post was abolished, as was the post of secretary general, which had been vacant since the arrest of Rudolf Slánský in 1951. Very soon, however, Antonín Novotný, the first secretary of the Central Committee, seized decisive power. In 1953–56 there was a political struggle between Zápotocký and Novotný. While President Zápotocký represented the "liberal" wing in the Presidium of the Central Committee, Novotný was a conservative hardliner. This internal conflict had a profound impact on the policy of collectivization.

In the spring of 1953, the pressure on the peasants to join JZDs gradually relaxed. On August 1, 1953, Antonín Zápotocký announced in his speech at the construction of the dam in Klíčava near Zbečno that the process of collectivization would be carried out voluntarily.[41] Zápotocký repeated his idea about the "new course" in his speech in Stěžery near Hradec Králové on September 27, 1953. Here, Zápotocký said that JZDs should be organized only in regions where they could be economically effective.[42] Those who did not wish to stay in collective farms were allowed to leave them. Before that announcement, Zápotocký delivered a similar speech in the Central Committee of the Communist Party on September 4, where a discussion about the further development of collectivization had taken place.[43] His idea was not accepted by all members, however, and it is not by chance that it was this Central Committee which recommended to increase compulsory deliveries for the *kulaks* by 10 percent in December of the same year. On March 14, 1954, Zápotocký again repeated his idea in a speech at the National Theatre on the occasion of the first anniversary of Gottwald's death.[44] Zápotocký did not oppose collectivization as such, but he proposed a more moderate course. His concept was soon criticized by Antonín Novotný, who demanded a con-

---

[41] *Pravda*, August 3, 1953.
[42] *Pravda*, September 9, 1953.
[43] NA ČR, f. ÚV KSČ, 01, zasedání ÚV KSČ 4. 9. 1953.
[44] *Pravda*, March 15, 1954.

tinuation of collectivization and also to return to the methods of 1950–53, if necessary.[45] Novotný's attitude soon started to prevail. The Tenth Party Congress, which took place on June 11–15, 1954, confirmed Novotný as the first secretary and did not make any changes in agricultural policy.[46]

The relaxation and Zápotocký's "new course" led to a stagnation of collectivization. Between 1953 and 1956, almost one half of the existing JZDs were dissolved, because most of the peasants left them and returned to private farming. From 1953 to 1955, the size of agricultural land in the possession of JZDs dropped by about 10 percent in the Czech lands and 4 percent in Slovakia.[47] The party leadership was alarmed. After some discussions in the highest echelons of the party, Novotný's followers began to turn the tide, and the plenary session of the Central Committee of the Communist Party on June 29–30, 1955 decided to renew collectivization.[48] On June 30 the Political Bureau of the Communist Party decided on new measures against private farmers, which were legally formalized on September 21 in the form of governmental decree 50/1955. The decree allowed local authorities to confiscate the plots of *kulaks* if they did not work on it effectively. According to the governmental decree 50/1955, every owner of arable or other agricultural lands who was not using his property for production was obliged to "let it for free and unpaid usage" to the local collective farms or state farms. In fact, the decree could be used against any peasant who did not want to join "voluntarily" the collective farm. However, the political events in 1956 in the USSR and other East European countries and, mainly, the impact of the uprisings in Poland and Hungary, forced the party leadership to postpone the next wave of collectivization until the following year.[49]

The Second Wave of Collectivization

During the year 1956 the methods of collectivization were criticized in Czechoslovakia, which like most East European countries was shaken by political crises.[50] But the political decision of the Central Committee of June 30, 1955 remained in force and the plan for further collectiviza-

45 Barnovský, "Prestávka medzi," 238–239.
46 *Protokol 10*, ÚV KSČ, 1954. See also: NA ČR, f. ÚV KSČ, 00/10, sv. 17, a. j. 257.
47 Průcha, "Vývoj zemědělské," 41.
48 NA ČR, f. ÚV KSČ, f. ÚV KSČ, 01, ÚV KSČ 29.–30. 6. 1955.
49 Barnovský, "Prestávka medzi," 239.
50 Rychlík, "Wahrnemung des 20."

tion was not revoked. The Party Conference in Prague on June 11–15, 1956 confirmed the course against *kulaks* in the countryside.[51] The action started next year. On July 31, 1957 governmental decree 49/1957 reintroduced the possibility of raising the compulsory deliveries for *kulaks* by 10 percent and this was enforced in contracts for the year 1958. Zápotocký, an opponent of forcible collectivization, was politically marginalized. On November 13, 1957, Zápotocký died and on November 19, Antonín Novotný was elected as the new president. Being now both president and the Communist Party first secretary, he concentrated a great deal of power in his hands. The collectivization process started again, accelerating in 1958 and reaching its peak one year later. According to Antonín Novotný's plan, full collectivization was to be achieved by 1960.[52]

Again, the propaganda campaign on the advantages of collective farms started. This time, however, more sophisticated economic rather than administrative pressure was applied and the ideological zeal was partly abandoned. The party leadership now moved its propaganda from smallholders and landless laborers to middle peasants and sometimes even to large farmers. They were offered the jobs of chairmen in the restored collective farms, and the authorities also promised them support from the state. Thus, the committees of the collective farms were now mainly composed of middle and sometimes even bigger farmers who commanded considerable authority among the peasants.

There were again some show trials against *kulaks* in the late 1950s but the system had fundamentally changed. Most of the big farms had collapsed or were about to go bankrupt. The JZDs now started to work mainly on lands that no longer belonged to its members. New members in fact did not differ much from hired workers because they had no property in the cooperative farm and felt no responsibility for it. The new law 55/1956 on social insurance, adopted on November 30, 1956, also had a significant impact. The law was extended to private and collective farmers, but the conditions for the latter—if they were members of a collective farm of type III or IV—were better. From January 1, 1957, members of the JZDs were included into the state health and social insurance system, which entitled them to medical care and pensions. Members of the family were also included in the system.[53] A farm which was indebted could be

51 *Celostátní konference*, 1956, 17.
52 Blažek and Kubálek, *Kolektivizace venkova*, 130.
53 Decree of government 56/1956 of December 18, 1956.

transferred to a JZD or a state farm in exchange for a state pension. Many *kulaks* took the chance to get rid of their farms for compensation.[54]

The Eleventh Congress of the Communist Party, which took place on June 18–21, 1958 in Prague, issued a resolution demanding the completion of collectivization.[55] After the congress, the combined economic pressure increased. Peasants realized that further resistance would be both difficult and useless. The process of collectivization was completed by and large in 1960, when JZDs held already 68.5 percent of overall agricultural land in the Czech territory and 65.8 percent in Slovakia. State farms comprised 19 percent of the land in Czech territory and 9.9 percent in Slovakia.[56] In the most productive regions with fertile land, the percentage of JZDs was even higher, for instance in the region of Olomouc. In the Haná region (with high soil quality), 81 percent of land was farmed by JZDs or state farms in 1959–60.[57] This was certainly a big achievement for the regime.

Communist rulers officially claimed that middle farmers had realized the advantages of collective farming. However, the reality was different. After the suppression of the Hungarian uprising in 1956, Czechoslovak farmers evaluated the political situation and came to the conclusion that there was no chance for any policy change, thus realizing that they would not be able to maintain their farms. Peasants were therefore joining JZDs massively, but with no enthusiasm. There were still some regional differences. While in areas with fertile land, the collectivization of land was practically absolute, in the highlands and mountain areas some small private farms still survived. The government did not organize JZDs there because the mountainous lands were not as suitable for modern agriculture.

In the 1960s, Czech and Slovak villages changed almost completely. Due to considerable state subsidies, the collectives became quite productive. The young generation accepted the transformation of the countryside, which for them was the reality of the day. Thus, there were no attempts on the part of former peasants to take back the land during the "Prague Spring" reform movement in 1968. The "Program of Action" adopted on April 5, 1968 by the new reform leadership of the Commu-

54 Blažek and Kubálek, *Kolektivizace venkova*, 134.

55 *Protokol XI*, 1958. See also: NA ČR, f. ÚV KSČ, 00/11, sv. 12, a. j. 257.

56 Průcha, "Vývoj zemědělské," 41

57 Blažek and Kubálek, *Kolektivizace venkova*, 134.

nist Party led by Alexander Dubček,[58] which was later denounced by the Soviet leadership as "revisionist," did not include any reference to private farming. The program merely called for more independence for JZDs in the economic life of the countryside.[59] Indeed, the leaders of cooperative farms tried to escape the rigid state and party control and produce commodities that could be sold in the market. After the Soviet invasion of Czechoslovakia on August 21, 1968, these reform attempts were stopped and, one year later, denounced.

The "Third Stage" of Collectivization

On April 17, 1969, the conservative Gustáv Husák replaced Alexander Dubček as first secretary of the KSČ. Husák consolidates power further six years later by becoming president of the republic. During Husák's rule, collective farms were gradually transformed into big agricultural complexes, sometimes covering several villages. These big new JZDs often employed hundreds of people including various specialists. More agricultural land was now needed. On November 13, 1975, a new law on agricultural land was passed.[60] According to this law, the "productive socialist organizations," i.e. state farms and collective farms, could use the land of private owners for their own production if owners were not willing or not able to cultivate it. In fact, the law renewed governmental decree 50/1955 and extended it. Even now, the land remained in possession of the original owner, but if he or she died, the heirs often came under pressure to pass their land to the state authorities. The law was occasionally used against some remaining private peasants still sporadically existing in mountain areas where cooperative farms were not established. Thus, some historians speak of a third wave of collectivization. The aim of the new law was not to force the remaining private peasants to join cooperative farms but rather to obtain more land where necessary. Most of the remaining private land in the mountains was now taken over by the state and attached either to big cooperative farms or, more often, to state farms where they served usually as pastures. The law did not have much impact on the number of

[58] Alexander Dubček succeeded Antonín Novotný on January 5, 1968. On March 22, 1968, Novotný resigned also as the president of the republic. He was replaced by general Ludvík Svoboda on March 30.

[59] *Rok šedesátý osmý*, 1969, 128.

[60] Law 123/1975.

JZDs. The size of the land in possession of JZDs and state farms increased only by about 2 percent between 1970 and 1980.[61]

In the 1970s and 80s, collective farmers secured a high standard of living due to state subsidies, accommodating themselves within a social stratum that was loyal to the rulers. On the other hand, many managers of these cooperatives started to run their own businesses that were usually disconnected from agriculture. This shift was due to new legislation on agricultural cooperatives[62] that ordered collective farms to merge into even bigger complexes and also allowed them to produce non-agricultural goods (like furniture, machinery, and in the late 1980s, even personal computers).[63] Opportunities in non-agricultural production were further extended in the 1980s. This non-agricultural production was originally considered something exceptional but the new law on agricultural cooperatives—90/1988 of June 15, 1988—considerably extended this possibility. In fact, some chairmen of these cooperatives abandoned agricultural production altogether and specialized on more lucrative economic sectors. These new cooperatives were the first capitalist enterprises within the planned socialist economy.

On the other hand, the average size of JZDs rapidly increased—in the Czech Republic[64] the average went from 610 hectares in 1970 to 2,459 hectares in 1980. In 1989, the average size increased further to 2,574. In Slovakia, the average JZD collective had 832 hectares in 1970, 2,690 in 1980, and 2,720 hectares in 1989.[65]

In the second half of November 1989, massive demonstrations and a subsequent general strike brought the communist regime in Czechoslovakia to the end. It should be noted that the "collective farmers" did not take part in the protests and remained passive. While farmers constituted some of the strongest resistance to the communist regime in the early years, they remained loyal to the regime during its final dissolution. The reasons were not ideological or political, but rather economic. The agricultural policy of communist Czechoslovakia was based on the doctrine of food self-sufficiency, which required heavy subsidies to agricultural production. This system brought a relatively high standard of living for the

[61] Průcha, "Vývoj zemědělské výroby," 41.

[62] Law 122/1975.

[63] This was the famous case of *Agrokombinát Slušovice* in Central Moravia.

[64] The Czech Republic was established on January 1, 1969, together with the Slovak Republic as a result of the federalization of Czechoslovakia adopted on October 27, 1968 (143/1968).

[65] Průcha, "Vývoj zemědělské výroby," 41.

members of many JZDs and even employees of the state farms. In addition, there was very little enthusiasm to start private farming immediately after 1989: the original farmers, who so strongly resisted collectivization in the 1950s, were in 1989 either dead or too old, and most of the younger members of JZDs had no personal experience with private farming. They were aware of the risks and prefer to stay in JZDs.

CONCLUSION: LOCAL TRADITION IN CZECHOSLOVAKIA AND THE SOVIET MODEL

Forced collectivization was an integral part of postwar political developments in all communist countries in Central and Eastern Europe, but it took different forms in each of these countries. In the USSR, collectivization in the late 1920s and early 1930s was based on the Russian tradition of *mir*, i.e. the collective usage of land and pastures by villagers, and on the *obščina*, the collective responsibility of peasants in some aspects of public life (mainly taxation), which had existed until the Stolypin agrarian reform in 1907.[66] The Bolshevik revolution nationalized all land in the country according to the decree on land in 1917.[67] In this decree, land was then divided and distributed to peasants who were not owners but only users of the land in legal terms. Thus, from a legal point of view, collectivization in the USSR did not entail the confiscation of land because it already belonged to the state.[68]

In Central and Eastern Europe, collectivization took place in late 1940s and early 1950s. Although the Soviet model was employed from the very beginning, the main aims of the communist policy in agriculture were kept secret for tactical reasons. The primary reasons for collectivization in the USSR as well as Eastern Europe, in general, were political, not economic. The communists wanted to liquidate the private sector in agriculture. There was a similar pattern in all communist countries after

[66] This decree from November 9, 1906 was the legal basis of Russian Prime Minister Pyotr Stolypin's reforms from 1906–11, confirmed later by the law of June 14, 1910. "Dějiny světa VII," 1965, 458–59.

[67] "Dějiny Všesvazové komunistické strany," 1951, 213–14.

[68] The agricultural land, which before November 8 (October 26) 1917 belonged to smallholders, was not formally confiscated, but the owner could not dispose with it and had to work on it. The formal difference between this land and all other lands confiscated under the decree disappeared and all land was considered to be state property.

World War II. The first step was the implementation of agricultural land reforms, which very often were not necessary. The peasants obtained land but this usually did not help their economic situation, due to the lack of machinery, existing compulsory deliveries, and low prices, as well as other factors. Such a situation led to economic failure, and that was later used as an argument for collectivization. There are some similarities in all Eastern European countries: after the first phase of collectivization in the early 50s, there was a stalemate in collectivization in 1953–56, after Stalin's death. After 1957, a second wave of collectivization was initiated and that was, generally speaking, successful. Yet, collectivization was interrupted definitively in Yugoslavia in 1953, and the second wave did not take place in Poland after 1956. The reasons for these peculiarities can be found in the decision-making of local political elites who, at that particular moment, considered collectivization politically harmful for their rule and legitimacy.

The Soviet model of collectivization was also introduced in Czechoslovakia in 1949. There were, however, some differences mainly due to Czech (rather than Slovak) legal and agricultural traditions. In the Czech lands there existed a long tradition of cooperative movements, while in Slovakia there were relatively fewer cooperatives. The communists could take advantage of this tradition. However, it soon became evident that the JZDs had very little in common with real cooperatives, as self-government was totally absent, and JZDs could not make decisions on what they would produce nor on what they could sell in the open market.

While there were many similarities between JZDs and the Soviet *kolkhoz*, there were also three basic differences. First, the peasants who joined—either voluntarily or forcibly—the so-called socialist "cooperatives" were able to partly retain their private property over land. This included their houses, gardens, small animals, and all personal belongings. While machinery had to be "sold" or even "voluntarily donated" to the JZD, the land formally remained in the possession of "cooperative farmers." Second, the JZDs had their own property, which belonged to the cooperative as a corporate body, not to the state. Third, unlike in the USSR, the ideological zeal was abandoned during the second wave of collectivization (1957–60) in Czechoslovakia. Smallholders and agricultural workers were generally no longer allowed to monopolize the leadership of Czech and Slovak cooperative farms. On the contrary, the Czechoslovak regime integrated middle farmers into these cooperatives, and sometimes even *kulaks*. This policy helps explain why some JZDs were run quite efficiently.

While in the 50s there was a strong resistance to the JZDs, in the 60s and 70s the situation changed, as cooperative farmers did not oppose the communist regime. This, of course, did not mean that cooperative farmers would now believe in the ideals of a future "classless communist society," which was—at least in theory—the final goal of KSČ. Rather, cooperative farmers simply lost interest in private farming because the heavy state subsidies brought them a relatively high standard of living without much worries. The chairman of cooperative farms, in fact the professional managers, were primarily interested in the relaxation of state control over production and prices. Their demand was generally reached in 1988. In the last years of the communist dictatorship, some cooperatives became capitalist enterprises that produced for the market. The chairmen of cooperative farms became *de facto* new big landowners using modern methods for cultivating land that still belonged, formally, to the cooperative. On the other hand, production was mainly still being bought by the state, now for relatively good prices, which meant a big advantage for cooperative managers.

It is very difficult to evaluate the complex and arduous process of collectivization in Czechoslovakia in its entirety; it is certain, however, that, on the long run, the process brought about radical changes for the Czech and Slovak agriculture. Its impact and legacy are still visible in the Czech and Slovak Republics. After the fall of communism, relatively few farmers reclaimed their land and engaged in private farming. Most of the former cooperative farms have been transformed into agricultural business companies. It is not incidental that after the fall of communism former managers of cooperative farms purchased abandoned lands and failed state agricultural projects to become entrepreneurs in a new society.

BIBLIOGRAPHY

ARCHIVAL SOURCES

Archiv Národního zemědělského muzea [Archives of the National Museum of Agriculture], Prague.

Národní archiv České republiky [The National Archive of the Czech Republic] (NA ČR) in Prague.[69]

[69] The Archive of the Central Committee of the Communist Party of Czechoslovakia was taken over by the National Archive of the Czech Republic in 1990 and now exists as a special fund (f. ÚV KSČ). According to the classification of the Archive of the Central Committee of KSČ, the numbers of the fund are as follows: 01 stands for meetings of the Central Committe, and 02/1 for meeting of the Presidium of the Central Committee.

NA ČR, f. ÚV KSČ, 02/1, sv [svazek]. [file] 2, a. j. [archivni jednotka], [archive unit]125. PÚV KSČ 28. 6. 1948. [Meeting of the Presidium of the Central Committee of the Communist Party of Czechoslovakia on 28 June 1948]
NA ČR, f. ÚV KSČ, sv. 4, a. j. 138, PÚV KSČ 14. 10. 1948.
NA ČR, f. ÚV KSČ, zasedání ÚV KSČ 17.–18. 11. 1948. [Meeting of the Central Committee of the Communist Party of Czechoslovakia]
NA ČR, f. ÚV KSČ, 02, zasedání ÚV KSČ 4. 9. 1953.

ONLINE SOURCES

Sbírka zákonů [Collection of laws], http://aplikace.mvcr.cz/sbirka-zakonu
Sbírka zákonů a nařízení státu československého [Collection of Laws and Decrees of the Czechoslovak state], laws 215/1919, 81/1920.
Sbierka nariadení Slovenskej národnej rady [Collection of Decrees of the Slovak National Council], decrees No 4/1945, No 104/1945 and No 64/1946.
Slovenský zákoník [Slovak Code], law 46/1940.

PRESS

Pravda (Bratislava), August 3, 1953; September 29, 1953; March 15, 1954.

EDITIONS OF DOCUMENTS

Juněcová, Jiřina and Pšeničková, Jana eds. *Zemědělské družstevnictví: kolektivizace zemědělství. Vznik JZD 1948–1949* [Agricultural cooperatives: The collectivization of agriculture. The Formation of JZDs 1948–1949]. Prague: Státní ústřední archiv, 1995.
Pšeničková, Jana, ed. *Zemědělské družstevnictví: kolektivizace zemědělství: vznik JZD—1950.* [Agricultural cooperatives: The collectivization of agriculture. The Formation of JZDs— 1950]. Prague: Státní ústřední archiv, 1998.
Pšeničková, Jana, ed. *Zemědělské družstevnictví: kolektivizace zemědělství: vznik JZD—1951.* [Agricultural cooperatives: The collectivization of agriculture. The Formation of JZDs—1951]. Prague: Státní ústřední archiv, 1999.
Pšeničková, Jana, ed. *Zemědělské družstevnictví: kolektivizace zemědělství: vznik JZD—1952.* [Agricultural cooperatives: The collectivization of agriculture. The Formation of JZDs—1952]. Prague: Státní ústřední archiv, 2000.
Pšeničková, Jana, ed. *Zemědělské družstevnictví: kolektivizace zemědělství: vznik JZD—1953.* [Agricultural cooperatives: The collectivization of agriculture. The Formation of JZDs—1953]. Prague: Státní ústřední archiv, 2002.
Pšeničková, Jana, ed. *Zemědělské družstevnictví: kolektivizace zemědělství: vznik JZD—1950.* [Agricultural cooperatives: The collectivization of agriculture. The Formation of JZDs—1954]. Prague: Národní archiv, 2008.

BOOKS AND ARTICLES

Barnovský, Michal. "Prestávka medzi prvou a druhou etapou kolektivizácie" [The break between the first and the second view of collectivization]. *Osudy zemědělského družstevnictví ve 20. století. SSM* 7 (2002): 235–42.
Blažek, Petr and Michal Kubálek, eds. *Kolektivizace venkova v Československu v letech 1948–1960 a středoevropské souvislosti* [The collectivization of the countryside in Czechoslovakia, 1948–1960 and its Central European connections]. Prague: Dokořán, 2008.

Březina, Vladimír and Pernes Jiří, eds. *Závěrečná fáze kolektivizace zemědělství v Československu 1957–1960* [The final phase of collectivization in Czechoslovakia, 1957–1960]. Brno: Stilus, 2009.
Cambel, Samuel, ed. *Kapitoly z dejín socialistického poľnohospodárstva v Československu* [Chapters from the history of socialist agriculture in Czechoslovakia]. Bratislava: Pravda, 1982.
———, Vasiľ Skrip, and Augustin Vanko. *Roľnícka politika v období výstavby socializmu na Slovensku* [Agricultural policy in the period of building socialism in Slovakia]. Bratislava: Pravda, 1986.
———, Želmíra Kratochvílová, and Jan Pešek. *Dějiny kolektivizace československého zemědělství.* [The history of collectivization of the Czechoslovak agriculture]. Prague: Svoboda, 1989.
———. *Päťdesiate roky na slovenskej dedine. Najťažšie roky kolektivizácie* [The fifties in the Slovak village. The most difficult years of collectivization]. Prešov: Universum, 2005.
"Celostátní konference Komunistické strany Československa" [The all-state conference of the Communist Party of Czechoslovakia], *Nová mysl* [New Thought], June 1956, 17.
Dějiny novověku [A history of modern times], vol. III. Prague: Státní pedagogické nakladatelství, 1973.
Dějiny světa [A History of the world], vol. VII. Prague: Nakladatelství politické literatury, 1965.
Dějiny Všesvazové komunistické strany (bolševiků) [A history of the All-Union Communist Party (bolsheviks)]. Prague: Svoboda, 1951.
Gottwald, Klement. *Spisy* [Works], vol. XV. Praha: SNPL, 1961.
Hlavová, Viera. *Kulak—triedny nepriateľ "Dedinský boháč" v kontexte kolektivizácie na Slovensku 1949–1960* [Kulak—a class enemy. "A Village Rich" in the context of collectivization in Slovakia, 1949–1960]. Bratislava: Veda, 2010.
Jech, Karel. *Soumrak selského stavu* [The sunset of the peasantry]. Prague: ÚSD AV ČR: 2001.
———. *Kolektivizace a vyhánění sedláků z půdy* [Collectivization and expulsions of farmers from their land]. Prague: Vyšehrad, 2008.
Protokol IX. řádného sjezdu Komunistické strany Československa 25–29. 5. 1949 [Shorthand Report of the Ninth Congress of the Communist Party of Czechoslovakia]. Prague: ÚV KSČ, 1949.
Průcha, Václav. "Vývoj zemědělské výroby v Československu v období socializace vesnice." *Osudy zemědělského družstevnictví ve 20. století. SSM* 7 (2002): 31–41.
———, ed. *Hospodářské a sociální dějiny Československa 1918–1992.* [Economic and social history of Czechoslovakia 1918–1992], vol. II. Brno: Doplněk, 2009.
Rok šedesátý osmý v usneseních a dokumentech ÚV KSČ [Year 1968 in the resolutions and documents of the Central Committee of KSČ]. Prague: Svoboda, 1969.
Rychlík, Jan. "Pozemková reforma v Československu v letech 1919–1938" [The Land Reform in Czechoslovakia 1919–1938]. *Vědecké práce Zemědělského muzea* (VPZM) [Scientific works of the Museum of Agriculture] 27 (1987–1988): 127–48.
———. "Príprava a priebeh pozemkovej reformy na Slovensku v rokoch 1938–1945" [The preparation and realization of the agricultural reform in Slovakia, 1938–1945]. *Agrikultúra* 22 (1989): 143–65.

———. "Pozemková reforma na Slovensku v rokoch 1945–1950" [Land reform in Slovakia, 1945–1950]. *Historický časopis* [Historical Journal] 41, no. 4 (1993): 394–413.

———. "Pozemková reforma v českých zemích v letech 1945–1948" [The land reform in the Czech lands, 1945–1948] in *Zemědělství na rozcestí 1945–1948* [Agriculture at the crossroads, 1945–1948]. *Studie Slováckého muzea* [Studies of the Slovácké museum—SSM] 3 (1998): 7–19.

———. "Wahrnemung des 20. Parteitages der KPdSU in den Ländern Mittel- und-Osteuropas. Besonderheiten und Ähnlichkeiten" in Joanna Szymoniczek and Eugeniusz Cezary Król, eds. *Das Jahr 1956 in Polen und seine Resonanz in Europa*. Warsaw: Institut Szudiów politicznych PAN, 2010, 141–47.

THE COLLECTIVIZATION OF AGRICULTURE IN CZECHOSLOVAKIA: CHRONOLOGY OF EVENTS, 1948–1975

Jan Rychlík

1948

February 25: The *coup d'état* in Prague. Political power is taken over by the Communist Party.

February 29: Klement Gottwald, communist prime minister and chairman of the Communist Party (president of Czechoslovakia from June 14, 1948) addresses the Congress of Peasant Committees held in Prague. He promises the peasants that the Soviet-style *kolkhoz* will not be established in Czechoslovakia.

March 21: The Constitutional National Assembly passed the law on New Agricultural Reform (46/1948). Private ownership of the land was limited to 50 hectares providing that the owner is really working on the land.

1949

February 23: The National Assembly passed law 69/1949 on United Agricultural Cooperatives (*Jednotné zemědělské družstvo* or JZD in Czech; *Jednotné rolnícke družstvo* or JRD in Slovak). JZDs were the Czechoslovak version of Soviet *kolkhoz*. Unlike in the Soviet *kolkhoz*, the members of a JZD remained *de jure* the owners of the land.

May 25–29: Ninth Congress of the Czechoslovak Communist Party (*Komunistická strana Československa*—KSČ) adopts a new program that includes the collectivization of agriculture. The program also mentions the liquidation of "capitalist elements" in the village (the Russian word *kulak*, unknown in Czech and Slovak language, started to be used only in 1950).

Increased compulsory deliveries are imposed upon *kulaks*.

Attempts at voluntary collectivization bring no success.

1950

The beginning of forcible collectivization. Increased compulsory deliveries are imposed on farmers who do not want to join JZDs and the *kulaks*. Management of the JZDs is left mainly smallholders or landless agricultural workers with no experience and no authority, leading to bad economical results.

1951

October 22: Secret joint decision of the minister of security, minister of interior and minister of justice on *kulaks* is passed (known as the "decision of three ministers"). It allows the administrative authorities to deport *kulaks* and their families from the place of their residence to the border areas or any other place where they are forced to live. The deportation action was known as "Action K," e.g. *kulak*.

1952 to beginning of 1953

High peak of forcible collectivization. Property is confiscated from the *kulaks* who could not fulfill the compulsory deliveries. Many of them are imprisoned for "sabotage" and their families deported from the native villages. *Kulaks* were simultaneously forbidden to join JZD.

In 1953 JZDs (collective farms) and state farms farmed on 45 percent of all arable land in Czechoslovakia.

1953

March 14: President Klement Gottwald dies nine days after the death of Stalin. Gottwald is succeeded by Antonín Zápotocký, who decides to adopt a milder course on collectivization. On the other hand, Antonín Novotný, the new first secretary of the Communist Party, is an adherent of further forced collectivization.

April–May: Because the administrative pressure eases, peasants start to leave the JZDs.

July: Deportations of *kulaks* from the villages is temporally suspended.

August 1: In a public speech in Klíčany, President Antonín Zápotocký announces that nobody will be forced to join a JZD in the future and those who were forced to join can leave. This "new policy" was confirmed by another Zápotocký speech in Stěžery near Hradec Králové on 23 September. Both speeches speed the departure of peasants from JZDs.

September 4: Discussion on collectivization in the Central Committee of the Communist Party between the adherents of a "milder" and "harder" course.

1954

January: The "decision of three ministers" on *kulaks* was repealed and "Action K" formally finished. The *kulaks* and their families are allowed to return to their homes.

March 14: In a speech delivered in the National Theatre in Prague on occasion of the first anniversary of Gottwald's death, President Zápotocký confirms the policy on collectivization. His speech is denounced in the Political Secretariat of the Central Committee of KSČ by Antonín Novotný. In April, Moscow also denounces Zápotocký's attempts.

1955

June 29–30: The Central Committee of the Communist Party decides to start the second phase of collectivization.

1956

New discussions about collectivization. Due to the internal political tension and also due to developments in Poland and Hungary, the beginning of the second wave of collectivization is postponed. The number of peasants in JZDs decreases by about 40 percent compared to 1953.

1957

November: President Antonín Zápotocký dies on November 13 and is succeeded by Antonín Novotný (19 November), who was already the first secretary of the Communist Party. This opens the way to the second wave of collectivization.

1958

June 18–21: Eleventh Congress of the Czechoslovak Communist Party takes place in Prague and proclaims the policy of the final liquidation of *kulaks* as a "capitalist class in the village." The course for new collectivization is officially approved.

1958–1960

The second wave of collectivization in Czechoslovakia. Compared with the first wave, the leadership of JZDs is now in the hands of middle farmers or even former *kulaks*. The second wave is relatively successful.

1961

By the beginning of 1961, the process of collectivization is completed. Only several hundred private peasants remain, mainly in isolated areas in the mountains.

1968

The reform movement in Czechoslovakia has practically no impact on agriculture. JZDs remain stable during the period and no attempts by the farmers to leave the JZDs are observed.

1975

November 19: The Federal Assembly (the Parliament) approves the law on agricultural cooperatives (No 122/1975) and on the use of agricultural land (No 123/1975). The JZDs are gradually transformed into big enterprises. The law 123/1975 allows the JZDs or the state farms to use the land of all owners. This law is sometimes called the "third wave of collectivization."

The Forced Collectivization of Agriculture in Hungary, 1948–1961[1]

JÓZSEF Ö. KOVÁCS

Despite the Hungarian Communist Party's repressive anti-peasant policies and the suppression of the 1956 Revolution, more than three quarters of the Hungarian peasantry refused to join the collective system until the forced collectivization campaign from the end of 1958 until early 1961.[2] Although this fact alone is important—as it involved more than 1.2 million peasants—this essay looks more closely at the social and structural factors that shaped the experience of collectivization in Hungary. Specifically, it addresses the following questions: How traumatic was the social experience of collectivization? What structural changes did it generate with respect to proprietorship, agricultural management, and peasant lifestyles?[3] How did collectivization (re)shape society as a whole? What forms of individual and collective behavior developed during the collectivization campaign?

---

1 This paper is part of OTKA (Hungarian Scientific Research Fund) research program T49424.

2 In 1958 statistical sources reported about 1,580,000 agrarian workers in the private sector. In 1962, only 111,000 of these workers remained. Orbán, *Két agrárforradalom*, 218. During the same period, the number of workers in collectives increased from 169,000 to 1,200,000. Valuch, *Magyarország*, 199. The proportion of new members without property was 53.3 percent; therefore their loss in land value is difficult to estimate. Pető and Szakács, *A hazai gazdaság*, 439.

3 Weber, *Wirtschaft*, 1; Suter and Hettling, "Struktur und Ereignis," 2001, 24–25; Alexander, "Toward a Theory;" Grenier, "A 'hosszú," 1–2, 306–12.

In approaching these questions, some methodological issues must be addressed in order to establish on what criteria written historical sources should be evaluated, as there are several forms of historical observation, each with a specific viewpoint and methodology. According to French historian Marc Bloch, "the variety of historical evidence is nearly infinite. Everything that man says or writes, everything that he makes, everything he touches can and ought to teach us about him."[4] This study gives priority to the experiences and memories of contemporaries, which are important sources of historical observation because dictatorships attempted to suppress those memories by imposing forgetfulness.[5]

In Soviet regimes, the massive expropriation of private properties took place by acts of terror. This procedure was justified by the argument that these properties were a basic obstacle to industrialization and the seizure of political power. The agents of dictatorship considered such an approach a precondition of controlled property acquisition and its partial distribution. Party leaders also regarded it as an instrument for minimizing peasant independence, both material and psychological. In manner and in timing, the collectivization campaign impacted all aspects of economic, social, and cultural life. At the same time, it is important to consider forced collectivization as one element of communist socialization. The term "socialization" refers to an interactive social structure in which people were necessarily involved by way of state control, propaganda, political mobilization and personal interests. This particular social situation was given the official designation of an "issue of social importance."

Both in methodology and subject matter, this study is primarily a political social history of collectivization from the bottom up. In the context of the Soviet system, the first sustained period of collectivization in Hungary lasted from 1948 to 1953. Stalin's death in 1953 ushered in a brief period of liberalization, but party leaders initiated a new collectivization campaign just prior to the 1956 Revolution. Following an intermediate period from 1957 to 1958, party leaders undertook the final phase, which lasted from 1958 to 1961. The first section of this chapter reviews the main characteristics of the Communist Party's agrarian policies, man-

[4] Bloch, *Historian's*, 66.

[5] My study is based on written historical documents that at the time were generally classified as secret. I use documents in the archival collections of the national party, ministries, and court system, as well as of each county-level archive of these organizations. I was unable to utilize information based on approximately 40 interviews due to the divergent narrative contents and space limitations.

agement, and officially designated collectives. The second part of the study presents the agrarian situation following the Revolution of 1956, the concluding campaign of collectivization, and issues relating to the social effects of state terror, as well as reactions to it and its consequences.

THE PROPAGATED IMAGE OF THE FUTURE AND THE ROLE OF AGRARIAN POLICY

Important changes in Hungarian population and social structure occurred after the Second World War. The total population of the country increased from 9 million at the end of the war to 9.96 million by 1960. By the 1980s, there had occurred a dramatic decrease in rural population,[6] as a result of German and Slovak migration following the war,[7] and anti-peasant agrarian policies in the 1960s that drew an estimated 90,000 to 100,000 people annually from villages to cities in search of industrial work. In that decade alone, the population of rural communities fell from 5.7 million to 5.3 million. Despite this trend, more than half of the Hungarian population still lived in rural communities. In 1949, nearly 52 percent of total employment was in the agrarian sector. This figure dropped to 38 percent in 1960, to 27 percent in 1970, to 20 percent in 1980, and finally to 8.5 percent in 1995. Moreover, one million wage earners, 20 to 25 percent of gainfully employed workers, left the agricultural sector between 1950 and 1970. Among the most detrimental change was the steep decline of able-bodied workers due to an aging population. In 1949, the proportion of those over 60 was 10.7 percent, and in 1995, this was 19.4 percent. From 1959 on, as a result of industrial decentralization—which dispersed industrial production in numerous Hungarian urban areas—there was some measure of equalization, but the number of those commuting to cities was 636,000 in 1960 and this number had increased to 977,000 by 1970. One of the most visible elements of intervention in rural life was the policy directed against the outlying farm system.

After the Second World War, as a result of the radically changed property structure, agrarian policy became the centerpiece of the government's

[6] Valuch, *Magyarország*, 33–34, 199.
[7] The Germans were relocated and the Slovaks were part of a Hungarian-Slovak population exchange.

broader campaign to transform the value systems of village life.[8] From a historical perspective, the issue of landed property at that time can be interpreted as a paradox.[9] While a significant segment of the peasantry wished to receive land at the time of the 1945 land reform, property ownership and the inequities it generated became the root of many problems several years later. The manner in which the Communist Party implemented the land distribution program also gave rise to new conflicts. According to the March 1945 land reform regulation, small or "dwarf" plots became the dominant form of farm land: small landowners accounted for 80 percent of the agrarian population, large peasant holdings amounted to 3 percent, and 17 percent were without property.[10] As a result of the land distribution, a significant segment of the peasantry developed a strong sense of land ownership, which in turn played a significant role in the resistance to forced collectivization that began in 1948, as peasants regarded the campaign as a betrayal of the party's earlier promise of independent land ownership.

The new structure also substantially transformed traditional village customs. Thus, the party's agrarian policies were met with strong resistance. The overall objective starting in 1948 was to promote the "building of socialism" in the country through collective agriculture with the expressed aim of raising living standards.[11] To that end, the proposed method was to change the prevailing property structure through state ownership. Those who stayed in villages were to participate in the management and production of large-scale farms. Taxation and compulsory requisition systems were introduced to fund the campaign and, ultimately, to discipline the peasantry. Although given their wartime experiences these aggressive new systems were by no means a novelty for peasants,[12] the party's agrarian policies were indeed traumatic, involving violence, terror, and attacks on the church and religion.[13] The Sovietization and occupation of villages began with the violent takeover of church schools by state authorities. Radical agrarian policies followed. This factor is generally overlooked in Hungarian historical research, although it is of great importance when considering the destruction of contemporary village networks and social structures.

---

8 Bauerkämper, ed., *Junkerland*; Schier, *Alltagsleben*, 281; Ö. Kovács, "Egy politikai."

9 Szakács, "A földreformtól;" Valuch, *Magyarország*, 189–91.

10 Valuch, *Magyarország*, 190.

11 Habuda et al., eds., *Magyar Dolgozók*, 15–32, 55–65.

12 Nagy Varga, "Alkalmazkodás."

13 Habuda et al., eds., *Magyar Dolgozók*, 164–66.

THE FIRST COLLECTIVE FARMING UNITS

After the Second World War, existing cooperatives based on voluntary accession were reorganized. About 100 cooperatives were terminated in the period before collectivization began in 1948. They did not receive state support. Subsequently, collective farms were organized on farms where a minimum of 171 hectares of pasture or 34.2 hectares of vineyard or orchard had been transferred to private owners under the 1945 land reform law. Local Communist Party members exerted decisive influence over these "village cooperatives," which had a significant social and economic impact, with 2,199,673 members in a total 7,753 cooperatives.[14] Internal official Communist Party documents from February 1948 onward contain references to the party's intention to transform small peasant properties, but no public notice of the collective farming campaign appeared before July 1948.[15] After this, party statements included the familiar anti-*kulak* Stalinist doctrines regarding the need to restrict the activities of "village exploiters." The first groups were organized in accordance with the government's August 1948 decree to confiscate leased properties exceeding 14.3 hectares. The first peasants to join collectives were those who had received no land at the time of the 1945 land reform—landless peasants, and returning prisoners of war. Properties were frequently taken over arbitrarily or via state-supported force.[16] As in the Soviet Union and Bulgaria, three types of organizations were established: Type I, in which land was cultivated communally but resources were distributed according to the size of land contributed; Type II, in which land was cultivated communally and resources were distributed according to the average crop yield irrespective of the size of land contributed; and Type III, in which resources were distributed according to the quantity of labor expended.

Collective management applied to all properties owned or leased by members, except for the 0.86 hectare household plots.[17] Initially, collectives required a minimum of ten members and a land area of 17.1 hect-

[14] Honvári, "Magyarország," 514–25.

[15] The First Secretary of the party, Mátyás Rákosi, spoke about the radical changes in his speech of August 20, 1948 in Kecskemét.

[16] Honvári, "Magyarország," 514–25; MOL, M-KS-276. f. 93. cs. 147.ő.e. March 25, 1949, 44.

[17] Honvári, "Magyarország," 524.

ares. At the end of 1949, 98.5 percent of members were poor farmers or landless peasants. At the end of 1950, 59 percent of the land area and 65 percent of personal property were supported by the state, in the form of grants, investment, or credit. On December 31, 1950, 2,185 collectives were registered, of which the vast majority were of the third type.[18]

The Communist Party achieved these results primarily by disintegrating individual properties through economic, administrative, and violent measures in order to force the peasantry to join collectives.[19] Between 1949 and 1953 the party undertook a particularly devastating campaign to integrate isolated homesteads, special villages, and estates into collectives, a program that involved 2,280 communities and 2.3 millions hectares of land.[20] In half of these communities the membership drive occurred twice, and in others even more frequently. Despite relentless state pressure, many peasants were unwilling to cede their land to collectives. As a consequence, the state faced several difficulties, including challenges over property rights, out migration from farming communities, labor shortages, and a significant decrease in productivity in comparison to private farming.[21] A shortage of technical expertise also exacerbated the agricultural crisis.[22] Effective management and intensive labor was instead limited to household plots, which provided an estimated 70 percent of peasant income in the 1950s and more than half of the peasant income in the 1960s.[23]

It is worth considering the quantitative results of the first waves of collectivization, from 1948 through July 1953, after which Imre Nagy's gov-

---

18 Honvári, "Magyarország," 525–26.

19 Nádasdi, *Tagosítások*. Although there was a multi-party system in the post-1945 years, a gradual Sovietization of Hungary took place in this period, leading to the establishment of full political domination by the Communist Party beginning in 1947–48. There was a one-party system in Hungary until the late 1980s.

20 MOL, M-KS-276. f. 85.cs. 9.ő.e. January 14, 1949, 2–5; *Magyar Közlöny* 17, January 22, 1949; *Magyar Közlöny* 33, February 12, 1949. At the same time, dummy collective farms were formed in western and eastern Hungary with equipment twice the amount required, in order to get the impression of being much more developed than they generally were. MOL, M-KS-276. f. 85. cs. 10.ő.e. January 21, 1949, 49–51.

21 Through so-called "offers of land," 857,142 ha of land were transferred to state ownership, the greater part of which the collectives were unable to cultivate. Honvári, "Magyarország," 529.

22 On December 31, 1953, 2,203 of 3,307 collective farm directors had no qualifications at all. Honvári, "Magyarország," 533; MOL, M-KS-276. f. 85. cs. 10.őe. 1949, 7–16.

23 Honvári, "Magyarország," 535; Valuch, "Községünkben."

ernment came into power, ushering in an era of relaxation during which the number of collectives decreased. Before 1959, collective units had reached their greatest numbers. In mid-1953, there were 5,224 registered collectives with 376,000 members, which accounted for 1.6 million hectares, or 26 percent of the total agricultural land. Private land accounted for 54 percent of the total land area, and the remaining balance was otherwise classified.[24]

According to a party report in 1949, there were several dimensions of social segregation in collectives: new landowners, traditional small and medium peasant owners, former prisoners of war, Hungarian settlers expelled from Slovakia, Communist Party loyalists, and relatives.[25] In some instances women or older people were excluded from collectives. According to party reports, members were recruited almost exclusively from poor, propertyless social groups who had joined the Communist Party.[26] Regarding the composition of collective members, one report stated: "The role of the party was predominant both in taking the initiative and in asserting its leadership role." In this respect the composition of the third type groups in western Hungary was typical, since they belonged to this group almost without exception.[27] The high proportion of new property owners in this region was striking. Membership up to the beginning of 1949 consisted primarily of family heads, and very few women or youths joined. Contrary to state propaganda, only 26 percent of collectives included women, and only an insignificant number were members.[28]

Regarding the social composition of the collective farms, the proportion of poor farmers was highest in eastern Hungary, while in northern Hungary, where the concentration of industrial areas was higher, new property owners comprised the majority of collective farmers. In Hungary, the officially calculated average size of land for a collective farm was less than 228 hectares.[29] At the time of the formation of collective farms in 1949, just 4.1 percent of ploughland belonged to collectives; by 1956, that figure had increased to 12.4 percent. However, many of these farms were unable to successfully cultivate lands, and by the spring of 1952, more

[24] Honvári, "Magyarország," 536.
[25] MOL, M-KS-276. f. 85. cs. 10.ő.e. 1949, 7–16.
[26] MOL, M-KS-276. f. 85. cs. 10.ő.e. 1949, 7–16.
[27] MOL, M-KS-276. f. 85. cs. 10.ő.e. 1949, 7–16. The classification of collectives can be found on page 217.
[28] MOL, M-KS-276. f. 85. cs. 10.ő.e. 1949, 7–16.
[29] Honvári, "Magyarország," 530.

than half of the officially designated collectives faced a critical shortage of workers.[30] This problem was officially attributed to the fact that women were not permitted to be members of collective farms. Moreover, many members had been recruited to industrial plants or state farms. In addition, members of the large collective farms in eastern Hungary reportedly refused to work.[31]

Hence, such changes in Hungary's demographic and social structure after the Second World War played an important role in the collectivization campaign. In 1952, a ministry official commented on the demographic changes:

> Labor force out-migration is partially due to the lack of perseverance of the membership. The membership largely consists of destitute wage earners or poor peasants, who had been seasonally employed agricultural wage labor prior to joining the collective farms. Last year, they were unable to earn sufficient income to subsist until the early summer when they were to receive the first significant new income. We receive daily requests for advance payments especially from eastern Hungary. There are also requests from western Hungary to a more limited extent as well as from cooperatives with declining income. The majority of cooperatives have not made advance cash payments since last September.

By 1955, an estimated 20 percent of collective farm members were between the ages of 50 and 60, and 15 percent were above 60—yet despite this, most collective farms failed to adequately support the elderly.[32] In many cases, the reason was not the inability to establish the Social Fund, a public fund to support older citizens that collective farms were required to finance, but was due to:

---

30 MOL, XIX-K-1-ah. 18.d. March 25, 1952, no. 2451/IV. For each member in 1953 the average size of land was 4.4 ha. In 1954, it was 4.7 ha. Statistically, membership consisted mostly of "workers of limited work capacity and advanced age," 20% were between the ages of 50 and 60, about 15% were over 60. The collective directors attempted to get rid of the old workers. MOL, XIX-K-1-ah. 19.d. January 26, 1955, no. 9. 7; MOL, XIX-K-1-ah. 19.d. January 22, 1955, no. 3. 2.

31 MOL, XIX-K-1-ah. 18.d. March 25, 1952. no. 2451/IV.

32 MOL, XIX-K-1-ah. 18.d. March 25, 1952. no. 2451/IV.

> the cooperative leaders' tendency to get rid of members unable to work by expulsion and other methods. This rigid policy increases the economic insecurity of the membership and represents a further threat since a significant part of the village population consists of workers above 50 whose children have departed, left properties to their elderly parents and got employed in industrial or management positions.[33]

Collective farms, as a rule, failed to function according to their operating procedures. According to a report, the leadership:

> frequently violated the most fundamental rights of the members, expelled them, usually without the approval of their council and without justification. There were thousands of legitimate grievances concerning the unjustified distribution of income and the illegal retention of earned dividends. The abuse of membership rights in numerous collective farms undermined the confidence of the members, weakened their economic livelihood and discouraged new members from joining. The uncertainty of livelihood and the use of violence as a management practice gave rise to the prevalence of individual production methods. The expansion of household plots to more than one acre in size, the increase in the number of farm animals or the practice of partial cultivation could be mentioned here. Especially in cooperatives near the capital, an increasing practice was industrial and commercial activity using the cover of cooperatives and state assistance to engage in prohibited activities and concealing financial mismanagement.[34]

The minister of agriculture, Ferenc Erdei, a renowned known sociologist of rural society in the prewar period, submitted a report to the Communist Party leadership on January 26, 1955 concerning the status of agricultural collectives.[35] The poor condition of agriculture was obvious, and Erdei did not try to conceal it, which was unusual at the time. According to the report, only a third of the collectives (about 1,000) showed positive results.[36] Financial reports revealed that out of the 3,255 collectives,

[33] MOL, XIX-K-1-ah. 19.d. January 22, 1955.
[34] MOL, XIX-K-1-ah. 19.d. January 22, 1955.
[35] MOL, XIX-K-1-ah. 19.d. January 26, 1955, no. 9.
[36] MOL, XIX-K-1-ah. 19.d. January 26, 1955, no. 9.

23 percent were debt free, and the remaining were in operating in debt.[37] According to Erdei's ministerial report, party and state authorities were said to tolerate wasteful management of common property. The term "waste" was in fact a self-regulatory mechanism in the socialist system. In fact, waste took place on a major scale, as indicated by the data on state-owned land areas and the growing fallow areas, for example.[38]

According to official reports by the agricultural ministry, production by collectives was low due to extensive soil utilization. Moreover, the collectives suffered from shortage of manpower and the progressive expansion of grain cultivation areas, which covered 38.3 percent of the ploughland in 1955.

Table 1. Plant Cultivation for Principal Acreage and Crop Yields in Hungary, 1955.

Crops	% of Plough-land Plan	% of Plough-land Actual	Total Boxcar Plan	Total Boxcar Actual	Average Production[2] (q) Plan	Average Production (q) Collective	Average Productio (q) Individua
Bread grain	39.1	38.3	42,079	30,511	8.3	6.1	6.5
Fodder	15.1	10.9	17,011	13,238	8.6	7.1	8.0
Corn	10.6	10.9	18,699	16,872	13.5	11.9	13.0
Potatoes	2.0	1.9	15,229	12,528	57.8	48.0	56.0
Rice	1.5	1.4	3,057	1,803	15.3	9.6	-
Sugar beets	2.2	2.3	37,202	31,890	127.9	107.0	110.0
Sunflower	2.6	2.3	2,333	1,974	6.9	6	6.8
Rough fodder	10.7	9.1	45,015	31,263	27.5	26.4	29.2

Source: MOL, XIX-K-1-ah. 19.d. January 26, 1955, no. 9, 6.

37 The average value of farm property for 0.57 ha (1 *hold*) was HUF 309, while in the case of "active" collectives it was HUF 1,036. This reveals the significant differences between the several collective units. MOL, XIX-K-1-ah. 19.d. January 26, 1955, no. 9.

38 MOL-KS-276. 93. cs. 407. ő.e. December 27, 1952.

Table 1 data are revealing if agrarian production processes are viewed as a symbiotic system. These adjusted figures demonstrate that unrealistic production plans could not be realized even with state assistance. One of the major consequences of extensive farming was the qualitative deterioration of farm animals due to fodder shortage and the resulting decline of production income.[39]

Table 2. Number, Area, and Membership of Collectives in Hungary, 1953–57.

Date	Size in 1,000 ha	Ploughland size 1,000 ha	Ploughland % of national area	Number	Membership
30 June 1953	1,337	1,107	20.3	3,768	300,370
31 Dec. 1953	932	739	13.6	3,307	193,328
31 Dec.1954	885	682	12.6	3,239	174,583
31 Dec.1955	1,133	868	16.2	3,759	253,355
30 Sept.1956	1,275	982	18.3	3,907	292,837
31 Dec. 1956	517	388	7.2	1,617	96,126
31 March 1957	669	498	9.3	2,536	119,739

Source: Honvári, "Magyarország," 551.

The data from Table 2 (distortions notwithstanding) indicate the changing levels of state intervention. It is evident that the Imre Nagy government, realizing in the second half of 1953 that forced collectivization and violence were no longer acceptable, introduced a new soft-line approach, which resulted in the decline in the number of collective farms and membership. In retrospect, 1952 seems to be the most repressive year of collectivization since 70 percent of collectives dissolved subsequently were formed in that year. 80 percent of those who left at this time had been small- and medium-sized landholders.[40] After the dismissal of Imre Nagy in March 1955, a new wave of violent collectivization started, which coincided with the largest decline in collective farms.

[39] Although the number of farm animals per 57 ha (100 *hold*) increased, the number of cows (3.7) was low compared to the private sector (11.5).

[40] Honvári, "Magyarország," 551.

Silent Battlefield and Segregation after 1956

The fact that the majority of the pre-1956 collectives disintegrated under the impact of the 1956 Revolution demonstrates the significance of peasant resistance. The majority of collectives did not disintegrate in the course of the revolution, but subsequently, during the post-revolutionary period in November and December. The process of withdrawals and dissolutions reached its peak on January 15, 1957 when, according to official records, only 37 percent of the 3,954 collectives still functioned.[41] The formation of cooperative ventures in 1957–58 were "voluntary," comprising just 5 to 10 percent of the agrarian sector, with the great majority of farming enterprises operating independently.[42] These statistical figures indicate the broad-scale failure of the Socialist project.

Post-1956 collective membership reflected the stratified economic positions and management skills of the agrarian population. This gave way to a new pattern of segregation within village society, which fuelled various internal conflicts. The group designated by the party as "agrarian proletarians" were the first to join the collectives.[43] Agrarian proletarian membership increased from 37.6 percent to 57.4 percent in the year period between March 1957 and March 1958. By contrast, the proportion of medium-sized peasants fell from 26.6 percent to 15.3 percent, and the number of small peasants from 35.8 percent to 27.3 percent over that same time period.[44] The two years after the 1956 Revolution can be regarded as the only period in postwar Hungarian social history when young age groups chose to remain and perform agricultural activities instead of leaving. Likewise, in the two years following the revolution there was a reversal of the out-migration trend among rural youth—a trend detectable in the composition of membership in collectives. At the same time, older persons were excluded from collective farms. In addition, party members formed an absolute minority of village society. According to available data, just 14.8 percent, of the 122,296 members of collectives

---

41 MOL, M-KS-288.f. 11.cs. 84.ő.e. May 29, 1957, 2. In the case of other types of agricultural units, which provided greater individual activity, the rate of disintegration was smaller. These figures show the main trend.

42 Valuch, *Magyarország*, 195; Varga, "Az agrárium," 304.

43 "Agrarian proletarians" can be defined as landless peasants.

44 MOL, M-KS-288.f. 11.cs. 84.ő.e. May 29, 1957, 2.

held party membership cards in 1957.[45] Party officials regarded agrarian proletarians and those of similar status as primary candidates for party support. Those with low economic resources and social standing considered party membership as an attractive solution.

In the spring of 1957, labor shortages became a critical problem on collective farms, with 425, or 16.5 percent, of 2,570 units reporting constant shortages. The collective farms also faced scarcity of organic fertilizers due to a major reduction in animal stocks. The number of cattle decreased by 63 percent and of pigs by 66 percent. There was also a shortage in administrative labor: approximately 300 units, or 11.5 percent, had no bookkeepers, and in 382 units there was no basic administrative staff. As a result of peasant demands and disorganization by party officials, the actual size of household plots exceeded the officially authorized allotment. In 397, or 15.4 percent, of the collective farms the system of work units was not observed. In 227 of these units, partial cultivation was the standard practice, with 40 to 60 percent of members working on a part-time basis. Accordingly, expenditures for common funds (seeds and fodder) took precedence over their wage payments.[46]

In addition, collective management failed to ensure equitable compensation among different work groups, which triggered dissatisfaction and conflicts between workers. Those engaged in animal breeding received lower monthly wages. The size of the cooperative land area was decisive, since members assigned the cultivation of work-intensive plants (corn, beets, and pasture) to specific workers or external wage earners. As a result, members who performed partial work did not contribute their share to cooperative activities.[47] Nor did the income-distribution system function according to the rules and directives of collective farming. Typically, the least stable cooperatives with elderly members and ineffective leadership paid wages by the hour. Another compensation method was based on a combination of wages by the hour, performance, and partial cultivation. An additional source of dissatisfaction was the failure to compensate owners for properties incorporated into collective farms. Such compensation was based on the size and quality of the properties. In fact, the majority of collective farms "did not assume the obligation to pay compensation." These collective farms were usually based on state properties in reserve.[48]

[45] MOL, M-KS-288.f. 11.cs. 84.ő.e. May 29, 1957, 2.
[46] MOL, M-KS-288.f. 11.cs. 84.ő.e. May 29, 1957, 2.
[47] MOL, M-KS-288.f. 11.cs. 84.ő.e. May 29, 1957, 2.16.
[48] MOL, M-KS-288.f. 11.cs. 84.ő.e. May 29, 1957, 2.16.

Managers and members of cooperatives that had not disintegrated or were restored after 1956 attempted to assert their personal interests in a variety of ways. For example, new members were required to provide a financial or in-kind contribution. In the east Hungarian county of Hajdú, for instance, new members were obliged to make a specific payment or to contribute farm animals or equipment equal in value. These collective farms were usually run by middle peasants. They seemed to prefer "collective enterprises" that primarily utilized state resources instead of private farming. The formation of such collective farms was determined by local-regional conditions and personal-social relationships. However, even in western Hungary similar efforts were taken in order to receive money or in-kind contributions from would-be members, as for example in the village of Újrónafő in Győr county.[49]

The modified operating procedures after 1956 generally did not address social or retirement issues, which had become major problems in previous years. At the same time, the elderly were generally not accepted as members.[50]

It is necessary to address occurrences of physical violence in the 1940s and 50s, and especially after 1956. Collective physical violence and reprisals and their impact following the 1956 Revolution have not yet been comprehensively evaluated by scholars, but personal notes indicate that this was not a new experience, especially during the persecution of the *kulaks*.[51] Meanwhile, confidential party reports from the spring of 1957 refer to "illegal excesses" by Győr county security forces. According to a party report:

> [S]ecurity forces in the district of Mosonmagyaróvár carried out beatings of the poor and peasants without justification. While carrying out organizational work in a village, the reporter encountered people who had traces of beatings on them. When security forces entered a village, they displayed a red flag on the village council building as their first action in order to give public notice of their presence. In Dévény they placed a red flag in the hands of a Christ statue.[52]

49 MOL, M-KS-288.f. 11.cs. 84.ő.e. May 29, 1957, 2.16.
50 MOL, M-KS-288.f. 11.cs. 84.ő.e. May 29, 1957, 2.16.
51 MOL, M-KS-276.f. 93. 142. d. April 11, 1949, 214–16.
52 MOL, M-KS-288.f. 22. 1.ő.e. March 26, 1957, 43–44.

Unpublished letters received by party daily *Népszabadság* also document the countrywide practice of physical violence. Out of 6,041 letters mailed between February 15 and March 31, 1957, 94 concerned violence by security forces, the police, and the Workers' Guard. According to those letters, the latter organization, the Workers' Guard, was responsible for numerous indiscriminate beatings, including a clash involving 27 teachers in the Mór district.[53] This is in keeping with the tactics of the Kádár dictatorship, which used physical violence to intimidate social groups, reduce opposition and secure power.

The Final Collectivization Campaigns, 1958–1961

In May 1958, leaders of the collectivization campaign ordered Workers' Guard units to be stationed in each district across the country to carry out the party's final collectivization campaigns.[54] By the second half of that year, reprisals against supporters of the 1956 Revolution had mostly come to an end with the execution of the former Prime Minister Imre Nagy—who had been welcomed by peasants—and his associates. Party leadership issued new political instructions to its apparatus and the police organizations, of which the collectivization campaign was an important element.[55]

As part of the final campaign, leaders evaluated the regional factors contributing to the success or failure of prior efforts at collectivization. They considered counties in eastern Hungary—Békés, Szabolcs, Szolnok—and the western Hungarian counties of Fejér as areas of "above average development potential." They gave priority to ploughland as they found the collective farms in these areas comparatively productive and well functioning. The large number of poor peasants in these areas was taken into consideration as was "[t]he significant revolutionary traditions of the peasantry," along with the acknowledgment that "the events of 1956 led to fewer disturbances in these regions than elsewhere."[56] The campaign's

---

[53] MOL, M-KS-288.f. 11. 46.ő.e. April 3, 1957, 1.

[54] MOL, M-KS-288.f. 11.cs. 275.ő.e. May 17, 1958.

[55] The county and district prosecutors and courts unusually rendered heavy sentences for each minor verbal comment that criticized the collective campaign. The great number of agents circulating in agrarian society regularly reported on such events. MOL, XIX-B-14. ORFK 387.d. 1961.

[56] MOL, M-KS-288.f. 11.cs. 348.ő.e. August 7, 1958, 34.

decision-makers and executive staff consisted of 50 to 100 local party members and special activists in each county, along with an additional 20 to 50 members at the county level. The total area to be collectivized was set at 4.9 million hectares—or 80 percent of the country's agricultural land. The three-year program included two alternative plans: one that projected a collectivization rate of 70 percent, and another at 60 to 65 percent. But it was clear that both alternatives lacked sufficient funding.[57]

An important initial strategy was to control the churches, which harbored opposition to party policies in the villages. According to a September 1958 party report, the political police and party authorities had regained control over key church positions.[58] Soon thereafter local council elections took place, which were regarded as an important test of support for the party and collectivization. It is interesting to note that some participants in the council elections associated the event with collectivization. "In several places the enemy spreads the view that those who sign the election registers will be required to join the collective."[59]

Top party leadership designated Győr county, close to the Austrian border, as the first area to be fully collectivized, a decision influenced by regional economic, social, and political factors. The Győr campaign launched in late 1958 was conceived as a model for the post-1956 collectivization drive. Party leadership believed the "most effective solution" was "the incorporation of the entire community into the collective," as opposed to the former practice of establishing small collectives.[60] They referred to experiences in the neighboring county of Veszprém, where, according to their opinion, the results had confirmed total collectivization was possible. Although the public was mostly aware of the brutalities the party resorted to during the campaign, leaders attempted to conceal their actions. This is demonstrated in a letter written by a female radio reporter, according to whom leaders were unwilling to provide information, who wrote: "Since this property owner refused to join, he was beaten, his head was smashed against the wall and he was forced to lick the wall."[61]

---

[57] MOL, M-KS-288.f. 11.cs. 348.ő.e. August 7, 1958, 34.

[58] It is well known that the maintenance of church parishes depended on state support. MOL, M-KS-288.f. 11.cs. 373.ő.e. September 22, 1958.

[59] MOL, M-KS-288.f. 11.cs. 376, 425.ő.e. September 27, 1958, January 24, 1959.

[60] MOL, M-KS-288.f. 11.cs. 427, 436.ő.e. February 2, 13, 1959.

[61] Győr-Sopron County Archive, MSZMP, 1.f. 13.f.cs. 17.ő.e. February 4, 1959. Nationally, local council and party leaders were divided on collectivization since some were also property owners.

By the spring of 1959 as much as 55.1 percent of all ploughland was collectivized. The first actions took place in western Hungary, in Győr, Veszprém, Komárom, and Fejér, and in Szolnok counties in central Hungary—regions to which the party had given preference from the beginning.[62] The campaign continued in predominantly agrarian regions in the second phase in the autumn of 1960 and in 1961. By 1962, some 75 percent of agrarian workers had become collective members.

Table 3 reflects the crisis and structural change regarding the development of individual property ownership. It is noteworthy that in 1957–58, prior to the final phase of collectivization, the number of individually owned properties increased substantially.

Table 3. Number of Individually Owned Properties in Hungary, 1949–62.

	0–0.57 ha	0.57–2.86 ha	2.86–5.71 ha	5.71–11.4 ha	11.4–14.3 ha	14.3 ha –	Total
1949	219,200	664,700	458,000	240,300	30,600	47,200	1,600,000
1950	211,600	638,800	452,400	234,000	33,700	34,000	1,604,500
1951	207,000	601,100	436,500	224,500	23,400	25,000	1,517,500
1952	193,900	567,400	399,600	207,700	17,700	14,000	1,400,300
1953	230,600	553,500	317,400	143,800	14,200	10,500	1,270,000
1954	234,600	663,500	373,200	169,100	15,500	9,100	1,465,000
1955	247,600	702,000	380,700	159,800	12,800	7,100	1,510,000
1956	268,400	655,500	352,300	150,900	11,400	6,500	1,445,000
1957	276,600	738,800	417,200	173,700	11,400	4,900	1,622,000
1958	267,739	756,185	423,923	172,361	10,800	3,840	1,634,848
1959	290,000	780,000	350,000	119,000	5,500	2,100	1,446,600
1960	300,000	367,859	95,081	69,469	11,681	1,302	845,392
1961	0	135,000	15,000	8,500	1,500	0	160,000
1962	0	123,580	13,670	6,800	950	0	145,000

Source: Mezőgazdasági adattár, *1972; Valuch,* Magyarország, *199.*[63]

[62] MOL, M-KS-288.f. 11.cs. 446.ő.e. February 28, 1959.

[63] In 1961 and 1962 agricultural units larger than 14.3 ha were tabulated jointly. After 1961 the land categories were changed.

Social Reactions and Consequences to Post-1956 Collectivization

Along with the capacity to purchase more goods, one of the widespread results in the period prior to the completion of collectivization was the increase in private savings by the public. According to an April 1958 report, party leadership was already aware of this problem. But in December 1958, personal savings that amounted to HUF 2.3 billion had increased to 3.1 billion by mid-April 1959. Party leadership considered the increase "substantially greater" than expected. The deposits were larger in Budapest than in the countryside.[64]

Under pressure, the party made radical changes in the spring of 1959, decisions which affected all social groups. According to party reports, in many communities it was widely believed that a new push to organize collectives would be launched in the autumn, as in other socialist countries—a view based on Khrushchev's statements.[65] In regions close to the capital, there was a widespread tendency among members to leave collectives or to refuse to perform inventory assignments. According to official statements, 121,000 members of collectives established in the spring of 1959, covering 35 percent of new ploughland and representing more than one-third of new members, would start work in the autumn.[66]

Following the most brutal phase of collectivization in the winter of 1959–60, the party dispatched several hundreds of observers to the countryside to report on the "mood of the village." According to a commonly reported story: "Following the reorganization in winter people did not initiate common activities, and therefore the majority of working peasants became uncertain and started to drink."[67] Some observers reported "this is a good system, but the basic flaw is that people are forced to join. Collectives do not have a good impact on peasants because as individual owners they were efficient workers but in the collectives they have lost their dedication."[68] There were repeated accounts of several problems: the failure to meet wage and other payments, broken promises, wide-

[64] MOL, M-KS-288.f. 11.cs. 469.ő.e. April 17, 1959.
[65] At the same time it was generally known in villages that in Yugoslavia and Poland "no collectives existed."
[66] MOL, M-KS-288.f. 11.cs. 485.ő.e. May 25, 1959.
[67] MOL, M-KS-288.f. 11.cs. 625.ő.e. March 18, 1960, 1–11.
[68] MOL, M-KS-288.f. 11.cs. 625.ő.e. March 18 1960, 1–11.

spread desire to move to urban areas, conflicts between poor and wealthy members, and ethnic and nationality conflicts. Peasants refused to accept discrimination, compared to other social groups like urban workers who received family support. Visits to general practitioners increased—also a recurring topic in these reports—and physicians were often forced to treat five to six people simultaneously because of the multitude of patients. Some doctors were unable to cope with the increase in patients and left their districts. Women complained about shortages in nurseries and kindergartens that were overloaded with infants, as their mothers were forced to seek employment.[69]

Complaints of "Prisoner Status" and Resistance with Adjustment

The harsh reality of social conditions during the final phase of the collectivization campaign is confirmed by internal party documents.[70] Several cases came to the attention of officials after peasants submitted complaints to state authorities or radio stations and newspapers, or submitted group petitions, signed by thousands. Common patterns of the communist system and its representatives are highlighted by an example from Fót, in Pest county, where a group of five to ten party representatives remained in the village for three months in 1960. They intimidated and physically assaulted more than 50 village residents who refused to join the collective. These actions included beatings, starvation, interrogations and questioning before the village council. Party agitators often victimized peasants, even though in some cases they originated from the same social class, including Workers' Guards, industrial workers, city employees, and teachers.[71] In the judicial proceedings related to these atrocities, prosecutors dismissed the charges on party instructions. High officials protected the agitators who had committed these atrocities.

---

[69] MOL, M-KS-288.f. 11.cs. 625.ő.e. March 18, 1960, 1–11. At this time the social security system did not provide full coverage to agricultural workers. Full and equal coverage to agricultural workers was provided only after collectivization was completed.

[70] MOL, M-KS-288.f. 28. cs. 12.ő.e. December 13, 1961, 22–25.

[71] In addition to the central reports in the Hungarian National Archives, I did research in all county archives. I will account for my findings in a study completing this research. Ö. Kovács, "Ekkora gyűlölet."

Victims who appealed to *Népszabadság* and state authorities in either oral or written complaints often adopted the viewpoint of "prisoner status."[72] The majority of these letters were written by members of the older generation, who made up the dominant segment of rural society. One such villager from Püspökmolnári in Vas county in western Hungary reported: "One day about fifty agitators arrived at the village and by means of beating and physical violence they forced people to join the collective farm. Some inhabitants were arrested and taken away."[73] Another anonymous author of a letter from Nyírbogát in Szabolcs county in eastern Hungary claimed that several people who had played an active role in support of the 1956 Revolution (now) acted as executors of the collectivization campaign as a form of compensation and accommodation.

Several private farmers from Tiszakécske in the Danube-Tisza region in east-central Hungary signed their letters stating: "Peasants are cited to appear at the collective organizing office at night and are persuaded by truncheons to join."[74] The testimony of older farmers in Albertirsa reports of how collectivization had created a split between generations that had altered traditional peasant life: "They (elderly farmers) are in despair since they cannot work, their land is confiscated and therefore their living is uncertain. Their children took care of them in the past in return for their land, but no one is willing to do so for their annuity."[75]

By this time, most of those capable of working had left the villages or were commuting between rural communities and the urban workplace. As a result, women and old people became the dominant members of collective farms. Letters from these members describe poverty, hunger, lack of firewood, and the desire to leave the collectives.[76]

In the month of February 1960, *Népszabadság* received 3,008 letters, 2,021 of which, or 67 percent, had come from the countryside. Of these, 648, or 22 percent, reported on village problems related to the violence that occurred while the cooperatives were being established. One letter, however, which refers to Sztálinváros, the "Hungarian City of Stalin," the country's new steel center, which starkly contradicts the widely promoted party myth of "worker-peasant" relationships:

---

[72] On the concept of "prisoner status," and the mental problems in the Soviet systems, see Sztompka, *Sociology*, 246–7, 281.

[73] MOL, M-KS-288.f. 11.cs. 605.ő.e. 1960.

[74] MOL, M-KS-288.f. 11.cs. 605.ő.e. 1960.

[75] MOL, M-KS-288.f. 11.cs. 605.ő.e. 1960.

[76] MOL, M-KS-288.f. 11.cs. 605.ő.e. 1960.

> I wish to let you know about the events that took place for five weeks in the southern part of Fejér county, in Cece, Alap, Sárszentmiklós, and Sárbogárd. First the miners appeared, but they did not succeed, then the workers of Sztálinváros, who seemed to have lost all human qualities. They failed as well. Therefore other methods were applied: tax assessments and personal humiliation of the elderly. Those who resisted were obliged to appear before the council chambers. They were forced to sign by verbal and physical violence. Many people left their homes in the morning, but when they returned they were taken by force. They then reported that the cooperatives were fully organized.[77]

To what extent did peasants resist collectivization? Official reports document mass demonstrations by peasants—and in several cases led by women.[78] Throughout the collectivization campaign, independent farmers protested in numerous ways, from flight to verbal and physical threats. Occasionally, there were organized actions. On one occasion, a hand grenade was thrown on the council building.[79] On another occasion, on March 4, 1960, a group of 200 people forced their way into the council building in Ruzsa, in the county of Csongrád in southern Hungary, and presented the council president with a declaration of their intention to withdrawal from the collective. The police and the military dispersed the crowd. Official documents cite frequent cases of individuals and their family members becoming physically ill as a result of the persecutions. There were suicides, which authorities attempted to conceal. Police reports provide accounts of two suicide cases related to the conduct of party agitators in November 1960 in Orosháza in southern Hungary—one ending in suicide by hanging, the other was prevented.[80]

The following methods of terror and violence were used during collectivization by party agitators: First, there was *administrative coercion*, which included: citations; dismissing individuals, relatives, or family members from their jobs; military conscription; and threatening to exclude or excluding children from school. Second, there was *verbal, psychological pressure*, and *intimidation*, which included: threats by the council president over loudspeakers; misrepresenting agitators as a "Soviet officer,"

[77] MOL, M-KS-288.f. 11.cs. 614.ő.e. March 10, 1960.
[78] MOL, M-KS-288.f. 11.cs. 615.ő.e. March 5, 1960; 615. ő.e. March 10, 1960.
[79] MOL, M-KS-288.f. 11.cs. 705.ő.e. August 24, 1960.
[80] MOL, XIX-B-14. ORFK 387.d. January 28, 1961.

"policeman," or "prosecutor" followed by orders to sign membership; forcing minors to sign membership agreements; searching dwellings; destructing personal property such as furnishings, vehicles and equipment; cutting the tails off horses; and confiscating food and beverages. Third, there was *physical violence*, which included: arrests on the street or at night by police, the Workers' Guard, armed civilians and other "people educators"; torture by standing up on one foot; stepping on feet; kicking, pulling of hair and ears, banging heads on tables and on walls; pulling of noses and genitals; use of boxer and truncheon; running in an overcoat; sweating by the stove; beatings of the candidate and his family; humiliation of women and removal of their clothing; armed threats to wives while beating their husbands; beating of feet by rubber hose; and forcing victims to take glowing embers from the stove in their bare hands.

Following the physical-psychological torture in the council building or the house of culture, the second phase of the process was a friendly, polite conversation with a familiar person, usually a local administrator, who assisted in signing the declaration of membership. Although atrocities were widespread in all counties, even reports of such acts were enough to force people join collectives.

The Interpretation of Forced Collectivization

The following section reviews forced collectivization from the perspective of contemporary experiences. Communist ideology and its role in agricultural collectivization can then be formulated based on this perspective.

A highly important contemporary experience was that the violent intervention of the communist power structure resulted in the destruction of the centuries-old, generally well-functioning coping strategies and traditions of the peasantry, shaped around preservation and preparedness.[81] But, with a third of collective members over 60, collectivization did not completely destroy rural traditions and their experiences are also worth examining.[82] Most older villagers felt their fate was once again decided by others. The process of identifying supporters or opponents of "the Socialist reorganization of agriculture" was a function of propaganda.[83]

[81] Márkus, "Az ismeretlen," 296–97.
[82] Valuch, *Magyarország*, 198.
[83] Edelman, "Politikai," 105, 120.

Those involved in this confrontation were not so much concerned with the battle against enemies, but with survival. This is indicated by statistical data: as a result of *kulak* persecutions, more than 70,000 people were expelled from their traditional social environment.[84] Repressive agrarian policies resulted in the convictions of more than 400,000 peasants on charges of "endangering public provisions" from 1949 to 1955.[85]

Forced collectivization was a source of additional social deprivation: it destroyed traditional relationships, led to a decline in social cohesion and created new sources of conflicts as well as political control over church organizations. The complex impact of collectivization has been documented by contemporary sociological research. In a study of an eastern Hungarian village, for example, scholars have traced the emergence of a "collective-neurosis."[86] Among the adult population, alcoholism, for instance, increased from 30 to 42 percent from 1960 to 1970.[87] Subsequent research after 1990 confirmed the enduring negative consequences of collectivization, including the lack of social relationships, weak social solidarity and civil society, lack of confidence, anomie—all of which were cited as typical village forms of alienation as a result of collectivization.[88]

Another contemporary experience consisted of dramatic effects on rural life both at the macro and micro levels. At the macro level, one million wage earners, one fourth to one fifth of the labor force, left agriculture between 1950 and 1970. This change supports the view among scholars that the social contexts of collectivization should be examined in future studies as a comprehensive social-urban issue, a research agenda which has not yet been addressed. The social impact of collectivization includes a basic change in the approach to work, which also involves the radical transformation of personal and family interests. For many, the traditional value of land and its cultivation was lost.[89] Wages of workers in collectives had declined by the end of the 1960s.[90]

---

84 Valuch, *Magyarország*, 192.
85 Valuch, *Magyarország*, 192.
86 It is worth mentioning that they pursue their life course in a manner divergent from customary peasant traditions. For those peasant traditions see Fél and Hofer, *Proper Peasants*.
87 Juhász, "A neurózis."
88 Andorka, "A falusi."
89 Valuch, *Magyarország*, 197–99.
90 Valuch, "Községünkben."

The employment of violence and coercion was a third essential contemporary experience. Agitation for collectivization and atrocities undermined the possibility of normal adult life.[91] Agrarian proletarians or poor peasants did not care where they worked. Thus, the majority accepted membership in collectives, partly motivated by promised social benefits. Those who were forced to join were unable to utilize the first defense mechanism, which excluded agitation. The second opportunity, which involved confrontation, was a choice only to a very limited extent. The great majority experienced the third alternative when they left the village or signed the membership agreement under the threat of violence. The basic message of this period was that whatever had been restored in the course of the 1956 Revolution, would be taken away at any cost, even if the result was chaos and irrationality. Contemporaries experienced this reality. At the same time, repression produced some advantages. Leaving aside the utility of the household plot, it is estimated that 30 percent of the managers of collectives were of *kulak* origin in the post-collectivization period.[92]

Finally, the fourth contemporary experience can be defined as the acceptance of a cultural compromise with the Soviet system. Hungarian society was transformed into one that was based on three behavioral categories: forbidden, tolerated, and supported. The official principle was: "Whatever was not specifically forbidden, was permitted." People were forced to live within these limits. This can be designated as "the rationalization of subjection."[93]

The official communist ideology characterized the party's agricultural policies as a program of modernization. Its most important goals were rapid and comprehensive industrialization and the dissolution of traditional structures. This is the reason why the idea proved to be attractive to many people. Contemporary experiences, as outlined above, substantiate these program goals. This program was based on the idea of "proletarian dictatorship," however, and not democracy, which served as an ideological justification for the party's campaign of mass terror and the liquidation of *kulaks*. Theoretically, terror was embodied within the totality of the communist program. The Communist Party centralized all areas of life activities and attempted to control every aspect of society—a program that could be realized only through imposing fear and terror on society.[94]

91 Berger and Berger, *Sociology*, 3–10.
92 Hamar, ed., *Agrárátalakulás*, 11; Swain, *Collective Farms*, 114–29.
93 Elster, *A társadalom*, 166.
94 Langewiesche, "Fortschritt."

At the same time, individuals were not simply passive objects of political power in the Soviet system, but they actively shaped, multiplied, and reinforced it. One of the basic differences between modern and traditional authoritarian systems is that the dictatorships of the twentieth century demanded enthusiasm and participation from their subjects. The Soviets envisioned the state as a gardener, whose task was to tame and develop the wild human society, to build a "new Socialist man," who reflected Soviet ideology in dress, manners and customs.[95] Although this fundamental framework does not explain all aspects of communist Hungary, it is important to view the development of the country's sociopolitical culture in the communist era from this perspective. Furthermore, as our review of contemporary accounts indicate, these experiences provide historical testimony for the propagation and attempted realization of this program.

CONCLUSION

The first phase of collectivization, which began in 1948, faced a major challenge during the Revolution of 1956. During the two weeks of the revolution, the majority of collective farms were dissolved. This is an important indication of the repudiation of forced collectivization by the Hungarian peasantry and their support for independent agrarian production. The prevalence of this viewpoint played an important role in the subsequent resistance to forced collectivization following the suppression of the revolution. In the second phase of collectivization, from 1958 to 1961, the use of terror and intimidation as practiced in other Soviet regimes resulted in the elimination of a great majority of privately owned farms. The significance of the revolution can be understood in two respects: full collectivization was not initiated until two years after 1956, and the revolution was a key factor in the party's decision to introduce a model of agrarian management that diverged from the traditional Soviet system. Although collectivization campaigns were initiated in 1948–1949 in Bulgaria, Romania, and Czechoslovakia, no open peasant revolts comparable to those that took place in Hungary occurred in Bulgaria and Romania. The Revolution of 1956 was unique, because it was as much an urban revolution as it was a rural one.

The history of forced collectivization in Hungary indicates that, as in all socialist states, not only were the organizations of power—party orga-

[95] Scott, *Seeing Like a State*, 147–53; Baberowski, *Der rote Terror*, 8–13.

nizations, secret police, state police, Workers' Guards, and in part, party councils used to mobilize peasants—but other social and occupational groups (industrial workers, members of collectives, teachers, officials, and professionals) were deployed as well. As an elementary feature of modern dictatorship, party leadership propagated and executed its policies by mobilizing the participation and support of all social groups. The scope of such mobilization and participation was more obvious when the first stage of collectivization was completed. In many cases, peasants forced into membership were dispatched as agitators the following week. This policy was designed to "conceal" and share responsibility. As a consequence, the memory of and responsibility for past events was dispersed across society.

Violence by party agitators was typically a last resort, used when all other methods failed. Optimal productivity from farmers could not be expected under these conditions.[96] In fact, forced campaigns meant qualifications, experience, and motivation were frequently lacking. Although the party-state had confiscated the means of production, it was unable to control the production process, since the most important element—human resources—were less dedicated to productivity, having been forced to cede to the state their property, and with it, their traditional agricultural system and practices. In later years, this was compensated by the household plot—a concession well understood party leaders.

Following the official completion of collectivization in the 1960s, communist party and state leadership in all countries faced the same problems: low productivity of collective farms, economic losses, and debt.[97] The basic contradiction of the socialist system was also evident in property relations, as members were owners only in principle. This situation had a major impact on interest issues—for example, it meant that the state received its share of annual income on a priority basis, with the balance distributed to members.

Hungarian agrarian policies in the end meant an increase in prices for agricultural producers. Moreover, a basic directive of Kádár's 1960s-era New Economic Mechanism—a policy which Kádár explained as "whatever was not specifically forbidden, was permitted"—meant collective farmers could pursue additional economic activities. They were allowed to raise more farm animals on their personal property. They could commit

[96] Durkheim, *Division of Labor*, 353–57; Weber, *Wirtschaft*, 19–20, 32, 86–89; Kornai, *A szocialista*; Dahrendorf, *Reflections*, 16–17.

[97] Varga, "Az agrárium," 305; Schöne, *Frühling*, 244–51.

themselves to work on a part-time basis, and they could receive payment-in-kind. Social-support policies were extended to members, which contributed significantly to the acceptance of the system. Under a party-controlled interest group system, a lobby was formed which could veto price and financial decisions. In 1967, a law was passed requiring collective members to receive priority compensation. Hungarian decision-makers did not support the practice of other Socialist countries, including East Germany, of creating excessively large production units in the 1970s. As a consequence, Hungarian agriculture continued to prosper. In the 1960s and even more so in the 1970s, infrastructural improvements expanded significantly, especially in rural areas. Modernization increased productivity and consumerism. This was the other side of Kádár's policies, which ushered in an era of so-called "self-exploitation." It meant that, in effect, personal property of the collective member made possible an intensive family economy. In addition, the composition of small producers in Hungary underwent a basic change. Their majority was no longer of peasant, but of industrial worker origin.[98]

The socialist experience includes several elements that are clearly reflected in our earlier review of the contemporary experiences of forced collectivization. These are, for example, the employment of violence and its impact on human life, resistance, and adjustment as responses to overwhelming pressures on individuals. In addition, if we consider Socialism in the context of the systems approach, we observe the process of waste as an integral element of this economic and social system.[99] Dissatisfaction with the shortages of vital necessities was not only a matter of concern for the top leadership, but also for the common man subjected to political and economic controls. The concept of a "moral economy" has been used to characterize those dissatisfied with the shortage of material goods and consider protest and disobedience as legitimate actions.[100] The observation of everyday life in a socialist society substantiates the prevalence of a moral economy. A possible explanation is that the population was deprived of their property and their traditional social environment. Frequently, the probable consequence was disobedience to dictatorship.[101] Research on collectivization reveals numerous reports of theft by both communist

[98] Varga, "Az agrárium," 311, 314, 328; Hann, "Die Bauern."
[99] Kemény, "A szocializmus."
[100] Thompson, "Economy," 77–79; Thompson, *Az angol munkásosztály*, 74–75; Viola, *Peasant Rebels*, 206–9.
[101] Cerutti, "Norma és gyakorlat," 153–56.

leaders and peasants, as well as various actions that violated "socialist legality." The latter did occur in previous social systems, but their frequency and form changed considerably. Another difference is that the previous categories of "poor" and "rich" have been replaced by "my" and "common" as opposites.[102]

In our evaluation of forced collectivization, we have emphasized the few but notable concessions granted by authorities to the collective farm members, such as the household plot, the allowance of non-agricultural economic activities, participation in agricultural marketing, and infrastructural assistance. But it is important to emphasize that forced collectivization resulted in the destruction of traditional peasant social and economic institutions, as well as social cohesion, and dispersed a substantial proportion of peasant population to urban and industrial areas. In addition, the use of violence, persecution and subjection to governmental controls imposed an overwhelming pressure especially on those peasants who were primarily independent, land-owning farmers. One of the essential features of forced collectivization was the destruction of traditional ways of life. Land collectivization was an agricultural system that may have produced economic benefits (in Hungary, at least), but at an incalculable cost to individuals living within it.

BIBLIOGRAPHY

ARCHIVAL SOURCES

Győr-Sopron County Archive:

Fond: MSZMP (*Hungarian Socialist Workers' Party), files:* MSZMP, 1.f. 13.f.cs. 17.ő.e. 4 February 1959.

Magyar Országos Levéltár, Hungarian National Archive, (MOL):

Fond: M-KS-276. MSZMP, files: M-KS-276. f. 85. cs. 9.ő.e. 14 January 1949.; M-KS-276. f. 85. cs. 10.ő.e. 21 January 1949.; M-KS-276. f. 85. cs. 10.őe. 1949.; M-KS-276. f. 93. 142. d. 11 April 1949.; M-KS-276. f. 93. cs. 147ő.e. 25 March 1949.; M-KS-276. 93. cs. 407. ő.e. 27 December 1952.

Fond: M-KS-288. MSZMP, files: M-KS-288.f. 11. 46.ő.e. 3 April 1957.; M-KS-288.f. 11.cs. 84.ő.e. 29 May 1957.; M-KS-288.f. 11.cs. 275.ő.e. 17 May 1958.; M-KS-288.f. 11.cs. 348.ő.e. 7 August 1958.; M-KS-288.f. 11.cs. 373.ő.e. 22

102 Szilágyi, "Törvények," 750. After 1956, the majority of the forests in the countryside were cut down, for example 80% of the forests in Pest county. Pest County Archive, XXVIII. 750. 1. f. 3. cs. 1.ő.e. December 17, 1956.

September 1958.; M-KS-288.f. 11.cs. 376, 425.ő.e. 27 September 1958.; M-KS-288.f. 11.cs. 376, 425.ő.e. 24 January 1959.; M-KS-288.f. 11.cs. 427, 436.ő.e. 2, 13 February 1959.; M-KS-288.f. 11.cs. 446.ő.e. 28 February 1959.; M-KS-288.f. 11.cs. 469.ő.e. 17 April 1959.; M-KS-288.f. 11.cs. 485.ő.e. 25 May 1959.; M-KS-288.f. 11.cs. 625.ő.e. 18 March 1960.; M-KS-288.f. 11.cs. 605.ő.e. 1960.; M-KS-288.f. 11.cs. 614.ő.e. 10 March 1960.; M-KS-288.f. 11.cs. 614.ő.e. 5 March 1960.; M-KS-288.f. 11.cs. 615.ő.e. 10 March 1960.; M-KS-288.f. 11.cs. 705.ő.e. 24 August 1960.; M-KS-288.f. 22. 1.ő.e. 26 March 1957.; M-KS-288.f. 28. cs. 12.ő.e. 13 December 1961.

Fond: XIX-K-1-ah. FM (Ministry of Agriculture), files: XIX-K-1-ah. 18.d. 25 March 1952. no. 2451/IV; XIX-K-1-ah. 19.d. 26. January 1955. no. 9.; XIX-K-1-ah. 19.d. 22 January 1955. no. 3.; XIX-K-1-ah. 19.d. 26 January 1955. no. 9.

Fond: XIX-B-14. ORFK (Police), files: XIX-B-14. ORFK 387.d. 1961.

Pest Megyei Levéltár, Pest County Archive:

Fond: MSZMP, files: XXVIII. 750. 1. f. 3. cs. 1.ő.e. 17 December 1956.

OTHER SECONDARY MATERIALS

Magyar Közlöny [Hungarian Gazette): *Magyar Közlöny* 17, January 22, 1949; *Magyar Közlöny* 33, February 12, 1949.

Magyar statisztikai évkönyv 1957. [Hungarian statistical yearbook 1957). Budapest, 1958.

Mezőgazdasági adattár I–II. [Agricultural data I–II.] Budapest, 1972.

BOOKS AND ARTICLES

Alexander, Jeffrey C. "Toward a Theory of Cultural Trauma" in J. C. Aleaxander et al. eds. *Cultural Trauma and Collective Identity.* Berkeley: University of California Press, 2004, 1–30.

Andorka, Rudolf. "A falusi társadalom változásai" [Changes of Village Societies]. *Agrártörténeti Szemle* 1–4 (1994): 3–26.

Baberowski, Jörg. *Der rote Terror. Die Geschichte des Stalinismus* [Red Terror: History of Stalinism]. Munich: DVA, 2004.

Bauerkämper, Arnd, ed. "*Junkerland in Bauerhand?*" *Durchführung, Auswirkungen und Stellenwert der Bodenreform in der Sowjetischen Besatzungszone* [Junkerland in Peasant Hand? Execution, Consequences and Rating of the Land Reform in the Soviet Zone of Occupation]. Stuttgart: Steiner, 1996.

Berger, Peter Ludwig and Brigitte Berger. *Sociology. A Biographical Approach.* New York: Basic Books, 1972.

Bloch, Marc. *The Historian's Craft.* New York: Vintage, 1964.

Cerutti, Simona. "Norma és gyakorlat, avagy szembeállításuk jogosságáról" [Principles and Practice, or Justification of their Contrast] in Gábor Czoch and Gábor Sonkoly, eds. *Társadalomtörténet másképp. A francia társadalomtörténet új útjai a kilencvenes években* [A Different Social History. New Ways of the French Social History in the 90s]. Debrecen: Csokonai, 1995, 151–68.

Dahrendorf, Ralf. *Reflections on the Revolution in Europe.* London: Chatto and Windus, 1990.

Durkheim, Emile. *The Division of Labor in Society.* New York: The Free Press, 1964.

Edelman, Murray. "Politikai ellenségek konstruálása" [The Construction of Political Enemies] in Márton Szabó, ed. *Az ellenség neve* [The Name of the Enemy]. Budapest: Jószöveg, 1998, 84–123.

Elster, Jan. *A társadalom fogaskerekei. Magyarázó mechanizmusok a társadalomtudományokban* [The Cogwheels of Society. Explanatory Mechanisms in the Social Sciences]. Budapest: Osiris, 1997.

Fél, Edit and Tamás Hofer. *Proper Peasants. Traditional Life in a Hungarian Village.* Chicago: Aldine Publishing Company, 1969.

Grenier, Jean-Yves. "A 'hosszú időtartam' új értelmezése" [New interpretation of the long duration]. *Aetas* 1–2 (1999): 306–12.

Habuda, Miklós et al., eds. *A Magyar Dolgozók Pártja határozatai 1948–1956* [Resolutions and Documents of the Hungarian Workers' Party, 1948–1956]. Budapest: Napvilág, 1998.

Hamar, Anna, ed. *Agrárátalakulás Magyarországon a kilencvenes években I* [Agricultural Transformation in Hungary in the 90s]. Szolnok: KGF, 1998.

Hann, Chris. "Die Bauern und das Land. Eigentumsrechte in sozialistischen und postsozialistischen Staatssystemen im Vergleich" [The Peasants and the Land. Proprietorship in Socialist and Post-Socialist State Systems Compared] in Hannes Siegrist and David Sugarmann, eds. *Eigentum im Vergleich internationalen Vergleich (18–20. Jahrhundert)* [Property in International Comparison (18th–20th Centuries)]. Göttingen: V/R, 1999, 161–84.

Honvári, János. "Magyarország gazdasági fejlődése a II. világháború után (1945–1958)" [Economic Development of Hungary after World War II, 1945–1958] in János Honvári, ed. *Magyarország gazdaságtörténete a honfoglalástól a 20. század közepéig* [Economic history of Hungary from settlement to the mid-20th century]. Budapest: Aula, 1997, 427–576.

Juhász, Pál. "A neurózis morbiditásának emelkedése egy magyar falu közösségében" [Increase in Morbidity of Neurosis in the Community of a Hungarian Village] in Rudolf Andorka et al., eds. *A deviáns viselkedés szociológiája* [Sociology of deviant behavior]. Budapest: Gondolat, 1974, 504–27.

Kemény, István. "A szocializmus a pazarlás rendszere. (Orosz István levélinterjúja)" [Socialism is the system of waste. (Letter interview of István Orosz)] in István Kemény, ed. *Közelről s távolról* [From Near and Far]. Budapest: Gondolat, 1991, 85–95.

Kornai, János. *A szocialista rendszer. Kritikai politikai gazdaságtan* [The Socialist System: The Political Economy of Communism]. Budapest: HVG, 1993.

Kovács, József Ö. "Egy politikai diktatúra társadalmi korlátai. Jelenkortörténeti kutatások az NDK-ról" [Social Limits of a Political Dictatorship. Contemporary Historical Analyses of the GDR]. *Korall* 15–16 (May 2004): 179–204.

———. "'Ekkora gyűlölet még nem volt a falunkban, mint most.' Szövegek és kommentárok az erőszakos kollektivizálás befejező hullámáról" ["There was never such hatred in our village." Documents and commentaries on the final waves of forced collectivization]. *Századvég* 1 (2008): 37–69.

Langewiesche, Dieter. "Fortschritt als sozialistische Hoffnung" in Klaus Schönhoven and Dietrich Staritz, eds. *Sozialismus und Kommunismus im Wandel. Hermann Weber zum 65. Geburtstag.* Cologne: Akademie, 1993, 39–55.

Márkus, István. "Az ismeretlen főszereplő: a szegényparasztság" [The Unknown Actor: The Poor Peasants] in István Márkus, ed. *Az ismeretlen főszereplő. Tan-*

ulmányok [The Unknown Actor. Studies]. Budapest: Szépirodalmi, 1991, 286–97.

Nádasdi, József. *Tagosítások és birtokrendezések Magyarországon a XIX. század közepétől 1956-ig* [Redistribution of Landed Property in Hungary from the Middle of the 19th Century to 1956]. Nyíregyháza: Stúdium, 1996.

Nagy Varga, Vera. "Alkalmazkodás és szembenállás: volt kulákok egy mezővárosban" [Accommodation and Resistance: Former *Kulaks* of a Market Town] in Gyula Kocsis, ed. *Alkalmazkodás, ellenállás, mobilitás* [Accommodation, Resistance and Mobility]. Cegléd: Kossuth Múzeum, 1995, 49–110.

Orbán, Sándor. *Két agrárforradalom Magyarországon. Demokratikus és szocialista agrárátalakulás 1945–1961* [Two Agrarian Revolutions in Hungary: Democratic and Socialist Agrarian Transformation, 1945–1961]. Budapest: Akadémiai, 1972.

Pető, Iván and Sándor Szakács. *A hazai gazdaság négy évtizedének története 1945–1985. I. Az újjáépítés és a tervutasításos irányítás időszaka 1945–1968.* [Economic History of Hungary, 1945–1985. Period of the Reconstruction and the Planned Economy, 1945–1968]. Budapest: KJK, 1985.

Schier, Barbara. *Alltagsleben im "sozialistischen Dorf." Merxleben und seine LPG im Spannungsfeld der SED-Agrarpolitik 1945–1990* [Everyday Life in the "Socialist Village." Merxleben and its Cooperative in the Conflicts of Communist Agrarian Policy, 1945–1990]. Münster: Waxmann, 2001.

Schöne, Jens. *Frühling auf dem Lande? Die Kollektivierung der DDR-Landwirtschaft.* [Springtime in the Province? The Collectivization of Agriculture in the GDR]. Berlin: Ch. Links, 2005.

Scott, James C. *Seeing Like a State. How Certain Schemes to Improve the Human Condition Have Failed.* New Haven: Yale University Press, 1998.

Suter, Andreas and Manfred Hettling. "Struktur und Ereignis—Wege zu einer Sozialgeschichte" [Structure and Event: Ways to a Social History] in Andreas Suter et al., eds. *Struktur und Ereignis* [Structure and Event]. Göttingen: V/R, 2001, 7–32.

Swain, Nigel. *Collective Farms Which Work?* Cambridge: Cambridge University Press, 1985.

Szakács, Sándor. "A földreformtól a kollektivizálásig 1945–1956" [Land Reform and Collectivization, 1945–1956] in Péter Gunst, ed. *A magyar agrártársadalom a jobbágyság felszabadulásától napjainkig* [Hungarian Rural Societies from the Liberation to the Present]. Budapest: Napvilág, 1998, 288–344.

Szilágyi, Miklós. "Törvények, szokásjog, jogszokás" [Laws, Customary Law, Legal Tradition] in Mihály Sárkány and Miklós Szilágyi, eds. *Társadalom. Magyar Néprajz VIII* [Society. Hungarian ethnography]. Budapest: Akadémiai, 2000, 693–759.

Thompson, Edward Palmer. *Az angol munkásosztály születése* [Making of the English Working Class]. Budapest: Osiris, 2007.

———. "Economy of the English Crowd in the Eighteenth Century." *Past & Present* 50 (February 1971): 76–136.

Viola, Lynne. *Peasant Rebels Under Stalin: Collectivization and the Culture of Peasant Resistance.* Oxford: Oxford University Press, 1996.

Sztompka, Piotr. *The Sociology of Social Change.* Oxford: Blackwell, 1993.

Valuch, Tibor. *Magyarország társadalomtörténete a XX. század második felében* [Social history of Hungary in the second half of the 20th century]. Budapest: Osiris, 2001.

———. "'Községünkben nagy előrehaladást értünk el a szocializmus építése terén.' A történeti parasztság és az életmód változásai Magyarországon a hatvanas években" ["We have achieved great progress in building socialism in our community." Historical peasantry and way-of-life changes in Hungary in the 1960s] in János M. Rainer, ed. *Múlt századi hétköznapok* [Everyday in the Last Century]. Budapest: 56-os Intézet, 2003, 129–76.

Varga, Zsuzsanna. "Az agrárium 1945-től napjainkig" [Agriculture from 1945 to the present] in János Estók et al., eds. *Agrárvilág Magyarországon 1848–2002* [The Agrarian World in Hungary, 1848–2002]. Budapest: Argumentum, 2003, 261–339.

Weber, Max. *Wirtschaft und Gesellschaft. Grundriss der verstehenden Soziologie* [Economy and Society. An Outline of Interpretive Sociology]. I. Halbband. mit textkritischen Erläuterungen herausgegeben von Johannes Winckelmann. Tübingen: J. C. B. Mohr, 1976.

LAND REFORM AND THE COLLECTIVIZATION OF AGRICULTURE IN HUNGARY: CHRONOLOGY OF EVENTS, 1945–1962

József Ö. Kovács: 1945–1951

Zsuzsanna Varga: 1952–1961

1945

March 17: The Provisional National Government issues decree 600/1945 on the abolition of the large-estate system and the distribution of land to the agrarian population. The stated goals of the decree are: "the termination of the large estate system" in order to "ensure the country's democratic transformation and future development," and "transferring the manorial estates and intellectual development of the peasant population oppressed for centuries." It indicated the general direction of agrarian policy strategies: "Following the termination of the large estate system, Hungary's agriculture will rest on strong, healthy and market oriented smallholdings which shall constitute the registered private property of smallholders."

March 29: The official distribution of land begins. Almost 60 percent of the 3,222,800 hectares of expropriated land is distributed among 642,342 claimants—predominantly agricultural workers, farmhands, and the owners of dwarf estates. The remaining 40 percent (mainly forests and pasture land) becomes the property of the state, villages, or cooperatives. Over 300,000 people do not receive the land to which they were entitled. Land reform transforms the characteristically large- and medium-sized, estate-centered Hungarian agrarian structure to a characteristically small scale farming system.

September 14: The land reform is enacted by Law VI:1945.

1947

January 14: Law V: 1947 concludes the land reform.

1948

The establishment of a network of state-owned machine stations takes place under pressure to follow the Soviet model. In May 1948 there are 11 machine stations in the country; a year later the number rises to 221.

Summer: The economic repression of the *kulaks* begins. In addition to land and compulsory deliveries, a new tax ("agricultural-development duty") is imposed on landowners whose estates exceeded 8.5 hectares and whose net cadastral income is above 150 gold crowns. This graduated contribution had to be paid in the form of produce, for which the producer received no remuneration. Larger leaseholds are expropriated, and *kulaks* are banned from trading land freely. A maximum size of 22.8 hectares is imposed for the privately owned and rented estates of *kulaks*. Compared to the prewar period, the peasant population's tax liabilities increases. In 1948, each farm type was subject to more taxes than in 1938. Between 1948 and 1953, almost 400,000 peasants (among them, 40,000 "wealthy peasants") are convicted, primarily for public supply offences.

July 25: The Ministry of the Interior issues a communiqué concerning a conspiracy uncovered in the Ministry of Agriculture. The show trial—on the threshold of collectivization—represents a warning to the whole farming society.

August 20: Mátyás Rákosi, the party's secretary general, announces plans for the rapid "Socialist reorganization" of agriculture. As part of this process, under the pretext of democratising the cooperatives, the old (pre-1945) cooperatives are forcedly amalgamated with the farmers' cooperative network in 1948–1949.

November 27: Mátyás Rákosi announces the liquidation of the small-scale farm system: "Within three to four years, this issue must be advanced to a point where 90 percent of the Hungarian peasant population will cultivate their land in proper Socialist collective farming."

December 18: A government decree is issued on the organization and regulation of the operation of agricultural cooperatives.

1949

A new term was added to the vocabulary of class struggle—the "*kulak* list." A *kulak* was a person whose lands reached or exceeded 14.25 hectares, or whose net cadastral income reached or exceeded 350 gold crowns.

March 3: The Politburo of the Hungarian Workers' Party passes a resolution stating that, in the course of the First Five-Year Plan (1950–1954), transition is to be accomplished from small peasant farming to large-scale farming. According to the plans, 60 percent of arable land is to be cultivated by cooperatives and 6 percent by state farms, by 1954.

The vinicultural and viticultural inspectorates and the wine communities are dissolved. Until the end of 1949 approximately 1,300 cooperative farms (or production cooperative groups) are established. Their average membership is 27 people, the average area 232 cadastral yokes. These are primarily the agrarian proletariats' and dwarf estate holders' crop producing associations. The sudden and forced increase does not represent development; the established new collective farms' production conditions are unsatisfactory and the majority of farms do not meet the various conditions for successful production. The "failed" organization also entails the destruction of traditional and small peasant farms in catastrophic proportions. As of 1949, fleeing from the land became characteristic.

1951

Late January to early February: The forced organization of cooperatives begins. Only the landless and certain strata of the poor peasantry show any interest in collective farming. Three types of cooperative groups are differentiated: Type I is known as the "general cooperative group;" type II is the "average distribution cooperative group;" and type III is called the "joint cooperative group." While a total of 2,185 cooperatives are formed up to December 1950, their number increases to 5,110 by the end of 1952.

September 6: Decree issued by the Politburo of the Hungarian Workers' Party on the "anti-*kulak*" policy.

Throughout 1951, the proportion of land owned by *kulaks* falls by 70 percent compared to 1948, and by 1952 more than one-third of the *kulaks* have become entirely landless. According to official figures from the time, at least 40 percent of members do not perform collective work, and in 1951, and especially 1952, tens of thousands of members leave the cooperatives.

1952

January 5: Establishment of the State Ministry of Agriculture and Forestry, and the Food Industrial Ministry.

September 16: The Council of Ministers tightens the official measures against those who contravene the regulations on agricultural production and the compulsory delivery system.

According to the data of Central Statistical Office, cooperatives and state farms cultivate 37 percent of the arable land during 1952. As a result of the ruthless collecting of compulsory delivery quotas—so-called barn sweeping—on the one hand 800,000 of the 1.2 million peasant families are left without sowing seed or sufficient grain to make bread, while on the other hand tens of thousands of cooperative members are left without their "portion."

1953

March 5: Stalin dies.

June 13–14: Talks between Hungarian party and government delegation and Soviet leaders in Moscow. The Hungarian agrarian policy is criticized on the following counts: coerced industrialization (especially the overdevelopment of heavy industry), the over-enlargement of the army, forced collectivization, the lowering of living standards, the disruption of supply, excessive bureaucracy and authoritarianism, show trials, and the cult of personality. Mátyás Rákosi is instructed by Moscow to retain his position as the party's secretary-general, but to relinquish his prime-ministership to Imre Nagy.

June 27–28: The session of the MDP's Central Leadership accepts a resolution of critical tone on the problems and tasks of the party. ("June resolutions")

July 4: Imre Nagy forms a new government, whose policy of the New Course aims at improving living standards. The leadership's main emphasis is on agrarian policy. Accordingly, the Council of Ministers passes almost 50 decrees on agriculture within a few months in the second half of 1953.

July 12: Significant concessions are introduced to the compulsory delivery system.

August 23: A government decree reduces the tax burdens on the peasantry.

September 25: A decree is issued by the Minister of Agriculture regulating the procedures for leaving and dissolving cooperatives.

December 13: Delivery obligations on agricultural producers are set for three years in advance.

In the second half of 1953, some 688 of the 5,224 cooperatives and cooperative groups are disbanded. The combined membership of the cooperatives and cooperative groups fall from 376,000 to 126,000 by the end of 1953. The 1,620,000 hectares of collectively cultivated land registered in mid-1953 falls to 477,000 hectares by the end of the year. At the same time, around 200,000 private farms returned to individual farming.

1954

October 31: Establishment of the Ministry of State Farms.

255 more cooperatives and cooperative groups are disbanded throughout the year. Cooperative membership falls by a further 20,000. The collectively cultivated land falls by 61,000 hectares during 1954. Agriculture receives 400 million forints more in investments than in the previous year. The series of agrarian political measures not only improves the situation of the peasantry but, with the expansion of the food supply, also results in favorable changes in every stratum of the population. Public consumption increases by almost 20 percent compared to 1953.

1955

January 8: Talks between Hungarian and Soviet party leaders in Moscow. The policies of Imre Nagy are sharply criticized. The main element of the criticism is that the policy of living standards had been overemphasized, thus the development of industry, and of heavy industry in particular, has been neglected, while in the villages the organization of cooperatives is ignored and support is given to the stabilization of small peasant farms. Khrushchev clarifies this point: "Whether or not there is a war, we must develop industry. There may be enough bacon, but if there are no airplanes, we will be in trouble."

April 14: The session of the Central Leadership of the MDP discharges Imre Nagy from all his functions since he is unwilling to undertake Socialist reform.

April 18: The National Assembly removes Imre Nagy from his post as Prime Minister and elects András Hegedűs in his place.

May 14: Hungary joins the newly established Warsaw Pact.

December 14: Hungary is accepted as a member of the United Nations.

In the course of the year: The delivery and tax burdens on agricultural producers increases, and the three-year ban on leaving cooperatives is reintroduced.

1956

February 14–25: The Twentieth Congress of the Soviet Communist Party in Moscow.

July 18–21: At the session of the MDP's Central Leadership, Mátyás Rákosi resigns from his position as First Secretary, and is replaced by Ernő Gerő.

October 23: A mass demonstration organized by Budapest university students, in sympathy with events in Poland, turns into armed revolt in the evening hours. During the night, Soviet tank units enter Budapest.

October 24: The Presidium appoints Imre Nagy as head of the Council of Ministers. Armed clashes break out all over Budapest, partly between the revolutionaries and troops loyal to the government, and partly between the revolutionaries and Soviet troops.

October 25: The MDP leadership relieves Ernő Gerő of his post and appoints János Kádár in his place as First Secretary.

October 26: Clashes and demonstrations spread throughout the entire country.

October 28: In his speech, broadcast on radio in the afternoon, Imre Nagy refers to the revolution as a national democratic movement and announces the removal of Soviet military units from Budapest.

October 30: The compulsory delivery system is abolished.

October 31: The Presidium of the MDP announces the disbanding of the party, which is re-formed as the Hungarian Socialist Workers' Party (HSWP).

November 4: The second Soviet intervention is launched; Imre Nagy and his entourage take refuge in the Yugoslav Embassy; a Revolutionary Workers' and Peasants' Government is formed, headed by János Kádár.

November 12: Expansion of the Kádár government. Imre Dögei is appointed as head of the Ministry of Agriculture. The compulsory delivery system—already abolished under the Imre Nagy government—is abolished once again.

1957

March 20–28: A delegation of government and party officials led by János Kádár travel to Moscow and signs an agreement on the continued stationing of Soviet troops in Hungary.

May 9: The National Assembly elects the new government and confirms János Kádár as its head.

June 27–29: National Conference of the HSWP.

July: Publication of the HSWP's "Agrarian Political Theses," spelling out the party's agrarian political program. In order to distance themselves from the previous agrarian policy, the program declares that both cooperative and private farming should be supported.

December 28: A law decree (65/1957) is issued by the Presidium on old age, disability, and widows' pensions for cooperative farm members.

1958

June 16: Execution of Imre Nagy, Miklós Gimes, and Pál Maléter after the handing down of death sentences on the previous day by the Supreme Court's Council of People's Tribunals.

December 7: Communiqué on the December session of the HSWP's Central Committee concerning agrarian political issues. Relaunch of collectivization.

1959

March 19: The Presidium passes law decree (7/1959) on agricultural cooperatives and cooperative groups.

October 22: The session of the HSWP's Central Committee passes a resolution on the tasks of the cooperative movement.

The number of cooperative members doubles throughout 1959 (from 200,000 to 500,000).

1960

January 15: Changes to the composition of the government: Imre Dögei is dismissed from his post as Minister of Agriculture and replaced by Pál Losonczi.

February 2–3: Representatives of the European Socialist countries' Communist and Workers' Parties hold talks on Socialist agriculture in Moscow.

February 14: Government decree (6/1960) on old age and disability allowances for cooperative members.

February 16: On the session of the Political Committee, the party announces the essence of the correction. The party leadership tolerated and acknowledged the idea that agricultural cooperatives temporarily should apply methods adapted to their conditions, differing from the Soviet *kolkhoz* model.

In the course of the year, the number of cooperative members rises from 500,000 to 900,000.

1961

February 19: The HSWP's Central Committee issues a communiqué to inform the public that "a turning point has been reached in the transformation of Hungarian agriculture: following industry, Socialist relations in agriculture have also become dominant." The statistical data reflect the turnaround as follows: in 1961 there were 271 state farms, approximately 4,200 cooperatives, and almost 165,000 individual farms registered. The cooperatives and state farms cultivated 93% of the country's arable land.

August 26: The last executions in relation to the 1956 Revolution are carried out.

December 15–16: The National Assembly passes a law on the protection of arable land, and on forests and gamekeeping.

PART III

Land Collectivization in Southeastern Europe

The Collectivization of Agriculture in Romania, 1949–1962

CONSTANTIN IORDACHI
DORIN DOBRINCU

"You cannot escape death or the *kolkhoz*."
— Gheorghe Gheorghiu-Dej[1]

INTRODUCTION

The collectivization of agriculture in Romania (1949–1962) was one of the longest and most arduous campaigns of social engineering in the countryside launched in post-1945 Eastern Europe, involving a war against the peasantry lasting more than 13 years. The length of this campaign was caused by a number of structural as well as contingent social and political factors. In 1945, Romania was the second most populous country (after Poland) in Eastern Europe, with 18 million inhabitants. Despite the relative success of the sustained efforts of industrialization conducted mostly since the mid-1930s, Romania had one of the largest agricultural sectors in Eastern Europe, with approximately 80 percent of the population living in rural areas. The Romanian Communist Party (*Partidul Comunist Român*, or PCR),[2] which had been outlawed since 1924, had emerged in 1944 as a new force on Romania's political scene. Although in 1945 the Communist Party managed to seize political power, for many years it had a weak

[1] Statement attributed to Dej by Marin, witness account in *În serviciul*, 109.

[2] Following its fusion with the Romanian Social Democratic Party, in 1948 the Romanian Communist Party was renamed the Romanian Workers' Party (RWP). In 1965, RWP returned to its pre-1948 name, the Romanian Communist Party. In this chapter, we employ either the generic term the "party" or the RWP acronym for the period up to 1965, and the Romanian Communist Party or the RCP acronym for the post-1965 period.

presence in rural areas. The party also suffered from factionalism and its cohesion was constantly marred by an internal struggle for power that was also manifest in a lack of party consensus over the country's agricultural policy. In Romania, just as in the USSR's interwar period or in other postwar popular democracies, the crystallization of the party's collectivization strategy polarized communist elites in rival factions, forcing them to articulate their priorities and loyalties. The internal power struggle within the party finally ended in 1952–53 with the elimination of the group led by Ana Pauker and her close collaborators, Vasile Luca and Teohari Georgescu. Overall, due to the political hesitation of the party elites, and to the peasants' passive and active resistance against collectivization—who remained staunchly devoted to their traditional way of life—the process of collectivization was especially protracted, marked by violent offensives followed by tactical withdrawals and concessions by authorities.

Based on a wealth of archival materials, this chapter provides a critical overview of the main stages and features of the process of collectivization in Romania, in an effort to compare it with similar campaigns that took place in the former Soviet-dominated Eastern Europe.[3] We argue that the evolution of the collectivization process in Romania illuminates both strong political motivations as well as the obstacles and limitations communist political elites faced in collectivizing the rural economy. At the beginning of the collectivization campaign, Romania's communist elites lacked strong political support at home and acted as subordinates to an external center of power. During the collectivization campaign, the communist authorities managed to establish an authoritarian structure of power based on two main pillars, the party

[3] For the history of collectivization in Romania one should consult the following archival funds: ANIC, fonds: CC al PCR—Cancelarie; CC al PCR—Secţia Administrativ-Politică; CC al PCR—Secţia Organizatorică; various county branches of the National Archives; ACNSAS, fonds: "Documentar" and "Penal;" Arhivele Direcţiei Instanţelor Militare, including the Archives of Territorial Courts in Bucharest, Iaşi, Cluj and Timişoara; Romanian Military Archives (Arhivele Militare Române) in Bucharest and Piteşti. Witness accounts can be found in archives of oral history testimonies at the Centrul Internaţional de Studii asupra Comunismului, Memorialului Victimelor Comunismului şi al Rezistenţei, Bucharest; the research center *The Third Europe* (A treia Europă), Timişoara; Centrul de Istorie Orală al Radiodifuziunii Române, Bucharest; and Institute of Oral History [Institutul de Istorie Orală], Cluj-Napoca. See also oral interviews published in the following studies: Costea, "Procesul colectivizării," 1998; Diaconescu, "Drăgoeşti," 1992; Folea, "Transformarea socialistă," 1995; Roşca, "Rezistenţa comunei," 1995; Cioşan, "Colectivizarea agriculturii," 1998; Ivaşcu, "Noi nu am fost oameni," 2000; Rujan, "Procesul colectivizării," 2001.

and the state—amalgamated in a heavy centralized party-state edifice—to consolidate its grip on rural areas and perfect its repressive apparatus. The party also established the basis of its autonomous policy toward Moscow, which would culminate in its policy of autarchic industrialization and its semi-independent foreign policy. In retrospect, the collectivization campaign appears not only as a crucial dimension in the evolution of the communist economy toward urbanization and industrialization, but also as an essential component of the communist strategy of political domination in rural areas.

The Romanian Communist Party and the Agrarian Question

Communists and the private property over land

The Romanian Communist Party was founded in 1921, during the reorganization of Romania's political system following World War I. In 1924, the party was outlawed for its revolutionary activity, its total political subordination to Moscow, and its open campaign for the territorial dismemberment of Greater Romania. The party remained outside the political system during the interwar and wartime periods. In 1944, after Marshall Ion Antonescu's removal from power as a result of a *coup d'état* in which the Romanian Communist Party actively participated, the party regained its legal status.

In the early postwar years, at a time when the Communist Party made sustained efforts at gaining internal legitimacy in order to capture the political power, the party's position on the agrarian question was always changing—as a result of both political pragmatism and a deliberate strategy of concealing its long-term goals. By judging the Soviet model that the Romanian Workers' Party (RWP) emulated, it was evident that the party's final aim was the collectivization of agriculture. Initially, the Romanian communists strategically argued nevertheless for the necessity of completing the "bourgeois-democratic revolution" in order to create the objective conditions for the socialist transformation of agriculture. To this end—and notwithstanding the fact that in 1921 Romania implemented one of the most thorough agrarian reforms in Europe (in terms of both amount and percentage or redistributed land)—Romanian communists pleaded for a new land reform to expropriate large landowners.[4] To destabilize the

[4] *Scânteia*, September 21, 1944.

political situation of the country and weaken resistance to the communist takeover, the party openly encouraged peasants to illegally occupy large estates even before the new agrarian reform was passed by the parliament.

The agrarian reform was adopted on March 23, 1945 soon after the establishment of the first communist-dominated government, led by Petru Groza. The reform was motivated more by political imperatives than economic reasoning: the aim was not to consolidate the small peasant property, as was claimed by the official governmental propaganda, but rather to abolish large estates and eliminate the aristocracy as a class, thus removing a major obstacle to communist rule in the countryside.[5]

After the implementation of the agrarian reform, the communists continued to conceal their plans to collectivize agriculture, strategically presenting themselves as defenders of peasants' land property.[6] However in 1947, the RWP launched a tacit but well-concerted campaign of confiscating large agricultural estates. In December 1947, after the abolition of the monarchy, the property of the royal family was nationalized by the state; on August 3, 1948 the state confiscated the land of confessional schools; on November 3, the goods of private sanitary institutions were also nationalized; and on December 2, the properties of the Greek-Catholic Church came under state ownership.[7] At the same time, at the official level the Communist Party continued to pledge support for small peasant property. On February 21–23, 1948 a resolution of the RWP argued for the first time in favor of cooperation as a means of alleviating the economic situation of the peasantry, but tactically emphasized that the process was to be voluntary and not state-driven or coercive. Likewise, in the spring of 1948 Secretary General Gheorghiu-Dej continued to deny rumors that the state intended to collectivize private lands.[8]

In March 1949, communist authorities took the first concrete steps toward eliminating private property in agriculture. According to Decree 83, adopted in March 1949, the state confiscated all private land properties above 50 hectares (many of which had already been expropriated once under the law on agrarian reform, no. 187 of March 23, 1945). Although

---

5 Şandru, *Reforma agrară*.

6 Şandru, "Colectivizarea agriculturii."

7 ANIC, fond Reforma agrară din 1945—Centrala, file 71/1948, ff. 1–2; "Monitorul Oficial," August 3, 1948, 177; ANIC, fond Ministerul Afacerilor Interne—Direcţia Administraţiei de Stat, file 8/1949, f. 4; "Monitorul Oficial," December 2, 1948, 281, cited in Şandru, "Colectivizarea agriculturii," 51.

8 Roberts, *Rumania*, 322.

officially the law targeted the so-called *moşieri* (a derogative social label used by communists to stigmatize large landowners as decadent pseudo-aristocracy, originating from the word *moşie*, meaning landed estate), the decree also eliminated a prosperous stratum of middle peasantry who cultivated the land by using mechanized technology.[9] Under Decree 83, the state confiscated the land, the agricultural inventory, cattle, and all the bank credits or revenues generated by agricultural activities (art. 2). The decree was implemented overnight, and came into effect on March 2, 1949.[10] Authorities in the "field" (this being the conspiratorial term employed in official documents) operated in secrecy in order to take landowners by surprise.[11] Targeted estates were occupied by military and paramilitary troops, which were prepared in advance. These properties would later serve different purposes, as head offices for state farm enterprises or for future collective farms, or as city halls or police stations. Numerous abuses and acts of theft were committed in the process.[12] Owners who resisted these requisitions or illegally hid assets were severely punished with 5 to 15 years of forced labor and their property was confiscated.[13] In addition to having their land and houses confiscated, landowners were deported together with their families and placed under house arrest in various locations across the country. Those who were not found at their homes at the moment when the decree was enacted were put under house arrest in the location where they were identified. Estimates of the number of deported persons under Decree 83 vary considerably. According to partial statistics compiled by Securitate organs, 2,000 families, comprising a minimum of 3,000 persons, were deported.[14] Historian Henry Roberts estimates the number of deported persons in the night of March 2–3, 1949 at 17,000.[15] According to historian Dumitru Şandru, at least 9,000 great landowners and rich peasants were deported,[16] an estimate he bases on

---

9 Ionescu, *Comunismul*, 218.

10 ASRI, fond Documentar, dos. 7778, vol. 36, ff. 44–45. See also Şandru, "Decretul 83," 1993. Ionescu ironically referred to the action of March 2–3, 1949 as "the last agrarian reform in Romania." See Ionescu, *Comunismul*, 218.

11 ANIC, fond CC al PCR—Secţia Organizatorică, file 144/1949, ff. 1–5.

12 For details, see Râpeanu, "Deportări şi confiscări;" Ţârău, "Aplicarea decretului;" Ionescu-Gură, "Decretul 83."

13 *Buletinul Oficial*, no. 1, March 2, 1949. See also Iancu, Ţârău, and Traşcă, eds., *Colectivizarea agriculturii*, 86–8.

14 ASRI, fond Documentar, file 7778, vol. 27, ff. 81, 89; vol. 32, f. 3.

15 Roberts, *Rumania*, 323.

16 Şandru, "Colectivizarea agriculturii," 52.

the number of 7,703 private agricultural estates over 50 hectares officially registered with the Ministry of Agriculture in November 1948. The deportation of landowners was illegal, as Decree no. 83/1949 did not contain any stipulations enabling authorities to arrest landowners and place them under house arrest. In order to justify these deportations (post-factum, as it were) and to legalize the status of detainees, the government issued HCM no. 1154/October 26, 1950 (later modified by HCM no. 344/March 15, 1951) which partially regulated the status of "certain categories of persons" placed under house arrest.[17] During the inter-party power struggles that took place in the mid-1960s, party leaders would later denounce the deportations of March 1949 as illegal acts of settling political scores and purging political rivals.[18]

In the following years, land confiscations continued uninterrupted, with the final aim of eliminating aristocracy as a social class. These confiscations lacked any legal basis, and as such, were not publicized by authorities.

Forced collection quotas and collectivization

Between 1945 and 1949, the most important change in the agrarian policy of the Communist Party was the introduction of a mechanism of forced requisitions, or agricultural collection quotas. At a time of a pronounced economic crisis and a dramatic decline in agricultural production as a result of disruptions caused by the war, these collection quotas offered the state an immediate solution for providing food supplies to urban areas and for paying heavy war reparations to the Soviet Union. Above all, however, collection quotas were an instrument of "class struggle" in rural areas. Through this campaign, the state was confiscating an important part of the farmers' harvest of cereals, sunflowers, potatoes, hay, milk, meat, wool, etc. This policy followed the model developed by the Soviets after the Bolshevik revolution and then applied in the 1920s and 30s throughout the rural areas of the Soviet Union. In Romania, quotas were introduced for the first time in 1945, and were forced upon a peasantry already impoverished by years of war, as well as by the requisitions made by the occupational troops and the drought and famine in an important agricultural area of the country in 1946–47. The system of collections was organized by

[17] ASRI, fond Documentar, file 7778, vol. 36, ff. 36–7.

[18] ASRI, fond Documentar, file 7778, vol. 27, f. 82.

means of an extensive legal framework that stipulated in detail the obligations of all producers and introduced severe penalties and punishments for failing to comply. At the local level, the system was organized and implemented by local party leadership with the support of the administration, the militia, and the justice system. Through the bureaucratic apparatus of the system of collections, the state also implemented new means and forms of control over the rural population. As expected, the collections led to great discontent among the peasants.[19] Quota collections lasted until the beginning of 1957, when, in order to release social tensions in the countryside and to thus diminish the potential destabilizing impact of the 1956 Hungarian Revolution, forced requisitions were mostly abolished, except for the requisition of meat which was maintained until 1962.

The history of the system of collective quotas can be divided into four distinct periods: 1945–47; 1948–52; 1953–56; and 1957–62.[20] The first "transition" period, from 1945–47, was marked by a relative prudence and moderation by the RWP in implementing this system. The second period, from 1948–52, was characterized by legal instability, with sudden policy reversals. These major oscillations reflected the transition from a market economy to a planned socialist system, as well as the fight for power among rival factions of the Communist Party. The period 1953–56 was marked by a vigorous and steady implementation of the system of forced collections in order to compensate for the proven weakness of the already sizeable but inefficient socialist agricultural sector. In the last period, from 1957–62, the system of forced quotas was gradually replaced by production contracts between the state and individual producers. The new pragmatic policy of production contracts contrasted sharply with the new wave of violence in rural areas that accompanied the last (post-1958) stage of collectivization.

The system of forced collection prepared the ground for the collectivization of agriculture, as it destroyed the economic standing of middle and rich peasant households. Numerous confidential documents issued by the party leaders involved in shaping the agrarian policy highlight the close connection between the imposition of the forced collections, the campaign of collectivization, and the party's overall economic strategy. For example, Ana Pauker, who was in charge of the Agrarian

[19] Dobrincu, "Colectivizarea," 285–86.

[20] For a comprehensive treatment of the system of forced requisition, summarized here, see Roske, "Colectivizarea şi mecanismul," 113–35.

Commission of the Central Committie of the RWP [*Comisia Agrară a Comitetului Central al Partidului Muncitoresc Român*], pointed out on May 29–30, 1951, at the meeting of prime secretaries and secretaries in charge of agrarian policies, that "collection quotas are a tax that we impose on peasants, on peasant households."[21] Parts of the quotas were used to provide food supplies for workers and state bureaucrats in urban areas and for the army. Another significant portion of the requisitioned goods was exported in order to generate capital for developing industry (mostly for the acquisition of industrial machinery and raw materials). Pauker argued that private owners continued to be, several years after the establishment of the new communist regime, "a small capitalist entrepreneur, in his own way, a petty-bourgeois individual peasant."[22] She portrayed collection quotas as a patriotic "duty to the state," arguing that peasants "should understand that it is a small sacrifice from their part that they have to make, that of providing our country with the food they need to deliver through the system of quotas. The one who tries to escape, in one way or another, from this basic duty, should be humiliated and stigmatized by his own village."[23] Pauker concluded by reiterating her conviction that Romania's agricultural crisis can only be solved by collectivization.

THE COLLECTIVIZATION CAMPAIGN: STAGES AND DIMENSIONS

Collectivization and the principle of class struggle

After seizing power in March 1946 and establishing a new political regime, the Romanian Communist Party (RWP) launched in 1949 what was to be its longest and most difficult campaign: the collectivization of agriculture (1949–1962). The plenary meeting of the Central Committee of the RWP that took place on 3–5 March 1949 marked a critical turn in the agrarian policy of the communist regime in Romania. Party leaders unanimously decided to launch collectivization, which was to lead to the "socialist transformation of agriculture." Although in the previous years, communist

---

21 ANIC, fond CC al PCR—Secţia Organizatorică, file 61/1951, ff. 1–6.
22 ANIC, fond CC al PCR—Secţia Organizatorică, file 61/1951, ff. 1–6.
23 ANIC, fond CC al PCR—Secţia Organizatorică, file 61/1951, ff. 1–6.

officials formally guaranteed private property, the new policy of collectivization directly opposed, and forcefully nullified, the results of the 1945 agrarian reform.[24]

The Communist Party deployed a multitude of arguments in favor of collectivization, all motivated by strategic political considerations. From an economic viewpoint, the need for collectivization was justified on the grounds that small peasant property was not viable and could not adapt to the conditions of a modern economy due to its lack of technological progress. According to the Marxist-Leninist dogma, only large-scale agricultural production owned by the state—as the only entity able to produce and successfully implement modern agriculture technology—could produce efficient results, making possible the process of socialist urbanization and industrialization.[25] However, these economic justifications were secondary to the party's ideological and political agenda. The existence of a free peasantry outside the socialist economic system and party control was unacceptable to communist authorities. On the one hand, the Marxist-Leninist dogma argued that small property in agriculture unavoidably generates capitalist production in the countryside. On the other hand, the free, unregimented peasantry controlled vital resources that the party wished to employ for the payment of war debts, the survival of the urban population, and most importantly, for the industrial development of the country. Although some of these goals were partially fulfilled by way of the system of forced collections in the countryside, it was clear that this ad hoc, temporary solution could not serve as basis for long-term economic policies of urbanization and industrialization. The social reorganization of the rural world through the expropriation of the peasantry appeared as a fundamental condition for the implementation of the communist project of forced industrialization.

The social policy of the RWP in rural areas was fully detailed in the resolution of the party plenary meeting adopted in March 1949. As political scientist Ghiţa Ionescu pointed out, this resolution is a "fundamental political text in the collectivization campaign (...) the clearest declaration of intention in the agrarian policy of the Romanian communist regime."[26] During the plenary meeting, Secretary General Gheorghe Gheorghiu-Dej described the social policy of the RWP in rural areas as follows: "We rely

[24] Frunză, *Istoria Stalinismului*, 43.
[25] See Mitrany, *Marx Against.*
[26] Ionescu, *Comunismul în România*, 219.

on the poor peasantry, we forge a closer alliance with the middle peasantry and we wage a continuous fight against the *chiaburi*."[27] The final resolution delineated five social categories of rural society: 1) landless peasants, the so-called agrarian proletariat, comprising approximately 2.5 percent of the total rural population; 2) poor peasants, comprising 56.5 percent of the rural population; 3) middle peasants, comprising 34 percent of the population; 4) the *chiaburi*, a small but influential, and for communists, a dangerous social group; and 5) the remnants of the old aristocracy, the former landowning class of the country (together, the last two strata made up an estimated 7 percent of the rural population).[28] In contrast with the resolution of the 1929 Plenary Meeting of the Central Committee of the Communist Party of the Soviet Union, which had explicitly called for the liquidation of the *kulaks*, the Romanian Workers' Party demanded only their economic exploitation, a policy officially and euphemistically called to neutralize the *chiaburi* (*limitarea* or *îngrădirea chiaburilor*). This did not mean that the Romanian Workers' Party wanted to deviate from the Soviet model or intended to tolerate the *chiaburi* in the social structure of socialist villages. Nevertheless, the party was aware of the weak political infrastructure of the new regime in rural areas and the essential economic role the *chiaburi* provided; hence, party leaders argued that "the complete elimination of the exploitation by *chiaburi* in rural areas will be done later, when conditions will be ripe, and with the assistance of the working class."[29]

The strategy of the communist authorities for transforming rural society was based on dividing peasant communities and breaking local ties by introducing the principle of class struggle with the aim of dismantling local networks and discontinuing forms of collective solidarity.[30] The elimination of large landowners had already greatly affected the social fabric of rural communities, depriving poor peasants of social patrons and opportunities for employment, thus making them vulnerable to state administrative pressures. Once it had confiscated agricultural estates, the "pater-

---

[27] "Rezoluţia şedinţei plenare," 95; Gheorghiu-Dej, *Articole şi cuvântări*, 261. The word *chiabur* is of Turkish origin (see *chibar*, meaning rich, noble); in Romanian it means "rich, well-to-do peasant." See *DEX*, 1996, 168. In the communist political language, the term *chiabur* was used as a synonym for the Russian word *kulak* and invested with political-ideological connotations.

[28] Gheorghiu-Dej, *Articole şi cuvântări*, 261.

[29] Gheorghiu-Dej, *Articole şi cuvântări*, 302.

[30] Kligman and Verdery, "Social Dimensions," 127–146.

nalist" socialist state could thus replace large landowners and take over the roles traditionally fulfilled by the aristocracy.[31]

With limited economic prospects, landless peasants were forced to work for the socialist state and act as its allies in the local collectivization campaigns. The next policy step was to pressure poor peasants to turn against the wealthier farmers, stigmatized as *chiaburi*. A March 1949 party resolution introduced the policy of containment (*îngrădirea*) of the *chiaburi*, which aimed to constrain, by all means, their economic development and possibilities to hire workers. The main techniques of this containment were "a class-based fiscal policy, which was to ease the situation of poor or middle peasants, on the one hand, but to pressure the *chiaburi*, on the other hand" and "a class-based policy of credits, forced requisitions etc."[32] The containment of *chiaburi* was to be followed, according to party documents, by the "liquidation" of wealthier peasants, or *kulaks*. In addition to the economic measures taken against the *kulaks*, the party also used direct repression, confiscating goods, and empoloying violence, including deportations.[33] This regime of terror was key to the collectivization campaign's success in the countryside.[34]

Undoubtedly, there existed economic differentiation among the peasants; in fact, the five social categories delineated in the party's resolution of March 1949 were, by and large, accurate. But the poor paesants' collective identity and interests were mainly constructed in opposition to large "absentee" landowners and townsfolk, seen as "strangers," and not to the villagers' richer strata as the party argued. Poor peasants envied the wealthier peasants but at the same time respected them, because they were part of their social networks and because they were ready to offer solidarity in case of need, as part of the village community.[35] Communist authorities understood that the *kulaks* represented a model for the rest of the peas-

[31] For the concept of the "paternalist state" and its relevance for communist Romania, with reference to gender relations and policies of forced reproduction, see Kligman, *Politics of Duplicity*. In this paper, we extend the concept of the "paternalist state" to refer to the communist policy of eliminating large landowners from villages and taking over their paternalist functions of providing means of livelihood for the poor peasantry through their enrollment in the newly established collective farms.

[32] *Rezoluţia şedinţei plenare a C.C. al P.M.R. din 3–5 martie 1949*, 96.

[33] Tănase, *Elite şi societate*, 76.

[34] Cătănuş and Roske, eds., *Colectivizarea agriculturii*, 15.

[35] Cătănuş and Roske, eds., *Colectivizarea agriculturii*, 15.

ants. A depiction of the *kulaks* and their social role by Gheorghiu-Dej is suggestive in this respect:

> The *kulaks*—the rural bourgeoisie—are a very special individual, an individual who made his household working, exploiting, and speculating, he is a combative individual, an element who is not very much different from the rest of the peasants if you look at their clothes, at the fact that they work. This is the reason why in the peasants' eyes the *kulaks* are still seen as worthy fellows, good hardworking family men. How strong this point of view is in the village upon the rest of the peasants, about their physiognomy of good hardworking men, is also proved by the fact that this mentality and the way of considering the good hardworking men has even penetrated inside our party.[36]

But the attempt to introduce the principle of "class struggle" in the countryside by setting poor peasants against the rich ones did not yield the results the party had anticipated. In many instances, peasants manifested their solidarity with great landowners or middle peasants, who were generally regarded as important to the local economy and to the village communities. In certain regions, numerous peasants opposed the deportation of great landowners, to which they were linked by clientelist or even blood relations. These actions of solidarity forced authorities to temporary renounce the deportation and to postpone it for several days, organizing it mostly during the night. In other cases, village communities helped those peasants who were classified as *chiaburi* with material support or labor to enable them to fulfill the mandatory quotas.[37] In doing so, peasants confronted the official propaganda to divide the village community with an alternative view, symbolically dividing their society into "us," the village community, and "they," the communist authorities and their collaborators, the latter being stigmatized as "traitors," "lazybones," "pushers," or "thugs." More efficient tools of social transformation proved to be administrative measures of systematic discrimination and repression targeting those rich peasants officially classified as *chiaburi*. The *chiaburi*

---

[36] ANIC, fond CC al PCR—Secţia Cancelarie, file 21/1949, f. 82.

[37] See Open Society Archive, Budapest, RL 2966/1959, "Persecutions, Illegal arrests, deportations. Source Athens, 68 years old Greek-Romanian refugee-expatriate merchant," f. 2. See also Kligman, "Creating Communist Authority."

were demonized by the communist official propaganda and discriminated against by the local and national administration.

The most important form of discrimination was the agricultural tax, which was applied progressively "in the spirit of the class struggle."[38] Numerous cases of total confiscation of the wealth of rich peasants also occurred. Thus, a special commission set up by the party in 1950 reported that the confiscation of property was a method widely used by the leadership of many collective farms as a means of increasing the size of their agricultural land or for obtaining buildings necessary for administrative-bureaucratic purposes. In most cases, such measures were initiated and approved by party organs at the county level, and were implemented by local county councils, assisted by the Ministry of the Interior and the Ministry of Justice. But in many cases local councils or even the mayors themselves took the liberty to order ad-hoc confiscations of property. The report of the special commission pointed out numerous abuses in the implementation of these measures: many confiscations were done illegally and numerous peasants were arrested as a result of administrative set-ups orchestrated by local authorities. For example, guns were secretly placed by agents in peasant households, and their subsequent "discovery" by the local militia resulted in the peasants' imprisonment and confiscation of property, although the law forbade this practice. Moreover, in numerous cases, even those classified as "poor" or "middle peasants" were subjected to this policy of confiscation. In some other cases, confiscations also took place in villages where collective farms had not been yet set up.[39]

Collectivization, deportations, and inter-ethnic relations

Collectivization served not only as a main form of "class struggle" in the countryside, but also as an instrument of repression against certain ethnic (Serbs or Germans) or religious (Greek-Catholic) groups, who became undesirable especially in border areas that were regarded by authorities as unsafe. Thus, according to the decision no. 200/1951, 10,099 families were deported from regions near the Romanian–Yugoslav border (Sânnicolau Mare, Timişoara, Deta, Reşiţa, Oraviţa, and Mehadia regions) as well as from the Băneasa and Constanţa regions, which comprised a

[38] ANIC, fond CC al PCR—Secţia Organizatorică, file 161/1949, f. 15.
[39] ANIC, fond CC al PCR—Secţia Administrativ-Politică, file 110/1950, ff. 57–61.

total of 43,899 persons. These were mainly peasant families, most relatively well off and of diverse ethnic origins: Romanians (most of them refugees from Bessarabia and Northern Bukovina territories occupied by the Soviet Union in 1940–41 and in 1944, or Macedo-Romanian refugees from Southern Dobrogea, a territory ceded to Bulgaria in 1940), as well as Serbs and Germans. The entire operation was coordinated by a special commission led by General Gheorghe Pintilie, deputy minister of Internal Affairs and the leader of the Securitate.[40] These persons were loaded into trucks and transported with a few belongings to the nearby railway stations and then deported to the Bărăgan plain in the Bucharest, Constanţa, and Galaţi regions. The deportees settled in 18 newly created villages (in the Securitate documents, these settlements were referred to as "special villages").[41] The conditions in these "special villages" or "special communes"[42] were drastic: abandoned in open fields, the deportees were forced to build their dwellings on the open soil and to improvise a means of subsistence in an unfamiliar and inhospitable environment. In view of this tragic experience of deportation, the Bărăgan plain has been often called "the Siberia of Romania."[43]

Stages in the collectivization campaign, 1949–62

The process of collectivization in Romania took place in two major waves: between 1949–1953 and 1953–1962, each divided into several stages and sub-stages.[44] The first wave in the "socialist transformation of the agriculture" (1949–1953) coincided with Moscow's attempt to impose rapid collectivization in the entire Soviet-dominated communist camp, while local

[40] ASRI, fond Documentar, file 7778, vol. 36, ff. 37–38; vol. 27, ff. 81, 89.

[41] These villages were: Dropia, Pelicani, Ezerul, and Olaru, in the Călăraşi region; Măzăreni and Zagna, in the Brăila region; Rubla, Schei, and Bumbăcari, in the Călmăţui region; Brateş, in the Galaţi region; Salcâmi, Valea Viilor, Răchitoasa, Moila Gâldăului, and Lăteşti, in the Feteşti region; Fundata and Viişoara, in the Slobozia region; and Dâlga Nouă in the Lehliu region. See ASRI, fond Documentar, file 7778, vol. 27, f. 127; vol. 36, ff. 37, 55.

[42] ASRI, fond Documentar, file 7778, vol. 27, f. 127.

[43] See published memoirs or oral history accounts by survivors in: Vighi and Marineasa, *Rusalii '51*; Vighi, Marineasa, and Sămânţă, *Deportarea în Bărăgan*; Vultur, *Istorie trăită*; Bălan, *Regimul concentraţionar*, 146–168; Vultur, "Deportări în Bărăgan."

[44] Levy, "The First Wave," 27–48; Oprea, "The Final Offensive," 49–79.

leaders opted for a more gradual process. Robert Levy has identified five distinct stages in the first wave of collectivization in Romania:

1) March 1949–February 1950;
2) February–June 15, 1950;
3) June 15, 1950–September 1950;
4) October 1950–January 1952; and
5) January 1952–April 1953.[45]

These stages were all marked by contradictory political directives from Bucharest, as a result of internal power struggles among rival political factions within the party, as leaders grappled with how closely to follow the Soviet model of collectivization. During this first wave, Romania's communist authorities were hesitant to follow Moscow's recommendation for full collectivization. This fact explains the party's shifts in its collectivization policy, marked by recurrent violent campaigns for forced collectivization but also by sudden and dramatic abandonments of this policy when confronted with violent peasant riots. Overall, efforts by local elites to fulfill Moscow's wish for full collectivization generated minor but costly progress, triggering violent peasant riots.

On May 29–30, 1951, during the fourth stage of Romanian collectivization, prime secretaries and secretaries in charge of agrarian policies held a meeting at which Ana Pauker acknowledged that "as it is well known, many collective farms were established without the peasants' consent, in some places even with the wild oppression of peasants' will, going as far as hanging them by their arms, threats of arrestments and other such practices."[46] Yet Pauker was nevertheless convinced that "there is no other road for us but the collectivization of the peasantry. There is no other socialist road. There can be other forms, but this is the goal: the collectivization of agriculture. There are different speeds, slower or faster, but that is the end point we have to get to."[47]

On September 18, 1951, still during the Stalinist period, the Central Committee of the Romanian Workers' Party published a decision criticizing the "errors" of the regime's agrarian policy, a political move that

---

[45] Levy, "The First Wave," 27–48.

[46] ANIC, fond CC al PCR—Secţia Organizatorică, file 61/1951, ff. 6–7. On Ana Pauker's attitude toward the collectivization of agriculture, see Levy, *Ana Pauker*. For another mention of this meeting, see above, fnt. 21–3.

[47] ANIC, fond CC al PCR—Secţia Organizatorică, file 61/1951, ff. 6–7.

exhibited striking similarities to the actions taken in the USSR in 1931. The document stipulated that the aim of total collectivization should be abandoned and replaced with an emphasis on farm associations or voluntary agricultural associations, in which peasants maintained their individual property.[48] In line with this decision, the plenary session of the Central Committee of the Romanian Workers' Party held on August 19–20, 1953, after Stalin's death, marked a change in the party's agrarian policy. The collectivization campaign was gradually slowed down, de facto bringing "the first assault the communist elite organized upon peasantry" to an end. Overall, the first wave of collectivization had thus ended in failure (1949–53); it was in fact the first openly acknowledged failure the communists had had since they took power in1945.[49]

At the political level, the most significant event during this initial wave of the collectivization campaign was the elimination of the faction led by Ana Pauker from the leadership of the agrarian section of the Central Committee of the Party. She was accused of "rightist deviation"and held responsible for the abuses and delays in the collectivization campaign. On July 3, 1952, after the "rightist deviationists" had been eliminated, Gheorghe Gheorghiu-Dej announced an open fight against the *chiaburi*, launching a new campaign of political purges, arrests, and show trials in Romania.[50]

The villages collectivized in the first wave were generally located in regions in which the rural economy and particularly the agricultural sector had suffered greatly as a result of World War II, and in those rural areas hardest hit by the severe drought of 1945–46. In these regions, poor peasants constituted a large and easy-to-manipulate category by the administrative pressures of the party-state, given the crises in social relations caused by the elimination and deportation of large landowners. It is interesting to note, however, that many villages which initiated the process of collectivization in 1949 by establishing the first collective farms were to be fully collectivized only in 1961–62. This was due to the fact that the initial enrollment of poor peasants allowed for rapid progress in the first phase of collectivization but, since the voluntary potential of collectivization was limited to the poorest peasants, the process stagnated after that initial drive. Since middle and rich peasants refused to join collective

[48] Ionescu, *Communism in Rumania*, 234; the text of the decision is published in Cătănuş and Roske, eds., *Colectivizarea agriculturii*, 246–64.

[49] Tănase, *Elite şi societate*, 77.

[50] Roske, "Colectivizarea agriculturii," 160.

farms, full collectivization was possible only by violent means in the last stage of collectivization.

A special category of collectivized villages were those located in regions considered "unsafe" by the party, such as the border areas or in provinces in which anti-communist resistance was particularly strong, like in Dobrogea or in Maramureş. In these regions, early collectivization was a form of political repression or even of punishing and disciplining the local population. This practice suggests that there was a direct correlation between the intensity of anti-communist resistance and the violence employed in the process of collectivization.[51]

The changes in the agrarian policy of the PMR from 1951–53 were not carried out for humanitarian reasons but primordially for economic ones: the need to supply the cities with much-needed food, as well as the requirement to export to USSR, led the regime to certain concessions.[52] At the same time, the central objective—to abolish private property—was consistently followed in order to gain control over society, which was the party's absolute priority.[53]

The second wave of collectivization (1953–62) confirmed the total subordination of Romanian communist political elites to Moscow.[54] The wave unfolded in several stages. The first stage took place during summer 1953 and December 1955. This period is usually regarded as "the years of stagnation," since after Stalin's death the campaign of collectivization was gradually halted and the party's emphasis shifted to consolidating existing collective farms. The second stage corresponds roughly with the year 1956. On the political level, party officials resumed vigorous debates and plans regarding full collectivization. Yet for strategic reasons, the implementation of this plan was postponed, mostly as a result of foreign policy considerations—the most important being the 1956 Revolution in Hungary and the fear of revolutionary unrest in Romania.

The third and final stage of the second wave of Romanian collectivization occurred between 1957–62. In 1957, communist authorities conducted a local experiment in the Galaţi region in order to explore the potential for collectivization. The results of this experiment were considered encouraging by authorities, who decided to resume collectiviza-

---

[51] See Iordachi, "Constanţa," 131; and Kligman, "Creating Communist Authority," 165–6.

[52] Ionescu, *Communism in Rumania*, 234; Tănase, *Elite şi societate*, 77–8.

[53] Tănase, *Elite şi societate*, 78.

[54] Oprea, "The Final Offensive," 49–79.

tion in full force.[55] The campaign restarted in Constanţa region, nearby Galaţi, where the party dispatched or mobilized 30,000 activists and agitators (one activist per every three non-collectivized peasants).[56] In November 1957, following a campaign of unprecedented violence led by Vasile Vâlcu, the local secretary general for the regional organization, Constanţa was declared the first collectivized region of Romania. This coincided with the 40th anniversary of the 1917 "Great Socialist Revolution," to which the accomplishment was dedicated.

Map 1: The collectivization of agriculture in Romania by administrative regions, according to the percentage of collectivized land, at the end of 1958.

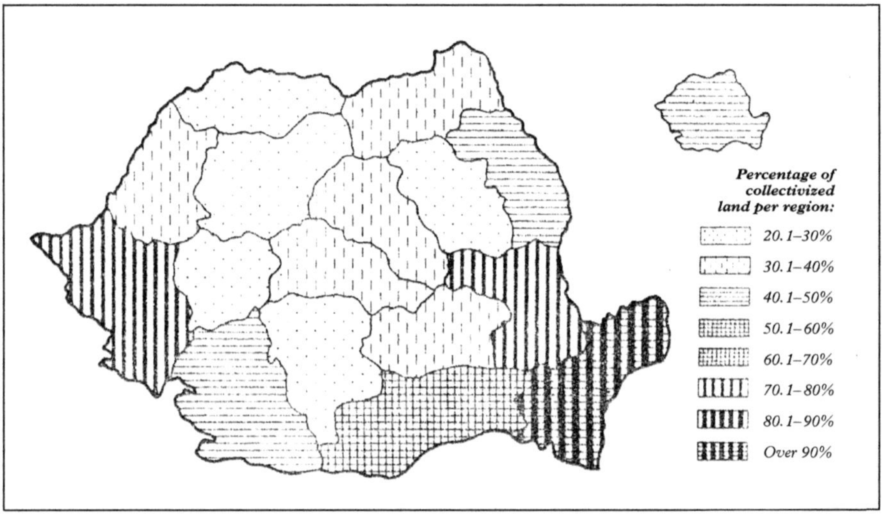

Instigated by Soviet advisors, the rapid collectivization of Dobrogea was part of a broader plan to achieve the complete socialist transformation of the Constanţa region. Due to its strategic geo-political position that allowed the Soviets to consolidate their control over the Balkans and the eastern Mediterranean region, the province was regarded by authorities as a pilot project in socialist modernization. Its success in the collectivization campaign was grossly exaggerated and embellished by official propa-

[55] ANIC, fond CC al PCR— Secţia Cancelarie, file 62/1956, "Protocol no. 25 al şedinţei Secretariatului CC din 11 iunie 1956," ff. 1–2; Oprea, "The Final Offensive," 59–63.

[56] *Scânteia*, October 19, 1957, 1.

ganda. Dobrogea was popularized in brochures and other materials "documenting" the miraculous transformation of the province from "a pile of ruin" during the bourgeois past to the region with "the highest standard of living of the rural population" in Romania.[57]

The example of Dobrogea generated a fierce competition among regional party organizations to complete the collectivization campaign in record time so that in the entire country the process ended officially in March 1962, three years ahead of the deadline set by the party. Each region was nevertheless collectivized following a political and geo-strategic logic (see Map 1 above)[58]. Priority was given to the multi-ethnic border provinces, such as Dobrogea and the Banat. These were followed by regions that had been considered unsafe from a political point of view, such as Maramureş, and then by the Bărăgan plain, the agrarian hinterland near the capital Bucharest, which was not only geopolitically important but also vital in terms of food supplies. Regions of less strategic importance, such as Suceava, which were ethnically homogeneous and had a limited economic significance, were among the last areas to be collectivized, even though they were predominantly rural regions.

On April 27–28, 1962, at an extraordinary session of the Grand National Assembly in Bucharest, Gheorghiu-Dej proclaimed the success of the campaign for the "socialist transformation of the agriculture." After thirteen years of systematic and persistent actions against the peasantry, the socialist sector in agriculture encompassed 96 percent of the total arable land of the country and 93.4 percent of the total agricultural land.[59]

Resistance and Repression in the Collectivization Campaign

The measures the authorities took in rural areas—especially economic ones—sparked protests among the peasantry, in the form of individual resistance and collective uprisings. During the collectivization campaign numerous riots occurred in different regions of Romania, some resulting in open confrontations with the troops of the Militia and Securitate,

[57] Iordachi, "Constanţa," 106.

[58] For discussion of the motivations behind the chronological succession of collectivization in various regions, see also Kligman and Verdery, *Peasants under Siege*, 143–45.

[59] Gheorghiu-Dej, *Raport*, 5.

which cost many peasant lives. Some peasants were executed, others were arrested, tried, and sentenced. The main target of repression were those peasants who held prestigious positions in local communities, and who had played a role during the anti-government resistance and were seen as obstacles in the collectivization campaign. In many cases, however, peasants became random victims of local activists' revenge. Resistant peasants were unmercifully repressed, as the communist authorities feared that the riots could spread.[60] Overall, during the collectivization campaign, Romanian villagers were subjected to a particularly brutal repression, and no appeals against these brutalities were accepted by the regime. In order to cope with the strains of the system of collecting quotas and the pressures of the collectivization campaign, Romanian peasants adapted various strategies of (unequal) negotiation, passive but also active resistance to these policies.[61] State violence and peasant opposition were such mutual conditions of the collectivization campaign that the two issues cannot be analyzed separately.

In 1949, mandatory collection quotas were so high that peasants could only preserve seeds for the next year's harvest but not food reserves. Such a situation was common in numerous counties, mostly in Hunedoara and Someş, according to party reports. In certain villages, local authorities went from household to household to confiscate all grains in order to fulfill quotas.[62] In that year, the harvest was especially poor in southern Romania, hence the authorities focused their campaign of forced requisitions on Transylvania and the Banat.[63]

These practices of spoliation stirred numerous violent revolts in the summer of 1949.[64] An enumeration of the locations of these revolts illustrates the magnitude of these protests, which occurred in the counties of: Botoşani (Roma commune);[65] Dorohoi (Vorniceni, Avrămeni, Grămeşti, Dersca, Cândeşti, Brăeşti, Cracalia, Dimăcheni, Havârna, Hilişeu, Mihăileni, Zamostea, Liveni, Adăşeni, and Mitoc);[66] Rădăuţi

60 Dobrincu, "Transformarea socialistă," 459–87.

61 See Iordachi and Dobrincu, *Transforming Peasants*.

62 ANIC, fond CC al PCR—Secţia Organizatorică, file 161/1949, ff. 13–14.

63 ANIC, fond CC al PCR—Secţia Organizatorică, file 161/1949, f. 10.

64 See Dobrincu, "Colectivizarea," 291.

65 ASRI, fond Documentar, file 1908, f. 358; ANIC, fond CC al PCR—Secţia Organizatorică, file 144/1949, ff. 52–58.

66 ASRI, fond Documentar, file 1908, ff. 353–357, ff. 384–386; file 7611, vol. 1, ff. 61–74.

(Calafindeşti, Frătăuţii Noi, Rogojeşti, Milişăuţi, Văscăuţi, Bălcăuţi, Negostina, Condeşti, Horodnicul de Sus, Morjinca, and Ruşi);[67] Suceava (Bălăceana and Stroieşti);[68] Bihor (Cheţ, Almaşul Mare, Almaşul Mic, Chenetea, Cociuba Mare, Tăuteu, Bicaciu, Giriş, Girişul Negru, Batăr, Belfir, Coroiu, Ucuriş, Ursad, Craiova [today Craiva], Susag, Diosig, Arpăşel, Tulca, Talpoş, Suiug, Marginea, Saldabagiul, Barcău, Barba, Bicaciu, and Marţihaza);[69] Arad (Şomoşcheş, Moţiori, Berechiu, Apateu, Şepreuş, Ţipar, and Vărşand);[70] Sălaj (Zalău area);[71] Braşov (Drăguş, Ucea de Sus, Arpaşul de Jos, and Sebeş);[72] Târnava Mare (Zorgheş and Chijapla);[73] and Făgăraş, Turda, and Mureş.[74] Communist authorities intervened promptly and brutally, alarmed by the intensity and scale of these revolts, especially of those from Bihor, Arad, and Rădăuţi counties, but also of the prospect of unrest spreading to other parts of the country. On orders from the authorities, Securitate and Militia troops or border patrols opened fire on the revolting peasants, causing numerous casualties. They also executed participants in the revolt (16 peasants in Bihor county and 12 in Arad county). Hundreds of peasants were arrested, sent to trial and sentenced by military tribunals to long prison terms. Authorities also deported the families of those killed or imprisoned, mostly in the province of Dobrogea. According to available data, 1,548 persons were deported from Bihor and Arad counties between August 3 and 12, 1949. On the night of August 8, 1949, over 100 persons were deported from Rădăuţi, Suceava, and Dorohoi counties.[75] Soon after pacifying these areas, agricul-

---

67 ASRI, fond Documentar, file 1908, ff. 156, 352; file 4638, ff. 93–102; ANIC, fond CC al PCR—Secţia Organizatorică, file 41/1959, f. 4; file 161/1949, f. 13; AN Suceava, fond Poliţia Oraşului Rădăuţi, file 10/1949, ff. 1, 2–3 bis, f. 9.

68 Dobrincu, "Transformarea socialistă."

69 ASRI, fond Documentar, file 1908, ff. 102–5, 107–8, 111–4, 144–50, 203–6, 240, 306–8; file 4638, ff. 45–6, 65 bis şi ff. 290–2; ANIC, fond CC al PCR—Secţia Organizatorică, file 41/1959, ff. 3–4; file 71/1949, ff. 3–8 bis; file 176/1949, ff. 1–20; Ardelean, "O răscoală ţărănească," 3; Breazu and Criste, "Revolta ţărănească," 698.

70 ASRI, fond Documentar, file 1908, ff. 61–5, 128, 144–150, 158, 170, 200, 270; file 4638, ff. 19–38, 66–9; ANIC, fond CC al PCR—Secţia Organizatorică, file 41/1959, ff. 3–4; file 71/1949, ff. 8 bis-10.

71 ASRI, fond Documentar, file 1908, ff. 144–150, 201–202 and ff. 209–10.

72 ASRI, fond Documentar, file 1908, ff. 435–6.

73 ASRI, fond Documentar, file 1908, f. 401.

74 ANIC, fond CC al PCR—Secţia Organizatorică, file 161/1949, f. 11.

75 ANIC, fond CC al PCR—Secţia Organizatorică, file 71/1949, ff. 22–33 and ff. 39–50; Dobrincu, "Colectivizarea;" Dobrincu, "Transformarea socialistă."

tural production was resumed by force and the pace of collectivization was intensified.[76]

Protests against forced requisitions and quotas were frequent in 1950 as well. Authorities ordered peasants to pay all their state taxes in agricultural products. But the price fixed by the state authorities for agricultural products was considerably lower than the price on the socialist market.[77] In reaction, peasants filed letters of complaints, either individually or in groups.[78] In numerous cases, in order to provide the taxes owed to the state, peasants had to sell their entire grain reserves, which left them without food or seeds for the following year.[79] In July 1950, the authorities' refusal to modify this policy led to violent peasant riots, mostly in Ilfov, Ialomiţa, and Vlaşca counties.[80] In many villages, numerous party members or bureaucrats in the local administration sided with the peasants, some even leading these revolts. Peasant riots spread rapidly through southern Romania to some other regions of the country,[81] a phenomenon documented by numerous party reports.[82] Fearing the prospect of

---

[76] ANIC, fond CC al PCR—Secţia Organizatorică, file 161/1949, f. 11. In the archives of the former Central Committee of the Romanian Communist Party there are numerous photos documenting the peasant revolts of July–August 1949 in Arad and Bihor counties, including some with peasants executed and then publicly displayed in their villages to deter revolts in the future. In the following months, authorities launched a propaganda campaign targeting the revolting villages, staging long convoys of peasant carts transporting grains to the state granary as proof of the great and instant success of socialist agriculture in the region. Among the convoys, happy peasants were dancing the traditional *hora*, allegedly satisfied for being able to successfully deliver the mandatory collection quotas to the socialist state. See photos in ANIC, fond CC al PCR—Secţia Organizatorică, file 43/1949, ff. 27–45.

[77] ANIC, fond CC al PCR—Secţia Organizatorică, file 7/1950, f. 50.

[78] ANIC, fond CC al PCR—Secţia Organizatorică, file 7/1950, f. 29.

[79] ANIC, fond CC al PCR—Secţia Organizatorică, file 7/1950, f. 31.

[80] See the communes of Ciuperceni, Siliştea, Berceni, Tudor Vladimirescu, Căminească, Baciu, Udeni, Cosmeşti, Blejeşti, Scurtu, Negreni, Sârbeni, Corbii Mari, Corbii Ciungi and Ghimpaţi, etc. ASRI, fond Documentar, file 4640, ff. 148–9, 151–9; Damian, Breazu and Bălan, eds., *Colectivizarea în Vlaşca.*

[81] ANIC, fond CC al PCR—Secţia Organizatorică, file 7/1950, ff. 2–3.

[82] ANIC, fond CC al PCR—Secţia Organizatorică, file 7/1950, 1–50. For example, a general revolt took place in 1950 in the village of Piscuri, Sărdăneşti commune, Gorj county, stirred also by the obligation of mandatory quotas. Peasants clashed with the Militia and Securitate troops sent to repress them. The heavy fighting resulted in casualties on both sides (ANIC, fond CC al PCR—Secţia Organizatorică, file 7/1950, ff. 42–43; ACNSAS, fond Documentar, file 7, vol. 13, ff. 3–4).

increasing regional uprisings, authorities repressed the revolts using violence, once again.[83]

The socialist sector in agriculture expanded in the years 1957–58, mostly through an increase in the number of TOZ cooperatives.[84] However, the authorities did not renounce the idea of full collectivization, but rather they were preparing the final stages, which lead to more violence in rural areas. The expansion of the socialist sector was thus accompanied by numerous revolts, the most important taking place in Suraia, Răstoaca, and Vadu Roşca, in Vrancea county. They resulted in tens of thousands of victims and numerous arrests among the peasants, followed soon by political repression.[85]

This wave of peasant revolts coincided with the last stage of the collectivization campaign. The plenary meeting of the Central Committee of the Romanian Workers' Party in November 1958 stipulated the simultaneous acceleration of the pace of both collectivization and industrialization.[86] Pressure for collectivization stirred peasant revolts all over the country—in Cluj county (Cuceu commune) in Transylvania, in Gorj county (Măceşul de Jos), Mehedinţi county (Ciupercenii Noi and Calafat), Olt county (Dăbuleni and Vâlcele), Dolj county (Cioroiaşi and Băileşti), Teleorman county (Dobroteşti, Negreni and Tătărăşti) in Wallachia, and Suceava county (Drăgoieşti) in Moldova.[87]

Although the collectivization campaign ended in 1962, the subsequent years were marked by grave tensions in the rural world. According to a report from the Ministry of Foreign Affairs on November 4, 1964, entitled *Buletin Informativ Special cu privire la unele manifestări şi acţiuni negative în cadrul gospodăriilor agricole collective* [Special information bul-

---

83 Gâdea, "Se trage;" Stoian-Gărăgăianu, "Clopotele;" Călin, "7 iulie 1950;" Bălan, "Mişcări ţărăneşti;" Breazu, "Pârjolul din Vlaşca."

84 TOZ—*Tovarishchestvo po sovmestnoi Obrabotke Zemli*, or Association for Common Tilling of the Land) were loose Soviet agricultural associations in which peasant members preserved their private ownership over land but worked it collectively and shared the harvest proportionally depending on the size of their parcels and their contribution. These associations spread in the earlier stages of the collectivization campaign, being generally seen by peasants as an alternative to full collectivization and by authorities as a step toward collective farms. In Romanian they were referred to as *întovărăşiri*.

85 See Ioan, "Dreptul la proprietate;" Ioniţoiu, "Subjugarea ţărănimii;" Stoica, "One Step Back."

86 See Oprea, "The Final Offensive," 103–4.

87 Roske, "Colectivizarea agriculturii," 165–8.

letin concerning some actions and negative manifestations in collective farms] peasants were largely discontent with the state of affairs in the socialist system of agriculture. In villages, rumors circulated about imminent radical changes to the socialist sector. Numerous collectivized peasants expressed their hope that collective farms would be dismantled and that their land, animals, and inventory would be returned. In some areas, such rumors led to open rebellion against collective farms. Peasants also protested against the low level of remuneration in collective farms. Some peasants used forms of non-violent resistance, refusing to work and participating in numerous spontaneous agricultural strikes that took place in various regions of the country. Official documents also prove the widespread practice of theft by peasants from the collective farms,[88] which was described by authorities as a grave form of sabotage but regarded by peasants as a "legitimate" form of compensation and redistribution for their unpaid work.

Violence and resistance: repression in the countryside

During the collectivization campaign, the party-state consolidated its repressive apparatus and strengthened its control over the rural world. The great number of peasants arrested and convicted during this period reveal the massive scale of state repression in the countryside. At the beginning of the 1960s, when the communist regime was reconsolidating its power, Gheorghe Gheorghiu-Dej, the uncontested leader of the Romanian Workers' Party, denounced early abuses in the process of "building socialism" in Romania, claiming that these were due to the treacherous activity of the faction led by Ana Pauker. The Ministry of Justice provided the Central Committee with statistical data on the number of peasants imprisoned in the period from 1950–53 "for various offenses in the agricultural sector."[89] The report pointed out that in numerous cases these offenses were used against the peasants "with the aim of pressuring them to assist in the process of the socialist transformation of the agriculture."[90] From 1950–53, a total of 89,401 peasants were sentenced to prison for failure to fulfill production plans and to deliver collection quotas. In addi-

[88] ANIC, fond CC al PCR—Secţia Administrativ-Politică, file 17/1962, ff. 59–75.
[89] ANIC, fond CC al PCR—Secţia Administrativ-Politică, file 17/1962, ff. 59–75.
[90] ANIC, fond CC al PCR—Secţia Administrativ-Politică, file 17/1962, ff. 59–75.

tion, 80,077 peasants were punished for various minor infractions under Article 183/1949 (for violating rules against speculative or clandestine economic activities).[91]

Gheorghiu-Dej exploited these reports in order to document abuses in the Pauker-led collectivization campaign. In October 1961, in his speech to the Twenty-Second Congress of the Communist Party of the USSR held in Moscow, Gheorghiu-Dej stated that between 1949 and 1952, more than 80,000 peasants were arrested in Romania, most belonging to the poor and middle strata. Over 30,000 of these peasants were subjected to show trials.[92] Again, Gheorghiu-Dej blamed this campaign of terror on the repressive measures taken by the "deviationist" trio of Ana Pauker, Vasile Luca, and Teohari Georgescu.

More detailed data is available for the period September 1, 1948, to November 7, 1949. According to official data, during this period the Securitate detained 23,597 persons, among which there were 10,152 peasants (4,518 middle peasants, 2,979 poor peasants, and 2,655 *chiaburi*).[93] A report dated December 1, 1961, mentioned 34,738 peasants arrested by the Ministry of the Interior in 1951–52: out of this number, 22,088 were *chiaburi*, 7,226 were middle peasants, and 5,504 were poor peasants. During the same period the authorities staged 438 show trials against peasants.[94]

From a regional perspective, the highest number of arrests were made in the regions of Hunedoara, with 16,146 peasants (out of which there were 14,394 *chiaburi*, 1,254 middle peasants and 498 poor peasants), Oradea, with 9,760 peasants (2,919 *chiaburi*, 2,854 middle peasants and 3,987 poor peasants), Cluj, with 4,025 peasants (1,715 *chiaburi*, 1,811 middle peasants and 463 poor peasants) and Constanţa, with 1,484 peasants (1,210 *chiaburi*, 257 middle peasants, and 17 poor peasants). In most other regions of the country, there were less than 1,000 arrests. The lowest

---

91 ANIC, fond CC al PCR—Secţia Administrativ-Politică, file 17/1960, ff. 1–4.

92 *Scânteia*, no. 5371, December 7, 1961,1–2. See also Gheorghiu-Dej, *Articole şi cuvîntări*, 206.

93 ASRI, fond Documentar, file 9047, vol. 3, f. 43.

94 ASRI, fond Documentar, file 7778, vol. 27, ff. 9–10, 91. Other documents issued by the Ministry of Internal Affairs, Service "C," most probably also in 1960, mention that in 1951–52, some 4,644 persons from rural areas were arrested. Among them, 38 had participated in anti-communist armed gangs, while 4,606 were guilty of "agitation with hostile character against the process of socialist transformation of the agriculture" (1,305 persons in 1951 and 3,301 in 1952— ASRI, fond Documentar, file 7778, vol. 3, ff. 92, 99).

number of arrests occurred in Suceava, with only 4 *chiaburi* arrested (one condemned and three interned in labor camps), Ploieşti, with 98 peasants (65 *chiaburi*, 28 middle peasants and 5 poor peasants) and Bacău, with 99 peasants (55 *chiaburi*, 32 middle peasants and 12 poor peasants).[95]

Official figures documenting the harsh repression against the peasantry are extensive. Between 1950 and 1954, 4,865 peasants were interned in labor camps, joining an additional 17,212 prisoners who had already been in detention by 1950.[96] At the same time, 1,899 peasants were placed under house arrest (1,407 of which were former members of the Iron Guard, two former members of the National Peasant Party, two former members of the Liberal Party and 488 without political affiliation), together with other 1,734 of various professions.[97] Overall, out of a total of 11,526 persons on house arrest, there were 1,643 peasants (to which one has to add the deportees of the special villages of the Bărăgan plain, see above).[98] State repression of the peasantry intensified toward the end of the 1950s. Thousands of peasants were sentenced to prison and their property confiscated. Their families were obliged to pay high rent to the state in order to continue living in their own houses. These repressive activities were extensively documented by the bureaucratic machinery of the communist regime.[99]

---

95 ASRI, fond Documentar, file 7778, vol. 3, ff. 85–86.

96 ASRI, fond Documentar, file 7778, vol. 27, ff. 83–84.

97 ASRI, fond Documentar, file 7778, vol. 27, ff. 125–26.

98 ASRI, fond Documentar, file 7778, vol. 27, f. 127. An undated document, most probably issued in the mid-1950s, mentions that 166 peasants were sentenced to forced domicile for up to five years, one peasant for up to seven years, and 150 for an unlimited time. (ASRI, fond Documentar, file 7778, vol. 27, f. 128). For additional information on the repression against peasants in the first years of the collectivization campaign, see *Cartea Albă*, 1994–1997; Bocu, "Condamnarea ţăranilor," 85–9; Cătănuş and Roske, eds., *Colectivizarea agriculturii*; Dobrincu, "Dinamica represiunii;" Dobrincu, "Colectivizarea agriculturii."

99 Only at the end of 1950s and beginning of the 60s, when the collectivization campaign was to a great extent completed, did the leadership of the RWP decide to return the property of those prosecuted. See ANIC, fond CC al PCR—Secţia Organizatorică, file 15/1961, ff. 1–213; file 16/1961, ff. 1–249; file 50/1961, ff. 1–152; file 51/1961, ff. 1–154; file 52/1961, ff. 1–117; file 53/1961, ff. 1–155. For example, in Galaţi region, 167 peasants were pardoned under Decree no. 332/1959 (ANIC, fond CC al PCR—Secţia Organizatorică, file 52/1961, ff. 23–47); 112 peasants were pardoned under Decree no. 79/1960 (ff. 2–21), and 10 under Decree no. 100/1960 (ff. 21–2).

On March 18, 1959, 4,625 peasants were detained and were either sentenced to prison or put under investigation. Regarding their political affiliations, 744 had been members of the Legionary movement, 270 members of the National Peasant Party, 167 members of the National Liberal Party, 51 members of "other bourgeois parties," 241 had been members of the Communist Party or its youth organization, while 3,152 were placed in the category "diverse."[100] In the coming months, these figures were slightly modified: on May 30, 1959, 4,928 peasants were detained in Romanian labor camps, reminiscent of the Soviet "Gulag." According to their political affiliation, 870 had been members of the Legionary movement, 320 members of the National Peasant Party, 219 members of the National Liberal Party, 63 members of "other bourgeois parties," 263 had been members of the Communist Party or its youth organization. The remaining 3,194 persons were charged for various offenses: 2,556 for "conspiracy against the established political order and terrorism," 24 for "treason and espionage," 339 for "failure to denounce and for assisting infractions," 36 for "sabotage in industry, agriculture, transportations," 223 for infractions dealing with "religious sects," 58 for "spreading manifestos and possessing subversive or forbidden publications," 43 for "crime against peace and humanities," 177 for "thefts from the public wealth etc." Peasants represented by far the most important contingent of political prisoners, a clear indication that communist authorities were at war with the peasantry.[101] On August 18, 1959, 5,341 counter-revolutionary peasants had been interned in the Romanian labor camps. Among these, 705 peasants were arrested in 1956; 1,308 peasants were arrested in 1957; 1,829 peasants were arrested in 1958; and 1,499 peasants were arrested in 1959. Only 655 of these were proposed for amnesty, the rest continued to remain under state "surveillance."[102]

Gradually, due to the wealthy peasantry's pauperization and the change in the regime's agrarian policy, references to the *chiaburi* would eventually disappear from the official discourse. By the end of the collectivization campaign, the *chiaburi* were no longer targeted as a social category. Moreover, in order to enforce the complete collectivization of villages, the regime allowed and even encouraged rich peasants (who were formerly classified as *chiaburi*) to join collective farms.

---

[100] ASRI, fond Documentar, file 7778, vol. 3, f. 162.
[101] ASRI, fond Documentar, file 7778, vol. 3, ff. 160–1, 165, 168–71.
[102] ASRI, fond Documentar, file 7778, vol. 3, ff. 154–55, 157.

CONCLUSIONS: ROMANIAN COLLECTIVIZATION IN LONG-TERM PERSPECTIVE

The collectivization of the Romanian agriculture (1949–62) was part of the larger process of post-World War II campaigns of collectivization in the Soviet-dominated Eastern Europe that occurred at the initiative of and under Moscow's political control. While sharing general features with this process at a regional level, the campaign of collectivization in Romania exhibits nevertheless several particularities, shaped (among other factors) by the country's social structure and level of economic development, its political traditions and the strength of the Communist Party, its geopolitical position, and communist leaders' relation to Moscow.[103]

First, Romania was among the least urbanized and industrialized Eastern European countries, with eighty percent of its population living in rural areas and engaged in agriculture at the onset of collectivization. Second, the Romanian Communist Party was among the weakest parties of its kind in Central and Eastern Europe: outlawed in 1924 and forced to operate underground, it had had an estimated 1,000 members in 1944, many who were either placed under arrest by the Antonescu regime or who were active from Moscow. The task of transforming the countryside was tremendously challenging in Romania and, in view of the limited capabilities and expertise communists had at their disposal in 1944, highly unrealistic. This gap between political goals and capabilities explains why the campaign of collectivization in Romania was among the longest and most arduous in the Eastern Bloc, lasting for more than 13 years. It also explains why Romanian communists, highly vulnerable at home, were so heavily dependent on political assistance and support from Moscow.

The long history of the collectivization campaign highlights the inability of the Romanian Communist Party—consisting at the time of a small group of politically inexperienced activists with a limited understanding of rural society—to subordinate the peasantry. It was only though systematic efforts by the party for more than 13 years, employing large-scale and indiscriminate techniques of physical and psychological terror (relying on the socialist militia, the Securitate and paramilitary groups)

[103] For a more comprehensive comparison of Romanian collectivization with similar processes that took place in Soviet-dominated Eastern Europe, see Iordachi and Verdery, "Conclusions."

that the party succeeded in achieving full collectivization by 1962, enrolling the peasantry into the socialist sector of the economy.

The sustained efforts of the authorities and the huge human and material resources invested in this process proves that collectivization was a top political priority of the regime, on its success depended on the implementation of other key processes of socialist transformation, such as urbanization and industrialization. Although during this long campaign the party suffered numerous obstacles and drawbacks, in the end the outcome was celebrated, in a triumphalist manner, as a major constitutive success of the new regime. The elimination of the *moşieri* and *chiaburi* from the village social structure, and the quasi-complete incorporation of the peasantry into the socialist economy, allowed the party-state to launch evermore ambitious plans of socialist development. Three years after the end of collectivization campaign, Nicolae Ceauşescu, the new secretary general of the party, adopted a new constitution, changed the official name of the country (from the "Popular Republic" to the Socialist Republic of Romania) and of the party (from the Romanian Workers' Party to the Romanian Communist Party), and embarked on large-scale plans of industrialization, urbanization, and rural systematization meant to consolidate the country's economic independence and obliterate differences between the rural and urban regions and between the peasantry and the working class.

In economic terms, although the main goal of the campaign was the modernization of Romanian agriculture, socialist economic policies largely failed to achieve their aim of bringing prosperity to the rural society. True, notable progress was achieved in several regions of the country, due mostly to massive investment from the central budget. Overall, however, the economic gaps in agriculture increased as compared not only to Western European countries but also to neighboring socialist countries. With few exceptions, such as the model collective farm of Sântana in Arad county,[104] collective farms did not bring peasants the material prosperity promised by the official propaganda. On the contrary, the system contributed to the pauperization of the peasantry,[105] particularly in

[104] Goina, "Ce poţi face azi."

[105] Exceptions were provided by a few showcase collective farms, which managed to provide their members above-average material benefits. Also, in certain regions, male peasants were able to find jobs in nearby industrial centers, as daily commuters, and to thus get access to an industrial wage, while their wives worked in the local collective.

the late 1980s when, at a time of a general crisis of the socialist economy in Romania, collection quotas were reintroduced and food consumption was strictly rationalized, so that certain categories of peasants were forced to buy basic products (such as bread) from urban centers. Symptomatic for the failure of socialist agriculture was that, even under these constraints, the "private" agricultural sector made a significant contribution to the country's food production. Nicolae Ceauşescu himself admitted in 1984 that, although the private plots owned by cooperativized peasants and the non-collectivized households accounted, together, for only 15% of the agricultural land of the country, they owned 49% of the total herds of cows, 49% of the total number of sheep, and 57% of the total number of laying birds in the country. Dissatisfied with the lack of efficiency in the socialist sector, which consumed more than it produced, in January 1986 Ceauşescu reinstated the system of forced requisitions that functioned at the beginning of the collectivization campaign, meant to systematically exploit the peasantry for the benefit of the state. New regulations demanded that "all holders of farmland contract with the state a production plan of crop and animal husbandry," on the basis of which they were forced to handle their agricultural production to the state, an activity presented as "a patriotic duty of every peasant."[106] To assure the extraction of food requisitions vital to the socialist economy, the system of agricultural production and distribution was placed under strict surveillance, with common offences such as stealing or negligence criminalized as forms of subverting the national economy and severely punished.

No wonder that in December 1989, upon the collapse of the communist regime, Romanian peasants rushed to spontaneously dismantle collective farms, claiming back their lost property. As a result, the system of socialist agriculture was quickly disbanded and the private land ownership, as it existed in the pre-communist times, was slowly but steadily reinstated by the post-communist legal-bureaucratic system. In the long run, the decades-long efforts of the communist authorities to eliminate peasantry as a class thus suffered a resounding defeat. It must be noted however that, while the landowning peasantry re-emerged as a class, it did so in a form that was deeply transformed by the communist experience. Due to decades of socialist internal social mobility, marked by forced

[106] Ceauşescu, *Cuvântare la Consfătuirea*, 6, 8; Roske, "Colectivizarea şi mecanismul colectărilor," 130.

industrialization and socialist-style urbanization, numerous landowners (descendants of the original proprietors) are now urban dwellers, and have a different relation to land property and agricultural activities. The collectivization of agriculture failed, but its long-term political, social, and economic implications have had profound effects on Romanian society, amply visible in the post-communist period.

BIBLIOGRAPHY

ARCHIVAL SOURCES

ACNSAS—*Arhiva Consiliului Naţional pentru Studierea Arhivelor Securităţii* [Archive of the National Council for Study of Securitate Archives], Bucharest:
Fond Documentar [Documentation], file 7, vol. 13.

ASRI—*Arhiva Serviciului Român de Informaţii* [Archive of the Romanian Information Service], Bucharest:
Fond Documentar [Documentation], files: 1908; 4638; 4640; 7611, vol. 1; 7778, vol. 3, 27, 32, 36; 9047, vol. 3.

ANIC—*Arhivele Naţionale Istorice Centrale* [National Central Historical Archives], Bucharest:
Fond CC al PCR—Secţia Organizatorică [Central Committee of the Romanian Communist Party—The Organizational Section] files: 43/1949; 71/1949; 144/1949; 176/1949; 161/1949; 7/1950; 61/1951; 41/1959; 15/1961; 16/1961; 50/1961; 51/1961; 52/1961; 53/1961; 52/1961.
Fond CC al PCR—Secţia Administrativ-Politică [Central Committee of the Romanian Communist Party—The Political-Administrative Section], files: 110/1950; 17/1960; 17/1962.
Fond CC al PCR—Secţia Cancelarie [Central Committee of the Romanian Communist Party—Chancellery] file 21/1949.
Fond Reforma agrară din 1945—Centrala [The Agrarian Reform of 1945—The Central Bureau], file 71/1948.
Fond Ministerul Afacerilor Interne—Direcţia Administraţiei de Stat [The Ministry of Internal Affairs—The Direction of State Administration], file 8/1949.
Fond Ministerul Afacerilor Interne, Service "C," Arhiva [The Ministry of Internal Affairs—Service "C," Archive], 1960, vol. 3; vol. 27.
Open Society Archive (OSA), Budapest:
Fonds 300: Records of Radio Free Europe/Radio Liberty Research, Subfonds 60: Romanian Unit, Series 1: 202 Agriculture, RL 2966/1959.

AN Suceava—*Serviciul Judeţean Suceava al Arhivelor Naţionale,* Suceava [County Branch of the National Archives, Suceava]:
Fond "Poliţia Oraşului Rădăuţi" [Police Rădăuţi City], dos 10/1949.

PERIODICALS

Monitorul Oficial [Official Monitory] 68, March 23, 1945.
Monitorul Oficial [Official Monitory], August 3, 1948.
Buletinul Oficial [Official Bulletin] 1, March 2, 1949.
Scânteia [The Spark], September 21, 1944.
Scânteia [The Spark], October 19, 1957.
Scânteia [The Spark] 5371, December 7, 1961.

BOOKS AND ARTICLES

Ardelean, Petru M. "O răscoală ţărănească din 1949" [A peasant revolt of 1949]. *Dreptatea* 492 (November 1991): 3.

Bălan, Ion. "Mişcări ţărăneşti în judeţul Vlaşca. 1949–1950" [Peasant uprisings in Vlaşca county. 1949–1950]. *Arhivele Totalitarismului* 2–3 (1996): 59–64.

———. *Regimul concentraţionar din România, 1945–1964* [The labor camp system in Romania, 1945–1964]. Bucharest: Fundaţia Academia Civică, 2000.

Bocu, Liliana. "Condamnarea ţăranilor—începutul dezastrului. Procese ale Tribunalului Regiunii a III-a Militare Cluj" in Romulus Rusan, ed. *Instaurarea comunismului—între rezistenţă şi represiune* [The conviction of peasants—the beginning of the disaster. Trials by the Tribunal of the Third Military Region, Cluj]. Bucharest: Fundaţia Academia Civică, 1995, 85–89.

Breazu, Florentin. "Pârjolul din Vlaşca." *Magazin istoric* 10 (October 1999): 17–21.

Breazu, Steliana and Eleonora Criste. "Revolta ţărănească de la Ucuriş-Bihor din anul 1949" [The Peasant uprising in Ucuriş-Bihor in 1949] in Romulus Rusan, ed., *Anii 1949–1953: mecanismele terorii* [The Years 1949–1953: The Mechanisms of Terror], series *Analele Sighet*, 7, Bucharest: Fundaţia Academia Civică, 1999, 697–701.

Călin, Cornel P. "7 iulie 1950—răscoala ţăranilor din Vlaşca" [7 July 1950—The revolt of the peasants in Vlaşca]. *Memoria* 8, s.a. (c. 1993): 120–25.

Cartea Albă a Securităţii [The white book of the Securitate], vol. I, *23 August 1944–30 August 1948*; vol. II, *August 1948–Iulie 1958*; vol. III, *1958–1968*. Bucharest: Serviciul Român de Informaţii, 1994–1997.

Cătănuş, Dan and Octavian Roske, eds. *Colectivizarea agriculturii în România. Dimensiunea politică* [The collectivization of agriculture in Romania. The political dimension]. vol. I, *1949–1953*. Bucharest: Institutul Naţional pentru Studiul Totalitarismului, 2000.

Ceauşescu, Nicolae. *Cuvântare la Consfătuirea de lucru privind activitatea în agricultură, dezvoltarea mai puternică a producţiei vegetale şi animale în gospodăriile personale ale membrilor cooperatori şi ale producătorilor particulari, 17 ianuarie 1984* [Speech at the Meeting concerning the activity in agriculture, the further development of the vegetable and animal production in personal households of cooperativized peasants and of private producers, 17 January 1984]. Bucharest: Editura Politică, 1984.

Cioşan, Florin. "Colectivizarea agriculturii. Studiu de caz (localităţile Teaca şi Ocniţa, jud. Bistriţa-Năsăud). Cercetare de istorie orală" [The collectivization of agriculture. Case study (localities Teaca and Ocniţa, Bistriţa-Năsăud county). Oral history research] *Anuarul Institutului de Istorie Orală* I (1998): 298–336.

Costea, Ionuţ. "Procesul colectivizării şi memorie generaţională. Studiu de istorie orală." [The process of collectivization and generational memory. A study of oral history]. *Anuarul Institutului de Istorie Orală* I (1998): 289–97.

Damian, Ancu, Florentin Breazu, and Ion Bălan, eds. *Colectivizarea în Vlaşca, 1949–1950. Documente* [Collectivization in Vlaşca, 1949–1950. Documents]. Bucharest: Vinea, 2002.

Degeratu, Claudiu and Octavian Roske, eds. "Colectivizarea agriculturii. Modelul sovietic: Ridicarea necontenită a nivelului de trai." [The collectivization of agriculture. The Soviet model. The continuous rise in the standard of living]. *Arhivele Totalitarismului* II, no. 1–2 (1994): 80–91.

———, eds. "Colectivizarea agriculturii. Modelul sovietic: Drumul belşugului" [The collectivization of agriculture. The Soviet model. The road to abundance]. *Arhivele Totalitarismului* II, no. 3 (1994): 54–68.

———, eds. "Colectivizarea agriculturii. Modelul sovietic: Ştiinţa biruitoare" [The collectivization of agriculture. The Soviet model. The road to abundance]. *Arhivele Totalitarismului* II, no. 4 (1994): 47–59.

DEX. Dicţionarul Explicativ al Limbii Române [DEX. The explicative dictionary of the Romanian language] 2nd ed. Bucharest: Univers Enciclopedic, 1996.

Diaconescu, Sanda. "Drăgoeşti—satul martir" [Drăgoeşti—the martyr village]. *Memoria* 7 (1992): 78–96.

Dobrincu, Dorin, "Colectivizarea, cote şi revolte ţărăneşti în vestul României (1949)" [Collectivization, requisition quotas, and peasant revolts in Western Romania (1949)]. *Anuarul Institutului Român de Istorie Recentă* I (2002): 282–318.

———. "Transformarea socialistă a agriculturii, răscoalele ţărăneşti şi deportările din nordul Moldovei (1949)" [The socialist transformation of agriculture, peasant uprisings and deportations in Northern Moldova (1949)]. *Anuarul Institutului de Istorie A.D. Xenopol* XXXIX-XL (2002-2003): 459–87.

———. "Dinamica represiunii comuniste împotriva ţărănimii din România (1951–1959)" [The dinamics of communist repression against peasantry in Romania (1951–1959)]. *Arhiva Istorică a României* I, no. 2 (2004): 253–67.

———. "Colectivizarea agriculturii şi represiunea împotriva ţărănimii din România (1949–1962)" [The collectivization of agriculture and the repression against the peasantry in Romania (1949–1962)] in Ruxandra Cesereanu, ed. *Comunism şi represiune în România. Istoria tematică a unui fratricid national* [Communism and repression in Romania. The thematic history of a national fratricide]. Iaşi: Polirom, 2006, 108–25.

Dobrincu, Dorin and Constantin Iordachi, eds. *Ţărănimea şi puterea. Procesul de colectivizare a agriculturii în România (1949–1962)* [The peasantry and the power. The process of collectivization of agriculture in Romania, 1949–1962]. Forward by Gail Kligman and Katherine Verdery. Iaşi: Polirom, 2005.

Folea, Ioan. "Transformarea socialistă a agriculturii, aşa cum este ea prezentată de ţăranii brăileni în declaraţiile lor" [The socialist transformation of agriculture as it is presented by the peasants of Brăila in their declarations], in Romulus Rusan, ed., *Instaurarea comunismului între rezistenţă şi represiune* [The Establish of Communism between Resistance and Repression], series *Analele Sighet*, 2, Bucharest: Fundaţia Academia Civică, 1995, 113–26.

Frunză, Victor. *Istoria stalinismului în România* [The history of Stalinism in Romania]. Bucharest: Humanitas, 1990.

Gâdea, Anghel. "Se trage fără avertisment" [They shoot without warning]. *Revista 22*, no. 36 (September 11–17, 1992): 10.

Gheorghiu-Dej, Gheorghe. *Articole şi cuvântări* [Articles and speeches]. Bucharest: Editura pentru Literatură Politică, 1953.

———. *Articole şi cuvîntări, iunie 1961–decembrie 1962* [Articles and speeches, June 1961–December 1962]. Bucharest: Editura Politică, 1962.

———. *Raport cu privire la încheierea colectivizării şi reorganizarea conducerii agriculturii, prezentat la sesiunea extraordinară a Marii Adunări Naţionale, 27 aprilie 1962* [Report concerning the completion of the collectivization and the reorganization of the leadership of the agriculture presented at the extraordinary session of the Great National Assembly, 27 April 1962]. Bucharest: Editura Politică, 1962.

Goina, Călin. "Ce poţi face azi nu lăsa pe mâine: cazul colhozului model Viaţă Nouă, Sântana," in Dorin Dobrincu and Constantin Iordachi, eds. Ţărănimea şi puterea. Procesul de colectivizare a agriculturii în România (1949–1962) [The peasantry and the power. The process of collectivization of agriculture in Romania (1949–1962)]. Iaşi: Polirom, 2005, 368–92.

Guran, Petre. "Colectivizarea, între mit şi realitate" [Collectivization, between myth and reality] in Lucian Boia, ed. *Miturile comunismului românesc* [The myths of Romanian communism]. Bucharest: Editura Nemira, 157–76.

Iancu, Gheorghe, Virgiliu Ţârău, and Ottmar Traşcă, eds. *Colectivizarea agriculturii în România. Aspecte legislative. 1945–1962* [The collectivization of agriculture in Romania. Legislative aspects, 1945–1962]. Cluj-Napoca: Presa Universitară Clujeană, 2000.

Ioan, Viviana. "Dreptul la proprietate. Noi nu avem de dat pământ la colectiv" [The right to property. We have not given land to the collective farm]. *Revista 22*, no. 46 (November 30, 1990): 14.

Ionescu, Ghiţa. *Communism in Rumania, 1944–1962*. Oxford: Oxford University Press, 1964.

Ionescu-Gură, Nicoleta. "Decretul 83. Începutul sfârşitului la sate" [The decree no. 83. The beginning of the end in the countryside]. *Magazin istoric* 10 (Oct. 2000): 27–31; 11 (Nov. 2000): 40–44.

Ioniţoiu, Cicerone. "Subjugarea ţărănimii. Revolta din comuna Vadu Roşca" [The subjugation of the peasantry. The revolt in Vadu Roşca commune], in Romulus Rusan, ed. *Anii 1954-1960: fluxurile şi refluxurile stalinismului* [Years 1954–1960: Flows and Ebbs of Stalinism], series *Analele Sighet*, 8, Bucharest: Fundaţia Academia Civică, 2000, 309–14.

Iordachi, Constantin. "Constanţa, the First Collectivized Region: Soviet Geo-Political Interests and National and Regional Factors in the Collectivization of Dobrogea (1949–1962)" in Constantin Iordachi and Dorin Dobrincu, eds. *Transforming Peasants, Property and Power: The Collectivization of Agriculture in Romania, 1949–1962*. Budapest: CEU Press, 2009, 103–40.

——— and Dorin Dobrincu, eds. *Transforming Peasants, Property and Power: The Collectivization of Agriculture in Romania, 1949–1962*. Budapest: CEU Press, 2009.

——— and Katherine Verdery. "Conclusions" in Constantin Iordachi and Dorin Dobrincu, eds. *Transforming Peasants, Property and Power: The Collectivization of Agriculture in Romania, 1949–1962*. Budapest: CEU Press, 2009, 455–71.

Ivaşcu, Lavinia. "Noi nu am fost oameni că am fost chiaburi. Mărturii privind colectivizarea în Maramureş" ["We were not human beings because we were chiaburi." Eye-witness accounts of collectivization in Maramureş] in Romulus

Rusan, ed., *Anii 1954–1960: fluxurile şi refluxurile stalinismului* [Years 1954-1960: Flows and Ebbs of Stalinism], series *Analele Sighet*, 8, Bucharest: Fundaţia Academia Civică, 2000, 305–8.

Kligman, Gail. *The Politics of Duplicity: Controlling Reproduction in Ceauşescu's Romania*. Berkeley: University of California Press, 1998.

———. "Creating Communist Authority: Class Warfare and Collectivization in Ieud (Maramureş Region)" in Constantin Iordachi and Dorin Dobrincu, eds. *Transforming Peasants, Property and Power: The Collectivization of Agriculture in Romania, 1949–1962*. Budapest: CEU Press, 2009, 165–201.

——— and Katherine Verdery. "Social Dimensions of Collectivization: Formenting Class Warfare in Transylvania" in Vladimir Tismăneanu, Marc Morje Howard, and Rudra Sil, eds. *The World Order after Leninism: Essays in Honor of Ken Jowitt*. Seattle: University of Washington Press, 2006, 127–46.

Levy, Robert. *Ana Pauker: The Rise and Fall of a Jewish Communist*. Berkeley: University of California Press, 2001.

———. "The First Wave of the Collectivization Campaign: Central Policies and their Regional Implementation, 1949–53" in Constantin Iordachi and Dorin Dobrincu, eds. *Transforming Peasants, Property and Power: The Collectivization of Agriculture in Romania, 1949–1962*. Budapest: CEU Press, 2009, 27–48.

Marin, Gheorghe Gaston. În serviciul României lui Gheorghiu-Dej. Însemnări din viaţă [In the Service of Gheorghiu-Dej's Romania: Life Recollections]. Bucharest: Evenimentul Românesc, 2000.

Mitrany, David. *Marx Against the Peasant: A Study in Social Dogmatism*. Chapel Hill: University of North Carolina Press, 1951.

Oprea, Marius. "The Final Offensive: 'The Socialist Transformation of Agriculture' from Slogans to Reality (1953–1962)" in Constantin Iordachi and Dorin Dobrincu, eds. *Transforming Peasants, Property and Power: The Collectivization of Agriculture in Romania, 1949–1962*. Budapest: CEU Press, 2009, 49–79.

Râpeanu, Mihai. "Deportări şi confiscări în baza decretului 83/1949" [Deportations and confiscations under decree 83/1949] in Romulus Rusan, ed. *Instaurarea comunismului între rezistenţă şi represiune* [The Establish of Communism between Resistance and Repression], *Analele Sighet* 2, Bucharest: Fundaţia Academia Civică, 1995, 406–10.

Rezoluţii şi hotărîri ale Comitetului Central al Partidului Muncitoresc Român [Resolutions and decisions of the Central Committee of the Romanian Workers' Party], vol. I: *1948–1950*, ediţia a II-a, s.l. Bucharest: Editura pentru Literatură Politică, 1952.

Rezoluţia şedinţei plenare a comitetului central al P.M.R. din 3–5 Martie 1949: asupra sarcinilor partidului în lupta pentru întărirea alianţei clasei muncitoare cu ţărănimea muncitoare şi pentru transformarea socialistă a agriculturii [Resolution of the Plenary Session of the Central Committee of the RWP (Romanian Workers' Party) of 3–5 March 1949: on the party tasks in the fight for strengthening the alliance of the working class with the working peasantry and for the socialist transformation of agriculture]. Bucharest: Editura PMR, 1949.

Roberts, Henry. *Rumania. Political Problems of an Agrarian State*. Hamden, Conn.: Archon Books, 1969.

Roşca, Nuţu. "Rezistenţa comunei Bârsana la colectivizarea forţată" [Resistance to forced collectivization in Bârsana commune] in Romulus Rusan, ed., *Instau-*

rarea comunismului între rezistență și represiune [The Establish of Communism between Resistance and Repression], *Analele Sighet* 2, Bucharest: Fundația Academia Civică, 1995, 133–37.

Roske, Octavian. "Colectivizarea agriculturii în România, 1949–1962" [The collectivization of agriculture in Romania, 1949–1962]. *Arhivele Totalitarismului* I, no. 1 (1993): 146–68.

———. "Colectivizarea și mecanismul colectărilor: istorii paralele" [Collectivization and the mechanism of forced collections: parallel histories] in Dorin Dobrincu and Constantin Iordachi, eds. *Țărănimea și puterea. Procesul de colectivizare a agriculturii în România (1949–1962)* [The peasantry and the power. The process of collectivization of agriculture in Romania (1949–1962)]. Iași: Polirom, 2005, 113–35.

Rujan, Rodel. "Procesul colectivizării în Mândruloc și Cicir, jud. Arad" [The process of collectivization in Mândruloc and Cicir, Arad county]. *Anuarul Institutului de Istorie Orală* II (2001): 329–50.

Șandru, Dumitru. "Decretul 83/1949" [Decree no 83/1949]. *Arhivele Totalitarismului* I, no. 1 (1993): 133–45.

———. *Reforma agrară din 1945 în România* [The 1945 Agrarian Reforms in Romania] Bucharest: Institutul Național pentru Studiul Totalitarismului, 2000.

———. "Colectivizarea agriculturii și problema agrară: Repere social-politice" [The collectivization of agriculture and the agrarian problem: Social-political reference points] in Dorin Dobrincu and Constantin Iordachi, eds. *Țărănimea și puterea. Procesul de colectivizare a agriculturii în România 1949–1962* [The Peasantry and the power. The process of collectivization of agriculture in Romania (1949-1962)]. Iași: Polirom, 2005, 45–65.

Stoian-Gărăgăianu, Maria. "Clopotele din 7 iulie...." [The bells of July 7th]. *Memoria* 8, s.a. (c. 1993): 118–20.

Stoica, Cătălin Augustin. "One Step Back, Two Steps Forward: Institutionalizing the Party-State and Collective Property in Two Romanian Villages (Galați Region)" in Constantin Iordachi and Dorin Dobrincu, eds. *Transforming Peasants, Property and Power: The Collectivization of Agriculture in Romania, 1949–1962.* Budapest: CEU Press, 2009, 423–53.

Tănase, Stelian. *Elite și societate. Guvernarea Gheorghiu-Dej, 1948–1965* [Elites and society. The rule of Gheorghiu-Dej, 1948–1965]. Bucharest: Humanitas, 1998.

Țârău, Augustin. "Aplicarea decretului no. 83/1949 în județul Bihor (Noaptea moșierilor)" [The implementation of decree no. 83/1949 in Bihor county (The night of the great landowners)]. *Crisia* XXVI–XXVII (1999): 223–44.

Verdery, Katherine. "Exploiters Old and New: Making and Unmaking 'Rich Peasants' in Aurel Vlaicu (Hunedoara Region)" in Constantin Iordachi and Dorin Dobrincu, eds. *Transforming Peasants, Property and Power: The Collectivization of Agriculture in Romania, 1949–1962.* Budapest: CEU Press, 2009, 307–28.

Vighi, Daniel and Viorel Marineas, eds. *Rusalii '51. Fragmente din deportarea în Bărăgan* [Pentecost '51. Fragments from the deportation in Bărăgan]. Timișoara: Marineasa, 1994.

———, Viorel Marineasa, and Valentin Sămânță, eds. *Deportarea în Bărăgan, 1951–1956. Destine–Documente–Reportaje* [Deportations in Bărăgan. Destinies–Documents–Reportages]. Timișoara: Mirton, 1996.

Vultur, Smaranda, ed. *Istorie trăită—istorie povestită. Deportarea în Bărăgan (1951–1956)* [Written history—narrated history. The deportation to the Bărăgan plain, 1951–1956]. Timişoara: Amarcord, 1997.

———. "Deportări în Bărăgan. Limbaj, morală şi moravuri" [Deportations in Bărăgan. Discourses, morals and mores]. *Revista 22*, no. 814 (2005): 16–18.

THE COMMUNIST TAKE-OVER AND LAND COLLECTIVIZATION IN ROMANIA. CHRONOLOGY OF EVENTS, 1945–1962[107]

Constantin Iordachi and Dorin Dobrincu

1945

March 6: The establishment of a communist-dominated coalition government led by Dr. Petru Groza (1884–1958), the leader of the Ploughmen's Front.

March 23: The Law on the Agrarian Reform expropriates without any compensation all properties over 50 hectares of land, as well as the land of war criminals, ethnic Germans who collaborated with Nazi Germany, and the absentees. Part of the land is redistributed to poor or landless peasants in small lots of 5 hectares. The aim of the reform is the destruction of great land estates.

April 12: Adoption of the Regulation for the implementation of the agrarian reforms.

April 19: All private tractors and agricultural machines are confiscated by the state.

May 15: The establishment of county economic offices for the requisition of agricultural products to be sent to USSR as part of the war debt due by Romania.

July 16: The State institutes a monopoly over the circulation and commercialization of agricultural products. In the next months, it would also establish mandatory requisition quotas on all agricultural products.

October 26: The establishment of the Supreme Council of National Economy, led by Gheorghe Gheorgiu-Dej.

1946

January 8–9: Plenary of the Central Committee (CC) of the Romanian Communist Party (RCP). The establishment of peasant party cells in rural areas.

November 19: National parliamentary elections. After an electoral campaign marked by violence and obstruction, the Bloc of Democratic Parties, controlled by the communists, claims a sweeping victory. Archival evidence confirms that the results were grossly falsified by the ruling communist-led coalition.

[107] An earlier version of this chronology was published in Constantin Iordachi and Dorin Dobrincu, eds. *Transforming Peasants, Property and Power: The Collectivization of Agriculture in Romania, 1949–1962*. Budapest: CEU Press, 2009, 485–91.

1947

December 30: Forced abdication of King Mihai I. Abolition of the Monarchy and the proclamation of the People's Republic of Romania.

July 29: The Groza government bans the National Peasant Party, the most important oppositional party with large electoral support in rural areas.

October 30–November 12: After a summary public trial, the leaders of the National Peasant Party are sentenced to long-term imprisonment and the confiscation of property.

1948

February 21–23: Congress of unification of the Romanian Communist Party and the Romanian Social-Democratic Party. Creation of the Romanian Workers' Party (RWP) as a "unique party of the working class."

April 13: The Grand National Assembly adopts the first communist Constitution of the People's Republic of Romania.

June 11: The Law on the nationalization of main industrial, mining, insurance, transport, and banking units is adopted.

June 1: The establishment of the State Commission for Planning.

July 6: Decree no. 121 establishes a system of forced requisition of cereals.

August 30: Decree no. 221 establishes the General Directorate of People's Security (*Direcţia Generală a Securităţii Poporului*, or *Securitatea*). As the main organ of the communist political police, the Securitate is to be responsible for massive actions of repression against the anti-communist opposition and of terror against the population at large. It is also actively involved in the campaign of collectivization.

October 7: Decree for the organization of the Machine and Tractor Stations.

1949

January 25: Establishment of the Council for Mutual Economic Aid (CMEA), an organization meant to facilitate and coordinate economic cooperation of socialist countries. Romania is a founding member.

March 2–3: Decree no. 83 concerning the expropriation of the last remnants of former land estates. Large landowners are deported to labor camps together with their families.

March 3–5: The collectivization campaign is formally launched by the Resolution of the Plenary of the CC of the RWP. The resolution sets up an Agrarian Commission in charge of organizing and supervising the collectivization campaign.

June–August: The first collective farms are established.

July–August: Peasant revolts against forced requisition of grains and against collectivization. The revolts are brutally repressed, with numerous peasants being killed, wounded, or deported.

1950

January 23–25: Fifth Plenary on the CC of the RWP. The Agrarian Commission is replaced with an Agrarian Section of the CC, also led by Ana Pauker, with expanded responsibilities.

June 15 to September 1950: Massive coercion is used in the creation and consolidation of collective farms throughout the country. Campaigns of confiscating all *chiaburs'* land; their partial land donations (or "renunciations") are no longer tolerated.

July 23: Decision of the CC of RWP on the new administrative reorganization of the country in regions and districts, copying the Soviet model. Approved by the Grand National Assembly on 6 September.

July: Peasant uprisings against forced requisition throughout Romania, in Ilfov, Ialomiţa, and Vlaşca counties. Authorities respond with more repression, dispatching the militia, Securitate and the army, conducting summary executions and deportations.

December 12–13: The First Five-Year Plan is launched (1951–1955). The plan stipulated that 62% of the country's arable land is to be collectivized by 1955. An additional 8% is to be organized in State Farms.

1951

January 13: The regulations for mandatory milk quotas are published. Failure to deliver quotas results in harsh punishment.

March 2: Decision of the CC of the RWP and the Council of Ministers for the economic and organizational consolidation of the existing GACs. Each member of the collective farm is obliged to execute annually at least 80 days of work.

September 18: Decision of the CC of the RWP concerning the work on the establishment of collective farms and associations. Excesses, abuses and coercion in the collectivization campaign are openly acknowledged.

December 31: The regulations for mandatory meat quotas are published. According to official information, at the end of the year there were 1,089 GACs containing 301,690 ha of land and 75,379 families.

1952

May: Purge of the so-called "right-wing deviationist" faction from the party: Ana Pauker, Vasile Luca, and Teohari Georgescu.

September 24: The Grand National Assembly adopts a new Constitution, modeled on the 1936 Soviet Constitution.

September 27: Decree on the new administrative reorganization of Romania. The number of regions is reduced from 28 to 18.

1953

January–February: 252 new collective farms are established.

March 5: Joseph Stalin dies.

Soviet Prime Minister Georgy Maximilianovich Malenkov launches the "New Course" Policy.

Spring 1953: Romanian leaders temporary halt the collectivization campaign.

June 11: The Council of Ministers cancels overdue requisition quotas for 1951, and exempts from payment peasants living in areas affected by floods.

August: Plenary of the CC of the RWP adopts a series of measures relaxing the taxation and requisition quotas system, with the aim of giving incentives to private farmers.

September 10: The payment of overdue quotas for 1952 is cancelled, with the exception of wheat.

October: Massive transfer of 448,000 hectares of land from state reserves to collective farms.

December: Active policy to bolster the GACs and TOZs. Decree no. 505 offers preferential tax cuts to collectivized peasants as an incentive for joining the GAC.

1954

January: The Agriculture Department of the CC led by Moghioroş unsuccessfully tries to resuscitate the collectivization campaign. The government grants 10 to 20% tax cuts to GACs and TOZs.

January 9: Special educational institutions for GAC cadres are established.

January 30: Aggravation of penalties for failures to deliver mandatory quotas.

November 8: The Politburo ordered the revision of the lists of *chiaburi*, signaling a more *nuanced policy on class warfare.*

December 26: The government scraps the food retailing system of rations and grants additional tax cuts, as an impetus to incipient forms of free trade in agriculture.

End of 1954: There are 2,070 GACs containing 178,561 peasant families and 884,194 ha. In addition, there are 2,833 associations containing 139,125 peasant families and 315,119 ha.

1955

January 27–February 7: Following Khrushchev's recommendations on providing incentives for small private farm production, Romanian authorities encourage state contracts for animals and poultry and reduce requisition quotas.

March 1: At the Conference of Agricultural Workers, Dej praises the contribution of privately owned farms to a record corn harvest.

May 14: Signing of the Warsaw Pact, a military organization of socialist countries. Romania is a founding member.

August 4: A new Ministry of Requisitions is created to supervise the requisition quota system.

September 13: The government bans free trade of foodstuffs in order to prevent speculation in cereal prices.

December 23–28: The Seventh Congress of the RWP. Launching of the Second Five-Year Plan (1956–1960). Gheorghiu-Dej reelected First Secretary of the CC. The party abandons the "New Course" and aggressively resumes the collectivization campaign. Timişoara region is criticized for lagging behind in collectivization with only 13.9% of peasant families enrolled in the socialist sector.

End of 1955: The number of GACs increases with 82 units, to a total of 2,152. The number of TOZs almost doubles, from 2,833 in 1954 to 4,471.

1956

April 4: The creation of the Ministry of State Agricultural Farms.

June 11: Meeting of the CC evaluating the outcome of the experiment on collectivizing the Galaţi region. Decision to resume collectivization in the whole country.

October–November: The Hungarian revolution and the Soviet military intervention. The events cause a delay in Romania's collectivization campaign.

December 27–29: Plenary of the CC of the RWP. Dej demands that the "socialist transformation of agriculture" continues.

End of 1956: There are 2,564 GACs containing 231,392 peasant families and 1,101,650 ha. In addition, there are 8,130 associations containing 452,117 families and 753,352 ha.

1957

January–August: Intensification of bureaucratic-administrative measures to force the peasants to give up their land, resulting in the escalation of rural tensions.

January 19: The Ministry of Requisition Quotas is abolished.

August 2: Politburo meeting adopts new measures to compel the peasants to enroll in collective farms.

August 6: Decree no. 380 stipulates higher levels of taxation for private farms and TOZs, with retroactive effects, in order to force private owners to join collective farms.

September 21: Decree no. 446 punishes thefts of grain with prison terms of up to 12 years and the confiscation of property.

September–November: Growing tensions in rural areas as peasants put up stubborn resistance to the final assault of collectivization. Violent protests in Suraia, Vadu Roşca, and Răstoaca (today in Vrancea county) against forced collectivization. The government uses coercive methods.

October 18: Official press announces the completion of collectivization in Constanţa region, at a time when collectivized land encompasses only 51% of the country's total land and 52% of the total number of households.

1958

March 14: The Regional Party Committee and the Regional People's Council Constanţa publish a development plan for 1958–1960, aimed at turning Dobrogea into the first region with fully socialist agriculture.

April 3–6: Party meeting in Constanţa. Gheorghiu-Dej approves the plan for Dobrogea's agricultural development and promises substantial state investments with the purpose of achieving 80 to 90% mechanization in three years.

November 26–28: Plenary of the CC of the RWP. The relaunching of the campaign of collectivization triggers numerous peasant revolts.

1959

February–August: Peasant revolts against collectivization.

March 6: Gheorghiu-Dej claims that 3.6 million peasants are enrolled in GACs, making up 75% of the total rural population of the Craiova region, 72% of the Galaţi region, 56% of the Timişoara region, etc.

March 30: The Grand National Assembly issues the Decree for liquidating all forms of exploitation. Rich peasants are obliged to join collective farms and to donate their land to the state; in exchange, their mandatory quotas and their official classification as *chiaburi* are abolished. This also entails a final coercive campaign for *chiaburi* to join the collective farms.

July 13–14: Plenary of the CC of the RWP announces that 67% of the country's arable land is collectivized.

1960

March 1: Official statistics claim that GACs possess 76.6% of the country's arable land and 76.4% of the total rural population.

June 20–25: The Eighth Congress of the RWP. Launching of the Third Five-Year Plan (1960–1965). Gheorghiu-Dej is reelected First Secretary of the CC. He announces that 81% of the rural population is enrolled in collective farms or associations.

December 24: Adoption of a new law on the administrative reorganization of the country. The number of regions is reduced from 18 to 16.

December: Peasant revolts in Vadu Roşca (Vrancea county) and in Mozăceni Deal (Argeş county). Violent confrontations with the authorities, followed by brutal repression.

1961

June 30–July 1: Plenary of the CC of the RWP orders the consolidation of the newly established GACs. Report on the full collectivization of the Bucharest district.

September 1: The socialist sector in agriculture encompasses 82.8 percent of the total arable land; collective farms own 66.6 percent of the total arable land.

December 18–22: Conference of Collective Farmers in Bucharest celebrates the completion of collectivization. Collectivized farmers from all corners of Romania give laudatory speeches.

1962

January–February: Peasant uprisings in Drăgoieşti (Suceava county) and Mărceşti (Dâmboviţa county).

March 16: The campaign of collectivization is officially declared complete. The socialist sector in agriculture encompasses 96% of the arable land of the country and 3,201,000 peasant families. Mountainous regions are exempt from collectivization.

March 18: The region of Suceava is declared the last region of Romania to complete collectivization.

March 27–April 3: Extraordinary session of the Grand National Assembly proclaims the completion of collectivization. The session is attended by 11,000 peasants.

Collectivization in Yugoslavia: Rethinking Regional and National Interests

Melissa K. Bokovoy

With the acceptance of individual peasant production and land ownership in 1953, the Yugoslav Communists (Communist Party of Yugoslavia, CPY) ended the long struggle for control of the countryside and the hearts and minds of the Yugoslav peasants. The CPY's Commission for the Village admitted the party's collectivization program had failed, and wrote to the Central Committee, "we have lost the peasantry, now they will never trust us again." Vladimir Dedijer, Tito's official biographer, acknowledged defeat, but argued it was a defeat of Stalinist ideology rather than a failure of the CPY's policies. That ideology had dictated that the socialization of countryside could only be achieved through mandatory agricultural quotas (*otkup*), large-scale agricultural enterprises, collectivization, and class struggle, i.e. the liquidation of rich peasants (*kulaks*). In Dedijer's opinion, which was the party's official position, the major sin of the CPY was its acceptance of Stalin's ideology, the main tenants of which the party had followed blindly until 1953.

In Dedijer's drama, there were villains and victims: the villain was Stalin and his ideology; the victims were the Yugoslav communists who had been duped into believing that collectivization and the liquidation of the *kulak* were the only methods of bringing socialism to the countryside. His assessment gave the CPY a justification for its failed agricultural policies, and obscured from view the party's own miscalculations.

Six years later, in May 1959, Edvard Kardelj, communist Yugoslavia's leading theorist and Josip Broz Tito's longtime deputy, delivered a speech to delegates at the Ninth Plenary session of the Socialist Alliance of the Working People of Yugoslavia (SSRNJ) on the situation of Yugoslav agriculture, in which he attempted to redeem Yugoslav leaders by arguing they had "refused to accept the Stalinist type of collectivization" because it had caused "serious economic and political disturbances."[1] He assured the peasant deputies in the audience[2] that they need not fear any "Stalin-type" collectivization, and promised that the party and the state would help peasant agriculture become more productive and labor efficient.[3] He declared that Yugoslav leaders would follow "painless measures" to achieve "socialist relationships in the countryside" by helping peasants without exerting "bureaucratic pressures," to bring about changes in their beliefs.[4]

Kardelj would later acknowledge the more complicated history and experience of the CPY's attempt to socialize the countryside.[5] From 1941 to 1953, party leaders had fought amongst themselves, with agricultural experts, with peasants, and with the Soviets and their satellites, over how to convert the countryside to socialism. Those battles took place on every front—ideological, political, economic, and social—and each front involved a different configuration of actors. Dedijer's assertion that the party had blindly followed Stalinist ideology did not account for the immense complexity of the collectivization effort, or the roles individuals played in its implementation. According to party documents and Yugoslav historians, CPY leaders played pivotal roles in formulating and implementing plans to socialize the countryside. Federal and republican ministers, their assistants, state and local party bureaucrats, and the peas-

[1] Kardelj, *Problemi socijalisticke.*

[2] At the end of World War II, peasants made up close to 50% of party membership; by the late 1950s this number had dwindled to less than 15%.

[3] Kardelj, *Problemi socijalisticke*, 7.

[4] Kardelj intended his remarks to be an evaluation of the state's socialist agriculture policies against the backdrop of a broader debate on agrarian reform among socialist states after the Kremlin in late 1957 signaled to its satellites it was time to start collectivizing again. Kardelj noted that throughout the early months of this new campaign, which now included campaigns in the People's Republic of China and the Democratic Republic of Vietnam as well as its European satellites, Khrushchev made a series of statements that indirectly justified the decisions and experiences of Yugoslavia during its ill-fated collectivization campaign of 1949. Kardelj interpreted Khrushchev's statements about the necessity of providing material incentives and significant means for the democratization of collective farm management to mean that the Soviets had acknowledged that socialization of the village dealt with "living people" rather than "dead mechanisms" or "dogmatic Marxism." Kardelj, *Problemi socijalisticke*, 27.

[5] Kardelj, *Problemi socijalisticke.*

ants—whose property, production autonomy, and lifestyle were at stake—all had a significant impact on Yugoslavia's collectivization campaign.[6] This is not an unusual phenomenon for party-states, or for state socialism.[7] State structures and party elites have too often been portrayed in the literature as monolithic, immutable, impermeable, and impervious to external political and societal forces. In studies of state socialism in communist Yugoslavia, scholars often present early communist elites as if they had little connection to, or were influenced very little by, outside forces. But we now know that policy debates and decisions made by the CPY and by communist party leaders across Eastern Europe were in fact heavily influenced by external forces—Stalin, the Red Army, the United Nations Relief and Rehabilitation Agency, the Communist Information Bureau, and the events of the Cold War. Yet internally, the CPY and other communist elites appeared untouched by any outside forces.

As we have seen elsewhere throughout this volume, the collectivization of agriculture in Eastern Europe was constantly buffeted by various actors—peasants, veterans, agrarian experts, party members—which forced communist leaders to rethink, reformulate, and sometimes retreat from Soviet-style agrarian policies. With the exception of Poland and Yugoslavia, communist parties in 1950s and the early 1960s followed Soviet-style collectivization policies. In Yugoslavia, the most intriguing question concerning collectiviza-

6 J. B. Allcock argues that repeated reference to peasant resistance in the narrative of Yugoslav history assumes that peasant resistance forced the CPY to abandon collectivization. He finds structural and economic factors for the CPY's retreat. For him, peasant resistance is a myth. See Allcock, *Collectivization*, 39. In the last several years, historians and social scientists have begun exploring societal responses to the party-state's policies. See specifically for work on the peasant response to the CPY's agrarian policies in western Slavonia and eastern Bosnia: Kržišnik-Bukić, *Cazinska*; and on Croatian peasant resistance see McCarthy, *Peasant Revolutionaries*. For a general treatment of the subject see Bokovoy, Lilly, and Irvine, eds., *State and Society*.

7 In the creation and consolidation of communist Yugoslavia, scholars have focused primarily on the subject of the party-state, its leaders, its elites, its structures and institutions, its ideologies, its state-building processes and strategies, and the place of the Yugoslav state in international politics. Studies which specifically focus on the consolidation and build up of party-state power in Yugoslavia are: Rusinow, *Yugoslav Experiment*; Petranović, *Politička i ekonomska*; Bilandžić, *Borba za*. For more recent work see Pavlović, *Srpsko Selo*; and Popov, *Drama*. Both detail peasant noncompliance and resistance to the *otkup* and peasant work cooperatives but conclude that it was the ability of the state to breakthrough the Comecon's embargo, Western military and economic aid to Yugoslavia, and internal politics that steered the CPY away from Soviet practices in the villages.

tion and its abandonment is why the Yugoslav communists deviated from the Soviet model by altering two main tenets of Stalinist orthodoxy—collectivization and the destruction of the small independent farmers—in the face of accusations by the Soviets and the Communist Information Bureau in 1948. Peasant resistance, which is present in all of the other cases within this book, played a key role in the abandonment of collectivization in Yugoslavia. However, as the other chapters demonstrate, peasant resistance did not always force party leaders to abandon collectivization. Why, then, did party leaders in Yugoslavia abandon collectivization in the face of peasant resistance when other communist regimes did not?

The answer, in part, can be sought by examining the ideological and political strains within the party over the "peasant question," along with the pace and methods of modernization and industrialization. Top CPY leaders agreed on the necessity of centralized planning and rapid industrialization and modernization, but disagreed over what methods should be applied to bring about the socialist transformation of the countryside. Slovenian Communist Party leader Boris Kidrič supported rapid industrialization at the expense of the peasantry, but influential Croat communist Andrija Hebrang steadfastly urged for a more moderate and gradual approach. Hebrang found a powerful ally in the agrarian minister and head of the Agrarian Council, Vaso Čubrilović, who believed ignoring "the basic psychology of the peasant" would prove counterproductive.[8] In the face of peasant resistance to mandatory quotas in 1946, the CPY immediately yielded to peasant opposition to the procurement regimen.[9] But pressure to relax its program also came from the regional and national leadership within the CPY, as dissension often broke out when the national CPY leadership's plans threatened the interests of local and regional party cadres. Hence, when examining reasons why the party abandoned collectivization, regional and national factors warrant additional analysis.

This chapter focuses on the constellation of social and political groups that both shaped and challenged the CPY's collectivization policies: federal and republican ministers and their assistants, state and local party bureaucrats, and a heterogeneous peasantry. The first group directly implemented and administered the decisions at the top. State and local party bureaucrats, on the other hand, often argued with federal and

[8] *Zakondavni Rad*, 481. Vaso Čubrilović made these comments at the 21st meeting of AVNOJ.

[9] "Zakon o opštem," 342–43.

republican ministers over the merits of the plans imposed from above, and expressed their discontent by issuing formal complaints, practising bureaucratic obstructionism, or by simply refusing to implement CPY policies that ran counter to a variety of ideological, provincial, or nationalist interests. The peasantry was a diverse group, divided by region, ethnicity, religion, material circumstance, and wartime experiences. Nonetheless, the peasantry as a collective frequently challenged the state when the latter sought to undermine property ownership and threatened production autonomy and independence. Both groups, party bureaucrats and peasants, resisted or refused to comply when the national-level policy directives threatened regional and local interests and traditions. This not only undermined the party's broader collectivization efforts but also damaged the "brotherhood and unity" model of the Yugoslav socialist state.

In the period under consideration, 1945 to 1953, the CPY's collectivization program was met with varying degrees of resistance and compliance among national and regional mid-level officials, their advisors, state and local party bureaucrats, and peasants. Over the past decade, several studies have revealed regional and national variations in reactions to collectivization.[10] In light of this new scholarship, this chapter will examine how party-state functionaries and peasants in Croatia and Serbia, and in particular in Vojvodina, resisted or complied with bureaucratic directives concerning the creation of peasant work cooperatives (*Seljačka Radna Zadruga*, SRZ). Whether these groups resisted or complied depended on various factors: early interactions with the party-state,[11] interwar and wartime experiences, regional and/or national background, social and economic status, political affiliation, and economic self-interest—a conclusion implicit in Kardelj's critique of CPY policy and reinforced by subsequent studies.[12]

---

[10] Pavlović, *Srpsko Selo*; Popov, *Drama*; and McCarthy, *Peasant Revolutionaries*.

[11] These groups had already encountered the state during the CPY's earlier effort to redistribute land, colonize confiscated land, implement the mandatory deliveries of agriculture goods and farm equipment, and push the peasants into state farms, general agricultural cooperatives, and the first handful of SRZs.

[12] Kardelj, like others in the party, believed opposition to their policies was dictated by economic relationships and material interests. In a previous work, I focus attention on ethnic tensions that arose from CPY policies, but I argue that national and regional differences played a lesser or non-existent role in responses to the *otkup* and collectivization and that material and economic motives were dominant. See Bokovoy, *Peasants and Communists*. For a criticism of this approach, see Nielsen, "Review."

In Yugoslavia, conflict erupted between the communist party elite, state and local party bureaucrats, and peasants soon after the communists came to power. In 1945, plans for agrarian reform and colonization sparked conflict and disagreement in numerous arenas. Top party leaders clashed over the type of policies needed to transform the Yugoslav countryside. One faction, which included Boris Kidrič, Milovan Djilas, Moša Pijade, and members of the party's Economic Council, supported the collectivization of agriculture and strict measures to bring the countryside under state control. Another faction, which included Andrija Hebrang, Edward Kardelj, Vaso Čubrilović (Minister of Agriculture)[13] and members of Agrarian Council, advised against rapid collectivization on the grounds that peasants were not psychologically, politically, or economically prepared for dramatic changes. This faction believed that some form of socialism was necessary in the villages but that the changes should be gradual.

These early years were a time of intense political negotiation and brokering, which took place privately in CPY offices, and in the chambers of the federal parliament and its various committees charged with managing the countryside. Adding to this policy cacophony were regional and local rank-and-file officials who argued over the merits of the plans and policies coming from federal-level party leadership.

Finally, there are two peasantries in this story—the real peasantry of Yugoslavia which were divided by region, ethnicity, material circumstance, and wartime experience—and the mythical peasantry of the Yugoslav communists. The peasantry of the CPY appears as a loyal member of the partisan struggle, undifferentiated, unwavering, and following the interests of the Yugoslav communists and their new state. Indeed, certain segments of the Yugoslav peasantry fought with the communists during the war and helped the CPY come to power, but other segments of the peasantry were not allied with the Yugoslav communists. Nevertheless, by war's end, the heroic peasant partisan of the National Liberation Struggle was fast

[13] Čubrilović was a member of the left wing of the Agrarian Union, a Serbian party whose ideology can be best described as agrarian socialism. While serving in the Ministry of Agriculture, he and many of his advisors, like the Croatian agronomists Maksim Goranović and Dr. Ante Starc, promoted principles from the interwar cooperative movements. The interwar cooperative movement included the Croatian Peasant Party and worked to develop a network of credit, technical, and producer organizations for the peasantry. Members of this movement saw the peasantry as an underrepresented class whose interests it should protect. In all of their proposals for agricultural legislation, Čubrilović and his advisors fought for raising the technical, financial, and educational level of the peasantry.

becoming a national icon. In reality, the peasant partisans, and those peasants who had remained outside the partisan ranks, soon found themselves at odds with the CPY's goals for post-war Yugoslavia. Peasants represented a complex mix of national, regional, local, familial, political, and economic interests, with which the CPY was forced to contend as it attempted to avoid sparking open conflict between the state and peasants.[14]

POLITICS AND IDEOLOGY IN THE YUGOSLAV COUNTRYSIDE[15]

Members of the CPY leadership had been very well schooled in Soviet thought and practice concerning peasant and agrarian problems. During the interwar period, the CPY loyally incorporated Soviet ideology into its party platform. In the early 1920s, the CPY accepted Lenin's New Economic Policy (NEP) and his gradual approach to socializing the countryside. In the 1930s, the party adopted Stalin's anti-*kulak* stance and his new hard-lined program of forced collectivization. Although the CPY entered the war in 1941 armed with Stalinist ideology, the party eventually turned more pragmatic than dogmatic, aware that it needed to form a practical alliance with some portion of the peasantry in order to raise an army. This made their revolutionary experiences fundamentally different from those of the Bolsheviks.

In 1945, the Yugoslav communists acknowledged the peasants' participation in the partisan struggle by rewarding them with ownership rights to land, and a role to play in the CPY's creation myth. These rewards were supposed to cement the peasants' loyalty and support for the new regime. The actual result was a peasantry which increasingly demanded that it had to be negotiated with if the CPY took action that threatened the peasants' property, independence, and production.

When the Yugoslav communists came to power, they adapted communist ideology to fit the party's immediate needs. Yugoslav leaders studied and acknowledged Soviet ideology but always mixed and matched policies to suit the needs of the new Yugoslav state, especially its policies regarding the countryside. Such was the case when the party attempted

14 Kardelj, *Problemi socijalisticke*, 27.

15 For a full discussion of issues explored in this section, see Bokovoy, *Peasants and Communists*.

to nationalize private property in 1946 and 1948. When peasants resisted mandatory quotas, or *otkup*, in 1945 and 1946, the CPY enacted a general amnesty, leaving the land in the hands of the peasantry. A turning point in the CPY's agrarian policy came in 1947 when the CPY began to debate the merits of developing industry over agriculture. The CPY's policy in the countryside called for large-scale, capital-intensive agricultural production that would be developed through gradual and moderate means. Although they decided on implementing a Soviet-style "Five-year Plan," communist leaders rejected the formation of peasant work cooperatives. Instead, the CPY decided to form general agricultural cooperatives (OZZ). Hence, the CPY formally adopted the Soviet principle and rationale of large-scale farms as a more efficient means of production, and the idea that collective farms were a necessary step to achieving socialism in the countryside. But the peasants' move into collective farms was to be gradual, with no mention of initiating class warfare, and based on methods theorized, debated, and accepted by the Yugoslav communists. The CPY chose its own distinct and middle path to the socialization of the countryside.

The Soviet-Yugoslav dispute of 1948 and the CPY's increasing diplomatic, political, and economic isolation in January 1949 made this path much more treacherous and unpredictable than the CPY anticipated. In 1949, the CPY attempted to quickly and efficiently centralize existing agricultural land, machinery, equipment, tools, seed, and labor into peasant work cooperatives, and at the same time, avoid a violent, Soviet-style collectivization campaign. The Yugoslav leadership wanted the ideal: large-scale collective farms which would increase agricultural production, and to socialize the countryside with a minimal amount of bloodshed and violence. CPY leaders all agreed that the process within Yugoslavia would replicate neither the scale nor the intensity of the Soviet model. Tito himself denied that the Yugoslav collectivization program represented a "Great Turn." Instead, the middle path was a moderate collectivization campaign, attenuated by the careful approach taken by Yugoslav party leaders.

Collectivizing the Yugoslav Countryside

When Yugoslavia's collectivization campaign began in early 1949, less than 3% of the arable land was in 1,318 SRZs. At the time, the state controlled less than 15% of arable land throughout Yugoslavia. So when the Commission for the Village of the Central Committee of the CPY reported that 1,949 work cooperatives were established during the first two months of the campaign, from March 20–May 31, this was a precipitous increase.[16] From June 1–September 31, the pace slowed, and in this four-month period an additional 678 were formed. Then in the space of one month, another 1,030 appeared on the scene. The average cooperative according to year end statistics consisted of 51.4 households and 277 hectares, meaning that the average number of hectares a peasant household brought to the cooperative was 5.3.[17] The amount of land under the state's control had only increased by 2.9%, from a little under 15% to 17.9%.

Table 1: *Peasant Work Cooperatives in Yugoslavia, December 31, 1949.*

Republic	No. of SRZs	No. of households	No. of members	No. of members per household	Amount of land in SRZs / Amount of arable land	% of arable land	Average land per household	Average land per SRZ	Average households per SRZ
Serbia	1,857	138,673	645,177	5	798,049/ 4,743,000	16.8	5.75	429	74.6
Croatia	1,589	57,295	235,704	4.1	267,998/ 2,230,000	12.1	4.6	168	36.1
Slovenia	353	8,770	33,506	3.8	75,121/ 736,000	10.2	8.5	212	24.8
B i H	1,518	56,751	329,667	5.4	303,906/ 1,751,000	17.3	5.3	200	37.3
Macedonia	833	53,470	339,014	6.24	284,042/ 609,00	46.6	5.3	344	64.1
Montenegro	475	25,780	124,503	4.72	110,862/ 176,000	62.9	4.3	233	54.2
FPRY	6,625	340,739	1,707,571	5.0	1,839,978/ 10,245,000	17.9	5.3	277	51.4

Sources: *Jugoslavija, 1945–1964*, 113; Arhiv CK SKJ, Komisija za selo, XV 1/27, "*Razvoj seljačkih radnih zadruga*, December 31, 1949–June 30, 1951," and XV 1/108, "Problemi seljačkih radnih zadruga," 1–2.

[16] See Bokovoy, *Peasants and Communists*.
[17] Ibid.

If one were to look only at the statistics, the Yugoslav collectivization drive appears as if it were successful: nearly one-fifth of the arable land had been brought under state control. On paper, the number of SRZs grew rapidly, but the CPY was a long way from making these cooperatives work. This is readily apparent when one considers that in Montenegro and Macedonia, local cadres had succeeded in transforming a high percentage of the land into the peasant work cooperatives, 62 and 46 percent respectively. But these republics were economically insignificant—neither were food-export areas, and together they possessed less than 8 percent of the arable land.[18] The compliance patterns in these republics is equally interesting: many of the poorer Montenegrin peasants joined cooperatives in order to avoid *otkup* obligations and to improve their standard of living. Before 1939, close to half of Montenegrin peasants possessed less than 2 hectares, and were primarily pastoralists, who brought minimal land and resources into the peasant collectives.

In the grain-rich regions of eastern Croatia and Vojvodina, collectivization had mixed results. In Croatia, local and republican officials faced consistent peasant resistance, and at its peak, succeeded in collectivizing only 12.5 percent of the cultivated land.[19] Recent scholarship shows that peasants and local party officials in these regions circumvented, obstructed, and resisted mandatory quotas during the program's initial four years. At the same time, the state had begun coercing, arresting, and convicting peasants for not complying with mandatory quotas, the CPY had also begun modifying and changing those quotas, often with pressure from high-level Croatian party officials who sought to balance the need to meet the state's quotas with the need to placate Croatian peasants—many of whom had been staunch supporters of the interwar Croatian Peasant Party and who, along with Serbian peasants, had formed "the backbone of the Partisan forces victory over the Independent State of Croatia, a fascist-style collaborationist regime, headed by the Croatian *Ustasha*. The Croatian communists had earned peasant support by promising them land reform, guaranteeing private property and explicitly renouncing collectivization."[20]

The situation was even more delicate in Croatia's ethnically mixed border regions, which were important areas for agricultural production

---

[18] Tomasevich, "Immediate Effects," 124.
[19] Tomasevich, "Immediate Effects," 124.
[20] McCarthy, *Peasant Revolutionaries*, 321.

but where ethnic fissures still ran deep as a result of the *Ustasha*'s brutal and violent campaign against Serbs during the war. As a result, Serb peasants often resisted *otkup* and collectivization efforts, waiting to see if their Croatian neighbors complied. Distrust toward Croatian authorities who carried out these campaigns was also apparent in ethnically mixed regions in Vojvodina, where Hungarians were suspected by other residents (the majority of whom were Serbs) of shirking their duties toward the state.

In Vojvodina, collective farms consisted of approximately 40 percent of the cultivated land—a high percentage compared to Serbia proper or Croatia, and likely due to the influence local and republican party leaders had over peasants, many of whom were former partisans. Nonetheless, there was widespread resistance to and discontent with peasant cooperatives. Similar to their reactions to the colonization process, during which peasants complained about poor housing stock, shortages of food, farm and household implements and tools, and the distribution of land parcels, SRZ members were often disgusted with the collectivization process and protested the lack of land, machinery, seed, and technological expertise.

Throughout the campaign, the Yugoslav peasantry showed remarkable tenacity and ingenuity in dealing with the peasant work cooperative campaign. Since direct confrontation with the party was not advisable given the strength of the Yugoslav army, peasants used a wide variety of strategies to express their opinions and desires to party leaders. The Commission for the Village[21] meticulously recorded dozens upon dozens of incidents directed against local officials and those peasants who dared cooperate with the state. Taken as a whole, these reports show the endemic nature of peasant resistance.

Peasant resistance to collectivization varied in its degrees of confrontation. What distinguished the reactions of the peasantry in 1949 from previous tactics is the extent to which they used violence. Within the first few weeks of the campaign, a commission report from Serbia remarked upon several instances of violence and public unruliness. For example, a local cooperative was conducting an educational meeting when a *kulak* interrupted by shouting at the SRZ members: "Down with the work coop-

[21] The Central Committee of the CPY sanctioned the creation of a watchdog agency for the village in all republics, the Commission for the Village. This commission, made up of members of the Central Committee, was entrusted with the duty of recording the activities of local cadres and the peasantry in the campaign for creating peasant work cooperatives. Arhiv CK SKJ, Fond II 1/48, "Directiva CK KPJ o osnivanju Komisije za selo, March 14, 1949."

eratives, your mother is a communist!"[22] The meeting ended in chaos and disorder, which was an embarrassment for local officials, as the meeting had been disrupted in the presence of a district committee secretary. In the same region, a group of peasants had murdered a *kulak* who had planned to join a peasant work cooperative.[23] When the police arrived to investigate, the police were stoned.[24] As reported later, the murdered peasant had consented to join a cooperative and the other peasants intended to "persuade" him otherwise.[25] The author of this report added his own assessment of the situation in April 1949: "the *kulaks*, in some places, have openly begun to carry out violence and sabotage."[26]

That the peasants resorted to violence is not surprising: with few avenues open for dialogue with the CPY, they struck out at local officials and other peasants who decided to cooperate and join the peasant work cooperatives. Reports of violence are documented in correspondences from villagers—one described how "enemies" had shot at the party secretary through her open window while she was dining. Luckily, she was only "lightly" wounded in the head.[27] Peasants often warned their neighbors about the consequences of joining collectives. Slogans heard in Croatia cautioned: "There will be war, and those who are found in the cooperative will be hung."[28] A similar statement was overheard by the village commission in Sutjeska, Serbia: "Soon there will be war."[29]

Violence and intimidation were directed not only at those peasants who supported the regime through membership in the work cooperatives, but also aimed at those peasants who complied with the *otkup*. When Committee member Milivoje Milićević asked a local peasant why he was

---

22 Arhiv CK SKJ, Komisija za selo, XV 5 I/1, "Izveštaj po pitanju zadrugarstva, April 11, 1949," 1–8.

23 Arhiv CK SKJ, Komisija za selo, XV 5 I/1, "Izveštaj po pitanju zadrugarstva, April 11, 1949," 5.

24 Arhiv CK SKJ, Komisija za selo, XV 5 I/1, "Izveštaj po pitanju zadrugarstva, April 11, 1949," 4.

25 Arhiv CK SKJ, Komisija za selo, XV 5 I/1, "Izveštaj po pitanju zadrugarstva, April 11, 1949," 4.

26 Arhiv CK SKJ, Komisija za selo, XV 5 I/1, "Izveštaj po pitanju zadrugarstva, April 11, 1949," 1.

27 Arhiv CK SKJ, Komisija za selo, XV 6 I/9, "Referat Druga Dragutina Saili-a na II Plenum KPH, 1949," 10.

28 Arhiv CK SKJ, Komisija za selo, XV 6 I/9, "Referat Druga Dragutina Saili-a na II Plenum KPH, 1949," 9.

29 Arhiv CK SKJ, Komisija za selo, XV 5 I/1, "Izveštaj po pitanju zadrugarstva, April 11, 1949," 3.

reluctant to deliver his grain to the state collection agency, the peasant told him: "If I deliver all the grain [to the state collection agency], I don't dare return to the village, and if I don't deliver, you will punish me."[30] This peasant was acutely aware of the *repercussions* from neighbors for complying with the demands of the state. He saw little difference between village justice and state justice. To him, they were equal.

Violence was not the only weapon used by the peasants. Some peasants stirred up fears by warning of consequences for anyone who joined the cooperatives or cooperated with the *otkup*. One slogan warned, "He who enters into the cooperative, renounces Christ."[31] In a village in southern Serbia, an observer overheard this counsel, "If you go into the cooperative, you will lose your property."[32]

Stories of property loss and deprivation circulated among peasants throughout Yugoslavia. A commissioner, reporting from Croatia, noted how "some say land would be taken away, no work would be available for them in the SRZ, and they [the peasants] would have to go into industry."[33] The commissioner explained that in other regions in Croatia, peasants warned of hunger and losing control over families.[34] In Kosmet, some peasants threatened their Muslim neighbors that the work cooperative leadership "will take the veils off the women, you will eat from the cafeteria; they will order your children around; and you will no longer have a mosque or celebrate Ramadan."[35] The Commission also described how in a mixed Serbian and Croatian village in South Slavonia, the Serbs refused to enter into the peasant work cooperative unless the Croats also entered. Serbian residents claimed they had been victims of the Croatian *Ustasha* in 1941, and they refused to suffer more unless their Croatian neighbors joined them.[36]

---

[30] Arhiv CK SKJ, XV 1/17, "Zapisnik, October 10, 1949," 2.

[31] Arhiv CK SKJ, Komisija za selo, XV 6 I/9, "Referat Dragutin Saili-a na II Plenum KPH, 1949," 9.

[32] Arhiv CK SKJ, Komisija za selo, XV 6 I/9, "Referat Dragutin Saili-a na II Plenum KPH, 1949," 9.

[33] Arhiv CK SKJ, Komisija za selo, XV 6 I/9, "Referat Dragutin Saili-a na II Plenum KPH, 1949," 9.

[34] Arhiv CK SKJ, Komisija za selo, XV 6 I/9, "Referat Dragutin Saili-a na II Plenum KPH, 1949," 9.

[35] Arhiv Jugoslavije, SSOJ F9, Milorad Pesić, "O nekim pitanjima ideolosko-vaspitnog rada u organizacijama Narodne omladine," XIV Plenum, January 5–7, 1950.

[36] Arhiv CK SKJ, Komisija za selo, XV 1/46, "Izveštaj ostanju i razvoju zemljoradničkog zadrugarstva u Hrvatskoj," May 5, 1949.

Peasants who joined collectives were often ostracized and isolated by other villagers. A report from a village in Bosnia-Hercegovina describes how peasants who had formed a work cooperative lacked the respect of other peasants. One villager depicted cooperative members as "rude, crude and possessing too much bravado," and having attitudes that "negatively influenced the winning over of other peasants for the cooperative."[37] In his opinion, perhaps, the other peasants had good reasons to isolate these few. In some cases, families threatened to expel or shun their sons, daughters, sisters or brothers if they followed the state's directives. According to one account from Croatia, a widow of a war veteran had joined a peasant work cooperative, but was forced to withdraw because her future daughter-in-law refused to join.[38] Peasant cadres also faced familial pressure—some excused their parents from the peasant work cooperatives.

Intimidation, physical and psychological, proved to be an effective method of conveying dissatisfaction with the party's policies to local authorities. What is striking in all of these accounts is how the communiqués begin to describe the peasant's activities. In these confrontations, local authorities and members of the Commission see these direct attacks on the party and the state as challenges to their authority, power, and legitimacy in the countryside. Although many of these incidents were anonymous, the party believed the violence was driven by political or economic motivations. This was not simple self-interest. The peasants were wrestling with the party for control over the countryside. In acting against the state, peasants who resisted were soon being defined as "enemies," their actions as traitorous. They had not only made "unfriendly attacks on the cooperatives from the outside but attempted to break them up from the inside."[39]

How did the peasants see this struggle? Evidence shows that many peasants, regardless of nationality or region, believed that they were defending their most fundamental interests: land, property, family, and independence. The CPY threatened to take their land, control their fami-

[37] Arhiv CK SKJ, Komisija za selo, XV 7 I/2, "Izveštaj o sastanku Komisije za selo pri CK KPBiH održanog 6. April 1949," 3.

[38] Once the CPY had embarked upon this collectivization campaign, it asked the war veterans to write letters to friends, family, and families of deceased veterans urging them to join the SRZ. See Arhiv CK SKJ, Komisija za selo, XV 5 I/1, "Izveštaj po pitanju zadrugarstva, April 11, 1949," 2. The recounting of the war widow's tale is found in Arhiv CK SKJ, Komisija za selo, XV 6 I/9, "Referat Druga Dragutina Saili-a na II Plenum CK KPH, 1949," 11.

[39] Arhiv CK SKJ, Komisija za selo, XV 6 I/9, "Referat Dragutin Saili-a na II Plenum KPH, 1949," 9.

lies, force them into the work cooperatives, or worse, to drive them from their villages. Some peasants, both in Croatia and Serbia, even portrayed their confrontations with the party as war, and believed that violence and direct attacks were effective methods of defending their way of life. Those peasants who resorted to violence and intimidation believed this was the only way to exert their power as individuals. Collectively, these actions and slogans put the CPY on notice. Dragutin Saili, in his report to the Second Plenum of the Croatian Communist party (CPC), noted the various anti-collectivization slogans coming from the villages, and wrote: "These slogans are popular and aim to create uncertainty and non-confidence in our government."[40] Saili's observation was correct: inflammatory slogans and direct attacks on local authorities and peasant supporters effectively worked to undermine attempts to implement the government's collectivization policies.

Although direct attacks effectively challenged the CPY's authority, they were still risky. Yugoslav peasants therefore expressed their dissatisfaction or pressured local authorities using a variety of less-direct methods as well, such as vandalizing crops, livestock, machinery, and property. Indeed, acts of vandalism were far more common than violence or intimidation. Reports contain daily accounts of crop destruction, livestock slaughtering, arson, and sabotage of machinery, which were common tactics among peasants who were joining a cooperative. The law on SRZs required peasants to contribute land, livestock, farm equipment, seed, fodder and grain reserves, and any other materials related to agricultural production. In return, the peasants were to be compensated. Compensation depended on the type of cooperative the peasant joined and its leadership. Some peasants suspected that this compensation undervalued their property or was simply ignored. In many cases, peasants destroyed anything of value before joining in order to thwart the CPY's attempts to control their land and property.

A village commission report from the regions near Titograd and Cetinje in Montenegro described this trend, noting that when faced with joining the cooperative, peasants killed their livestock and burned their wheat.[41] The report detailed different types of vandalism directed at peasant work cooperatives. When faced with pressure to join the coop-

[40] Arhiv CK SKJ, Komisija za selo, XV 6 I/9, "Referat Dragutin Saili-a na II Plenum KPH, 1949," 9.

[41] Arhiv CK SKJ, Komisija za selo, XV 1/93, "Stanje i problemi zadrugarstva u Crnom Gori, 1949," 5.

erative in their *srez* (county), the peasants of Nikšičko tried to kill their livestock before entering. In a nearby village, the peasants burned their wheat in order to prevent the cooperative from gaining access to it. Some peasants who lived outside one cooperative raided the silo and picked fruit from the orchard.[42] While such occurrences were common throughout Yugoslavia, methods of vandalism varied according to the region, wealth of the local peasantry, and type of agricultural production. For example, slaughtering or hiding livestock was more common in areas where the dominant form of agriculture was livestock grazing. Records from village commissions in Hercegovina and Montenegro,[43] mountainous and soil-poor republics, highlighted how the destruction of livestock was increasingly common in those regions. In Trebinja, peasants slaughtered 50,000 sheep.[44] In the village of Sipani, peasants butchered 40 cattle. In Vračev gaj in southern Serbia, peasants tried to intimidate the cooperative's members by setting fire to some hay stacks which were alongside the cooperative's pig pen with some 300 pigs inside.[45] Officials in Hercegovina noted how the peasants hid their livestock and inventory before joining cooperatives.[46] In Croatia, a member of the Commission observed how some peasants sold their livestock before joining the collective or persuaded the cooperative to pay for their inventory.[47]

What was the motivation of these peasants to destroy their livelihoods? Some peasants who practiced such acts of vandalism did so out of desperation. They did not look at the long-term ramifications of such actions—they made sure that no one else could benefit from their labor and property. It became battle for control over production, property and village authority,

---

[42] Arhiv CK SKJ, Komisija za selo, XV 1/93, "Stanje i problemi zadrugarstva u Crnom Gori, 1949," 3.

[43] Arhiv CK SKJ, Komisija za selo, XV 1/25, "Zapisnik za sastanka aktiva Komisija za selo CK CPY, December 29, 1949," 1.

[44] Arhiv CK SKJ, Komisija za selo, XV 1/25, "Zapisnik za sastanka aktiva Komisija za selo CK CPY, December 29, 1949," 1.

[45] Arhiv CK SKJ, Komisija za selo, XV 5 I/1, "Izveštaj po pitnaju zadrugarstva, April 11, 1949," 4. In a similar incidence, the police arrested a peasant in Sinj Kotar, who had removed the roof from the warehouse of the SRZ. When asked why he did it, the peasant explained that the hay would get wet from the rain, making it unusable, and harm the cooperative. Arhiv CK SKJ, Komisija za selo, XV 6 I/9, "Referat Druga Dragutina Saili-a na II Plenum CK KPH, 1949," 9.

[46] Arhiv CK SKJ, Komisija za selo, XV 7 I/2, "Izveštaj o sastanku Komisije za selo pri CK KPBiH održanog, March 6, 1949," 4.

[47] Arhiv CK SKJ, Komisija za selo, XV 6 I/9, "Referat Druga Dragutina Saili-a na II Plenum CK KPH, 1949," 9.

and more so, over a way of life. Peasant resistance and vandalism signaled to the CPY that the villagers would not surrender the countryside without a fight. These acts of resistance, recorded by village commissions across Yugoslavia, sent a unified message from the villages to party leaders at the national and republican levels: the countryside would not be easily subdued.

Confusion and overzealousness among peasant and local officials often added to the chaos. Peasant cadres did not know how to form the proper type of work cooperatives for the terrain and population. Some refused to join the cooperative or granted exemptions to family and friends. Peasants who joined cooperatives were required to make only a three-year commitment, after which they could leave if they chose. Those who joined cooperatives unwillingly often found ways to undermine the project, such as infiltrating leadership positions in order to sabotage the collective. Such cases were reported in correspondences between villagers and party officials, who referred to these peasants as *kulaks* or speculators.[48] According to one report, "the SRZ pulled in different *kulak* elements with sabotage and speculative tendencies. In the Vojvodina, they have a considerable number of SRZ's without a party organization attached."[49] This remark revealed a series of problems that had come to plague the CPY in its campaign.

Some peasants who joined the SRZ were less than supportive of the new institution and tried to exploit it to their own advantage. The bulletin of the Commission in November described how in Bačka Topla:

> *Kulaks* enter with the intent to surrender land which they are unable to cultivate due to production obligations and delivery quotas. Or they enter with the intent to avoid obligations toward the state such as the *otkup*, taxes and other obligations. They wish to extract from this economy advantages for themselves. They blackmail the working peasant into working the land which they have brought in by withholding access to farm equipment necessary for the poor and middle peasant to work his plot ...

[48] This term is deceiving because of the poor accounting skills practiced in the formation of the peasant work cooperative. The only statistic available at the time was the average number of hectares, 5.5, brought into the cooperatives by each household. Keeping with the statistical definition put forward by the Yugoslavs, those peasants possessing 5.5 hectares are classified as middle peasants.

[49] Arhiv CK SKJ, Komisija za selo, XV 1/4, "Zapisnik sa prvog sastanka IV grupa aktiva pri Komisija sa selo CK KPJ, June 3, 1949," 1.

> There are cases in other republics as well, for this is the tendency of the rich peasant.[50]

In another case, the *kulaks* of Sutjeska in Serbia convinced the poor and middle peasants that they did not need to join the cooperatives since the poor and middle peasantry do not have large obligations to the state. This left the way clear for the *kulaks* to join, avoiding payment of taxes and the *otkup*, and determining the rental fees for farm equipment needed by the poor and middle peasants.[51] In the same account, the Commission described the process by which some *kulaks* came into a cooperative and wrestled the leadership from a group of unsure communist cadres. The *kulaks* formed a splinter group and convinced the poor and middle peasants to support them.[52] The commentator added that if the local OPO had been stronger, this coup would not have happened. In another report, a commissioner in Montenegro observed that in one village the party organization was not strong enough to destroy the *kulak* element since there was sympathy in the region for "the enemy."[53] A report from Serbia succinctly summarizes the problem: "The fact is that they [the party organizations] let in unfriendly elements into the SRZ. The party organization does not understand the seriousness of the class struggle within the SRZ and they do not take the proper measures."[54]

The weakness or lack of party organizations within the cooperatives created another problem for the CPY's collectivization drive. Infiltration by unfriendly elements was troublesome enough, but the reluctance of the peasant cadres to expel the offenders and/or the peasant cadres cooperation with the *kulaks* proved especially disconcerting. In the formation of some SRZ's, "some party members had hindered their formation,"[55] reported M. Jovović in a report to the Agit-Prop ministry. Jovović explained that the

---

50 Arhiv CK SKJ, Komisija za selo, XV 1/79, "Bilten, Komisija za selo," no. 4, November 2, 1949, 13.

51 Arhiv CK SKJ, Komisija za selo, XV 5 I/1, "Izveštaj po pitanju zadrugarstva, April 11, 1949," 4.

52 Arhiv CK SKJ, Komisija za selo, XV 5 I/1, "Izveštaj po pitanju zadrugarstva, April 11, 1949," 4.

53 Arhiv CK SKJ, Komisija za selo, XV 1/93, "Stanje i problemi zadrugarstva u Crnom Gori," 1949, 4.

54 Arhiv CK SKJ, Komisija za selo, XV 5 I/10, "Zapisnik sa sastanka Komisije za selo CK KPS, November 28, 1949," 1.

55 Arhiv CK SKJ, V KII/18, M. Jovović, "O nekim pitanjima dužnosti i prava članova partije," January 4, 1950.

cadres' obstruction was due to their loyalty to their parents and brothers. He also noted some resistance among Communist Party members in administering some of the basic tasks of the cooperative's leadership.[56] These basic tasks included creation of work brigades for the cooperative, establishing work norms, i.e. the number of work days demanded of each cooperative member, maintaining discipline, calculating payment in money and in kind to each member, land assessment, general accounting, and meeting the state's delivery quota. How the leadership accomplished these tasks depended on several factors: loyalty of the cadres to the party, their standing within the community, and their understanding of the regulations. Sometimes it was not clear which factor was the dominant one.

The local commissions' reports as the year progressed contained a litany of problems associated with the organization and administration of the peasant work cooperatives, often times describing the actions of the cooperatives' administrators. First, in simply choosing the number of cadres necessary for administrative positions, the commissions found too many administrators who had questionable "moral and political qualifications," and not enough workers.[57] Second, in the creation of work brigades, one report cited how the directional board (*upravni odbor*) unsystematically formed the work units on the cooperative farm. The report noted: "Most work brigades do not exist as formal working units, the cooperative brigade leader simply invites peasants who want and can work. There is no organized accounting of the work done or the days worked by each peasant."[58]

The report also complained that the intense physical work for the cultivation of SRZ crops was shunned by the cooperative members in order to work the plots around their homes.[59] Why was this the case? The cooperative members could keep and/or sell any of the agricultural products raised on these plots. It was not worth their effort to work on cooperative lands if compensation was inadequate and improperly disbursed. The problems of miscalculating work days, corruption in the sense that board members paid themselves or family members first or with bonuses, and

[56] Arhiv CK SKJ, V KII/18, M. Jovović, "O nekim pitanjima dužnosti i prava članova partije," January 4, 1950.

[57] Arhiv CK SKJ, Komisija za selo, XV 1/27, "Izveštaj za CK KPJ," December 20, 1949, 6.

[58] Arhiv CK SKJ, Komisija za selo, XV 1/64, "Izveštaj po nekim problemima iz rada Komisije za selo SK Titograd," September 23, 1949, 5.

[59] Arhiv CK SKJ, Komisija za selo, XV 1/64, "Izveštaj po nekim problemima iz rada Komisije za selo SK Titograd," September 23, 1949, 7.

poor access to the necessary farm implements and equipment discouraged the cooperative members from active and enthusiastic participation in the cooperative.[60] The commentator for the *srez* of Titograd contributed his own opinion, "The cooperative members still do not see in the cooperatives the source of their existence and path to a better life."[61]

VIOLENCE IN THE COUNTRYSIDE

The collectivization campaign in 1949 had effected the creation of over 5,000 new peasant cooperatives,[62] but it had not brought about the socialist transformation of the village. Facing entrance into the peasant work cooperatives, the peasants did not see in them a path to a better life. Instead, the peasants viewed communist policy as an usurpation of their production autonomy, land, and independence. Peasants knew that their activities would be more closely monitored and ultimately controlled in the cooperatives. Not only would the peasants have to make their land available to the state under the provisions of the different types of cooperatives, but the cooperative would gradually appropriate all decisions on land usage, cropping patterns, mechanization, the working day, and wages. The resistance practiced by the peasants attempted to deny the party this type of control over their lives, both productive and personal. By the end of 1949, the CPY realized that the collectivization campaign was failing to socialize the countryside, and decided to take a more hard-lined path, and to intimidate, expel, arrest, prosecute, and imprison those who neglected their duties and obligations in regards to both the collectivization and the *otkup*—whether they were local, regional or republican officials or peasants. It was the same tactic the party had used to squelch resistance to the *otkup*. Violence against the peasantry throughout 1949 and 1950 coincided with the persecution of Yugoslavs suspected of siding with the Cominform

[60] Arhiv CK SKJ, Komisija za selo, XV 1/64, "Izveštaj po nekim problemima iz rada Komisije za selo SK Titograd," September 23, 1949, 5–8. For other misuses and abuses of executive power within the cooperatives in each republic see Arhiv CK SKJ, Komisija za selo, XV 1/27, "Izveštaj za CK KPJ," December 20, 1949; and XV 1/68, "Izveštaj o izvršenom pregledu SRZ," October 12, 1949.

[61] Arhiv CK SKJ, Komisija za selo, XV 1/27, "Izveštaj za CK KPJ," December 20, 1949; and XV 1/68, "Izveštaj o izvršenom pregledu SRZ," October 12, 1949, 5.

[62] Arhiv CK SKJ, Komisija za selo, XV 1/27, "Izveštaj za CK KPJ," December 20, 1949, 1.

and the Soviets. As Banac has observed, "The anti-Cominformist purge was a unique opportunity for the KPJ leadership to cleanse the land of all potential troublemakers."[63] However, in the countryside, it was not always clear who the troublemakers were. Trouble could be interpreted as outright resistance or simple recalcitrance to the *otkup* and the peasant work cooperatives. Both responses were prevalent, and both were punished.

After the expiration of the 1947 general amnesty for peasants avoiding the *otkup*, the party-state adopted a stance of arresting, prosecuting, imprisoning or fining peasants who did not meet their delivery obligations. Anecdotal evidence from the Commission of the Village demonstrates how varied the state's response to non-compliance was in 1949. The type of reaction depended primarily on whether the local and republican officials were sympathetic or not to the peasants' plight. The most serious challenges to the *otkup* came from the grain growing regions of northern Yugoslavia—Vojvodina, Slavonia, and eastern Croatia.

Osman Karabegović, a supporter of the 1947 amnesty, noted how in the Vojvodina region, the peasants had the greatest burden of *otkup* obligations, and "it is here that the most uncontrolled repression exists."[64] What Karabegović meant was that due to the peasants' reluctance to deliver grain and other agricultural products, the local and republican cadres engaged in forms of intimidation, humiliation, and mistreatment throughout 1948 and 1949. Reporting 94 incidences of non-compliance, the regional committee (*Pokrajinski komitet*) for Vojvodina noted how these were dealt with by local cadres:

> Mistreatment ranged from cadres brandishing pistols, hurtful bodily injury (broken ribs, teeth knocked out, injured eyes, etc) to destruction of personal dignity by inducing the accused peasants to spit at each other. They were forced to play animals in one case, imitating a saddled horse bringing on its back to the secretary one member of the committee.[65]

According to the village commission reports, local committees carried out more than 1,000 confiscations in 1949. 220 of these confiscations were "total," meaning all of the peasants' harvest had been seized, even when it

63 Banac, *With Stalin*, 154.
64 Dautović and Vučelić, "Svašta," 52.
65 Dautović and Vučelić, "Svašta," 52.

was taken from people who could not afford it.[66] In this particular collection drive, there were reports of mistreatment of the peasants—slapping, pulling of beards, and other forms of physical abuse.

In the grain-growing regions of Croatia, similar reports caught the attention of the Croatian leadership. Croatian communist leader Vladimir Bakarić pointed out that close to 50 people had been wounded in Slavonia in a clash with authorities after peasants attempted to avoid the *otkup*.[67] More than 500 peasants also refused to participate in the spring sowing campaign, refusing to hand over their grain to the state collection agency or to plant their fields. Authorities prosecuted more than 400 peasants for non-compliance and confiscated the surplus from another 100. Yet, state pressure did not deter the peasants from resisting collectivization. Bakarić described the almost fatalistic attitude among peasants, recounting how some peasants said that "if one must go to prison, then go immediately, inevitably one must leave."[68]

Other peasants felt that if they continued to resist, the KPJ would eventually have to relent, an attitude that was reflected by one peasant who said: "they can't imprison all of us."[69] The state perhaps could not imprison all of the peasants, but they could intimidate, scare and coerce them into giving up their agricultural products and land. When villagers failed to meet the quotas, some local authorities sent search teams of local cadres, the police, or members of the state security force (*Uprava državne bezbednost*, UDBa) into the villages to find where the peasants had hidden their remaining harvests. In villages with stronger party support, local authorities used the power of the mob to intimidate those who had not complied.[70]

Dissatisfaction among the population often stemmed from simple bread-and-butter issues. In some regions, there was a scarcity of food, land, and work. In addition, drought had hit peasants hard, and only exacerbated the situation.[71] Peasants revolted in Macedonia, northern

---

66 Petranović, Končar, and Radonjic, *Sednice Centralnog*, 425.

67 Dautović and Vučelić, "Svašta," 52.

68 Dautović and Vučelić, "Svašta," 52.

69 Dautović and Vučelić, "Svašta," 52.

70 Arhiv CK SKJ, Komisija za selo, XV 1/27, "Izveštaj za CK KPJ, 20 decembra 1949," 2.

71 Tomasevich notes that the combined index of agricultural and livestock production (1947–56 average equals 100) fell from an average of 102 during the period of 1947–49 to 78 in 1950. The index of grain production fell from 112 for the period 1947–49 to 75 in 1950. He claims that the drought rather than collectivization was responsible for this decline in production. See Tomasevich, "Immediate Effects," 125.

Bosnia, and in the Banija and Kordun regions of Croatia. The rebellions in Croatia, which developed along the old military frontier with a long tradition of peasant uprisings, were quickly crushed.[72] In May 1950 a group of veterans and peasants in northwestern Bosnia near Bihać,[73] a partisan stronghold with a mixed population of Serbs and Muslims, mounted an insurrection against the local government in protest of the *otkup*. In early May 1950, a group of discontented, neglected veterans and peasants fused their grievances into a single act—an armed insurrection against the local government of the Cazinski *srez*, with its center at Bihać.[74] In the late evening and early morning of May 5–6, a group of nearly 40 marched toward Cazin from the village of Furjana.[75] Along the way, they set fire to the local committee's meeting place, and attacked a cooperative and divided up its goods and wheat reserves. This cooperative had long been a sore spot for the community because it had been established by colonists before the 1949 collectivization drive, and local peasants saw these outsiders as aiding the communists.[76] During their protest, peasant rebels also disarmed local policemen and cut telephone lines. The rebellion spread through five neighboring villages. When they reached Cazin, the rebels secured a hill above the town which sent a wave of confidence through the crowd and they began to chant "long live King Peter, long live the fatherland."[77]

On May 6, after receiving information about the riots, Alexander Ranković and Đuro Pucar called in the army to suppress the revolt. Although accounts vary over whether casualties occurred during or after the army intervened, nine people died as a result. The three organizers,

---

72 Wilson, *Tito's Yugoslavia*, 68.

73 Vera Kržišnik-Bukić, a Slovenian historian of Bosnia-Hercegovina, has detailed the 1950 rebellions in Cazin, Biha, Velika Kladuša (northwest Bosnia), and Slunj (central Croatia). She argues that these uprisings can be used as a window for looking at the dynamics of the other peasant rebellions in Macedonia and Banija. See Kržišnik-Bukić, *Cazinska*, 135.

74 In retelling the story of this revolt, I am relying on several secondary sources: Kržišnik-Bukić, *Cazinska*; Banac, *With Stalin*, 235–238; Dautović, "Cazinski krajina," 30–31; and Dautović and Vučelić, "Svašta." This latter source relies on documentation from the Commission for the Village.

75 May 6 is St. George's Day (Djurdjevdan), a Balkan peasant tradition symbolizing the beginning of spring and signaling the beginning of annual outlaw actions (*hajduk)* against authority. In Kržišnik-Bukić, *Cazinska*, 134–35; and Banac, *With Stalin*, 236.

76 Kržišnik-Bukić, *Cazinska*, 183. Cited in McCarthy, *Peasant Revolutionaries*, 375.

77 Dautović and Vučelić, "Hleba nema," 55.

Mile Devrnja, Nikola Pamaguša, and Mirko Radočaj, were eventually captured and executed after eluding authorities for several weeks with the help of local peasants who saw them as heroes and martyrs.[78] In many peasant cultures, these "social bandits" had become important symbols of resistance. For the peasants of northwestern Bosnia, the rebels were viewed in this way.[79]

The Cazinski Uprising reveals how national, economic, and rural issues merged around protests against collectivization. Initially, there were two separate rebel groups: Serbian partisans, led by Milan Bozić, and Muslim and Serbian peasants.[80] However, as the two groups moved through the countryside, they merged into a single force, targeting symbols of party and state power—the local people's committee, the police, the *otkup*, and agricultural cooperatives. According to records, the group did not destroy homes, stores, or attack ordinary citizens. Those in the crowd who shouted "long live King Peter" were probably Serbs, who connected the survival of their way of life to monarchy and nation.

Vladimir Bakarić downplayed the event to Western consulates, identifying the culprits as a small group of unhappy Serbs, *Chetnik* and *Ustasha* elements, and individuals who railed against the bureaucratic measures of the local authorities and cooperative officials.[81] In this interpretation, the rhetoric proved to be much more sophisticated when placed alongside actions taken by the state. After the suppression of the revolt, 714 people were arrested, including family members and neighbors who refused to help in the investigation. 288 individuals were brought to trial before military courts, of these 17 were condemned to death by firing squad and the others were given prison sentences of fifteen to twenty years. 426 individuals were tried in criminal court and all were sentenced to 2–3 years hard labor in the Brez mines.[82]

With his statement that the uprising had been carried out by Serbs and *Chetnik-Ustasha* elements, Bakarić had singled out specific groups that represented extreme nationalist positions. Nationalism indeed played a role in this uprising, and the CPY was made aware of the dangers that this group

[78] Kržišnik-Bukić, *Cazinska*, 183. Cited in McCarthy, *Peasant Revolutionaries*, 375.
[79] Eric Hobsbawm describes similar actions of peasant communities in southern Italy and Spain. He points to how these outlaws became important symbols of resistance to injustice and power. See Hobsbawm, *Primitive Rebels*.
[80] Dautović "Cazinski krajina," 30–31.
[81] Bekić, "Jugoslavija," 30–31.
[82] Banac, *With Stalin*, 233; and Dautović, "Cazinski krajina."

of Serbs—many who had served in the partisan army—now posed. They had marched against the local organs of state power, and along the way rallied other segments of the population, who were equally unhappy with the party's collectivization campaigns, the *oktup*, and the harsh bureaucratic measures taken by local authorities and cooperative officials.[83] The danger in this region, and elsewhere, was that former partisans with national sentiments would join with peasants against the party. The CPY leadership, faced with losing support of both groups, modified its policy in this region and reduced the amount required for the *otkup* as well as permitted peasants to sell the remainder of their produce at market value.[84]

Suppression and violent repression of any acts of resistance was but one way that the CPY chose to handle the situation in the countryside. Yet, this was not a viable solution, besieged as they were from the outside. A war in the countryside would make the new Yugoslavia vulnerable to attack, ideologically and militarily. Using violence to suppress the countryside would concede a crucial ideological point to the Soviets—the Yugoslav peasantry was not the pillar for building socialism as the CPY had thought. Pressing the peasants to join collectives or forcing them to comply with the *otkup* would consume too much of the Yugoslav state's military and police resources. The KPJ leadership had to decide between relaxing its policies in the countryside and trying to reassume its tacit alliance with the peasantry, or maintaining its assault on the countryside.

Conclusion

In March 1953, the Skupština passed the Law on Property Relations and Reorganization of Peasant Work Cooperatives, allowing members to withdraw from peasant work cooperatives, individually or in groups, and to liquidate or reorganize the collective if the majority of the membership so desired. With this, the CPY admitted defeat.[85] But this created another problem: as more prosperous peasants withdrew, poorer peasants were left with insufficient or low-quality land. In response, in May 1953 the party passed a law establishing 10 hectares as the maximum amount of arable land an individual could own, and excess land was deposited into a

[83] Dautović and Vučelić, "Hleba nema," 55–56.
[84] Dautović and Vučelić, "Hleba nema," 55.
[85] "Uredba o imovinskim odnosima," 1953.

general land fund that the government redistributed to the less prosperous farmers or state-created farms.[86] By limiting land parcels to 10 hectares, the party was hoping to decrease the potential for successful independent farming, thereby forcing peasants into joining the general agricultural cooperatives. However, the reorganization of cooperatives resulted in their disappearance by the late 1950s and interest in general agriculture cooperatives diminished as well.[87]

By the end of 1953, pressure from international powers, internal party differences, and peasant resistance had forced the Yugoslav communists to dismantle their collectivization campaign. Among these factors, the most important reason for the CPY's revision was its unwillingness to wage war against the peasantry in order to advance its Stalinist collectivization program. This decision was based on well-documented reports of peasant resistance by the village commissions, which were taken into consideration by the Politburo as they moved away from the Stalinist model. Thus, the peasants, speaking through local authorities and agrarian and political experts, had made their opinions and their resistance known to the top leadership. The CPY simply could not risk chaos and conflict in the countryside as it distanced itself from the Soviet Union, both diplomatically and ideologically, and sought help and aid from the West. After 1953, agriculture was freed from direct state control by decreasing the importance of peasant work cooperatives, abolishing the machine tractor stations, disassembling the collection points, and rescinding mandatory quotas. The KPJ placed greater emphasis on the general agricultural cooperative and hoped that gradual economic processes would eventually integrate peasants into the socialist society. Yet by 1957, the CPY admitted that cooperatives in general, even the general agricultural cooperatives, "failed to yield positive results under our conditions. The cooperatives in practice proved, over a short period of time, to have a negative effect in our country—the producers' loss of interest and the degradation of production."[88] Echoing the words of Kardelj when he addressed the situation of agriculture in a report to the Federal Assembly, the Minister for Agriculture and Forestry observed that "no one today or in the future counts on a solution in that form."[89]

---

[86] "Zakon o poljoprivrednom zemljišnom," 1953. This was the Law on the Creation of a National Land Fund and Regulation of its Distribution.

[87] For an overview of Yugoslav agriculture since 1957, see Stipetić, "Development."

[88] Skupština, *Stanje poljoprivrede*, 136.

[89] Skupština, *Stanje poljoprivrede*, 122.

The CPY never again attempted to socialize the countryside. Kardelj had envisioned that "the socialist society would allow the individual peasant freedom of action within the limits of his economic possibilities and also help him in his efforts to raise the productivity of his holding."[90] For Kardelj, as always, this meant creating some type of institution which would gradually and increasingly bind the peasant to the socialist economy. Such an institution never materialized. Vagueness and inconsistency continued to plague the CPY's policy on agriculture for years and by the time of Yugoslavia's disintegration in 1991, nearly nine-tenths of all agricultural land was in the hands of private producers.

BIBLIOGRAPHY

ARCHIVAL SOURCES

Arhiv CK CPY, Komisija za selo [The Commission for the countryside], XV 1/46, "Izveštaj ostanju i razvoju zemljoradničkog zadrugarstva u Hrvatskoj" [A Report on Conditions and Development of the Agricultural Cooperative Movement in Croatia], May 5, 1949.

Arhiv CK KPJ, XV 1/17, "Zapisnik" [A transcript], October 10, 1949.

Arhiv CK SKJ, Komisija za selo [The Commission for the countryside], XV 1/4, "Zapisnik sa prvog sastanka IV grupa aktiva pri Komisija sa selo CK KPJ" [A transcript of the First Session of the Fourth Group of the Commission for the countryside of the CC CPY], June 3, 1949.

Arhiv CK SKJ, Fond II 1/48, "Directiva CK KPJ o osnivanju Komisije za selo" [A CC CPY Directive on foundation of the Commission for the countryside], March 14, 1949.

Arhiv CK SKJ, Komisija za selo [The Commission for the countryside], XV 1/25, "Zapisnik za sastanka aktiva Komisija za selo CK CPY" [A Transcript of the meeting of the members of the Commission for the countryside of the CC CPY], December 29, 1949.

Arhiv CK SKJ, Komisija za selo [The Commission for the countryside], XV 1/27, "Izveštaj za CK CPY" [A Report for the CC CPY], December 20, 1949.

Arhiv CK SKJ, Komisija za selo [The Commission for the countryside], XV 1/27, "Razvoj seljačkih radnih zadruga" [The development of the peasants' labour collective farms], December 31, 1949–June 30, 1951.

Arhiv CK SKJ, Komisija za selo [The Commission for the countryside], XV 1/64, "Izveštaj po nekim problemima iz rada Komisije za selo SK Titograd" [Report on some problems faced by the Commission for the countryside in Titograd], September 23, 1949.

[90] Kardelj, "Dalji razvitak," 251. Cited in Johnson, *Transformation*, 189.

Arhiv CK SKJ, Komisija za selo [The Commission for the countryside], XV 1/68, "Izveštaj o izvršenom pregledu SRZ," October 12, 1949.
Arhiv CK SKJ, Komisija za selo [The Commission for the countryside], XV 1/79, "Bilten, Komisija za selo" [Bulletin, The Commission for the countryside] no. 4, November 2, 1949.
Arhiv CK SKJ, Komisija za selo [The Commission for the countryside], XV 1/93, "Stanje i problemi zadrugarstva u Crnom Gori" [Situation and problems of collective farms in Montenegro], 1949.
Arhiv CK SKJ, Komisija za selo [The Commission for the countryside], XV 1/108, "Problemi seljačkih radnih zadruga" [Problems faced by peasant collective farms], 1–2.
Arhiv CK SKJ, Komisija za selo [The Commission for the countryside], XV 5 I/1, "Izveštaj po pitanju zadrugarstva" [Report on collective farms], April 11, 1949.
Arhiv CK SKJ, Komisija za selo [The Commission for the countryside], XV 5 I/10, "Zapisnik sa sastanka Komisije za selo CK KPS" [A transcript from the meeting of the Commission for the countryside of the CC CPS]. November 28, 1949.
Arhiv CK SKJ, Komisija za selo [The Commission for the countryside], XV 6 I/9, "Referat Druga Dragutina Saili-a na II Plenum KPH" [An expose of Comrade Dragutin Saili at the Second Plenary Session of the CP of Croatia], 1949.
Arhiv CK SKJ, Komisija za selo [The Commission for the countryside], XV 7 I/2, "Izveštaj o sastanku Komisije za selo pri CK KPBiH održanog" [A Report on the meeting of the Commission for the countryside of the Central Committee of Bosnia and Hercegovina], April 6, 1949.
Arhiv CK SKJ, V KII/18, M. Jovović, "O nekim pitanjima dužnosti i prava članova partije" [On Some Aspects of Obligations and Rights of the Party Members], January 4, 1950.
Arhiv Jugoslavije, SSOJ F9, Milorad Pesić, "O nekim pitanjima ideolosko-vaspitnog rada u organizacijama Narodne omladine" [On some aspects of ideological-educational work in the organizations of the People's Youth], XIV Plenum, January 5–7, 150.

BOOKS AND ARTICLES
Allcock, John B. *The Collectivization of Yugoslav Agriculture and the Myth of Peasant Resistance*. Bradford: University of Bradford, 1981.
Banac, Ivo. *With Stalin Against Tito: Cominformist Splits in Yugoslav Communism*. Ithaca: Cornell University Press, 1988.
Bekić, Darko. "Jugoslavija u američkim arhivama, 1950–1954" [Yugoslavia in the US Archives] in Sava Dautović, "Cazinski krajina godine 1950" [Cazinska Krajina in 1950]. *Nin*, April 20, 1986, 30–31.
Bilandžić, Dušan. *Borba za samoupravni socijalizam u Jugoslaviji 1945–1969* [A struggle for a self-governing Socialism in Yugoslavia 1945-1969]. Zagreb: Institut za historiju radničkog pokreta Hrvatske, 1969.
Bokovoy, Melissa. *Peasants and Communists: Politics and Ideology in the Yugoslav Countryside, 1941–1953*. Pittsburgh: University of Pittsburgh Press, 1998.
———, Carol Lilly, and Jill Irvine, eds. *State and Society Relations in Yugoslavia, 1941–1992*. New York: St. Martin's Press, 1997.
Dautović, Sava. "Cazinski krajina godine 1950" [Cazinska Krajina in 1950]. *Nin*, April 20, 1986.

——— and Milorad Vučelić, "Svašta se događalo" [Anything could happen]. *Nin*, April 26, 1987.
——— and Milorad Vučelić, "Hleba nema za sve" [There is no bread for all]. *Nin*, May 3, 1987.
Hobsbawm, Eric. *Primitive Rebels*. New York: Norton, 1965.
Johnson, A, Ross. *The Transformation of Communist Ideology: The Yugoslav Case, 1945–1953*. Cambridge: MIT Press, 2002.
Jugoslavija, 1945–1964: Statistički Pregled [Statistical Review]. Belgrade, 1965.
Kardelj, Edvard. "Dalji razvitak našeg privrednog sistema" [Further development of our economic system]. *Komunist* (April 1953): 251.
———. *Problemi socijalisticke politike na selu* [Problems faced by the Socialist policy in the countryside]. Belgrade: Kultura, 1959.
Kržišnik-Bukić, Vera. *Cazinska buna, 1950* [Cazin uprising, 1950]. Sarajevo: Svjetlost, 1991.
McCarthy, Kate. *Peasant Revolutionaries and Partisan Power: Rural Resistance to Communist Agrarian Policies in Croatia, 1941–1953*. Phd dissertation, University of Pittsburgh, 1995.
Nielsen, Christian. "Review of Melissa Bokovoy, *Peasants and Communists: Politics and Ideology in the Yugoslav Countryside, 1941–1953*." H-Net Reviews, September 1998. http://www.h-net.org/reviews/showrev.php?id=2359 (accessed on November 25, 2012).
Pavlović, Momčilo. *Srpsko Selo: Otkup 1945–1952* [Serbian Village: Requisition 1945–1952]. Belgrade: Institute za savremenu istoriju, 1997.
Petranović, Branko. *Politička i ekonomska osnova narodne vlasti u Jugoslaviji za vreme obnove* [The political and economic foundations of the People's Government in Yugoslavia in the period of (postwar) reconstruction]. Belgrade: Institut za savremenu istoriju, 1969.
———, Ranko Končar, and Radovan Radonjić. *Sednice Centralnog Komiteta KPJ (1948–1952)* [Sessions of the Central Committee of the CPY (1948–1952)]. Beograd: Izdavački centar Komunist, 1985.
Popov, Jelena. *Drama na vojvodjanskom selu, 1945–1953* [Drama in the Vojvodina countryside, 1945–1953]. Novi Sad: Planoteum, 2002.
Rusinow, Dennison. *The Yugoslav Experiment, 1948–1974*. Berkeley: University of California Press, 1977.
Skupština, Narodna. *Stanje poljoprivrede i zadrugarstva i perspective njihovog razvoja* [The situation in agriculture and of the cooperative movement and perspectives of their development]. Belgrade: Kultura, 1957.
Stipetić, Vladimir. "The Development of the Peasant Economy in Socialist Yugoslavia," *Eastern European Economics* 20, no. 3/4 (1982): 166–99.
Tomasevich, Jozo. "Immediate Effects of the Cominform Resolution on the Yugoslav Economy" in Wayne Vucinich, ed. *At the Brink of War and Peace: The Tito-Stalin Split in a Historic Perspective*. Brooklyn: Brooklyn College Press, 1982, 89–100.
"Uredba o imovinskim odnosima i reorganizaciji seljačkih radnih zadruga od 30 marta 1953" [Regulation on property relations and reorganization of the Peasant's Labour Collective Farms from March 30, 1953]. *Službeni list*, no. 14 (1953).
Wilson, Duncan. *Tito's Yugoslavia*. Cambridge: Cambridge University Press, 1979.

Zakondavni Rad Predsedništva Antifašističkog veća narodnog oslobodjenja Jugoslavije i Predsedništva Privremene Narodne skupštine DFJ (19 novembra 1944–27 oktobra 1945) [The Legislative Activities of the Presidium of the Anti-fascist Committee of the People's Liberation of Yugoslavia and the Presidium of the Temporary People's Assembly of DFJ: November 19, 1944–October 27, 1945]. Belgrade: Prezidium Narodne skupštine FNRJ, n.d.

"Zakon o opštem pomilovanju krivica iz uredaba o otkupu poljoprivrednih proizvoda" in *Prvo redovno zasedanje Saveznog veća i Veća naroda: Stenografske beleške 15 maj–20 jula 1946* [The First Session of the Federal Council and of the Council of the Nations: Stenographic Notes, May 15–July 20, 1946]. Belgrade: Narodna Skupština, n.d.

"Zakon o poljoprivrednom zemljišnom fondu opštenarodne imovine i dodeljivanju zemlje poljoprivrednim organizacijama od 27 maja 1953" [Law on national agricultural land fund property and its allotment to agricultural organizations, May 27, 1953]. *Službeni list* no. 26 (1953).

COLLECTIVIZATION AND THE AGRARIAN QUESTION IN COMMUNIST YUGOSLAVIA CHRONOLOGY OF EVENTS, 1945–1953

Melissa K. Bokovoy

1945

March 7: The National Committee for the Liberation of Yugoslavia announced the formation of a Provisional Government under the authority of a Regency, in which Josip Broz Tito was Prime Minister and Dr. Ivan Šubašić was Minister of Foreign Affairs.

February-December: The provisional government issued numerous laws and decrees which regulated the agricultural sector and created communal agricultural organizations. They set pricing and production targets, established credit policies discriminating against private landholders, allocated capital goods and services through machine tractor stations, state agricultural farms and the state and cooperative sectors, set goals for state agricultural production, and granted federal and republican budget advantages to the state agricultural sector.

February: The "Law on the Transformation of Enemy Property into State Property," provided for the confiscation of all property of the German Reich and its citizens, of persons of German nationality (unless they had supported the NOB), war criminals, and collaborators. Under this law 636,847 hectares of agricultural land were confiscated.

April: The "Decree on the Organization of Agricultural Machine Stations" introduced the creation of agricultural machine stations.

April 23: The "Law concerning speculation and economic sabotage," aimed at limiting black market activities. The law established penalties for peasants who damage, destroy, or slaughter animate inventory, and defy requisition orders that went into effect during the war.

August 6: The "Law on State Agriculture Farms establishes state farms" was adopted. Two weeks later, the "Law on Agrarian Reform and Colonization" distributed 288,000 hectares of confiscated land to these farms.

August 23: The "Law on Agrarian Reform and Colonization" ordered the complete expropriation of land held by banks, enterprises, churches, monasteries, and other religious institutions without compensation. The state also confiscated land from well-to-do peasants who owned surfaces above the legal maximum of 25-35 hectares; from non-peasants who possessed more than 3-5 hectares of arable land; from owners who had abandoned their land during the war; and from German nationals and from those accused of war crimes. Out of the 1.56 million hectares confiscated, 797,000 hectares (51%) were distributed to over 316,415 peasant families (a little over 2.5 ha per family), among which 136,454 were landless peasants or colonists.

September 6: The Economic Council initiated a system of compulsory deliveries or collections of certain agricultural products (*otkup*) to be transported to state agencies. Following the Soviet model, the Economic Council directed each region and locality to deliver a fixed amount of certain agricultural goods to regional warehouses. The state agricultural agency, *Poljopromet*, was to direct the collection of grain, set the price of grain, and the quotas for each region.

December 15: Delegates of various republics discussed the progress of the agrarian reform and colonization, proposing measures to improve the image of the Ministry of Colonization.

December 31: 31 peasant work cooperatives (SRZs) were founded predominantly by former partisans and colonists, owning 96,000 hectares of arable land.

1946

January 10: The KPJ propaganda assaulted peasants who had shirked their responsibilities by hoarding supplies, speculating with prices for their produce, and black-marketing. Blagoje Nešković, a Serbian communist party leader, claimed that resistance to governmental rural policies was incited by the rich peasantry. He asserted that, without the *kulaks*' influence, poor and middle peasants would support the regime's policies.

May 22: The new, communist-dominated Federal Parliament, *Skupština,* adopted the "Law on the General Economic Plan and State Planning Bodies," introducing centralized planning and leading to the preparation and adoption of the first Five-Year Plan less than a year later.

June: The "Law concerning a General Amnesty to the Violators of the Orders on the Obligatory Collections of Agricultural Produce" pardoned peasants who had not yet completed or had not yet begun their sentences for violating the collection orders.

July: The "Basic Law on Cooperatives" designated four organizational types: 1) buying-selling cooperative; 2) processing-producing cooperative; 3) credit cooperative; and 4) peasant work cooperative. The latter was the most advanced form of rural socialist organization.

September 16: The state introduced production contracts (*kontrahiranje*) drawn up between individual peasant producers and state or cooperative enterprises before the spring sowing, regulating the conditions of production, selling, and the delivery of agricultural products. The peasant and the state were to estimate the peasant's yield for the contracted crop and after the harvest the peasant would deliver the predetermined amount.

Mid-1946 (to 1948): Skupština passed directives for the compulsory delivery of wool, cattle, tobacco, and grain in order to insure the supply of goods to the cities.

1947

February 6-8: Heads of the Federal Planning Commission met to discuss the Five-year Plan. The relationship between agriculture and industry was seen as the main point of the Plan.

March 15: The "Grain Collection Regulations in Yugoslavia, 1947–1948" limited local and republican official's discretion in determining the size of an individual's *otkup* obligation. The government in Belgrade determined a state purchase plan for each region and made the republican authorities responsible for the fulfillment of the plan.

April: The "Law on the Adoption of the first Five-Year Plan," provided for the establishment of the Federal Planning Commission for the preparation and implementation of a federal economic plan. The plan followed the Soviet model, being inspired by Soviet advisers and Soviet planning drafts. However, the Yugoslav plan rejected the Soviet corollary to industrialization—namely collectivization and dekulakization—as the means to rid the countryside of capitalist elements.

June: "Agricultural Cooperatives in a Planned Economy." In his seminal treatise, "Agricultural Cooperatives in the Planned Economy," Edvard Kardelj, the leading theoretician on agrarian question, proposed the creation of a new form of cooperative, the general agricultural cooperative (*opšta zemljoradnička zadruga*, OZZ).

November 21: A Central Committee meeting defined the *kulak*. According to Kardelj, "The *kulak* is a political concept—a man who does not fulfill his obligation to the state".

1948

The number of agricultural cooperatives grew from 5,041 in 1945 to 8,666 in 1948. Membership increased six-fold in that time, from 493,800 cooperative members in 1945 to 3,127,464 in 1948.

March (to June 1948): Soviet and Yugoslav communist leaders exchanged correspondence concerning Yugoslavia's domestic and foreign policy initiatives that countered Soviet interests.

March: The Soviets accused the Communist Party of Yugoslavia (*Komunistička partija Jugoslavije*, KPJ) of failure to promote "the spirit of class struggle" and noted that the leadership of the Party was not taking any measures to check "the capitalist elements".

May 4: In correspondence to the KPJ, Stalin and Molotov chastised the Yugoslav leaders for "their denial of the strengthening of capitalist elements, and in connection with this, the sharpening of the class struggle in the villages under the conditions of contemporary Yugoslavia."

June 5: At a meeting of the Economic Council, Boris Kidrič claimed that the KPJ did not want to "hit the *kulaks* frontally." Instead, "the party had to find new, indirect ways, to squeeze them economically." The aim was to undercut their market power—their ability to monopolize rural trade and labor markets—without eliminating them as individual producers.

June: A new cereals decree stiffened *otkup* regulations, reducing the peasants' subsistence allowance from 250kg per household member to 200kg. The peasants who did not comply faced stiffer enforcement from the state. The decree allowed the sale of grain and flour by the peasantry only after the completion of the *otkup*.

June 28: A Resolution of the Communist Information Bureau stated that the Yugoslav leaders are pursuing an incorrect policy in the countryside by ignoring class differentiation and by regarding the individual peasantry as a single entity, contrary to the Marxist-Leninist doctrine class struggle and to the Lenin thesis that small individual farming continually gives birth to capitalism on a mass scale.

1949

January: The Resolution entitled "The Basic Tasks of the Party in the Area of Socialist Transformation of the Village and Advancement of Agricultural Production," put greater emphasis on establishing peasant work cooperatives (SRZ), closely resembling Soviet collective farms. However, Kardelj pointed out that, unlike the Soviets, the Yugoslavs would not seek to create collective forms of agriculture simultaneously with a strike against the *kulak*.

February: The Central Committee of the KPJ sanctioned the creation of a watchdog agency for the village in all republics, the Commission for the Village. This commission, made up of members of the Central Committee, had to record the activities of local cadres and the peasantry in the campaign for creating peasant work cooperatives.

April: First reports of peasant resistance to the SRZ campaign were submitted by members of the Commission for the Village to the Central Committee. An eight-page report from Serbia mentioned instances of violence and public unruliness, arguing that "the *kulaks*, in some places, have openly begun to carry out violence and sabotage."

May: The "Basic Law concerning Agricultural Cooperatives" defined four types of work cooperatives differing in the degree to which the peasant surrendered his ownership rights. In the first type, the peasant retained ownership of the land and received a fixed rent from the cooperative. In the second, the peasants retained ownership of the land and received interest on the value of the land from the cooperative. In the third, the peasant retained the ownership of the land but renounced any benefits from interest or rent. In each of these types, members could leave the cooperative after three years and reclaim their land. Only the fourth and "highest type of cooperative" could be equated to the Soviet-style *kolkhoz*: the peasant fully transfers his land to cooperative ownership (except for his household).

December 29-30: At the Third Plenum of the Central Committee of the KPJ, Boris Kidrič acknowledged opposition to the collectivization campaign from within the party, from republican authorities and local cadres, and from peasants. He recommended that party cadres should not focus on the number of peasant work cooperatives established, but should "concentrate on internal organization and economic strengthening of the peasant work cooperatives."

December 29: The Commission for the Village pointed out that the Plenum resolution had started a race for the creation of peasant work cooperatives, and criticized the way in which local cadres implemented the Plenum's directives.

December 31: Between Dec. 31, 1948 and Dec. 31, 1949, 5021 new peasant work cooperatives were created, bringing the total to 6,625. 18.3% of agrarian households and 21.4% of Yugoslavia's arable land had been collectivized. The average cooperative consisted of 51.4 households and 277 hectares. The average number of hectares a peasant household brought to the cooperative was 5.3 hectares, yet considerable regional variation existed.

1950

January: The Executive Committee of the People's Front endorsed the general agricultural cooperative over the SRZ, in an effort to control agricultural production.

January: The KPJ leadership granted local and republican cadres permission to reorganize the peasant work cooperatives. Their major task was to consolidate the gains made in 1949 and make the SRZ the most viable and desirous institution in the countryside.

Spring: Reports of the Commission for the Village pointed out that the gravest crisis facing the leadership was the non-productivity of the SRZ and their "spontaneous" dismemberment, and the continued circumvention of the *otkup*. E.g.: In March, the Osijek region reported the disbanding of 36 SRZ's and the loss of 578 households.

Spring/Summer: Drought in Yugoslavia. Uprisings occurred in Macedonia, northern Bosnia, and in the Banija and Kordun regions of Croatia.

May: Discontented veterans and peasants channeled their grievances into an armed insurrection against the local government of Cazinski *srez*, with its center at Bihać. After the suppression of the revolt, 714 people were arrested; 288 individuals were brought to trial before military courts, of these, 17 were condemned to death by firing squad and the others were given prison sentences of fifteen to twenty years. 426 individuals were tried in criminal court and sentenced to 2-3 years hard labor in the Brez mines.

July: A group of 600 peasants from Smederevo "marched on Belgrade" asking for a reevaluation of their region's *otkup* obligations. In response, the authorities arrested 23 local functionaries; many were convicted, accused of being Cominformists. Overall, between 1948–1952, circa 5626 peasants were arrested and sentenced as Cominformists. In addition, in the first 10 months of 1950 the KPJ expelled 10,796 village cadres.

1951

January: Vladimir Bakarić criticized the SRZs organizational model, arguing that the SRZs should operate under more rigorous economic principles. In an article published in the March issue of the monthly for agricultural affairs *Socijalistička poljoprivreda* (Socialist Agriculture), Bakarić argued departing from the Soviet model in how work was organized (i.e. abolish the brigade system) and compensated, proposing to introduce cost accounting.

June: The Economic Council suggested reforming the "bureaucratized" and "incompetent command structures" of the SRZ. It recommended that the directors were selected from the most resourceful local members, able to draw up production plans and negotiate contracts with state enterprises. Absent from this description was a warning not to appoint a "*kulak*" to this position.

Mid-1951: The KPJ leadership sanctioned a series of legislative acts which began to dismantle the most hated institutions and practices which characterized the *otkup*.

Fall: Speaking at the Federal Council on Agriculture, the Minister of Agriculture, Mijalko Todorović, proposed corrections in the relationship between the state and the work cooperatives to make sure that the peasant work cooperatives had the necessary means to develop. He took special issue with the fact that the SRZs were grossly undercapitalized.

November: The KPJ resolution, "On the Future Pattern of Socialist Transformation of the Countryside," stated that the peasant work cooperatives were in need of reorganization and that the principle of voluntarism should be strictly enforced.

December 31: Slowdown in collectivization: only 350 new SRZs were created since Dec. 1949. The number of work cooperatives increased in Macedonia, Montenegro, and Serbia, but it remained static in Croatia, Slovenia, and Bosnia-Hercegovina. At the end of 1951, a little more than a fifth of the arable land and peasant households were collectivized.

1952

February: Minister of Agriculture Todorović proposed a bill for changing the method of accounting for work-days and payment, in response to peasant malcontent. Instead of payment for specified tasks, the cooperative was to pay its members by result; members could specify the type of work for which they were most suitable.

December 31: more than 2,700 SRZs had been dismantled and 600,000 hectares returned to private production. In addition, due to drought, grain production fell below 1950 and 1951 levels.

1953

March 30: The "Law on Property Relations and Reorganization of Peasant Work Cooperatives" allowed members of the peasant work cooperatives to withdraw from the cooperative, singly or in groups, or to liquidate or reorganize the collective if the majority members so desired.

May 27: The "Law on the Creation of a National Land Fund and Regulation of its Distribution" limited individual land holdings to a maximum of ten hectares. The excess land was deposited into a general land fund and redistributed to poor peasants or turned into state farms.

Collectivization and Social Change in Bulgaria, 1940s–1950s

MIHAIL GRUEV

In September 1944, the Soviet army occupied Bulgaria, a middle-sized Balkan country with 111,000 square km of territory and a population of about seven million inhabitants (after the return of the territories formerly under Bulgarian control in Macedonia and Thrace to Yugoslavia and Greece, respectively). At that time, Bulgaria was among the most "under-developed" European economies with a predominantly agrarian structure.

The *coup d'état* of September 9, 1944, coordinated with the Soviet advance, was designed as a new coalitional government of the so-called *Otechestven front* (Fatherland Front—FF), which included the dominant communists, the left wings of the Agrarian Union and Social Democrats, and the political organization *Zveno*, the latter with a strong following in military circles. The political influence and the rank and file of the Bulgarian Communist Party (BCP) witnessed a rapid growth—from about 14,000 members during the illegal period to 100,000 at the end of 1944. One year later, at the end of 1945, there were about 250,000 members of the BCP. Behind the facade of the so-called "people's democracy" (an official multiparty system), the BCP enjoyed a political monopoly, fully established in 1948.

Since that year—marked by the Fifth Congress of the Bulgarian Workers' Party/BWP—Bulgaria began emulating the Soviet model, initiating thorough political, economic, social, and cultural changes. Among these changes was the Bulgarian version of collectivization which became the closest possible analogue of the *kolkhoz* system. The transformation in the village, side by side with the Bulgarian communist version of radical modernization, was carried out by exercising the sheer force of the state machinery and in a relatively short period of time compared to the rest of the Soviet satellites. Collectivization started in 1948 and was fully accomplished by the end of the 1950s.

The main stages of the collectivization campaign roughly mirrored the respective strategy in the USSR of the 1920s–30s. Bulgaria's large-scale collectivization campaign started in 1948, after a brief period of political consolidation, and can be divided into three subsequent stages. The so-called "first massovization" of 1948–51 ended with peasant resistance on a large scale, especially in northwestern Bulgaria. 1951–56 was a period of relative stalemate without brutal force and pressure, the main efforts being directed towards stabilization of the existing TKZS (*Trudovo-kooperativno zemedelsko stopanstvo*—cooperative farm). The last stage—"the second massovization" of 1956–59—was marked once again by political and social oppression, thus completing the collectivization in the mountainous and peripheral regions.

The main task of this chapter is to analyze in chronological order this policy in the context of the general development of communist Bulgaria up to the 1960s. Structurally, the text is divided into several sections focusing on the preconditions of the changes in the agrarian system before 1944; the tools of the communist state policy in the village; the main stages of collectivization; and its long-lasting consequences. They outline the history of what can be regarded as the most radical and all-embracing social cataclysm in modern Bulgarian history.

The Bulgarian Village on the Eve of World War II

It is quite difficult to study the process of collectivization and to define and analyze its various direct and especially indirect effects. After all, collectivization was a giant project intended not only to change the structure of ownership but to implement overall social engineering, a genuine "genetic modification" of society. In Bulgaria, collectivization set the stage for a global economic, social and mental transformation in the rural world and in the lives of millions of people. It can safely be generalized that collectivization was an essential part of the policy of the BCP on the country's reorganization. The adoption of the "model" of collectivization imported from the USSR became possible after the huge political and social upheavals in the society in the initial post-World War II years when

Bulgaria was permanently included into the Soviet sphere of influence and an undisguised communist dictatorship was established in the country. However, it should be noted that the ground was already prepared, thus facilitating the implementation and consolidation of the process.

The *coup d'etat* of September 9, 1944 witnessed quite a sorrowful picture in Bulgarian villages and in the country's agriculture. By 1946 about 76 percent of the population was still rural and 68 percent relied on agriculture, stock-breeding, and forestry for their livelihood.[1] The processes of land fragmentation between the inheritors of different generations, which had already started in the years after the liberation from Ottoman rule in 1878, intensified. This trend was obvious after World War I when it gradually became common practice for women to inherit up to 50 percent of their brothers' shares from their parents.[2]

Consequently, by the 1940s the arable land was divided between approximately one million individual holdings scattered over approximately 13 million pieces of land (fields). The majority of them were quite small: according to the official statistics, 712,000 holdings owned up to 5 hectares, which constituted 37.5 percent of the total arable land; 254 holdings owned between 5 and 10 hectares and only 200 holdings owned more than 50 hectares.[3] Therefore, the modernization of agriculture, such as the introduction of machinery and scientific farming, was much impeded. Generally, Bulgaria's interwar governments attempted to ease the condition of Bulgarian peasants by freeing many farmers from their debts, imposing state protectionist measures in agriculture, and, in particular, pursuing partial land consolidation in some villages. These reforms yielded modest results and did not lead to a huge leap in agricultural modernization.

However, the endeavors of politicians, such as Andrey Lyapchev, Grigor Vassilev (conservatives) and Ivan Bagryanov (right-wing agrarian), as well as scholars like Yanaki Mollov, Anastas Totev, and Hristina

---

1 Zhivkova, *Bulgarskoto selo*, 46.

2 The first attempts at correcting this inequality were made during the first regime of Stefan Stambolov (1887–94). However, the reforms were in drastic contrast to centuries-old traditions, so these provisions were not applied in practice. Therefore, a new Heritage Act had to be ratified in 1906, whereby daughters were entitled to inherit property twice as small as that of sons. In the majority of cases, however, they continued to waive their right of shares in the coming decades.

3 Mateev, *Dvizhenieto*, 12.

Mocheva, had succeeded in outlining the main streams and tendencies toward improvement in the conditions of the Bulgarian village. Their initiatives and studies promoted economic prosperity and the education of landowners. They encouraged the development of the network of free cooperatives while preserving ownership, share participation and fair allocation of returns, increasing credits for mechanized agriculture, and the introduction of the modern achievements of agronomy into practice. It is difficult to say from today's point of view whether this evolutionary project for the modernization of agriculture failed due to the painfully slow pace of its implementation or due to general unawareness. The evolutionary project for the development of rural areas and the collectivist approach in its latent form ran in parallel even before World War II. Generally, it is problematic to consider the latter as a unified system of concepts, at least upon its emergence in Bulgaria during the first post-World War I decade. Tolstoyists, cooperative promoters sharing different views, and some representatives of agrarian-like groups interpreted it in their own ways. Within this spectrum of opinion, the communist variety was, without doubt, the clearest expression of the collectivist project. The legal Bulgarian Workers' Party (BWP), which included the underground BCP (1928–34) promoted actively the Soviet model and experience through its propaganda apparatus which had penetrated deep into society. The Soviet model of collectivization played an important role in the official propaganda campaign, a model which was, directly or indirectly depending on the target audience, recommended as panacea for solving the problems of the Bulgarian village. Leading communist politicians, such as Petko Napetov, Lazar Stanev, and Todor Pavlov joined this type of campaign by authoring political pamphlets. Influenced by this propaganda, the first three experimental farms for collective cultivation of the land were established as early as 1939 in the villages of Totleben in the Pleven region, Stezherovo in the Svishtov region, and Vurbovka in the Sevlievo region (in north-central Bulgaria). The principles on which these cooperatives functioned resembled, to some extent, the Soviet *kolkhoz* model. The three cooperatives were unambiguously represented as forerunners of the labor-cooperative farms (TKZSs) in pre-1989 historiography and memoirs.[4] Though their organization clearly suggested the direct involvement of the Communist Party, it should be noted that they were, in effect, subject to free cooperation while preserving private ownership of the land.

[4] Mateev, *Dvizhenieto*, 55–57; Iliev, *Brazdi prez*, 88.

LAND REFORM

Soon after the *coup d'etat* of September 9, 1944, communists were able to push for the implementation of their planned revolutionary project in villages, promoting the collectivization of the Bulgarian agriculture following the Soviet model. The formation of the first TKZSs had started already in 1944 but until the autumn of 1948 their number in the country did not exceed 400, while their membership accounted for about 8 percent of the arable land (including the land expropriated during the reform).[5] In this stage, mainly the communists and their sympathizers joined the collectives, "voluntarily" giving up their land, livestock, and equipment. Although part of them were quite hesitant, they had to follow the policy of the party, which had a rich set of tools for coercion and persuasion.

The abbreviation for "labor-cooperative farm" appeared for the first time in political propaganda in mid-November 1944 when the decree on TKZSs was prepared. However, already before the ratification of that act the state had made it clear that it backed this model. In early 1945, the National Committee of the Fatherland Front declared a one-month "collective action" in the country aimed at bringing the masses of peasants into the TKZSs.[6] Communist were the most active in the campaign, loudly acclaiming the fact that, by the spring of 1945, peasants of 12 villages had already joined collectives.[7] Although this was quite an overstatement, it nevertheless reflected the attitude of many local functionaries who had started to force the process "from below." To support it, in March 1945 fifteen tractors were presented to "the best TKZSs in the country" on behalf of Soviet Marshal Tolbukhin in a propaganda campaign.[8] Following a fairly long delay and coordination, the Council of Ministers finally approved the decree on TKZSs on April 13, 1945.[9]

Up to the present, Bulgarian historiography argued that it was not until 1948 that the coercive measures were imposed. But it should be noted that pressure had been exerted even during that initial period, at least in the villages where the first collective farms were formed. When outlining the borders of land tracts and inducing farmers to join the col-

[5] Migev, *Kolektivizatsiata*, 23.
[6] Minchev, *Bulgaria otnovo*, 118.
[7] *Otechestven front*, no. 196 (April 25, 1945); Marinov, *Za dobroto*, 127–30.
[8] Trifonova, *BCP i socialisticheskoto*, 40.
[9] It was published in *Darzhaven vestnik* [State Gazette] on April 25.

lectives, the methods of pressure did not differ from those in the following years. Those peasants who refused to join collective farms had their land confiscated and, in return, they were given other pieces of land, usually scattered on the periphery of the village, with soil quality, as a rule, not equal to the confiscated properties. Those "replacements," as they were called, were one of the most common levers for pressuring peasants and forcing them into collective farms.

The BWP and the coalition led by it realized that with the uncertain international position of Bulgaria (before signing the Paris Peace Treaty in 1947), the time was not ripe for mass forcible collectivization and that any further hasty moves in that direction would only alienate the majority of peasants. Consequently, in the autumn of 1945, the first wave of pressure on private farmers was abated and plain agitation was given priority. In addition, one can detect continuity of the agrarian policies pursued by nonpartisan governments. Thus, in February 1946 the National Assembly passed the Act on Village Sanitation. Under the new legislation, standard designs of residential and public buildings were to be popularized, annual competitions for beautification of towns and villages were to be announced, the towns and homes were to be cleaned, and the health culture of their inhabitants was to be improved.[10]

In July of the same year, the Act on Agrarian Education and Improvement of the Lifestyle of Villagers was passed. This act established a special office for the improvement of village life within the Ministry of Agriculture. Its task was to coordinate the efforts of individual comprehensive cooperatives, social committees, and associations with those of the state to open bakeries, public works, vacuum-canning factories, and fruit and vegetable-drying plants.[11] The move toward the electrification of villages continued. Thus, after September 9, 1944, governmental policies with respect to the rural population were the result of a complicated interaction between the directly inherited economic and social concepts of leading politicians, and the still reticently concealed and not quite crystallized radical intentions of the Communist Party.

The next important step of the government in the field of agricultural policy was the preparation and the ratification of the Act on the Labor-Landed Property, whereby the intent on carrying out agrarian reform was declared. It was prepared at the end of 1945, and following

---

[10] *Rabotnichesko delo*, no. 21, February 1, 1946.

[11] Kanatzieva, *Socialno-ikonomicheskata*, 146.

a brief discussion at the 26th General National Assembly, was passed on March 12, 1946.[12] The title and many provisions of that act demonstrated a succession to the reform policies of Alexander Stamboliyski in the field of agriculture.[13] This can be traced both in the introduction of an upper limit for owning a labor-landed property (in the law of 1921 it had been 30 hectares and now it was 30 hectares for the richest region—South Dobroudja—and 20 for the rest of the country) and in the greater restrictions on the so-called non-labor-landed property.[14] The formation of a state land fund, as well as the types of land attached to it—those of owners exceeding the above mentioned limits, state-owned property, land owned by municipalities, the Church, *waqfs* (land owned by Muslim religious foundations), schools and community centers—were regulated in a similar manner.

It should be noted that the law had been passed under conditions of a surge in opposition and was literally a takeover of much of the membership of the pro-communist Bulgarian Agrarian National Union (BANU) by its rival, BANU-Nikola Petkov.[15] The second party formed the backbone of the anti-communist opposition, while the leftist agrarians remained in the FF. The two wings were engaged in a bitter struggle for the symbols and principles of agrarian ideology. The Communist Party was definitely interested in demonstrating that the "real" BANU was part of the coalition it dominated and that the party supported the principles defended by the Union. Yet, in order to clear up peasants' suspicions that the Soviet *kolkhoz* system would be soon introduced, the BWP evaluated the situation tactically and agreed on the adoption of such a law. Naturally, in the spirit of the times, it contained provisions that set forth a variety of terms and conditions for acquiring and preserving private property which were, in fact, aimed at creating the preconditions for the enlargement of the base and network of the TKZSs. These were, in effect, the essential differences between the agrarian reform contemplated and only partly realized by Stamboliyski and the communist version, the implementation of which was stopped in an initial phase.

---

[12] *Darzhaven vestnik,* no. 81, April 9, 1946.

[13] Alexander Stamboliyski was the leader of the Bulgarian Agrarian Union and prime minister from 1919–23. His government introduced an Agrarian Reform Act, aimed at restrictions for the rich landowners and allotments to landless peasants.

[14] Cf. *Darzhaven vestnik,* no. 31, May 12, 1921.

[15] The party was named after its anti-communist leader Nikola Petkov.

Along the same lines, those peasants who were given land under this act were to be released from the obligation to pay the value of the properties received (up to 5 hectares) when joining a TKZS. Moreover, it was guaranteed that they would be able to retain from 0.5 to 1 hectare of land for private use, for which they had to pay. At the same time, however, Article 40 stated that "those to whom land had been given and who were members of a cooperative for collective use of lands would receive ideal parts of the fund land set for them, without it being parceled out." This provision clearly shows that it was, in fact, a fictitious land-settlement, under which the only benefit for these TKZS members was the receipt of a rent (up to 40 percent of the returns per hectare) which was gradually but surely decreasing. More tangible was the benefit for the so-called landless peasants to whom up to 1 hectare was given for private use (subsequently reduced to 0.5 hectare) and, in fact, they were the only ones who acquired something real (though temporarily).

With such wording, however, the Act on the Labor-Landed Property could not lead to an actual and relatively fair reallotment of agricultural property in the country even if it was to be fully implemented. The arguments for the bill acknowledged that about 1 million hectares would be needed to carry out a complete land-settlement by giving up to 5 hectares to landless and poor peasants.[16] Approximately 56,400 hectares were taken from 3,600 owners under this act. The lands taken from schools, community centers, the Church, and *waqfs* were also added, thus reaching 243,781 hectares or approximately one-fourth of the necessary land.[17] Actually, this huge difference between the required and the available land was the reason for the failure of Stamboliyski's former project.[18]

Since it was evident that the agrarian reform was only a propaganda move, the government was slow to enforce the act. Thus, only 21,590 households had been given land by the end of 1946 and in 1947 their number was 80,039.[19] It turned out, however, that these official figures for the progress of the agrarian reform were quite fictitious because they represented households that had joined the TKZSs in the meantime. The number of peasants to whom land was actually given was fewer. As the architect of the agricultural policy pursued by the Communist Party, Titko Chernokolev admitted in September 1948 that the real number of those to

[16] Stenographic records, 26th Assembly.

[17] CSA, fund 1B, inventory 13, file 111 (1), folio 84–90.

[18] Penchev, *Agrarnata reforma*, 219–20.

[19] CSA, fund 1B, inventory 13, file 111 (1), folio 14–30.

whom land had been given was 11,000 landless peasants and 29,000 poor peasants. Announcing this information, he admitted that the government had failed in its attempt to solve the problem by that means.[20]

Similar agrarian reforms or attempts to carry out such reforms between 1945 and 1948 were common in all countries that had entered the Soviet sphere of influence.[21] Furthermore, their parameters were more or less the same—the maximum areas of land which any individual owner might possess and anyone might be given, who was entitled to receive land, the terms of payment, and the compensations to be paid.[22] These indicators show clearly that the Act on the Labor-Landed Property in Bulgaria was dictated not by the actual economic conditions and needs in the countryside but by a geo-political strategy for the establishment of Soviet control in Eastern Europe. Another common feature was the suspension of these large-scale undertakings due to initial implementation of collectivization. However, it is safe to say that Bulgarian communists have been more strictly following the Soviet model, which explains the speed and the relative success during the initial stage.

Creeping Collectivization

If the promise to carry out land reform and to allocate land was the core of the regime's propaganda strategy at this stage, the quota system (or the so-called obligatory state deliveries) constituted the actual enforcement of the new agrarian model and was the basic tool for pressure on the peasantry. This system was inherited from the war years but was gradually improved to turn into a form of ruthless robbery of unprecedented size. The quota system was introduced in its classical pattern for the autumn crops of 1947, thus establishing a kind of colonial regime of draining peasants of products and resources. By the decree of the Council of Ministers approved on June 13, 1947, amended by a special ordinance of the Ministry of Agriculture and State Properties, the whole amount of grain was to be requisitioned from the producers, leaving them with 600 grams per family member per day and 200 grams of maize. Since the regime had

---

20 Chernokolev, *Dokumenti*, 125.

21 See the respective chapters in this volume, especially the contributions of Jan Rychlík, Dariusz Jarosz, Arnd Bauerkämper, and Constantin Iordachi and Dorin Dobrincu.

22 Popov, *Agrarnite otnoshenia*, 82–84.

encountered difficulties with the collection of requisitions in the previous years and the practice of hiding of products were common, now an obligatory threshing on a common threshing floor was enforced. Moreover, agricultural producers were to pay back to the state first and then take what remained.[23]

The official name of this campaign was "obligatory buyout" but it was, in effect, "putting the squeeze" on the peasantry. The state purchase organization paid for grain, maize, milk, and meat at prices that were between one-fourth and one-third of their market value. In the next few years the system of obligatory state deliveries[24] was expanded gradually, covering new crops. This practice of the communist regime in Bulgaria, directly borrowed from Stalin's rural policy, would become the main instrument for forcing peasants in TKZSs. On May 21, 1948, the Council of Ministers approved a new decree for the buyout of crops that had been harvested in the previous year.[25]

Altogether, the quota system achieved its full scale by this act. It encompassed wheat, rye, barley, oats, vetch, beans, and sunflower seeds (a separate decree was issued for maize). Of forage crops, hay and straw were subject to obligatory buyouts. Potatoes were not left out either. The quotas on meat and animal products—such as hides, wool, and milk—were very heavy. Here a differentiation was made between "rich," "middle," and "poor" peasants. With respect to grain deliveries, the country's districts were divided into five categories for the first time, depending on their topography, soil, and climatic conditions. This division remained almost unchanged until the removal of deliveries in 1959. One basic principle in the decree was the gradual increase of taxation in kind depending on the area of arable land.[26] Thus, according to some rough calculations, the obligatory state deliveries deprived peasants of between 50 and 70 percent of their income. The state's perspective was often quite different from the will of local leaders, opening the way for violence and arbitrariness. Peasants had no alternative but to join the TKZS where only the lands for private use were subject to taxation, or

---

23 *Darzhaven vestnik*, no. 158 (July 11, 1947); Migev, *Problemi na*, 17–18.

24 This term replaced "the quotas" of 1950 but, in fact, it referred to the same thing.

25 CSA, fund 136, inventory 3, file 66, folio 143–150; *Darzhaven vestnik*, no. 126 (June 1, 1948).

26 Migev, *Problemi na*, 23.

to migrate to cities. The coming years proved that the former option was often the first step toward the latter.

The other lever closely linked to the quota system was the so-called crop plans. To a certain extent, they were also a heritage of the war years but under the new conditions they started to play the role of a specific guarantee that any private agricultural producer would fulfill their obligations towards the state. In accordance with the buyout plan, the state transmitted the crop plans to the districts through the Ministry of Agriculture and State Properties while the respective District Agrarian Services passed them on to the municipalities. They, in turn, conveyed them to the respective owners through the local *aktiv*, such as the chairmen of local people's councils (this was the definition of a village mayor at that time), party secretaries, and chairmen of the TKZSs. These plans served as basis for the preparation of the accounts on deliveries.

In principle, natural disasters and cataclysms (a significant risk in agriculture) were not taken into consideration and usually were left at the expense of peasants. Crop plans were an important indicator for the growing role of the state in agriculture. They contributed to the extinction of some traditional crops and local varieties and approaches in land cultivation in many places. However, many new ones appeared. Thus, in areas where only cereals had been grown, farmers and peasants started to plant crops like tobacco and cotton, independently of whether the soil and climatic conditions were suitable or not.

Another key instrument in the regime's policy on creeping collectivization was the strict social stratification and segregation. Obviously, there was no clear distinction between small, middle and large landed property under Bulgarian conditions. The differences between them did not lead to obvious contrasts in the way and quality of life. It should also be noted that there were different criteria and "yardsticks" by which a farm would be measured as "small," "middle" or "large" in different areas of the country, depending on whether it was lowland or highland. The actual social stratification in these three categories was largely an ideological construct, the outcome of the dogmatic adherence of Bulgarian communists to Lenin's teaching and the Soviet political practice of the 1920s and 30s. Usually, "a petty peasant" meant an owner of approximately up to 2 hectares of land. Any holding in excess of that conditional limit was considered to be "middle-sized" although both agrarian economy and the practice in Western Europe have shown that holdings of less than 5 hectares are not viable. This was, in fact, the lower limit for economically healthy and independent production units in agriculture. One of the most widespread doctrines in the Communist Party was the belief that the petty peasantry, together with middle-income peasants, could be relied on to carry out the so called "socialist transformation" in agriculture. The first move toward "agrarian reform" and the quota system therefore involved providing some assistance for the smallest rural holdings from which loyalty and support were expected. But in many cases the petty peasants were the ones who resisted the policy of coercive collectivization most strongly.

The next category of peasants was the so-called "middle peasants." Depending on the conditions, it applied to owners of land between 2 and 5 hectares and in some cases even up to 10 hectares. They were considered a tactical and temporary, but not long-term strategic ally of the Communist Party, through which political superiority would be achieved, which in turn would guarantee the non-reversibility of the process initiated by the rulers. But local functionaries usually looked down on these so-called middle peasants as people who had to be guided and controlled because of their alleged immanent susceptibility to "the wiles of the reactionaries."

Most negative was the attitude to rich peasants who were exposed to all sorts of economic, political, and psychological pressure. Since they were to be relied on for the fulfillment of state plans and the society benefited from them in the earlier years, a borderline category was introduced in the political language of that time—"a wealthy peasant" owning over 10 hectares of land and showing some loyalty to the government at the same time. In this case, the pressure on this category was mainly economic but there was a real possibility that its representatives would be permanently qualified as "enemies."

Undoubtedly, the greatest pressure was exerted on *kulaks*. This term appeared in the speeches of communist functionaries of that time as a direct borrowing from the Soviet political jargon. Initially, it was unfamiliar to those who were not trained in the USSR. That's why it often appeared as *kolaks* in the printed pamphlets of some ill-informed functionaries and propagandists of that time who probably associated it subconsciously with people who ate the well-known ritual bread, usually baked on traditional occasions and symbolizing abundance.

However, this term was not well defined and had fuzzy boundaries. The very ideologists of the Communist Party used it mainly with political connotations as "class and ideological enemies of the new state." The definition of *kulak* relevant for the Bulgarian conditions was the one given in May 1929 by the Council of People's Commissars of the USSR: "A *kulak* is anyone who uses hired labor, owns a mill, a sawmill or equipment which he rents out, is involved in trade, money-lending and generally, has other sources of non-labor income, including priests."[27] Given the obvious differences between the traditional Russian village and its Bulgarian counterpart, the very borrowing of the term mirrors the mechanical copying of the Stalinist strategy. It is clear that this image was constructed as an antithesis of the "working peasantry"—an exploiter, a rural capitalist, and a bloodsucker. In the course of time the definition had to be extended by including people who were labeled *kulaks* for various reasons. Initially, it included people who had not made the obligatory state deliveries. The dogmatic adoption of Lenin's formulation of small commodity production giving birth to capitalism constantly, from its very outset, increased the distance between the Communist Party and larger numbers of peasants classified unambiguously as *kulaks*. By default, the term included all followers of Nikola Petkov, functionaries of disbanded parties in the country, grocers, traders, and tavern-keepers.

The Bulgarian equivalent of this category was most clearly presented in a speech by Titko Chernokolev at the international meeting of representatives of people's democratic countries on the construction of socialism in rural areas held in September 1947: "Who is a *kulak*? What class group do *kulaks* constitute in our village? The *kulaks* in a village—these are capitalists, slave drivers who organize their farms with hired hand, exploit other man's labor and amass a fortune [...] Besides, we must take into consideration the various extra enterprises operated by *kulaks* in villages—taverns, mills, creameries, carding-machines and other side enterprises; the links of certain social classes to trade capital."[28]

Therefore, the term included huge numbers of people engaged in non-agrarian work against whom a systematic campaign was started. Even the Nationalization Act of December 1947 aimed to deprive them of creameries, carding-machines, and soda and lemonade workshops. Under the Act on Reduction of the Number of Taverns and Prevention of Drunkenness, one tavern was left to every 700 inhabitants and the first one had to be owned by the respective people's council (municipal property). The authorities were to decide how many other taverns could be opened and how many of them were to remain private. But the law can be looked at from another aspect. In addition to the introduction of a monopoly on alcoholic drinks, it was of great importance for the gradual driving out and

[27] Conquest, *Harvest of Sorrow*, 63.
[28] Chernokolev, *Dokumenti*, 114.

the subsequent liquidation of private trading in alcoholic drinks.[29] As a result, the state assimilated another important territory of economic and personal independence. Ultimately, it turned out that one could neither produce alcohol nor sell it, or even drink alcohol without the government's consent and surveillance.

When eventually there were no longer peasants who could fit into the classical definition of *kulak*, this label started haunting even the destitute. All too often their income was smaller than that of the clerk or the party bureaucrat who was working "for their destruction as a class."

The anthropologist Katherine Verdery, who has studied collectivization in Romania, has drawn attention to the fact that in village daily practice "the class struggle" was part of the constant process of constructing the image of *kulak* in the local community by attributing the term to certain peasants. On the other hand, the stereotype was often deconstructed as soon as the accused persons demonstrated active collaboration or willingness to cooperate with the government. There were also cases of repeated falling of an already "labeled" man or group into the category of *kulaks* under certain circumstances.[30] In fact, this process involved a permanent struggle between individualizing (or decontextualizing), on the one hand, and communalizing (or contextualizing), on the other.[31] In all cases it came to the imaginary construction of a status group alternative to the official government. As a rule, the *kulaks* were to blame for all economic hardships and failures of the governmental undertakings while the poor results of collective farms were attributed to "*kulaks* penetrating into them." One of the important aspects of party policy was the creation of social intolerance towards those people. It was achieved mainly by pointing an accusing finger at them—their names were on special lists prominently displayed, and condemning graffiti were painted on their houses. The second stage of the process of *kulakization* can be called "segregationist." It found expression in refusals to serve them in stores, taverns, and cafes, and a publicly demonstrated reluctance to talk to them.

Copying closely the Soviet terminology, the term *podkulachnik* ("*kulak*-friendly person") appeared in the political jargon of local communist functionaries in 1949. Usually, it referred to anyone who refused to adopt the communists' attitude as a social standard and maintained their

---

[29] Mihov, *Kapitalisticheskite*, 18.
[30] Verdery, *Dialogic Collectivization*, 4.
[31] Verdery, *Dialogic Collectivization*, 9.

normal neighborly and friendly relations. Since such behavior could jeopardize the construction of the new hierarchy in villages, the *podkulachnik* were also subjected to persecution.

Another form of resource confiscation from wealthier peasants was the tax policy pursued by the regime. Under the inherited system, the burdens fell jointly on the peasants who, according to some estimates, had to pay 83 taxes and charges.[32] From 1946 on, the graduated income tax was adopted, whereby any farmer who owned up to 3 hectares planted with cereals or 2 hectares planted with intensive crops, one cow, and no more than five sheep was to be exempted from taxation.[33] It is clear that this legislation stimulated small and middle agricultural landowners on whom the state relied heavily to fulfill their quotas and, in effect, to provide for the country's population. The tax regulations favored grain production that proved to be one of the few sources of foreign exchange funds for the country in the first postwar years. Under the new Income Tax Act, passed in 1950, the rate of taxation of private agricultural producers rose steeply. According to some estimates, the tax burden on petty peasants was reduced, but for the wealthier ones the tax burden was twice as high. At the same time, only "the lands for private use" of the TKZS members were symbolically taxable. The income tax was quite restrictive with respect to private owners, such as craftsmen and grocers. In fact, the aim of the new legislation was to persuade the peasants that their only choice was to join TKZSs.[34]

One of the important tools for pressuring the peasantry was price policy. The restriction of market mechanisms in this field went through the regulation of prices not only of agricultural products but of any product in the country. The so called "marginal prices" had already been introduced since 1945. Prices were very much governed by strict provisions and the sweeping powers of the state by virtue of the law against profiteering. The black market continued its parallel existence despite a great number of criminal proceedings against it. Thus, there were three price rates operating in agriculture. First, it was the "quota system." The value of agricultural products under it was largely symbolic—very often it was half of the rate of the "marginal prices" and a quarter of or even less than the actual value of the products. The system's existence can be rather explained

[32] *Darzhaven vestnik*, no. 234 (October 12, 1946); Chernokolev, *Dokumenti*, 84.

[33] Chernokolev, *Dokumenti*, 119.

[34] SA-Vidin, fund 180, inventory 1, file 31, folio 58–61; fund 332 B, inventory 2, file 2, folio 367.

as "common courtesy" to formally disguise the plundering of peasants behind some kind of "sale." But the fulfillment of quotas was the condition under which farmers could offer their surplus on the market.

They were to bear in mind the already mentioned "marginal prices," too. In addition, the "free buyout" of agricultural products, such as meat, hides, and eggs, by the state was affected at such prices. The term should be used conditionally because very often the buyout measures had nothing to do with the economy. As the whole propaganda and administrative apparatus was set to work for the fulfillment of state plans, the "free buyout" was forced in many cases. The category of "marginal prices" referred also to the so-called industry goods that were subject to rationing. No peasant could receive such goods unless he was issued a certificate verifying that he had fulfilled his deliveries to the state. Due to the chronic deficit of such commodities, it was a common practice to replace industrial goods with agricultural products calculated at "marginal prices."[35] Thus, peasants found themselves in a situation in which they were to sell cheap to the state and to buy from it at prices lower than the actual prices.

The third category was the actual market price. Formally, they were classified as "black market" but the government often shut their eyes to the existence of that market. The reason was that there were numerous cases when peasants were forced to buy from this very market in order to fulfill their deliveries to the state. Usually, this kind of transaction was effected "in the shade," with direct negotiation between the producer and the buyer.[36] Although there were occasional show trials against black marketers, the functioning of that market secured the quota system. Peasants who had produced less than they had to deliver to the state would compensate for the difference from that very market. However, they were again the innocent victims of the system.

The functioning of the ration system in the country was also based on the social and class differentiation, officially postulated by the Communist Party. According to its ideology, the working class was the "avantgarde" of the society that had to be built. It was therefore officially tolerated. This opinion appeared in the governmental documents to guarantee the minimum standards of living of different groups of Bulgarian citizens. For example, the amount of wheat flour per day, which a peasant had to be given, was calculated at 300 grams, while a worker was given a bread

[35] Kanatzieva, *Socialno-ikonomicheskata*, 130.
[36] Migev, *Problemi na*, 121.

card calculated for 1 kilogram.[37] Even as the peasants started leaving the villages and settled in the towns, initially as temporary and seasonal workers at industrial sites, the government ordered that they be discharged from obligatory state deliveries only for the period of time while they were engaged in construction and industry. They were to receive bigger rations respectively only for that period.[38] Those rations were, of course, of different values. For example, at the discretion of local figures of authority, the wealthier a farmer was, the less he needed rationed commodities. As for the *kulaks*—one of the regular measures against them was to deprive them of rations as a rule. Thus, peasants fell into another kind of dependence on their local leaders—they could "take the bread out of their mouth" and this was one of the most actively used tools in the course of collectivization. As the ration system was formally repealed partially in 1951 and wholly in 1953, it was no longer a factor in the final phase of collectivization at the end of 1950s. Scarcity of certain products and at different periods of time nonetheless remained a problem.

The next but not less important lever of the regime for forcing peasants into TKZSs was the direct psychological pressure and coercion. It took all sorts of forms and always accompanied other tools for enforcing policies. In the context of the arbitrary powers of raging militia and court officials who used every possible pressure, peasants lived in constant fear. The systematic assaults on those who were stubborn—arrests, internments, and show trials—struck terror in their hearts and ultimately, turned many resisters into reliable agents. According to official procurator's figures, in 1948–51 there were 43,465 cases of "crimes against the state and FF government," and the death sentences reached 250–280 yearly.[39] This was only a part of the arrested, many of them being peasants whose cases did not enter the procurator's offices. In the same period the official number of the deported peasants was 4,208 families with 18,315 members.[40]

Maybe the most effective psychological pressure on peasants was the binding of their collective enrollment to the future of their children. The fear that a peasant's son or daughter would not be able to continue their

[37] CSA, fund 172, inventory 1, file 15, folio 192 [Accounts of the Office of State Supply and State reserve].

[38] CSA, fund 136, inventory 5, file 2374 [Decree of the Council of Ministers no. 2454 of September 8, 1950].

[39] Ognyanov, *Politicheskata sistema*, 204, 206.

[40] Gruev, *Preorani slogove*, 207.

education (some were, in effect, prevented from studying), proved to be the most powerful weapon that paved the way for many firebrands.

THE FIRST COLLECTIVIZATION DRIVE: PRESSURE AND COERCION, 1948–51

At the beginning of 1948 the regime openly declared its intentions concerning Bulgarian villages, making it clear that "the socialist reorganization of agriculture" was inevitable and without alternatives. The next step of the government was the Act on the Buyout of Large Farm Equipment from Private Owners, passed on February 18, 1948.[41] Under this act, the heavy farm equipment was nationalized and the former owners were to be indemnified with securities, against which they would receive virtually nothing. This meant the end of the competition between private and collective sectors in the village because the former lost any chance of modernization at the expense of the latter. Farmers were left equipped only with a plough and a pair of oxen to compete against modern technology like tractors and combine harvesters. It was a shattering blow not only to the so called *kulaks* and capitalist landowners but mainly to the middle-income peasants who owned between 50 and 100 hectares. Obviously, they would not be able to cultivate such an area of land without equipment but only with manual labor and draught animals, irrespective of the size of their families. Therefore, the only alternative for the majority farmers was to join the TKZSs.

The regime took its next step toward the strengthening of TKZSs in parallel with the already organized delivery and acceptance committees which had to carry out the actual confiscation of large agricultural equipment. The act on amendment and modification of the decree on TKZSs was passed on March 4, 1948.[42] Under the law, the Ministry of Agriculture and Forestry acquired legitimating supervisory and disposing functions over collective farms in the country and was conferred the rights to approve newly established collectives, to issue binding instructions, and to overrule decisions of their managing bodies. The new amendments obliged members to contribute their whole land, animals, and farm implements to the collectives. Thus, a clear dividing line between members of

41 *Darzhaven vestnik*, no. 48 (February 28, 1948).

42 *Darzhaven vestnik*, no. 63 (March 18, 1948).

cooperatives and private farmers was established. Any peasant who had already joined the new units faced a new dilemma and there was little hope that they would be able to keep some of their property and to be both half-collective members and half-private producers.

Undoubtedly, all these changes were a crucial moment in bringing TKZSs in line with the Soviet *kolkhoz* system and if there were still any differences, they would soon disappear. Formally, there was no nationalization of the land as in the USSR, but Bulgarian peasants, in effect, lost ownership of their land. In addition, there was the introduction of a small rent, but it was abolished at the end of the 1950s.

The last, sixteenth plenum of the Communist Party prior to the Fifth congress timed for the end of the year was held on July 12–13, 1948. Bulgarian historiography has traditionally regarded this plenum as a turning point for the finalization of transforming the political state system that started in 1944 and led to the formation of open communist dictatorship along Stalinist lines. The higher party leadership intensified political agitation to end the transition period. In agriculture, measures were aimed at the strict collection of the obligatory state deliveries, forcing of cooperative agriculture, and social segregation of the so-called "*kulak* and capitalist elements." On August 26, 1948, the Politburo asked Titko Chernokolev to prepare a series of steps along these lines, whereby the government gave the green light to the next phase of the dogmatic following of the Soviet lead—*massovization* of collectivization. Notwithstanding the purposeful efforts of the state supported by the structures of the Communist Party at all levels, the initial figures from which the *massovization* started were discouraging. The total amount of land contributed to the collectives until that time was approximately 7 percent.[43]

In the very end of 1948, the Great National Assembly voted on the Act on Prohibition of Sharecropper Agreement and Other Contracts.[44] It, in effect, liquidated the private copyhold relations in the Bulgarian village and opened the way for TKZSs to be the sole and non-competition copyholder and, therefore, exploiter of farm hand.

Initially, the TKZSs constituted production units attached to village comprehensive cooperatives. However, they soon developed the well-known "cuckoo syndrome" and started seizing more and more businesses from the village, extending their powers over the auxiliary processes,

[43] Trifonova, *BCP i socialisticheskoto*, 103.
[44] *Darzhaven vestnik*, no. 304 (December 27, 1948).

trade, and services. Thus, other business units in rural areas were gradually pushed out. At first they took the existing village (municipal) stock-breeding farms into their hands, thus gaining control of cattle-breeding and forcing private farmers to turn to their services when impregnating the cattle. The TKZSs also seized village distilleries from consumer cooperatives, thus acquiring not only a new resource but also supervisory functions over the production of brandy by any household. Some collectives started opening special shops for their members like bakeries and public baths. They also granted loans in the form of money and materials for building new houses. The creation of the Central Cooperative Union in April 1947, whereby the cooperation, in effect, was socialized, made village comprehensive cooperatives meaningless. Under the new conditions, these cooperatives turned into local subdivisions of the state buyout and trade. In fact, the state nationalized the cooperative trade network and proceeded to establish its own one only where there was none.[45] Thus, in the late 1940s the state liquidated the authentic cooperative movement through doubling structures, the main one being the TKZS. The private sector, not only in agriculture but also in trade and services, became more and more dependent on this overwhelming economic unit.

The policy pursued by the regime in agriculture in the late 1940s was continued and expanded in scope and intent in 1949—the first year of the First Five-Year Plan, which was declared an accelerated plan. Increasing the amount of land contributed to the TKZSs, as well as the total amount of arable areas, seemed a mammoth undertaking. On the eve of the spring campaign, by a series of regulations, the regime ordered that private lands not sowed within the period of time specified in the crop plan should be expropriated.[46] Thus, many local governors were able to take physical control of a large number of properties. This policy was temporarily and partially corrected after May 1949 when private producers were able to catch their breath. It was prompted by the Soviet example which the government copied to the last detail and according to which *peredishka* or some rest should be given before the next phase of forced collectivization. Stalin himself gave his consent to this.[47]

Softening the course became an official party stand presented on the June plenum of the Central Committee (CC) of the BCP in 1949.

[45] Manolova, "Obedinenie," 28; Ganev, *Moyat zhiznen*, 178–225.
[46] Migev, *Colectivizacijata*, 41.
[47] Migev, *Colectivizacijata*, 48.

Though this party forum has been regarded as emblematic of the final Stalinization and of the establishment of the totalitarian state in other fields of social development, with respect to collectivization, it played the role of pointing an accusing finger at many ruthless local governors. Of course, this new feature of the party policy was not based on concern about "protection of people's democratic legality," in fact it rather implied an ambition to carry out the grand plans in agriculture during the start of the accelerated year of the First Five-Year Plan. The idea was to inspire peasants with some confidence, so as to start bringing products to the market, reduce the hiding of grain, and at the same time to stabilize the already created TKZS prior to the next phase. On these lines and by virtue of the relevant decrees of the Council of Ministers, the obligatory state deliveries were decreased to some extent and the buyout prices of agricultural products were increased.[48] Also, 86 governmental committees were created to investigate the arbitrary acts in the formation of TKZS tracts of land.[49]

Peasants saw it as some loosening of the noose and large numbers of them began to make complaints about fields wrongly taken away from non-members of collectives, unequal replacements, or simply about robbing them of their property. The total number of such complainants was 61,854 in 1949. Thousands of applications for leaving TKZSs were submitted.[50] Thus, the *peredishka* confirmed the conviction of the communist leaders that any further continuation of collectivization would not be possible if the pressure on peasants was to be relaxed. It was unrealistic to expect the process to develop in constant progression, without unnecessary forcing. It became clear that the only possible implementation of the process was by fits and starts and by force.

The rates of forced collectivization accelerated sharply from the early 1950s. The implementation of the large-scale plan for *massovization* began with the 138th decree of the Council of Ministers and the BCP CC, issued on January 20, 1950. It laid down the responsibilities of the Ministry of Agriculture, the amounts of deliveries and the moves toward the forthcoming *massovization*.[51] Under this decree, another 250,000 hectares were to be contributed to the cooperatives, whereby the total amount of collective land estimated to reach 800,000 hectares—approximately one half of

48 Migev, *Problemi na*, 109–11.
49 CSA, fund 1B, inventory 18, file 50, folio 3.
50 CSA, fund 1B, inventory 18, file 50, folio 3
51 *Rabotnichesko delo*, no. 18 (January 23, 1950).

the arable areas in the country. The estimated deliveries of milk and wool were particularly high.[52]

The next plenum of the BCP CC dedicated to the course of collectivization was held on April 4, 1950. Its aim was to finally approve the so called "Model Statute of TKZS" which would unify the functioning mechanisms of any farm, the amount of lands, and the number of animals left "for individual use" in different parts of the country. At the plenum and at the second national conference of TKZSs opening on the next day, party leader Vulko Chervenkov and the Minister of Agriculture Titko Chernokolev, appealed to the party *aktiv* not to act too hastily and to pay attention to the winning of the so called "middle-income peasant" to join the collectives.[53] Their speeches were aimed at promoting more moderate views and restraining the calls for limitation and even elimination of the rent.

The approval of the "Model Statute" laid down strict limits on the management of lands "for private use"—from 0.2 to 0.5 hectares (including the house with the household yard), depending on the geographical conditions of the district, as well as on the number of animals owned—one head of draught animals (a horse, a donkey, or a mule), one head of cattle (a cow or a buffalo-cow, a goat as an exception) and up to five sheep. It also regulated the maximum amount of rent to be paid to the members of the collective, as well as their commitments to the state. The tendency toward the smooth reduction of rents was clearly outlined—from 40 percent in the former "Standard Statute" to 30 percent, with the possibility of its gradual removal. The supervisory functions over the collective farms were transferred from the Central Cooperative Union to the Ministry of Agriculture and TKZSs, in effect, were turned into subordinate structures of the latter.

In April 1950 the regime made a new correction in the decree on milk delivery. Under it, cattle was to be subject to taxation "all down the line." In other words, the quota covered even barren cows and oxen.[54] The problem with barren cows and buffalo-cows was both funny and sad. According to another decree of the Council of Ministers, any farmer who had such an animal was to be exempt from delivery in May of the next year if he was able to present a certificate issued by the *municipal council* verifying that it had not calved and a second document issued by the dis-

[52] CSA, fund 136, inventory 5, file 439, folio 1–9.
[53] *Vtora nacionalna*, 1950; Migev, *Colectivizacijata*, 68–69.
[54] CSA, fund 136, inventory 5, file 599, folio 80–84.

trict veterinarian who had to confirm that it would not calve in the period from May to October.[55] All in all it came full circle—unless one got rid of the oxen, he would be forced to buy milk in order to pay off his debts to the state. And if one got rid of them, he would have nothing to cultivate the land with. In other words, there was again no other way out but joining the TKZS.

American anthropologist Gerald Creed has compared the gradual "softening" of peasants and the progressing *massovization* to "a rolling snowball resulting from the dense network of relations between them."[56] This interpretation aptly explains the social nature of the process. To start the rolling, however, a critical mass was badly needed—in this case, peasants who turned from core minority into dynamic majority. This critical mass was different depending on the local community, for instance, ethnic and religious groups, towns and villages, as well as areas and regions had strictly individual and subjective variations. In the months of *massovization*, the numbers and names of all those joining the collective were broadcast on radio day by day, so the average psychological line was between one fourth and one third of the peasants. Here personalities were of crucial importance—the poorer usually waited for the wealthier, hardworking and authoritative men in a village. Therefore, if we are to use Creed's metaphor, *massovization* might be compared to a stone talus or an avalanche, in which the top layer was controlled first and "weighed heavily" on those below. In turn, those who finally succumbed became a "burden" to the most stubborn and stable members of the local community due to the close-knit network of family and neighborly relations. This accelerated the process like an avalanche and multiplied its destructive power and the consequences for the hardest opponents.

After the mass had finally fallen down they were either totally ruined and/or rejected as unnecessary debris that did not belong to the village. They were usually frowned upon by the government and by their fellow villagers who were not like them. In this dynamic process the so-called "loyalists" played a special role in the village community. These were the majority of people who had no firm political convictions but backed the government by old tradition and well-developed instincts and proved to be a very important component in the accumulation of the "critical mass" concerned. Here we have to introduce another term which gives a fairly

[55] CSA, fund 136, inventory 8, file 78, folio 2.

[56] Creed, *Domesticating Revolution*, 67.

correct description of the social process of that time—the so-called skeptical collaboration. The majority had no alternative but to collaborate with the regime even if they did not have any personal or ideological motivation. The lack of motivation would facilitate the erosion of the classical system from the inside when the time was ripe.[57]

Many accompanying economic, social, and psychological phenomena ran in parallel with this process and formed a specific mentality and attitude toward the world. Thus, another problem emerged with public and personal identities and the behavioral models behind them. One had to speak and act outdoors in one way and when at home, in the intimate family environment, in quite a different way. Of course, this "double speak" was characteristic not only of the village, but was common to the social nature of the entire communist system as well. In rural areas, however, in the context of huge upheavals, it stood in stark contrast. The necessity of playing a role in the private and public spheres predetermined the discord in the social and political identity of ordinary people in whom hypocrisy had implanted itself permanently. Initially, the main agent of these processes was fear. In the first years it had particular physical and sensory roots, but little by little it was replaced by a vague and depersonalized feeling which gave birth to the new social and moral standards. These general tendencies presupposed numerous exceptions which, however, did not challenge their relevance in the majority of cases.

Divisions and Dissent in the Villages and in the Communist Leadership, 1950–51

The pressure on peasants was on such a scale that joining TKZSs became their only chance of survival. By the end of 1950, more than 50 percent of Bulgarian peasants were members of collective farms. However, the discontent among those who got "inside" and among those who were still "outside" was so strong that there was growing spontaneous unrest in many rural areas of the country. Interestingly, these were mostly women's riots. Women's resistance in Bulgaria had made history before, mainly through antiwar and anti-famine demonstrations in the period between 1913 and 1918. Also, it is a well-known fact that women have been guardians of tradition to much greater extent than men who often tried to make

[57] For more details see Gruev, *Preorani slogove.*

their family's living away from home. Since in the conventional view, collectivization would "blow up the tradition," women naturally opposed this process more emphatically. One of the main reasons, however, was related to the general (shared by men and women) fear of inevitable repressions and the hope that the government would have a more lenient attitude toward the resistance of "the fairer sex" purely for humane reasons.

Resistance began on July 3, 1950 in the large Catholic village of Burdarski Geran whose inhabitants were settlers from Banat. The riot in the Orthodox village of Stavertzi in the Oryahovo region was even better organized. Four days after the developments in Burdarski Geran, the inhabitants of Kozloduy rose in revolt. The women's riots in the three villages in the plain of Vratza, as well as in Butan, Lipnitza, Krushovene, Dolni Lukovit, and Glozhene, remained in effect the most publicly demonstrated discontent over the TKZSs and the government's policy. But the participants did not go beyond the invisible bounds of reasonable behavior and when it came to use of weapons, the majority of the population was paralyzed. Thus, the government gradually restored its control over the situation. The following show trials against the more active participants threw the peasants in hesitation and confusion and dampened their enthusiasm for any further resistance. The unrest in the towns and villages in the Vratza region in 1950 were not the only ones in the country. At the same time there were uprisings in the regions of Assenovgrad, Plovdiv, Parvomay, and in some villages in other regions. Interestingly enough, all rural communities involved were relatively rich, situated in the plains, whose inhabitants depended on the land for their livelihood but also had self-confidence and felt secure. They regarded the land as an added value, by which they measured themselves against their neighbors and their own strength.

New unrest broke out in dozens of villages in the spring of 1951, spurred on by a famine the previous winter that was caused by the state deliveries and the increasing pressure of joining TKZSs. Now the center of resistance resided the regions of Vidin and Kula, in northwest Bulgaria. The immediate cause of this new wave of rural unrest was the new line toward some relaxation of pressure adopted "from above." In fact, this was Stalin's next "advice" for his substitute in Bulgaria, Chervenkov. Comparing the situation in the USSR in the beginning of 1930s and in Bulgaria twenty years later, the Soviet leader stated that he had to speed up collectivization. However, according to his opinion, "shared" with Bulgarian comrades, it was not necessary right now.

The new line of policy was taken up by the entire communist leadership in Bulgaria and became the reason for overall softening of the course with respect to peasantry. One of the main initiatives on these lines was to distance the central leadership from the violence and arbitrariness "from below" and to put it in the role of an arbitrator and fair judge in the conflict between peasants (but not *kulaks*) and local upstart rulers. On the same lines, an extended plenum of the BCP CC was convened on March 12, 1951 in Sofia to discuss and condemn "the perversions of the party policy." In his main report, Chervenkov criticized mostly the measures taken in the course of *massovization* in the district of Kula and in the village of Yablanitza in the Teteven region, but his intent was to dampen the "enthusiasm" of the local functionaries throughout the country. A week later, on March 19 of the same year, a special joint decree was published by the Council of Ministers and the BCP CC, whereby "the perversions of people's democratic legality" were stigmatized. The new line was loudly proclaimed in the official organ of the party, in separate brochures specially printed, and on the radio. The expectation was that people in villages would be placated and unrest would cease.

However, the reaction was quite the opposite. Peasants saw it as giving the green light to leave the collectives and seek the return of the fields, gardens, vineyards, animals, and equipment of which they had been deprived. A real rural revolution started in many places. The fight between peasants and the government was quite dramatic in the areas of Plovdiv and Assenovgrad, where 13 and 12 villages respectively rose against the regime.[58] The situation was similar in the area of Pleven—the region which suffered the first casualties in a direct fight with the forces of order.[59] The districts of Kula and Vidin did not make exception. The unrest spread to the villages of Gramada, Chichil, Kosta Perchovo, Rabrovo, Izvor Mahala, Brankovtsi, and some others. As in the previous year, they were organized spontaneously and were aimed at jointly leaving the collective farms, and seeking the return of the taken cattle and equipment. A mass submission of applications for leaving of TKZSs ensued. From Kula, the riots shifted to the neighboring district of Vidin. Here the collectives were broken up in the villages of Novo Selo, Deleina, Tiyanovtsi, Negovanovtsi, and some others. Undoubtedly, the mass rural rising in the two districts was influenced by their proximity to the Yugoslavian

[58] Ministry of the Interior, fund 1, inventory 1, file 1795, folio 85–89; file 1873, folio 2 [Reports on the agents' and operative workers' operations at the division of the State Security with the district offices of the Ministry of the Interior in Plovdiv and Assenovgrad in 1951].

[59] Migev, *Colectivizacijata*, 123–125; Josifov, *Totalitarnoto*, 186.

border in a time when the bilateral relations were at the point of greatest hostility. Many underground groups were formed on Yugoslav territory and raised the spirits of the rebels, promising them armed support. Thus, hundreds of peasants fled across the border.

The government took all necessary measures to prevent the unrest from spreading to other parts of the country. Secretary of the BCP CC Todor Zhikov arrived on the scene, accompanied by his long-time political companion, Tsola Dragoicheva. Hundreds of militiamen, officers of the State Security, and full-time and part-time political workers were mobilized. Huge numbers of people were arrested. Some of them were brought to show trials in their home villages, others were sent to prison and concentration camps without trial. Also, mass campaigns were organized for forceful internment of family members, relatives, and friends of those who had participated in the riots or who had fled abroad. The total number of those interned from that part of the country was 3,500, plus roughly the same number of detainees. The government's measures, in effect, succeeded in paralyzing peasants' resistance and avoiding further unrest, but at this stage they were not able to return the peasants to the collective farms.

However, the regime decided not to do any further experiments and abandoned the *massovization* of collectivization for the time being. In line with Stalinist protocol, the latest developments had to be justified with "enemy activities" within the Communist Party. According to this distorted view, the leaders of the BCP's district committees in Vidin and Kula had to be replaced and some of their members brought to trial and sent to prison. A scapegoat was identified at the highest level—in the very Politburo of the BCP CC—in the Minister of Agriculture Titko Chernokolev. Though he had dedicated his political career to collectivization, he had failed in his attempt to win "the love of the party boss." Now, at various party forums he was repeatedly accused of the unrealistically high plans for the creation of new collectives espoused by the Ministry or of non-fulfillment of different kinds of obligatory state deliveries.[60] Though in the previous year he had been forbidden to issue orders in connection with the course of collectivization, on June 15, 1951, he released a "circular letter" on behalf of the Ministry of Agriculture and Forestry, whereby those peasants who had sowed their fields prior to joining the collectives were allowed to harvest privately.[61]

60 *Rabotnichesko delo*, no. 126 (May 6, 1950); Giurova, "*Kooperirane*."

61 CSA, fund 1B, inventory 5, file 160, folio 80 [Discussion at the Agrarian Department of the BCP CC].

The reaction of the party boss was very sharp. By his personal telegram, Chernokolev's "circular letter" was repealed and he was removed from all posts, expelled from the BCP and arrested.[62] The ex-Minister of Agriculture was to spend seven months under investigation at the State Security, whose officials were obviously hesitant about what charges to bring against him. The actions taken with respect to Chernokolev, as well as the assessments and epithets used, closely resembled the drama that had unfolded two years ago with the second ranking person in the Communist Party, Traicho Kostov. It is most likely that the Kremlin gave the green light to such developments. Chernokolev's case was emblematic of the logic in decision-making, the relationships and the vicissitudes at the top of the party pyramid that governed the state. It also marked the epilogue of the second stage of collectivization when the policy of pressure was temporarily abandoned.

The Deceleration of Collectivization, 1951–56

In all former attempts at dividing Bulgarian collectivization into periods, the abandonment of policy of pressure by communist leaders in the early 1950s has been taken into consideration. In late 1951 this third period started, characterized by "a go-slow policy" pursued until early 1956. The appeals for "observance of the people's democratic/socialist legality" resulted in actual suspension of pressure immediately after the unrest ended. After the events of March and April 1951, the government's efforts were aimed mainly at "standing their ground." It related both to striving for preservation of the total area of land contributed to collectives and strengthening economic, financial, and personnel measures for collective farms. The formation of new collectives was temporarily suspended, as was the pressure on new owners to join the existing ones. It brought about an odd reconciliation between the peasantry and the state, which turned into a dynamic equilibrium manifested mainly in "a game of nerves." However, the abandonment of *massovization* did in no way mean restriction of pressure on the private producer. As the regime was afraid that stabilization of the private sector would follow, pressure through the instrument of deliveries intensified. Though the grain quotas in 1952 remained unchanged as compared to those of the previous year, the increased

[62] Migev, *Colectivizacijata*, 141–42.

burden of fulfilling them was due to a severe winter and poor crop. The deliveries of animal products were drastically increased. It related mainly to milk, where an additional plan for "hidden sheep" was transmitted.[63] This "presumption of guiltiness" was rather an expression of the regime's despair in its inability to gain control over cattle, and was further evidence of the lack of change in the internal colonial policy of exhausting the peasantry. Coercion on private producers also intensified through the so-called replacements (of land plots) used since the early years after communists came to power. With the enlargement of the collectives and increase in their number, they began to affect more and more people.

One of the reasons for the poor condition of peasants during that period was associated with the repeated reduction of prices in the basic foodstuffs and industrial goods undertaken by the government. This cheap propaganda trick on behalf of the regime had a very negative impact on agricultural producers whose products were additionally devaluated. This affected both products brought to the free market and buyout prices fixed by the state. According to the estimates made by the historian Vladimir Migev, these steps reduced the agricultural commodity prices by half compared to 1950.[64]

In all spheres of Bulgarian society, as well as in the entire communist system, the death of Stalin on March 5, 1953 marked the beginning of furtive processes of liberalization and some "release of the grip." The "ration course" adopted in the USSR and transferred to Bulgaria opened the way for a partial "rehabilitation" of the private sector in agriculture and a relaxation of the quota pressure on peasantry. State deliveries as a whole were reduced by governmental decrees; some freedom was given to sales at free prices. This strengthened private rural producers though the state continued to exert reverse pressure.

Comparison between the private and collective sectors showed marked disparity in terms of economic results, output, and motivation. It was clearly illustrated even by the official statistics of the so-called auxiliary farms of collective members. They constituted 9 to 11 percent of the total land belonging to TKZSs but gave on average one-fourth of the total output in this sector, including about two-thirds of the output in stockbreeding.[65] These figures disclosed by the official statistics show that the

---

63 Migev, *Problemi na*, 129.

64 Migev, *Colectivizacijata*, 151.

65 Migev, *Colectivizacijata*, 198–99; Marcheva, *Todor Zhivkov*, 235.

auxiliary farms were at least three times more efficient than the other activities of TKZSs. However, it will be very far-fetched to say that this form of production and existence was "proto-capitalistic" and "market-orientated." It should be noted that it was a parasite on the classical system of cooperative agriculture, on which it drew resources (machines for cultivation, sowing seeds, breeding stock and offspring, etc.), contributed to the development and flourishing of misuses and thefts, and guaranteed the sale of its products, but did not take any risk and did not suffer any losses.

Full-scale Collectivization and the Creation of the "Socialist Man," 1956–60s

The fourth and final stage of collectivizing the Bulgarian village, also known as the period of the second *massovization*, started in early 1956, a year which was very important for the political system. This alone is indicative of the conditional dividing line in the regime's history and of the obvious succession between Chervenkov, who was sinking into oblivion, and Todor Zhivkov, who had his star on the ascendant in terms of long-term strategy, tactics, and modes of pressure. The Polish researcher of collectivization, Dariusz Jarosz, has paid attention to the fact that the directive on continuation and finalization of *massovization* wherever possible was transmitted centrally by Khrushchev to all satellites.[66] This hypothesis can be confirmed indirectly by experiences in Bulgaria, Hungary, the GDR, Romania, and Czechoslovakia.[67] An exception to this pattern is Yugoslavia, where the regime abandoned cooperation in 1953, and Poland, where the process stopped in 1956 with hardly 11 percent of the land contributed to collectives. However, the concurrence of events in all other countries clearly suggests that it was a directive from Moscow. Its implementation became somewhat of a test of loyalty and fitness in the new political assembly at the top of the respective countries. Thus, all

[66] See Dariusz Jarosz's contribution to this volume. Prof. Jarosz has worked in the archives of the CC of VCP(b) and presented this thesis orally at the conference on collectivization in Eastern Europe organized by Central European University in Budapest, June 22–23, 2007. He stated that he was not allowed to make notes and was given documents at officials' discretion but one of them was the minutes of such a meeting at the Kremlin.

[67] For more details, see Boyer, ed., *Zur Phisionomie*; and Wädekin, *Agrarian Policies*.

emerging groupings spoke as one on their intents to finalize collectivization on the eve of the April plenum in 1956.

Generally, the methods and the very technology of the campaign were quite familiar to the peasants from the first *massovization* and to the local functionaries who had to bear the same burden again. Once again, they were to rely on psychological pressure exerted by the newly formed "pressure groups" and when the eloquence of agitators gave no results, it was time for direct violence.

Shortly after the reorganization on the highest level, a new plenum of the CC was convened in July 1956 to consider only agricultural matters. The meeting decided that collectivization in the country should be completed in one, two, or a maximum of three years. A new system of measures would be planned, so as to visibly improve the living conditions of peasants in collective farms. First, "the land for private use" would be exempted from state deliveries. The most serious step towards the strengthening of the entire cooperative sector, however, was related to the so-called farm-workers' pensions, which the collective members would be granted from the beginning of the next year, 1957.[68] In fact, the money granted to this group was symbolic but peasants regarded it as a sign of great importance. Another important decision taken at the plenum was to allow *kulaks* to be members of collective farms, although without the right to hold management posts.[69]

These changes were aimed at publicly demonstrating the new face of the government, while preventing eventual resistance and calming down the most stubborn peasants. Thus, in 1958 and the early 1959 collectivization had already been set in motion in the last remote mountain villages and hamlets and encompassed all rural areas in Bulgaria. This dramatic finale to the process coincided with the culmination of mass depopulation in mountain regions and a radical change in means of livelihood. Thus, the separation from land and livestock caused less drama and turmoil than in the beginning. Many peasants had already said goodbye to their property in their thoughts and were heading for a fresh start.

All this resulted in a severe crisis in the Bulgarian village. Living there was an unattractive option against the possibilities offered by the city. These demographic trends and the attempts at purposeful state policy

---

68 *Izvestiya na Prezidiuma na Narodnoto Sabranie*, no. 1 (January 2, 1957); Tanov, ed., *Agrarnata politika*, 177–78.

69 *BCP v rezolyucii*, 1955, 31–40.

on their "softening" affected not only the village but also the towns and, in effect, the entire Bulgarian society. The trends can be summarized in two prevailing processes running in parallel and interacting with one another—*depopulation* of villages and *rustification* of towns. The process was described for the first time by two US sociologists, Irwin Sanders and his student Roger Whitaker, who conducted fieldwork in Bulgarian villages (Dragalevtzi in particular) from the 1930s to the early 1980s. They analyzed the gradual extinction of the traditional villager through his transformation into a member of a collective farm and ultimately, into a state day-laborer on his own land or proletarian within the Marxist understanding of the term.[70] However, depopulation has a larger meaning—losing the features of the old view of life and moral values, changing the lifestyle and means of livelihood, and last but not least, coming closer to living conditions in towns. The counter process of rustification ran in parallel with it. This demographic term could be roughly translated as "becoming countrified"—in the more generalized meaning of the term. Huge migration flow into towns resulted in gradual assimilation of rural stereotypes and the integration of traditional and urban economy and culture.

But what happened was quite the contrary—the migration flow from the village to the town was so huge that after the 1960s, it was evident that peasants were undergoing the reverse process of gradually absorbing city culture, adapting rather quickly to new living environments. Of course, this process did not apply to the peasantry in the classical meaning of the term. Torn from their natural environment and put in new economic conditions, the peasants had already lost some of their stereotypes and were susceptible to quick adaptation. As a result of these complicated processes, a new type of citizen emerged. He or she identified themselves as such rather because of the official terminology (including the law). The self defining as citizen did not mean specific means of livelihood, social characteristics or cultural adjustment. Behind most of them stood a peasant who was losing many of the traditional virtues of his or her former social category. The established new cultural model as a result of the disintegration of its forerunners in terms of the traditional rural and traditional city culture would prove to be twice as immunized and more sustainable than the system which gave birth to it. Thus, the parallel depopulation in villages and rustification of towns would bring the two former strata much closer together and would lead to the creation of the common "new

[70] Sanders and Whitaker, "Tradition and Modernization."

socialist man" who would combine features of both identities but would not be identical to either of them.

CONCLUSION

The collectivization in Bulgaria was the most radical and abrupt change in the process of establishing the new communist political and social system. Strictly following the Stalinist model, this process became crucial for the thorough Sovietization of the country. Accordingly, Bulgaria was among the new communist states where the transformation of the traditional village was accomplished relatively soon (together with Albania). It is well known that in Poland and Yugoslavia collectivization was halted in an initial stage, whereas in Romania, Hungary, Czechoslovakia, and the GDR it lasted up to the 1960s. The Bulgarian experience stands out for its rapid process of collectivization. Moreover, Bulgaria's agrarian model for collectivization, established in the 1950s, did not undergo substantial changes and remained relatively stable throughout the process. In contrast, in countries such as Hungary, a creeping decollectivization started even in the end of 1960s. Thus, Bulgaria's Soviet-like agrarian system remained virtually untouched.

Demographic figures clearly show the long lasting consequences of collectivization. The rural population constituted 75 percent of the total population in the late 1950s. By the end of the 1980s, only about 30 remained in the villages. By that time, the predominant part of the "socialist peasantry" consisted of aging retired individuals and families. Within two generations, the traditional Bulgarian village had rapidly disappeared. A completely new society emerged, marked by different social structure, economy, and culture. Collectivization was to a large extent the crucial tool in communism's version of forced modernization.

BIBLIOGRAPHY

ARCHIVAL MATERIALS

Stenographic records of the 26th National Assembly, 1 reg. session, 22 meeting, 125.

Central State Archive (CSA):

CSA, fund 1B, inventory 13, file 111 (1), folio 84–90.

CSA, fund 136, inventory 3, file 66, folio 143–50; *Darzhaven vestnik*, no. 126 (June 1, 1948).

CSA, fund 1B, inventory 13, file 111 (1), folio 14–30.
SA-Vidin, fund 180, inventory 1, file 31, folio 58–61; fund 332 B, inventory 2, file 2, folio 367.
CSA, fund 172, inventory 1, file 15, folio 192. [Accounts of the Office of State Supply and State reserve].
CSA, fund 136, inventory 5, file 2374 [Decree of the Council of Ministers no. 2454 of 8 September 1950].
CSA, fund 1B, inventory 18, file 50, folio 3.
CSA, fund 136, inventory 5, file 439, folio 1–9.
CSA, fund 136, inventory 5, file 599, folio 80–84.
CSA, inventory 8, file 78, folio 2.
CSA, fund 1B, inventory 5, file 160, folio 80. [Discussion at the Agrarian Department of the BCP CC].

Archive of the Ministry of the Interior, fund 1, inventory 1, file 1795, folio 85–89; file 1873, folio 2.

NEWSPAPERS

Otechestven fornt [Fatherland Front], no. 196 (25 April 1945).
Darzhaven vestnik: no. 31 (12 May 1921); no. 81 (9 April 1946); no. 234 (12 October 1946); no. 158 (11 July 1947); no. 48 (28 February 1948); no. 63 (18 March 1948); no. 304 (27 December 1948).
Izvestiya na Prezidiuma na Narodnoto Sabranie [Act on amendment and modification of the Pension Act], no. 1 (2 January 1957).
Rabotnichesko delo [Workers' Affairs], no. 21 (1 February 1946); no. 18 (23 January 1950); no. 126 (6 May 1950).

BOOKS AND ARTICLES

BCP v rezolyucii I resheniya na kongresite, konferenciite, plenumite na CC i zasedaniyata na Politburo [BCP in resolutions and decisions of the congresses, conferences, plenums and of the CC Politburo], vol. 4. Sofia: Partizdat, 1955, 31–40.
Boyer, Christian, ed. Zur *Phisionomie sozialistischer Wirtschaftsreformen. Die Sowjetunion, Polen, die Tschechoslowakei, Ungarn, die DDR und Jugoslawien im Vergleich.* Frankfurt am Main: Max Plank Institut für europäische Rechtsgeschichte, 2006.
Chernokolev, Titko. *Dokumenti, publitzistika i spomeni za nego* [Documents, Publications and Memoirs about Him]. Sofia: Partizdat, 1989.
Conquest, Robert. *The Harvest of Sorrow: Soviet Collectivization and the Terror-Famine.* Oxford: Oxford University Press, 1986.
—. *Chemerna zhetva. Sovjetska kolektivizacija I terror gladi* [The Harvest of Sorrow: Soviet Collectivization and the Terror-Famine]. Belgrade: Filip Višnjič, 1988.
Creed, Gerald. *Domesticating Revolution: From Socialist Reform to Ambivalent Transition in a Bulgarian Village.* University Park, PA: Penn State University Press, 1998.
Ganev, Hristo. *Moyat zhiznen put. Razvitie na kooperativnoto dvijenie v Bulgaria mejdu dvete svetovni voini* [My life Path. Development of cooperative movement in Bulgaria between the two world wars]. Sofia: Zahari Stoyanov, 2006.

Giurova, Sv. *"Kooperirane na selskoto stopanstvo v Plevenski okrug 1949–1951 g."* [Cooperation of agriculture in the Pleven district 1949–1951]. *Izvestia na museite v Severozapadna Bulgaria* [News of the Museums in Northwestern Bulgaria], 1989, v. 15, 144–89.

Gruev, Michail. *Preorani slogove. Kolektivizaciya i socialna promyana v Bulgarskiya severozapad, 40-te–50-te godini na XX vek* [Re-ploughed Boundaries: Collectivization and Social Change in Bulgarian Northwest, 1940s–1950s]. Sofia: Ciela, 2009.

Iliev, Bocho. *Brazdi prez vremeto* [Traces in Time]. Sofia: Trud, 2002.

Josifov, Kalin. *Totalitarnoto nasilie v bulgarskoto selo (1944–1951) i posleditzite za Bulgaria* [Totalitarian pressure in the Bulgarian village (1944–1951) and the consequences for Bulgaria]. Sofia: Universitetsko izdatelstvo "Sv. Kliment Ohridski," 2003.

Kanatzieva, Rimka. "Socialno-ikonomicheskata politika na BRP (k) po otnoshenie na seloto (9 septemvri 1944–1948 g.)" [Social and economic policy of BCP on the village (9 September 1944–1948)]. *Izvestiya na Instituta po istoriya na BCP* [Proceedings of the Institute of history of BCP], v. 48. Sofia: Partizdat, 1983, 120–52.

Manolova, Nadya. *Obedinenie i reorganizatzia na bulgarskoto kooperativno dvijenie v sredata na 40-te godini* [Integration and Reorganization of the Bulgarian cooperative Movement in the mid-1940s]. *Istoricheski pregled* (Historical Review) 2 (1990): 18–29.

Marcheva, Iliyana. *Todor Zhivkov—putyat kum vlastta. Politika i ikonomika v Bulgaria 1954–1964* [Todor Zhivkov—the Road to Power. Policy and Economy in Bulgaria 1954–1964]. Sofia: KOTA, 2000.

Marinov, Ivan. *Za dobroto na horata* [For the Good of People]. Sofia: Zemizdat, 1991.

Mateev, Boris. *Dvizhenieto za kooperativno zemedelie v Bulgaria v usloviata na kapitalizma* [The Bulgarian movement for cooperative agriculture under the conditions of capitalism]. Sofia: Partizdat, 1967.

Migev, Vladimir. *Kolektivizatsiata na bulgarskoto selo (1948–1958)* [Collectivization of Bulgarian Village (1948–1958)]. Sofia: Universitetsko izdatelstvo "Stopanstvo," 1995.

—. *Problemi na agrarnoto razvitie na Bulgaria (1944–1960)* [Problems of the agrarian development of Bulgaria (1944–1960)]. Sofia: K & K, 1998.

Mihov, Nikola. *Kapitalisticheskite selski stopanstva v Bulgaria i tiahnoto likvidirane (9 septemvri 1944–1958)* [The Capitalist farms in Bulgaria and their liquidation (9 September 1944–1958)]. Ph.D. thesis, University of Sofia, 1988.

Minchev, Mincho. *Bulgaria otnovo na krustoput (1942–1946)* [Bulgaria at the crossroads again (1942–1946)]. Sofia: Tilia, 1999.

Ognyanov, Lyubomir. *Politicheskata sistema v Bulgaria, 1949–1956* [The Political system in Bulgaria, 1949–1956]. Sofia: Standart, 2008.

Penchev, Pencho. "Agrarnata reforma na BZNS i razvitieto na kapiralizma v selskoto stopanstvo pri upravlenieto na purvia cabinet na Demokraticheskia sgovor" [The Agrarian Reform of the BANU and the Development of Capitalism in Agriculture during the Rule of the First Cabinet of the Democratic Union] in: *Zemedelskoto dvijenie v Bulgaria—istoria, razvitie, lichnosti* [The Agrarian Movement in Bulgaria—History, Development, Personalities]. Pazardjik: Bello Print, 2005, 222–36.

Popov, Nikola. *Agrarnite otnoshenia pri socialisma* [Agrarian relations during socialism]. Sofia: Partizdat, 1976.

Sanders, Irwin and Roger Whitaker. "Tradition and Modernization: the Case of Bulgaria" in Jesse G. Lutz and Salah S. El-Shaks, eds. *Tradition and Modernity*. Washington: University Press of America, 1982, 62–70.

Tanov, A., ed. *Agrarnata politika na BKP na suvremennia etap* [Agrarian policy of the BCP at the current stage]. Sofia: Partizdat, 1981.

Trifonova, Minka. *BCP i socialisticheskoto preustroistvo na selskoto stopanstvo* [BCP and the Socialist reorganization of agriculture]. Sofia: Partizdat, 1981.

Verdery, Katherine. *Dialogic Collectivization: "Rich peasants" and Unreliable Caders in the Romanian Countryside, 1948–1959*. University of Michigan, Working Paper, 2004.

Vtora nacionalna konferencija na TKZS. Materiali [Second National Conference of TKZSs. Manual]. Sofia: Partizdat, 1950.

Wädekin, Karl-Eugen. *Agrarian Policies in Communist Europe. A Critical Introduction*. Everett Jacobs, ed. The Hague/London: Rowman & Littlefield Publishers, 1982.

Zhivkova, Veska. *Bulgarskoto selo prez vekovete* [Bulgarian Village through the Centuries]. Sofia: Nauka i izkustvo, 1985.

THE COMMUNIST TAKE-OVER AND LAND COLLECTIVIZATION IN BULGARIA. CHRONOLOGY OF EVENTS, 1944–1959

Mihail Gruev

1944

September 9: One day after the Red Army invasion, the communist-dominated coalition People's Front takes power under the leadership of Kimon Georgiev (Prime Minister during the authoritarian military regime, 1934–1935).

September 17: The new government declares state support for cooperative farming.

September 30: Publishing of the decree for "People's Tribunal," the main targets of which are the "persons guilty for Bulgarian participation in World War II on German side" and "persecution of partisans" (communist resistance movement). 6 central and 90 provincial courts are established. 11,000 persons arrested between September 1944 and March 1945, among them policemen, army officers, other activists, mayors, writers and journalists, royal advisers, all ministers after 1944 and the three Regents of infant Tsar Simeon II.

1945

February 2: All persons convicted by the central courts put to death. 2,730 death sentences executed until the end of March 1945.

April 13: Publishing of the decree for establishing the "Working Cooperative Agricultural Farms" (WCAF).

September 7: Legalization of political opposition, the Agrarian Union under the leadership of Nikola Petkov becomes the most popular anti-communist party, supported by the so-called "right social-democrats" (Kosta Lulchev) and the Democratic Party (Nikola Mushanov).

November 18: The opposition boycotts the first parliamentary elections. People's Front "wins" 88.14% of the total number of votes.

1946

March 12: The Parliament passes the Agrarian Reform Act. Nationalization of all plots of arable land and estates bigger than 20 hectares (in South Dobrudja bigger than 30 hectares).

March 31: The second cabinet of Kimon Georgiev includes 5 communist ministers in key positions.

September 8: Referendum for Republic. According to official data 91.63% of the total number of registered voters cast their vote, out of which voters 95.63% voted for Republic.

September 15: Abolishing of the Monarchy. Vasil Kolarov becomes the first communist president.

October 27: Elections for Grand National Assembly. The communist-dominated coalition "People's Front" wins 70% of the votes.

November 23: The first cabinet with communist Prime Minister (Georgi Dimitrov).

1947

February 1–2: First National Conference of WCAF. Georgi Dimitrov declares that this form has no alternative.

February 10: The Peace Treaty between Bulgaria and the anti-Hitlerist coalition is signed in Paris.

March 6: Currency Reform. The state confiscates all private bank deposits.

August 16: Death sentence for Nikola Petkov. The oppositional Agrarian Union prohibited.

September 22: Execution of Nikola Petkov.

September 22–28: Bulgarian communist delegation takes part in establishing of the Cominform Bureau in Poland.

December 4: The Grand National Assembly approves the new "Dimitrov's" constitution, where WCAF becomes institutionalized as basic economic unit.

December 11: Second cabinet of Georgi Dimitrov, absolute domination of Bulgarian Workers' Party (communists).

December 23: Law for the nationalization of industry.

December 26: Law for the nationalization of banks.

1948

February 18: Law for compulsory buying-up by the state of the private "big farm equipment" (machinery etc.).

March 6: Law for the nationalization of forests.

July 12–13: Sixteenth plenum of the Central Committee of BRP (c.) marks the end of the agrarian reform implemented since 1946 and approves collectivization by force. By that moment collectivization encompasses 7% of the land and 8% of the peasants.

December 18–25: The Fifth congress of BRP. The party is renamed into Bulgarian Communist Party (BCP). The congress accepts the "Program for building up of socialism in Bulgaria."

1949

January 8–9: Bulgaria becomes co-founder of COMECON.

February 18: The First Five-Year Plan (1949–1953).

May 15: Only People's Front candidates are enlisted for the first local elections after World War II. According to the official data, they obtain 92.1% of the votes.

June 10–12: Following Josef Stalin's "advice," the June Plenum of the BCP decides to slow down the forced collectivization.

June: The opening of the largest labor camp in Bulgaria (on the Danubian island of Belene/Persin).

July 2: The death of Georgi Dimitrov, inherited by Vassil Kolarov.

December 17: Execution of Trajcho Kostov, the second figure in the party hierarchy, blamed for espionage for the British and the Americans.

1950

January: The beginning of a Turkish minority emigration wave on large scale, halted in the end of 1951. 155,000 Bulgarian Turks emigrate to Turkey and their lands are nationalized, becoming part of WCAF and the State Agricultural Farms (SAF).

January 23: The death of Vassil Kolarov, inherited by Vălko Tchervenkov.

April 4–5: Second National Conference of WCAF. According to the changed regulations, all members had to submit their entire land, livestock and farming equipment to the cooperative farms.

October 7–8: The October Plenum of the BCP declares a "turning point" in the process of collectivization (since more than 50% of the farmers and the arable land already included in WCAF).

October 11: Decree of the government and the Central Committee of the BCP for the "development and stabilization of WCAF," entitled "Course for collectivization on mass scale." Obligatory state deliveries take roughly two-thirds of the yields. A selective coupon system excludes the so-called *kulaks* (the term itself borrowed from the Soviet propaganda).

November: Decision on the "Experimental collectivization on a mass scale" in Vratsa region (northwest Bulgaria).

1951

January: Growing psychological and physical terror against the *kulaks*.

February 1: The new "Loan for the development of the people's economy" aiming at the extraction of cash from the population and the creation of state funds for the emerging heavy industry.

March: Peasant riots on large scale against collectivization in Vidin and Kula regions (northwest Bulgaria), Assenovgrad and Plovdiv regions (south Bulgaria).

March 12: The March Plenum of the BCP condemns "the deformations and distortions" during the collectivization in Kula district and Jablanitsa village, Teteven district.

March 20: Partial abolition of the coupon system.

April 11: Re-settlement of peasants from the most troubled regions. Some of them are from the northwest region, bordering Yugoslavia, others from villages and small towns in the interior, where protests and in some cases mass-scale revolts took place (more than 25,000 deported).

June 18: The agrarian minister Titko Chernokolev, accused of "deformations and distortions" during the collectivization, is fired and arrested. After Josef Stalin's intervention the investigation is stopped.

November 20: "Law for the abolition of all the laws adopted before 9 September 1944."

1952

May 12: Second currency reform aiming at complete extraction of cash resources by the state.

1953

January 1: Beginning of the Second Five-Year Plan (1953–1957).

August 30: Decree of the government and the Central Committee of the BCP for reduction of the prices and abolishment of the coupon system.

December 1–3: Third National Conference of WCAF finally abolishes the monetary rent for members. WCAF becomes a *kolkhoz*-like form.

1954

February 25–March 2: The Sixth Congress of the BCP. Following the Moscow course, Vậlko Tchervenkov is forced to hold only the post of Prime Minister. Todor Zhivkov becomes party leader.

1955

May 14: Bulgaria becomes co-founder of the Warsaw pact.

1956

February: Following a recommendation from Moscow the beginning of collectivization is announced also for the mountainous and semi-mountainous regions. This marks the beginning of the last wave of collectivization. Up to that moment WCAF includes approximately 60% of the arable land and 66% of all farmers.

April 2–6: The April Plenum of the BCP pronounces course for de-Stalinization. Vậlko Tchervenkov loses power and premiership, succeeded by Anton Yugov. Todor Zhivkov remains the most powerful figure.

December 30: According to a new law, the members of the WCAF begin to receive pensions for the first time.

1957

July 7: Dobrudja becomes the first region in Bulgaria with 100% collectivization of the arable land.

December 2–4: Fifth National Conference of WCAF declares 86.5% collectivization of the arable land in Bulgaria.

1958

January: The Third Five-Year Plan (1958–1962).

June 2–7: The Seventh Congress of the BCP. Todor Zhivkov declares that Bulgaria is already the second country in the world where socialism is victorious in the countryside.

October: Beginning of the consolidation of the existing WCAF in bigger structures (until the beginning of 1960).

1959

April 29: Abolishment of the obligatory state deliveries.

December: Collectivization in the high-mountain villages accomplished. According to the official data, 99% of the arable land is owned by the state and by the WCAF.

"Any Other Road Leads Only to the Restoration of Capitalism in the Countryside:" Land Collectivization in Albania[1]

ÖRJAN SJÖBERG

"Now look at what Albania is doing.
Albania is copying what we had done precisely,
only with greater, Asiatic brutality."[2]

After splitting with Moscow in 1961, Albania's communist party, the Albanian Party of Labor (APL),[3] proclaimed that its model of socialism was unique to the Eastern Bloc. This was indeed true, at least with respect to how rural areas and agricultural production were organized. Although collectivization in its earliest phases mirrored campaigns in other Soviet satellites,[4] the Albanian program soon took on an innovative character—or so the party claimed—which was expected to developed into a full-fledged state farming system before long. Only then, the official line suggested, socialism would be firmly established in the countryside. At this point, or so propaganda had it, cooperative forms of organization were replaced by true ownership and management by the people. The party's claim that Albania's path was unique warrants further scrutiny, in particular since it was based on ideologically charged propaganda. As we shall see, it was not exceptional, but it does contain features that can be considered sin-

---

1 I owe the quote which is part of the title to the work of Peter Prifti. It originates in a speech by Enver Hoxha at the Fifth Congress of the Albanian Party of Labor held in 1966. For context and bibliographic details, please refer to the penultimate section of this article and note 72 below, respectively.

2 "Speech of N. S. Khrushchev," 1962.

3 The Albanian Communist Party changed its name to the Albanian Party of Labor in 1948; for practical reasons, the latter name is used throughout this article.

4 Wädekin, *Agrarian Policies*; Sjöberg, *Rural Change*, chapter 4.

gular. For now, suffice it to note that Albania did consciously try to shape its own collectivization policies and solutions with a view to achieving the original aim as set at the very beginning of communist rule. Indeed, the fact that Albania held on to an orthodox line, also in face of an unyielding reality, helps explain some strange and also seemingly unrelated political events. Thus, the stubbornness with which Enver Hoxha pursued his goals has a bearing on the last major—and fatal—battle within the APL itself, the main victim of which was Hoxha's long time ally Mehmet Shehu. This event, which took place in late 1981, in turn marked amongst other things the beginning of yet another attempt to extend the reach of collectivized agriculture. As such it was the final wave of collectivization during the Cold War, but the consequences were little different from the experiences of the 1950s: retail markets for rural produce were outlawed, animals were slaughtered, the supply of eggs, milk, and meat dried up and rationing was reintroduced once again.

As for traits it shares with other communist states, Albania carried out land reforms before initiating collectivization. As in many Eastern European countries, collectivization proved a lengthy process that seldom benefitted the rural poor. The process also hinged on class warfare, which pitted smaller farmers against (at least nominally) wealthier *kulaks*. Once collectivization was achieved—which Albania did in a step-by-step fashion over several decades—there remained the problem of ensuring that collectives could deliver the food needed for a rapidly growing population, along with raw materials for processing and manufacturing industries, which at least in Albania's case were controlled by an increasingly stringent autarkic state planning model.

This contribution addresses the protracted process of collectivization in Albania, emphasizing developments during the second half of the 1940s and during the 1950s, but also following the process through up to the 1980s. The chapter builds on the established chronology of events, noting the measures taken by the APL to collectivize agriculture and to subdue rural areas in order to carry out its political and economic imperatives. Rather than relying on firsthand archival research or rehearsing established foreign secondary literature only, the aim here is to fill in some gaps about our understanding of the collectivization process by using the relatively limited post-socialist Albanian research. It is yet another step in our endeavor to understand the nature and role of collectivization in Albanian efforts to build socialism.

To appreciate why this might be of interest, it is useful to consider the current state of research on agrarian development in Albania during the communist period. Foreign researchers have typically not availed themselves of the opportunities that have been presented as a result of the collapse of communism two decades ago, at least not as far as rural change and agricultural development, including collectivization, is concerned.[5] The main exception is rural fieldwork focusing on the post-socialist period but also touching upon developments prior to the demise of the previous political system.[6] Albanian research, for its part, is also unexpectedly thin and surprisingly often hostage to the ideas, language, and tenor of research published prior to 1990–91. The main economic history covering the period of particular relevance here (e.g., 1940s and 1950s) is a case in point, adding new material and newly written segments to a text that is otherwise not all that different from work published by the senior author and colleagues during the 1970s and 1980s.[7] One does not, therefore, necessarily have to share the rather disillusioned view with respect to the quality and quantity of Albanian historical scholarship expressed by some respected historians of Albania to appreciate the problems at hand.[8] A number of major histories of modern Albania are of some use,[9] however, as are a small but growing number of articles—sometimes newspaper articles—that address critical events of relevance and attempt to understand why collectivization in Albania took the form and path it did. This includes more detailed studies done by the authors of the fourth and final volume of the history of Albania launched by the

[5] Tellingly, the final volume on modern Albanian history compiled by Pearson, *Albania in the 20th Century*, also contains rather few entries of relevance to the subject matter of this article.

[6] Examples of which include Kaser, Pichler, and Schwandner-Sievers, eds., *Die weite Welt*; de Waal, *Albania Today*; and Stahl, *Rent from the Land*.

[7] Fishta and Ziu, *Historia e ekonomisë*; Fishta and Toçi, *Ekonomia e Shqiërisë*. The former not only draws to a considerable degree on historical research conducted during the socialist period (including genuine archival research), but at times it also adopts a rather uncritical stance relative the then-mandated guidelines. It should be approached accordingly.

[8] Schmitt, "Vorwort;" Schmidt-Neke, "Zwischen Kaltem."

[9] Smirnova, *Historia e Shqipërisë*; Duka, *Histori e Shqipërisë*. In addition to such works, a collection of Russian archival documents translated into Albanian contains the occasional discussion on matters agricultural, the main virtue of which is to give a contemporary view of events as they took place in the late 40s and early 50s rather than from hindsight. See Lauka and Ymeri, eds., *Shqipëria në dokumentet.*

Academy of Sciences in the late 1990s.[10] These will be used here to fill in a few of the many issues that previous writings have left unaccounted for or have been forced to gloss over.

The Setting: Unique or Simply Stalinist?

In communist-era historiography, 1960 marked the year when, after a transitional period during which the foundations of socialism had been laid, socialism itself was firmly established. In agriculture, however, this was achieved only in part. During the 15 years following the communist takeover in November 1944, Albania followed the general pattern of land reform and collectivization policies that can be observed throughout the region. The main difference was that in Albania all agricultural land was (eventually) fully expropriated and no private ownership was permitted. In most other respects, including the deadline to collectivize agriculture no later than at the end of the Second Five-Year Plan (1960), Albanian collectivization appeared to be very similar to the experiences elsewhere.

However, although the drive to complete collectivization was in progress during the second half of the 1950s, this applied only to the lowland and highland basins, such as the Plain of Korçë in the south east, and not the mountain areas. Only in 1967 was collectivization in the highlands resumed, and ostensibly achieved. Furthermore, steps were taken to reduce the size and role of household plots, and in the 1970s, a new type of cooperative was introduced as a means for cooperatives to transition into state farms. Finally, as noted above, the last collectivization drive was launched in the early 1980s, a strategy with disastrous consequences and reversed in part only after the death of Enver Hoxha in April 1985.

Albania's radical collectivization program in some ways reflects the particular post-World War II political dynamics within the Socialist Bloc. As in Yugoslavia, where the role of Soviet military forces in liberating the country at the end of the war was limited,[11] Albanian leaders may have

---

[10] Although similar to a work published twenty years previously, Gjeçovi, ed., *Historia e popullit shqiptar*, does, unlike its predecessor, address the darker sides of communist Albania. The level of detail offered, however, is understandably not very different from other general histories that have been published during the past decade or so. The articles are a different matter, but they will be detailed as we move along.

[11] Swain and Swain, *Eastern Europe*, 8.

had greater room for maneuver. Yet, when it did not force particular strategies or solutions upon Albania, Yugoslavia figured prominently in intra-party debates and struggles. The Yugoslav card was played out first against those of a more liberal or nationalist persuasion and subsequently against those who preferred to lean against the Soviet Union. In fact, it appears that several members of the Politburo of the APL tried to make the Soviet Union interested in Albania, but were told that the Soviet Union would manage its affairs with Albania with Yugoslavia acting as an intermediary.[12] In 1948, as the faction siding with the Soviet Union triumphed, Tirana broke off relations with Belgrade. Because Tito's break with Stalin in part was a reaction against Moscow's agricultural polices, it stands to reason that Albania continued to favor the Moscow line.

Given the events that followed, perhaps it is more appropriate to say that post-1948 Albania remained true to Stalin rather than to the Soviet Union. As has long been held by outside observers, and more recently confirmed by Albanian scholars with access to party archives,[13] the new course adopted by party leaders throughout much of Central and Eastern Europe following the death of Stalin in 1953—and in particular the events unfolding in 1956—proved very unsettling to the communist factions that emerged victorious in 1948. As was also the case in some other countries of the communist camp,[14] Moscow's directives did not succeed in bringing about any liberalization. Instead, Soviet attempts to restore relations with Yugoslavia only made Albania more reluctant to follow the lead of the post-Stalin leadership in the Soviet Union. Thus, whether for ideological reasons or out of sheer opportunism, Hoxha out-maneuvered those in favor of staying close to Moscow.[15] With the party's final split with Moscow still a few years ahead, Hoxha turned to increasingly nationalist political rhetoric and autarkic policies following the demotion and purges

---

12 Lalaj, "The Soviet-Yugoslav," 123–27.

13 Among others see Dyrmishi, "Konferenca;" Lalaj, "1956." Foreign accounts picking up this theme include Pano, *People's Republic*, chapter 4; Pipa, "Party ideology."

14 Lankov, *From Stalin to Kim Il Sung*, 165.

15 It has frequently been argued that Hoxha throughout his career assumed positions that primarily served the end of retaining power, for himself or for his family (or the Hoxha-Shehu clan). There are also quite a few telling examples that would seem to bear this out. See Duka, *Histori e Shqipërisë*, 270; Blumi, "Politics of Culture."

of those in favor of de-Stalinization.[16] Khrushchev's visit to Albania in 1959 did little to alleviate the strains. The pronouncements made by the Soviet leader to the effect that Albania should focus its efforts on Mediterranean agricultural produce, natural resources and simple manufactured goods only served to strengthen the belief of the Albanian leadership that Khrushchev was not to be trusted.[17]

This is not to suggest that the APL's claims of a unique socialist path for Albania were entirely unfounded. As in other communist states, Albania's system took on traits unique to its particular social and political history and landscape. Some of these unique features deserve to be noted. Similarly, the literature on rural change and development, including that devoted to collectivization, have identified a number of traits that are typical of (but not necessarily unique to) Albania. Combined, these include the following.[18]

Despite the relatively modest size of the country—at 28,748 km² it is slightly smaller than Belgium—conditions for agriculture vary enormously across the territory of Albania. It is therefore difficult to design or implement policies without taking this diversity into consideration. This proved particularly so as a rapidly growing population had to be accommodated; the number of inhabitants grew from about 1.1 to 3.2 million during half a century of communist rule. At the early stages of this rapid population growth (1950s), which was predominantly rural in origin, large scale population relocation (typically from the interior to the western lowlands) took place. In part this was a result of Soviet-style industrialization providing job openings in urban or resource rich areas, but domestic migration was also linked to adverse conditions in the countryside. Overall, it implied a considerable flow of people leaving the mountainous interior in favor of the western lowlands where both urban and rural areas served as destination.[19]

---

16 Schnytzer, *Stalinist Economic*, 70–71. Here it is also recognized that this would become more pronounced as Albania shifted its allegiance to China. It must also be noted, however, that throughout the 1950s Albania remained critically dependent on aid from the Soviet Union and others.

17 A sample statement can be found in Pearson, *Albania in the 20th Century*, 556–57.

18 Sjöberg, *Rural Change*.

19 Not only was the population predominantly rural at the time when the communists assumed power (about 78%), nativity was notably higher in rural areas as well, with the great majority of the rural population engaged in smallholder agriculture or as tenants and agricultural laborers. While the natural rate of population

This in turn was followed, in the early 1960s, by a relatively strict control of domestic mobility and urbanization which of course only served to increase the population pressure in rural areas in general and, relative to the resource base, in the mountains in particular. In part this was a policy choice linked to an increased emphasis on autarky, both nationally and at the district level. The unfavorable outcome of such policies put pressure on the party and government agencies to find solutions to problems ranging from adverse incentives in agricultural production to a severe lack of funds for investment to convincing the urban population that agriculture was doing well despite the reintroduction of rationing in the 1980s.

Some of these features do have a bearing on the manner in which collectivization was implemented, but other issues stem from collectivization itself. None, however, warrants the conclusion that the experience of Albania with respect to land reform and collectivization as such was unique. This will become apparent as we now move on the particulars of agrarian reform instituted by the APL.

LAND REFORM

Agrarian reform was one of the first measures taken by the APL after establishing itself at the end of World War II. With the majority of the population living in rural areas and engaged in agriculture, this is not surprising. This is especially so since the land reform of the 1930s did not prove radically decisive in changing the livelihoods or social structure in the countryside.[20] Instead, a small number of large landowners dominated, but it bears pointing out that many rural dwellers owned or had access to land of their own or that was controlled by their kin. As

growth, the share of rural dwellers, and the contribution of rural nativity to the national total all declined during the period up to the collapse of communism, at the time of the 1989 census the rural population still made up 64.5% of the total, and the natural growth rate remained higher in rural areas compared to the towns. See for example Schmidt-Neke and Sjöberg, "Bevölkerungsstruktur."

[20] Before the war, several attempts at land reform were made. It was finally implemented in the 1930s, but most observers have drawn the conclusion that it was not at all a success and that it fell seriously short of its own not very demanding objectives; for a discussion, see Sjöberg, *Rural Change*, 29–33. It should be noted that politicians staunchly opposed to Communist Party economic policy were in favor of agrarian reform; see Gjeçovi, "Alternativat e zhvillimit."

an American historian noted at the time when Albania emerged from Turkish rule,[21]

> The basis of land ownership is the freehold, and the class of independent yeomen is very numerous. Most of the land belongs, however, to the great landowners, Beys and Pashas, who have received it as fiefs from the Sultan, especially for meritorious services rendered in war and peace. The land is leased by them to the peasantry, in the form of perpetual leases. The right of eviction has become obsolete, but the exactions of the landowners have proved disastrous to any agricultural development. Usually, the lessee is required to turn over to the landlord one-third of the produce; the result is that the peasant does not care to grow more crops, when he knows that his landlord will become richer and greedier thereby.

This picture is by and large confirmed by the work leading up to agrarian reform in the 1930s, the main architect of which suggested that 165 families or so controlled somewhere between one-third to two-thirds of the total, with the state also being a major landowner.[22] There is no doubt, therefore, that the distribution of land was very unequal across different segments of the rural population. Thus, starting in early 1945, land was registered, and by August of that year a new law on agrarian reform had been passed. Farm land greater than 20 hectares was to be confiscated, unless it was actively cultivated by modern means, in which case twice that area could be retained by the original owner. If the owner had non-agricultural sources of income, all agricultural land could be expropriated and redistributed to state farms without compensation. Landless or land poor households received 5 hectares of land, with additional land allocated to households exceeding a certain size. Enver Hoxha described this as the first revolution in socioeconomic relations of the countryside, with the slogan "*toka i takon atij që e punon*," or "land to the tiller."[23]

---

[21] Hazen, *Albania*, 174.
[22] Lorenzoni, *La questione agraria*, 15–16.
[23] Duka, *Histori e Shqipërisë*, 249. Also Fishta and Toçi, *Ekonomia e Shqiërisë*, 195.

The land reform program began in November 1945, and within a few months, by February 1946, party leaders decided its efforts were not radical enough. By May, it passed new legislation expropriating all agricultural land, olive groves, vineyards and fruit trees, and the upper limit of land allocated now set at 5 hectares.[24] Forest land and pastures, a critical resource throughout much of the country, were nationalized and the right to lease or sell land was withdrawn.

Standard Albanian socialist-era accounts tend to blur the measures taken as part of the agrarian reform and the early stages of collectivization. This is particularly so with respect to data on land that was expropriated and redistributed. In addition to the land confiscated from the largest land owners (or the *latifondiste*, of which there were seven, on average, owning about 2,000 ha each), data also lists land taken from 4,173 rich owners (who together controlled slightly in more than 91,000 hectares) that was redistributed to agricultural laborers, poor and middle peasants.[25] But in fact, four times as many owners were affected by partial or complete land seizures, a fact proudly acknowledged by party officials at the time. The number of rich peasants was a figure created as a result of the first wave of collectivization, which we will explore in more depth.

[24] Fishta and Toçi, *Ekonomia e Shqiërisë*, 199–200. From the point of view of establishing the status of property rights following the agrarian reform of 1946, it should be noted that it is not self-evident that all land was in fact nationalized. Existing legislation can be interpreted such that those who received land under the reform also received ownership (as opposed to usufruct) rights; see Durand, *Socialisme e développement*, 71–72. On the other hand, Durand also notes that access to land further implied an obligation to cultivate it. That peasants *de jure*, if not *de facto*, retained their status as owners is also pointed out by Fishta and Ziu, *Historia e ekonomisë*, 182. It was only later that public ownership of all land irrespective of type, use, or previous owner was formally nationalized, being firmly and finally confirmed by the Constitution of 1976, Art. 19. However, as also noted by Fishta and Ziu, to all intents and purposes by 1946 private land ownership as normally understood was a thing of the past.

[25] The data is available in *Anuari i statistikor*, 1962, 123, and often reproduced from this source, a prominent example of which is Fishta and Toçi, *Ekonomia e Shqiërisë*, 184–85. Note, however, that officially a larger number of owners still were affected by the provisions of the reform. Thus, more than 19,300 owners lost land while another 70,200 benefited from the redistribution. These numbers can be found in official pronouncements of the time, but are also reproduced with confidence (or uncritically?) in post-socialist scholarship, e.g., Fishta and Ziu, *Historia e ekonomisë*, 181–82.

EARLY COLLECTIVIZATION AND THE "ELIMINATION OF THE *KULAKS* AS A CLASS"

Land reform was seen as a useful vehicle for establishing, or at least reinforcing, support for the new regime in the countryside. Yet low productivity of newly created farms (and, subsequently, of the early collectives) was a continuing crisis for authorities,[26] as were the delivery quotas for individual farmers. In 1946, farmers were allowed to retain 250kg of grain per person, and any amount exceeding that was to be delivered to the state. Following the 1946 harvest crisis, the amount producers where allowed to retain was dramatically reduced.[27] Hence, many of the problems typically associated with the launching of collectivization—ranging from the slaughtering of animals[28] to a lack of incentives to produce at capacity—had already emerged prior to the collectivization drive.

In recognition of these problems, party and government set its sights on creating cooperatives. It has not been possible to establish to what extent this policy, and the details and timing of its implementation, was dependent on directives from Belgrade or Moscow. However, the fact remains that systematic collectivization followed the pattern of the other satellite states (aside from Yugoslavia) and the Cominform directive issued in June 1948.[29] But even prior to this, Albania had been dependent on foreign models, which may account for its continuing trend of following Soviet agrarian policy. After all, as Melissa Bokovoy notes in her research on Yugoslavia, the many "sins of the KPJ" included "an incorrect ideological and political line in the countryside and allowing 'capitalist elements' to flourish there,"[30] and Yugoslavia did serve as a role model for Albania up to this point. This was also true as far as policies on collectivization were concerned.[31]

---

[26] Kaser, "Economic continuities," 628.

[27] The party blamed this on the weather, but more likely, this was due to the lack of incentives and disorganization following land reform itself.

[28] Starting in 1946, peasants needed permission to slaughter their own stock, a regulation that presumably would not have been necessary if peasants willingly let go of their animals.

[29] Fishta and Toçi, *Ekonomia e Shqiërisë*, 284–301. This source provides a more detailed, yet no doubt officially endorsed, chronology of events and account of progress made, as opposed to general socialist-era works on the contemporary history of Albania, such as *Historia e Shipërisë*, 1983.

[30] Bokovoy, *Peasants and Communists*, 84–85.

[31] Fishta and Toçi, *Ekonomia e Shqiërisë*, 294.

In September 1948, after the break with Yugoslavia and after the Cominform directive had been issued, the party drew up a new model for cooperative farms.[32] A standard charter for cooperatives had been introduced already in 1946, which stipulated, for instance, that members should be enumerated based on the labor (60 percent) and land contributed (40 percent) and that they were allowed a household plot as big as 1–1.2 hectares.[33] This followed the original Yugoslav idea of a peasant work cooperative rather than Yugoslavia's later agricultural cooperative form, which the Cominform opposed.[34] Yet, given that Albania followed Yugoslavia's model at least up to late February 1948,[35] it should not come as a surprise that Albania settled on a more liberal collective form, as did the Yugoslavs. In 1948, however, the party adopted a version closely modeled on the Soviet *artel'* of 1935. This version was eventually replaced in 1956, although it took some time before it was universally adopted.[36]

A small number of cooperative farms had been created prior to these developments,[37] which lured members with lower quotas and taxes compared to individual farms. Although socialist era historiography suggests that these benefits did achieve their intended aim,[38] the productivity of cooperative farms was low, or at least lower than had been hoped for, and so were incomes. By 1950, more the 90 percent of the agricultural output originated outside the state and cooperative sectors, at which point there were an estimated 90 cooperatives with 4,500 member families controlling a total of about 20,000 hectares of land, including members' household plots.[39]

A key challenge was providing incentives to peasants in order to make collectives more attractive than private farming. The preceding

32 Fishta and Ziu, *Historia e ekonomisë*, 225 and 340, respectively.

33 Fishta and Ziu, *Historia e ekonomisë*, 228–29, detail these provisions, arguing that they followed Yugoslav models.

34 Bokovoy, *Peasants and Communists*, 46–48.

35 Duka, *Histori e Shqipërisë*, 239–43; Smirnova, *Historia e Shqipërisë*, 326–44.

36 For a summary see Sjöberg, *Rural Change*, 86–89; for a systematic comparison of cooperative statues see Wädekin, *Agrarian Policies*, 69–89.

37 On which, see Fisha and Ziu, *Historia e ekonomisë*, 223–26.

38 Fishta and Toçi, *Ekonomia e Shqiërisë*, 293.

39 Pano, *People's Republic*, 98. Official socialist-era statistics on the number of cooperatives, the number of members, and overall size are reproduced in Sjöberg, *Rural Change*, 88.

land reform had dismantled the largest estates of "the seven families,"[40] and redistributed the land; by 1947, the obvious "class enemy" had been driven from the countryside.[41] Indeed, many had been stripped of their assets, but in reality there was no clear division between "us" and "them," that is between large and small peasants.[42] Rather, following the agrarian reform, a large number of villagers gained land that they were not eager to give up to cooperatives, even with less heavy tax burdens and quotas.

The new farming structure suffered from low productivity—and became what historian Valentina Duka calls the "Achilles heel of the Albanian economy"[43]—which added to the urgency of introducing effective inducements. But instead of reducing the burden on cooperatives by taking measures to support the private sector, the party's strategy instead put increased pressure on individual farmers—despite later claims that positive measures were taken to support the private sector in an attempt to convincing the peasantry they were responsive to the sector's needs.[44] Some provisions applied equally to collective and private farmers, such as the penalties imposed on grain with high moisture contents. Others were first applied to the non-cooperative sector but then were extended to the cooperatives as well, as was the case with quotas. Rather than allowing peasants to retain a set amount of grain per capita, an absolute amount to be delivered was calculated—and this was calculated on the basis of some general capacity of the land without any regard for whether this was achievable or not.[45] At the same time, in 1947, the state brought most industrial users of agricultural produce under its control, while six region-

---

40 It should be noted that this did not necessarily imply that the land controlled by these families was enormous or that large amounts from these estates where redistributed; at about 11,000 ha it obviously was not. Land was also made available by confiscating holdings of "enemy elements and war criminals," while *çiflik* land under the control of the state along with major commercial farms previously owned by foreigners often was reorganized as state farms, the former tenants being turned into agricultural laborers employed by the state. See Sjöberg, *Rural Change*, 84–86; on the establishment of state farms, Fishta and Toçi, *Ekonomia e Shqiërisë*, 277–84, or Fishta and Ziu, *Historia e ekonomisë*, 219–22, 334–39.

41 Dyrmishi, "Cilësimi," 107.

42 Lange, *Die Agrarfrage*, 68. Lange contends that following the post-war land reform, it was hardly possible to identify any strata of rich peasants. See also, de Waal, *Albania Today*, 75–77.

43 Duka, *Histori e Shqipërisë*, 279.

44 Fishta and Toçi, *Ekonomia e Shqiërisë*, 303.

45 Sjöberg, *Rural Change*, 86.

ally based procurement companies were set up to act as intermediaries between agricultural producers and the food processing industry.[46] In effect, and as a parallel case to the machine and tractor stations that were also being set up starting in the late 1940s, by virtue of its monopsonist and monopolist powers this created a system that allowed for the extraction of resources from the countryside should the authorities so desire. Conceivably, this could be done for instance through the manipulation of the terms of trade between the agricultural sector and other parts of the economy.[47]

In January 1949, the Central Committee introduced a new rationing system referred to as the *tregu reciprok* (the reciprocal market). This was made up of three different parts: the first consisted of (nominally guaranteed) deliveries under a formal rationing system; the other two parts were reciprocal trade proper, under which exchanges between producers and consumers were conducted, and with the free market system, without any price controls. Thanks to these measures, the system supposedly allowed for lower prices than did the private market alone. This was so, or so it was claimed, because farmers were able to procure consumer goods and inputs at favorable prices in exchange for giving up (some of) their produce at similarly low compensation.[48] The system was said to guarantee deliveries of food to workers and essential products for cooperativists and small farmers, and also to strengthen the amicable relations between the city and countryside. In practice, the system did not perform particularly well and shortages of food and other consumer goods persisted. However, instead of adjusting its policies by improving incentives for farmers, the party in 1950 extended the system of compulsory deliveries and its control over retail and consumer services. Retail outlets were increasingly run by the state through a system of state trading companies, some of which were charged with supplying specific mining settlements or major industrial enterprises. Similarly, by 1952 the system of recip-

---

46 Hajdëri, *Industria ushqimore*, 122. The Albanian term for such an enterprise is *ndërmarrja e grumbullimit.*

47 In addition to the more general treatment offered by Schnytzer, *Stalinist Economic*, see Wildermuth, *Sich stützen auf*, 104–17. The latter source provides a useful discussion on the role of agriculture in providing the financial wherewithal for society as a whole. Thus, throughout the socialist period, direct and indirect taxation (including procurement prices biased against collective farms) played an important role in financing amongst other things industrial expansion.

48 Fishta and Ziu, *Historia e ekonomisë*, 395–400.

rocal deliveries was also extended, while at the same time the private retail sector was put under increasing pressure by increasing taxes and dwindling supplies.[49]

In March 1949, the APL set sight on eliminating the *elementi kulak* in the countryside. *Kulaks*—the enemies of the party, the people and fatherland—had of course been targeted earlier, but now that term extended to a broader strata of rich peasants. As historian Demir Dyrmishi notes, the actual term *kulak* had not been used previously, and now this group of villagers was equated with a capitalist plague in the countryside.[50] Authorities put considerable effort into defining criteria by which the *kulaks* could be recognized. This included employing obvious economic criteria such as monetary wealth but also the use of non-family labor, availability of draught animals, access to modern agricultural implements and inputs, and off-farm sales.

Once identified, the *kulaks* were subjected to various restrictions and discriminatory polices, including higher taxes and fees or barring them from using various services. For instance, quotas for grain, meat, wool, and industrial crops were higher for rich peasants, and they were denied access to the services of machine and tractor stations, and veterinary and extension services, where available, were open to them at higher prices. Similarly, *kulaks* had to pay to use pastures, and credits were unavailable to them. Those who did not fulfill their obligations toward the state could be confined to labor camps for up to five years, and their assets could also be confiscated.[51]

Similar to anti-*kulak* policies pursued in other socialist countries, the "class struggle in the countryside," and the effort to eliminate *kulaks* as a class became a priority for the APL. The first step was for local party members, youth, and state organizations to compile a list of those considered to be *kulaks*. In addition to the economic ones listed above, the standards applied also included political criteria, including an individual's

[49] Fishta and Ziu, *Historia e ekonomisë*, 400–11.

[50] Dyrmishi, "Cilësimi," 109.

[51] In addition to Dyrmishi, "Cilësimi," who discusses these and other provisions at some length, some of the decrees and (secret) directives have recently been made available by Elmas Leci in a series of articles in the daily *Shekulli*. See "Urdhri i Enver Hoxhës," 2008; "Enver Hoxha," 2008; "Enver ngarkon Mehmetin," 2008; "Kolektivizimi 1967," 2008; and "1967. Urdhri i Enver Hoxhës," 2008.

stance on the regime's policies, along with social standing and political activities prior to, during, and after the Second World War.

These criteria made it possible to categorize a large spectrum of individuals as *kulaks*, even those who were not particularly wealthy. But the class struggle in the countryside was primarily a political one, albeit one with pronounced economic implications. However, the extent to which this struggle turned violent is difficult to establish. There are quite a few instances in the literature, not least in the one issued during the socialist period, which suggest that the policies imposed on the countryside generated violent reactions.[52] Although the number of people sentenced to prison, forced labor camps, or into domestic exile was large,[53] how many of these were classified as *kulaks* is not possible to establish.

By the end of 1953, 2,877 households had been identified as being *kulaks*, up from 2,438 in 1952 and from 1,930 in 1951. When these families were not "liquidated economically," that is, put out of business, or treated as criminals, they were allowed to retain, in addition to the statutory amount of land, one cow and half a dozen small ruminants.[54] Once classified as a *kulak* there was little recourse for families. Although it was possible to appeal, most who did, did so without success. Only toward the end of the first collectivization drive in 1959, *kulaks* deemed by authorities as not posing a threat could be rehabilitated and allowed to join the cooperatives, which by then dominated the rural economy. According to Dyrmishi, the party's anti-*kulak* policies had created "one of the deepest wounds of the Albanian countryside."[55]

[52] *Historia e Shipërisë, Vëll. IV*, 1983; Shilego, *Lufta e klasave*. See also Prifti, *Socialist Albania*, 66–67.

[53] Sufaj, *Historia e burgjeve*, 106, 125. Sufaj notes that in January 1947, out of a total of 4,749 prisoners, 3,659 were sentenced on political grounds, with another 1,272 individuals held in labor camps for the same reason. Over time the proportions changed, and ten years later about two-fifths of all those held in prisons were classified as political prisoners rather than "ordinary" ones. The total number of prisoners (excluding the labor camps) had by then increased by about one-fifth. None of these figures include those sent into internal exile, a fate shared by many whose only crime had been their relative wealth.

[54] Dyrmishi, "Cilësimi," 113–14.

[55] Dyrmishi, "Cilësimi," 121.

The Drive to Complete Collectivization

The party's strategy of pitting different groups of villagers against each other did not initially increase cooperative membership. At the end of the First Five-Year Plan (1951–55), individual peasant farms still accounted for 87 percent of the cultivated area and 90 percent of agricultural output in the non-state agricultural sector. Cooperatives consisted of a mere 15,000 households, just a tenth of the total number of farm households. It was only at the end of the Second Five-Year Plan (1956–60) that these figures were reversed, with the cooperative sector by 1960 accounting for 84 percent of the area and 76 percent of production. As the cooperatives accounted for 70 percent of all deliveries to the state, we may draw the conclusion that the private sector was both more productive and more heavily taxed.[56]

In October 1958, *kulaks* who had managed to survive as independent farmers were subjected to confiscatory tax rates.[57] Lowland farmers were likely to have found it more expedient or necessary to join the cooperatives, and by 1960 the number of families registered as members reached about 115,000, an almost eightfold increase over the previous five years. Despite this increase, Albanian researchers have shown there was a grave disappointment in the leadership with the results of the First Five-Year Plan.[58] Growth rates were dismal and failed to meet established targets. In response, already in 1953 the APL introduced measures to redress the balance in favor of cooperative agriculture.[59] Yet, rather than reducing the emphasis on industrialization—which Enver Hoxha vehemently opposed[60]—or increasing investments in agriculture, the APL chose instead to rapidly expand its collectivization campaign. As the Second Five-Year Plan was introduced, additional efforts were made to increase the amount of land allocated to cooperatives, and a new model statute was

---

[56] Pano, *People's Republic*, 98; Sjöberg, *Rural Change*, 86, 89. The latter quotes evidence to the effect that quotas were 10 percent lower in cooperatives compared to the private sector, while tax rates were half of those outside cooperatives.

[57] Dyrmishi, "Cilësimi," 114.

[58] Fishta and Ziu, *Historia e ekonomisë*, 486.

[59] Schnytzer, *Stalinist Economic*, 68. It is unclear if Soviet pressure was a factor, but Albania would have been forced to address these circumstances nonetheless. More lenient terms, e.g., in procurement, were part of the attempts to do something about the lackluster performance of the collectivized sector.

[60] Schnytzer, *Stalinist Economic*.

introduced. The new charter imposed more stringent limits on household plots, which now could amount to at most 3 *dynym* (or 0.3 ha), but should be no less than half a *dynym*, the exact amount to be decided by the executive committee of the district level people's assembly. Similarly, privately owned livestock was reduced to a maximum of one or two cows, in addition to which 15 to 20 goats or sheep could be held, plus a number of fowl, again to be set by district.[61] This made joining collectives less attractive for independent farmers—indeed, many cooperatives failed to adopt the new statute because of the negative consequences it would have had for their members.[62] Yet the number of cooperatives would continue to increase, in line with the party's decision at the Third Congress of the APL (April–May 1956) that 85 percent of the arable land should be cultivated under cooperatives by the end of the Second Five-Year Plan.[63]

It is important to consider how such a rapid expansion of collectivized agriculture was implemented and achieved. Beyond the quota-calculation system and contracts imposed on the private sector,[64] which were key deterrents to private farming, a number of other factors shaped the pace and success of Albania's collectivization process. The massive relocation of people that took place before migration controls were strictly enforced in the 1960s, amounting to a flight from the interior,[65] meant that in the lowlands a larger proportion of the residents had less of an attachment to the soil than would have been the case if newcomers had not availed themselves of the opportunities to resettle. The extensive reclamation and drainage works in the western lowlands during the 1950s, some of them

61 These and other provisions, e.g., on organization, remuneration, and membership (which was explicitly not open to *kulaks*), can be found in the decision of the Council of Ministers, "Vendim Nr. 486 datë 6.12.1956 mbi aprovimin e statutit tip të kooperativave bujqësore."

62 Schwanke, "Die landwirtschaftlichen," 46–53.

63 Fishta and Ziu, *Historia e ekonomisë*, 500–1. As Prifti, *Socialist Albania*, 66, notes, the decision to speed up collectivization may have been initiated at the Party Plenum in December 1955.

64 Sjöberg, *Rural Change*, 89.

65 The migration that took place in particular during the 1950s and its consequences are discussed in Tirta, "Vështrim rreth popullsisë," and Tirta, "Lëvizje të popullsisë." As the scale of this relocation of people appears to have threatened the policies of the government, migration controls primarily aimed at rural-to-urban movements were introduced step-by-step starting in 1958. For details on these controls, which saw parallels in many other communist countries across the Eurasian continent, and the rationale behind them, see Sjöberg, "Rural retention," and Sjöberg, "Shortage," respectively.

begun already during the Italian occupation in the 1940s, also allowed for a colonization of new land where previous owners were not forced to give up their land. Similarly, some large estates could be turned into cooperative or state owned units without first sharing out the land to landless and land poor, all of which reduces tension and increases speed in the implementation of agrarian policies of the type that Albania embarked upon.[66]

As with the previous economic plan, the Second Five-Year Plan was revised midterm. Unlike the concessions made in 1953, however, the February 1958 revision reflected what Adi Schnytzer characterizes as the "revolutionary enthusiasm" of the day.[67] It has not been possible to establish whether directives from Moscow were instrumental in pushing collectivization forward. It remains a possibility, though, in need of closer investigation. In addition to the decisions made at the Third Congress of the APL, the one document that Albanian historians rely on is a February 1959 letter from the Central Committee to district party committees instructing them to "secure the completion" of collectivization.[68]

In the absence of additional primary or secondary documentation, it is only possible to offer a speculative assessment regarding the significance of this letter's timing. It followed a visit by Albanian high-ranking party members to Moscow in January 1959, when Soviet leaders promised economic concessions if Albania complied with Moscow's directives. It preceded a visit to Albania by Khrushchev himself in late May to early June 1959, during which the Soviet leader made clear he would not allow Albania as much flexibility on economic matters as the APL expected. While the strained relations with Yugoslavia was the main point of contention, Khrushchev's designs for Albania in the socialist division of labor also went against the preferences of the Albanian leadership. As the Soviet stance threatened to undermine the drive toward industrialization favored by Albanian leaders, the latter could presumably not accept the Soviet proposals for cooperation—despite that Soviet support for the Third Five-

[66] Some information on the expansion of agricultural land can be found in Fishta and Ziu, *Historia e ekonomisë*, 518–27.

[67] Schnytzer, *Stalinist Economic*, 70. The fact that Albania was richly rewarded by the Soviet Union for staying within the fold—the need to do so was strongly impressed on the Albanian leaders as they visited Moscow in the spring of 1957—may have something do with it. Soviet concessions included increased aid and the cancelling of debt, and the effects could be felt by ordinary people as rationing of food was removed towards the end of 1957. See further Pano, *People's Republic*, 126–27.

[68] Fishta and Ziu, *Historia e ekonomisë*, 499–500.

Year Plan was pledged and that this included some support for industrial expansion.

Therefore, at least two interpretations are possible. One is that Albanian leaders were following the same timetable issued from Moscow to other Bloc members, but high politics got in the way. The other is that the APL's directive to district committees was a response to domestic challenges, including the increasingly dire need to produce large quantities of grain to feed its rapidly growing population. If so, reduced support from Moscow was the price to be paid for holding on to Albania's favored course toward industrialization. While the reasons may have changed, the response from Tirana toward Soviet calls for a new and more benign stance on Tito's Yugoslavia was much the same as the reaction to de-Stalinization in the mid-50s: stay the course and implement the original Stalinist model of economic development. Leaders felt it was necessary to pursue a more autarkic course, bringing agriculture under central control given that sources of outside supply could not be relied upon. At the same time the traditional Stalinist model favored in Albania presupposed that industrialization could make use of resources extracted from the agricultural sector, no matter how unwilling the farming sector was in this process.

There is surprisingly little evidence that the rapid expansion of collectivized agriculture itself created problems, but at least in the mid-1950s it was still claimed by the party that *kulaks* sabotaged agricultural production, the work of the collectives in particular. By 1957, party sources quoted by Albanian economic historians suggest that there were still almost as many *kulaks* as had been the case five years previously, many of whom where intent on spreading anti-communist propaganda and disrupting the activities of the cooperatives. Hence the measures to increase the tax burden, while work was undertaken to improve the viability of cooperatives by increasing their size through amalgamations.[69]

By 1960, collectivization in the lowlands was nearly complete, and party leaders declared that the foundation for socialism in the countryside had been established. Government statistics at the time suggest that nowhere was the process fully completed, however, with only five of 26 districts achieving a 90 percent success rate.[70] A little more than 70 percent of all rural households had become members of cooperatives, and

[69] Fishta and Ziu, *Historia e ekonomisë*, 503–17.

[70] *Anuari i statistikor*, 1962, 126.

84 percent of the cultivated land area outside state farms had been collectivized. Variations in collectivization levels within the country were considerable, however, and in some mountain districts just a few percent of all rural households were cooperative members.

Developments Post-1960

In the 1960s, there were a number of dramatic changes to Albania's economic policies as a result of the split with Moscow. In the wake of the changing circumstances, Albanian leaders initiated some reforms in its economic planning, but few changes were made to its overall strategy. In fact, the party's pursuit of full collectivization remained central to its agenda. In what one Albanian historian recently has labeled "the greatest absurdity of the time,"[71] the party renewed a massive campaign to complete collectivization, which included controlling agriculture in the highlands and further reducing the importance of household plots.

Hence, the APL announced its plans to resume collectivization after the Fifth Congress in 1966. Its objective was not only to increase production—in which case, presumably, only those areas able to support large scale production would have been targeted—but also to eliminate capitalism entirely. "Any other road," Enver Hoxha claimed in his speech to the Fifth Congress, "leads only to the restoration of capitalism in the countryside."[72] Furthermore, the campaign was to be achieved with little delay. In a secret directive signed on February 28, 1967 by Secretary of the Central Committee of the APL, Hysni Kapo, the party declared its goal of creating 500 new cooperatives across 12 of the 26 districts within four to five months.[73] Whereas the struggle to achieve collectivization during the 1950s was relatively smooth, the post-1960s collectivization drive was fraught with challenges. Although the authorities claimed to have achieved their aim—490 cooperatives were established in the 12 districts and another 20 new ones elsewhere, and if not within 4–5 months then at least within a few years[74]—by offering attractive incentives, it was certainly not

[71] Leci, "Kolektivizimi 1967."

[72] *Kongresi V i PPSh* (Tirana: Naim Frashëri, 1967), 61; quoted by Prifti, *Socialist Albania*, 66.

[73] Reproduced in Leci, "Kolektivizimi 1967."

[74] Sadikaj, "Lufta për kolektivizimin," 12.

without conflict and violence.[75] While high level cadres often blamed this on the over-zealousness of local authorities, it appears that such occurrences were rather frequent. Peasants often resorted to classical acts of resistance, as seen elsewhere in the Eastern Bloc. Animals were slaughtered rather than handed over to cooperatives, and more passive forms of resistance, such as the retaining of boundary markers, were adopted. By and large, however, we do not know whether such acts of resistance were more or less common compared to previous waves of collectivization or whether they were similar or different in character. Anthropologist Clarissa de Waal, for instance, notes that in the central and more fertile parts of Mirditë, a district in north-central Albania, collectivization was met with "bitter resistance," as had the introduction of communism in the 1940s and 1950s, but in the mountains few were willing to risk "attracting the *sy keq*—evil/bad eye—of the government, leading to internment or prison."[76]

As for household plots, land was limited to 1 *dynym* (or 0.1 hectare) and one cow or 10 sheep or goats in the lowlands. Highlanders were allowed as much as 1 *dynym* of irrigated land, or a *dynym* and a half if unirrigated, and were also permitted to keep one cow plus ten small ruminants, with a somewhat more generous allocation to members of new cooperatives starting in March 1967.[77] The objective, however, was to eventually eliminate any vestiges of private ownership in agricultural production.[78] Household plots were further reduced, indeed minimized or done away with completely, as some cooperatives were transformed to a new type of cooperative introduced at the beginning of the 1970s—the agricultural cooperative of a higher type (*kooperativa bujqësore e tipit të lartë*, KTL). Within these cooperatives, and as a step toward the creation of new state farms, investments could be made by the government.[79] The idea was to boost productivity with increased input of mechanized production and physical infrastructure, without transferring resources in the (ideologically) wrong direction, namely from state to cooperative ownership.

The combined goal of improved productivity (or at least increased output) and the attempt to eliminate capitalism also seem to have been factors during the last phase of collectivization in the first half of the

---

[75] Sjöberg, *Rural Change*, 95–96.
[76] de Waal, *Albania Today*, 77.
[77] Leci, "1967. Urdhri i Enver Hoxhës."
[78] See for example: *Historia e popullit shqiptar*, 282.
[79] On this new type of cooperative, see Sjöberg, *Rural Change*, 97–98.

1980s.[80] It should be noted, too, that as external financial support dried up following Albania's break with China,[81] the APL increased its push toward complete centralization of control.[82] In the 1980s, party leaders targeted privately owned animals, collecting and transferring these animals to collectives, in order to satisfy the objectives of ideological purity and rational forms of production. Historians of contemporary Albania, however, think otherwise and have labeled the resulting measures as "extreme."[83] At this point, if not earlier, it became evident that the peasantry did not fully support the ideas that guided the Albanian leadership. Another wave of animal slaughter took place, and combined with widespread maltreatment and mismanagement of surviving livestock, the country was thrown into a food crisis. Rationing was reintroduced and once Enver Hoxha passed away, his successor Ramiz Alia introduced measures to redress the situation. Although the post-Hoxha leadership was not prepared to return in full to the provisions that prevailed before the last wave of collectivization was launched, he did at least concede that continuing collectivization might not be the answer to all the problems encountered in the countryside.

Conclusion

The experience of Albania shows that despite the effort to pursue its collectivization program, the regime was forced to make concessions in the face of unsupportive economic realities. Such concessions, however, were typically regarded as temporary in order to reach the party's long-term

---

80 Wildermuth, *Die Krise der albanischen.*

81 As Albania broke off relations with China in 1978, all technical and financial support was withdrawn by Beijing. As was the case when the Soviets left in the early 1960s, a chain of uncompleted investment projects were left behind, including prestige ones such as the hydropower station in Koman, the oil refinery in Ballsh and the steel works in Elbasan. Alternative sources of finance and technology became a major Albanian priority, but as these were not easy to come by, the pressure to mobilize domestic resources increased. As Schnytzer, *Stalinist Economic*, clearly shows, the manner in which this was done was in keeping with Stalinist prescriptions.

82 For a typical exponent of such sentiments, see Manoku, *Probleme të përqëendrimit.*

83 *Historia e popullit shqiptar*, 280.

goals. In the end, it seems that Albania went a step further than most of its peers simply by not giving up on the hope of one day completing collectivization and subsequently turning the cooperative sector into formally state-owned property and state-run production units. The results proved disastrous, but before the full consequences of the last round of collectivization had been allowed to run its course, communism collapsed and a new political order was established.

Since 1989, scholars have taken renewed look at developments during the socialist era. It goes without saying that during the communist period itself—in the case of Albania, from 1944 to 1991—there were clear limits on what scholars were able to research and what materials they could access. In those cases where access was allowed, there were strict limits on what could be published. Therefore the body of scholarship by socialist-era scholars remains fragmentary at best. Although Albanian historians have utilized the new research opportunities since 1989, the output of published research is so far rather limited, but increasing. There are quite a few opportunities of making important advances simply by providing basic information in order to fill in a number of gaps that still exist. Thus, although it is quite obvious that collectivization was an essential part of the strategy of the new rulers to consolidate power and to establish a new political order, the finer detail of this process is still not visible. In particular, it would be most useful if documentation could be found on how individual policies were designed, and on what grounds the decisions were arrived at. Similarly, what was the full impact on the ruling party, its top tier members and its rank-and-file? Although the former are likely to have been forced to take a stand on various matters relating to collectivization and collectivized agriculture, and in the process putting their career and at times perhaps their lives at risk, local party officials were also subject to a host of pressures, the consequences of which are still in need of more exploration.

It is also obvious that collectivization served to redefine the relations between state and local communities. Again, however, details of this relationship remain unclear. We can establish with a good measure of confidence that agriculture was seen as a resource base for forced industrialization and that measures such as the amalgamation of cooperatives went together with administrative reorganization. But what of the political role of rural areas in general, and of the agricultural sector in particular? Indeed, at various turns policies relative to rural areas in general and agriculture in particular were a source of intraparty contention, but did this happen primarily for ideological or political reasons or as a result of self-serving decisions, along the lines suggested by some historians?[84] Could rural policies be enlisted in the struggle for power or as a source of inspiration on how to proceed? Evidence of the strain on rural areas produced by these policies is also accumulating, but systematic accounts of intra-village strife and solidarity within the community, including between those that were singled out as *kulaks*, are still missing. It would be of more than passing interest, for instance, to know more about the relative importance of clan-based loyalties versus ideological preferences or political expediencies in the implementation of the anti-*kulak* campaigns.

Again, although the consequences in terms of livelihood and the quality of life are reasonably well documented, many details remain unclear. New research in these and related areas would allow us to better understand the logic and constraints under which the attempts by the Albanian Party of Labor to industrialize Albania occurred and how the agricultural sector fits into that process.

BIBLIOGRAPHY

Anuari i statistikor RPSh 1961 [Statistical yearbook of the People's Republic of Albania 1961]. Tirana: Dretoria e Statistikës, 1962.

Blumi, Isa. "The politics of culture and power: the roots of Hoxha's postwar state." *East European Quarterly* 31, no. 3 (1997): 379–98.

Bokovoy, Melissa K. *Peasants and Communists: Politics and Ideology in the Yugoslav Countryside, 1941–1953.* Pittsburgh: University of Pittsburgh Press, 1998.

de Waal, Clarissa. *Albania Today: A Portrait of Post-Communist Turbulence.* London: Tauris, 2005.

Duka, Valentina. *Histori e Shqipërisë, 1912–2000* [History of Albania, 1912–2000]. Tirana: Kristalina-KH, 2007.

Durand, Didier. *Socialisme e développement agricole: le cas d'Albanie.* Unpublished Thèse de Doctorat de troisième cycle. Grenoble: Université des Sciences Sociales, 1986.

Dyrmishi, Demir. "Konferenca e PPSh së Tiranës, shkaqe dhe rrjedhoja (prill 1956)" [The Tirana conference of the APL, causes and consequences (April 1956)]. *Studime Historike* 1 (2002): 115–30.

———. "Cilësimi dhe deklarimi kulakëve—formë e ashpër e luftës së klasave në fshat (1949–1969)" [The Qualification and Declaration of the *Kulaks*—a Savage Form of the Class Struggle in the Countryside]. *Studime Historike* 1–2 (2004): 107–23.

[84] Blumi, "Politics of Culture," and Duka, *Histori e Shqipërisë*.

Fishta, Iljaz and Veniamin Toçi. *Ekonomia e Shqiërisë në vitet e para të ndërtimit socialist 1944–1948* [The Economy of Albania during the First Years of Socialist Construction 1944–1948]. Tirana: Institut i Historisë, Akademia e Shkencave e RPSSh, 1984.

——— and Mihal Ziu. *Historia e ekonomisë së Shqipërisë (1944–1960)* [The Economic History of Albania (1944–1960)]. Tirana: Dita, 2004.

Gjeçovi, Xhelal, ed. *Historia e popullit shqiptar*, IV: *Shqiptarët gjatë luftës së dyte botërore dhe pas saj, 1939–1990* [The History of the Albanian people, IV: The Albanians during World War II and after, 1939–1990]. Tirana: Botimet Toena, 2002.

———. "Alternativat e zhvillimit të Shqipërisë pas Luftës së dytë botërore" [Albania's Development Options after World War II]. *Perla* 12, no. 2 (2007): 27–45.

Hajdëri, Dashamir. *Industria ushqimore në Shqipëri* [The food industry in Albania]. Tirana: Dita, 2005.

Hazen, Charles D. *Albania, Past and Present.* New York: Macmillan, 1919.

Historia e Shipërisë, Vëll. IV *(1944–1975)* [The history of Albania, vol. IV (1944–1975)]. Tirana: Instituti i Historisë, Akademia e Shkencave e RPSSh, 1983.

Kaser, Karl, Robert Pichler, and Stephanie Schwandner-Sievers, eds. *Die weite Welt und das Dorf. Albanische Emigration am Ende des 20. Jahrhunderts.* Vienna: Böhlau, 2002.

Kaser, Michael. "Economic Continuities in Albania's Turbulent History." *Europe-Asia Studies* 53, no. 4 (2001): 627–37.

Lalaj, Ana. "1956—tronditje në 'kampin' komunist" [1956—Tremors in the Communist "Camp"], *Studime Historike* 1–2 (2007): 114–25.

———. "The Soviet-Yugoslav break and Albania," *Studia Albanica* 2 (2005): 123–27.

Lange, Klaus. *Die Agrarfrage in der Politik der Partei der Arbeit Albaniens.* Albanische Forschungen 21. Munich: Trofenik, 1981.

Lankov, Andrei. *From Stalin to Kim Il Sung: The Formation of North Korea 1945–1960.* London: Hurst, 2002.

Lauka, Islam, Eshref Ymeri, eds. *Shqipëria në dokumentet e arkivave ruse* [Albania in documents from Russian archives]. Tirana: Toena, 2006.

Leci, Elmas. "Urdhri i Enver Hoxhës: Kulakët të trajtohen si armiq!" [Order of Enver Hoxha: *kulaks* to be treated as enemies!]. *Shekulli*, February 1, 2008.

———. "Enver Hoxha: Kulakët t'i rrjepim ekonomikisht, të pabindurit t'i eliminojmë" [Enver Hoxha: Strip the *kulaks* economically, eliminate those that disobey]. *Shekulli*, February 4, 2008.

———. "Enver ngarkon Mehmetin si 'dorë të forte' për Reformën Agrare e problemin kulakëve" [Enver assigns Mehmet as the "strong hand" to take charge of the agrarian reform and the *kulak* problem]. *Shekulli*, February 5, 2008.

———. "Kolektivizimi 1967. Absurdi më i madh i kohës" [Collectivization 1967. The greatest absurdity of the time]. *Shekulli*, February 9, 2008.

———. "1967. Urdhri i Enver Hoxhës për heqjen e oborreve kooperativiste" [1967. The order of Enver Hoxha to eliminate the cooperativists' household plots]. *Shekulli*, February 10, 2008.

Lorenzoni, Giovanni. *La questione agraria albanese. Studi-inchieste e proposte per una riforma agraria in Albania.* Bari: Laterza, 1930, 2nd edition.

Manoku, Ylli. *Probleme të përqëendrimit dhe specializimit të bujqësisë (Sipas përvjës së rrethit të Korçës)* [Problems of Concentration and Specialization of Agriculture]. Tirana: Shtëpia e Propagandës Bujqësore, 1985.

Pano, Nicholas C. *The People's Republic of Albania*. Baltimore: Johns Hopkins Press, 1968.

Pearson, Owen. *Albania in the 20th Century: A History*, 3. *Albania as Dictatorship and Democracy: From Isolation to the Kosovo War 1946–1998*. London: Tauris, 2006.

Pipa, Arshi. "Party ideology and purges in Albania." *Telos* 59 (Spring 1984): 69-100.

Prifti, Peter. *Socialist Albania since 1944: Domestic and Foreign Developments*. Cambridge, MA: MIT Press, 1978.

Sadikaj, Dilaver. "Lufta për kolektivizimin e zonave të thella malore të vendit dhe forcimin ekonomiko-organizativ të kooperativave të reja (1966–1970)" [The struggle to collectivize the deep mountain zone of the country and the economic-organizational strengthening of the new cooperatives (1966–1970)]. *Studime Historike* 1 (1987): 3–24.

Schmidt-Neke, Michael and Örjan Sjöberg. "Bevölkerungsstruktur," in Klaus-Detlev Grothusen, ed. *Südosteuropa-Handbuck*, Band VII: *Albanien*. Göttingen: Vandenhoeck & Ruprecht, 1993, 464–90.

———. "Zwischen Kaltem Krieg und Teleologie: das kommunistische Albanien als Objekt der Zeitgeschichtsforschung" in Oliver Jens Schmitt and Eva Anne Frantz, eds. *Albanische Geschichte: Stand und Perspektiven der Forschung*. Südosteuropäischen Arbeiten 140. Munich: R. Oldenbourg, 2009, 131–47.

Schmitt, Oliver Jens. "Vorwort" in Oliver Jens Schmitt and Eva Anne Frantz, eds. *Albanische Geschichte: Stand und Perspektiven der Forschung*. Südosteuropäischen Arbeiten 140. Munich: R. Oldenbourg, 2009, 7–12.

Schnytzer, Adi. *Stalinist Economic Strategy in Practice: The Case of Albania*. Oxford: Oxford University Press, 1982.

Schwanke, Robert. "Die landwirtschaftlichen Produktionsgenossenschaften in Albanien." Österreichische Osthefte 2, no. 1 (1960): 46–53.

Shilego, Turhan. *Lufta e klasave në Shqipëri (1948–1953)* [The class struggle in Albania (1948–1953)]. Tirana: Instituti i Historisë, Akademia e Shkencave e RPSSh, 1985.

Sjöberg, Örjan. *Rural Change and Development in Albania*. Boulder, CO: Westview, 1991.

———. "Rural retention in Albania: administrative restrictions on urban-bound migration." *East European Quarterly* 28, no. 2 (1994): 205–33.

———. "Shortage, priority and urban growth: towards a theory of urbanization under central planning." *Urban Studies* 36, no. 13 (1999): 2217–36.

Smirnova, Nina. *Historia e Shqipërisë përgjatë shekullit XX* [The history of Albania during the twentieth century]. Tirana: Ideart, 2004.

"Speech of N. S. Khrushchev at a friendly dinner in Yevksinograd (Varna), May 16, 1962," translated by Gary Goldberg, Cold War International History Project, Woodrow Wilson International Center for Scholars, Washington, DC. http://www.wilsoncenter/org [accessed on June 27, 2009].

Stahl, Johannes. *Rent from the Land: A Political Ecology of Postsocialist Rural Transformation*. London: Anthem, 2010.

Sufaj, Femi. *Historia e burgjeve të Shqipërisë gjate shek. XX* [The History of the Prisons of Albania during the Twentieth Century]. Tirana: Albin, 2000.

Swain, Geoffrey and Nigel Swain. *Eastern Europe since 1945*. Basingstoke: Macmillan, 2003.

Tirta, Mark. "Vështrim rreth popullsisë së ardhur në vendbanime të krijuara pas Çlirimit në mjedise bujqësore" [Observations on the population arriving in settlements created after liberation in agricultural settings]. *Etnografia shqiptare* 13 (1983): 5–30.

———. "Lëvizje të popullsisë dhe zhvillimi i qyteteve (1945–1985)" [Population movements and the development of towns (1945–1985)]. *Etnografia shqiptare* 15 (1987): 5–36.

Wädekin, Karl-Eugen. *Agrarian Policies in Communist Europe: A Critical Introduction.* The Hague: Martinus Nijhoff, 1982.

Wildermuth, Andreas. *Die Krise der albanischen Landwirtschaft. Lösungsversuche der Partei- und Staatsführung unter Ramiz Alia.* Munich: Hieronymus, 1989.

Wildermuth, Andreas. *"Sich stützen auf die eigenen Kräfte." Die Wirtschaftspolitik Albaniens nack dem Zweiten Weltkrieg.* Munich: Liliana Djeković Verlag, 1995.

THE COLLECTIVIZATION OF AGRICULTURE IN ALBANIA: CHRONOLOGY OF EVENTS

Örjan Sjöberg

1944

December: A commission for the registration of agricultural land and other assets is created within the Ministry of Economy.

1945

August 29: Land reform law.

September 3: Implementation of land reform begins in the Berat-Lushnjë zone.

1946

February: The land reform law under discussion, marking a radical turn.

April 20: Decree on cooperatives, a Yugoslav model charter adopted.

May 27: New land reform law.

October 24: First cooperative founded, in Krutjë ë Sipërme, Lushnjë district.

November 17: Land reform declared complete.

1948

October 26–27: A new model charter for agricultural cooperatives, closely modelled on the 1935 Soviet Model Charter, is introduced replacing the earlier (Yugoslav) one.

1949

January 1: Rationing of food introduced.

February 15–18: First Congress of Agricultural Cooperatives.

March: Measures to "eliminate the *kulaks* as a class" launched, and reiterated at the third plenum of the Central Committee in October.

1951

April–May: Decisions at the tenth ordinary plenum and eleventh extraordinary plenum of the Central Committee to consolidate the gains of collectivization and speed up the process in order to build socialism in the countryside; a "lower" form of cooperatives now permitted (only labor, not other resources, is pooled).

1954

October 13: The Politburo decides to approve new model statutes, formally recognising a type 1 and a type 2.

1955

December 27–28: Partial removal of food rationing.

1956

September 3–5: Third Congress of Cooperatives approves new model statute which implies, e.g. a further reduction in the size of the household plot.

1957

December: Rationing of food abolished.

1958

January 28: Decision by Council of Ministers to cancel debt of agricultural cooperatives.

1959

August 11: Following goal set by the Third Congress of the APL in April–May 1956, the Politburo decides to achieve the goal of 85% collectivization by end of Second Five-Year Plan.

1963

June 6–8: Tenth Plenum of Central Committee focuses on structural change in agriculture, which is to grow as fast as (or faster than) industry.

1966

November 1–8: Decision to complete collectivization of agriculture in remaining parts of the country.

1967

February 28: Directive issued by APL that collectivization should be completed in 4–5 months, i.e., well before end of 1967.

April 29: Decision by Central Committee and Council of Ministers on furthering the revolution, including measures to further reduce the household plot (and private gardens more generally; state farm workers no longer allowed to keep livestock).

June 20: Taxes on the household plot abolished.
July 19: Taxes on individuals and agricultural cooperatives are abolished.

1968
March 23: The model charter for agricultural cooperatives abolished.

1971
January 8: Statement by Central Committee on the status and problems of agriculture.
July: Decision to create cooperatives of a higher type, the KTL.

1972
Social benefits to be extended to members of agricultural cooperatives.
First KTL established.

1980
November 4: Further reduction in the size of the household plot.

1982
February 24: Decision by the Politburo to collectivize remaining private animals; members of lowland cooperatives had been "encouraged" to do so starting in 1980.

1985
Autumn: In a partial step back, formerly privately owned animals collectivized since 1982 now to be allocated to herds at the brigade level.

1989
February 1–2: Seventh Plenum of the Central Committee, devoted to agriculture. Decisions on remuneration and private markets, now again permitted.

1991
April 20: Decision by Council of Ministers to lift restrictions on private assets in agriculture. Independence of agricultural cooperatives as economic units is granted.

PART IV

Axes of Differentiation: Center and Periphery, "Class Struggle," Social and Ethnic Cleavages

Collectivization as Social Practice: Historical Narratives and Competing Memories as Sources of Agency in the Collectivization Campaign in the GDR

ARND BAUERKÄMPER

The transformation of agriculture and rural society in the GDR in the 1950s was initiated "from above" by the leadership of the Socialist Unity Party (*Sozialistische Einheitspartei Deutschlands*), or SED. At the same time, however, the process was implemented, interpreted and appropriated by various regional and local actors who either rejected collectivization or adapted to it and attempted to integrate the system into their ideologies. Often these reactions overlapped as much as the modes of conduct—from compliance to outright opposition. Individual and collective agency therefore served as either social or cultural resources supporting collectivization or became barriers to the process. How collectivization as a process of radical transformation was perceived, interpreted, and appropriated are important fields of historical investigations.

This chapter demonstrates how conflicting versions of history and different memories of past experiences, especially of Germany's past and of collectivization in the Soviet Union, respectively, had a particularly strong impact on individual and collective agency in the GDR's collectivization campaign. Yet these factors were far from uniform and static in rural society. In fact, opposing memories frequently clashed in rural society and the gulf separating peasants and farmers from the party and state functionaries proved to be almost insurmountable.[1] These conflicting memo-

[1] For remarks on the role of recollections in the GDR's collectivization of agriculture, see Bauerkämper, "Collectivization and Memory;" Bauerkämper, *Ländliche Gesellschaft*, 458–63.

ries coalesced around the notion of the "class war," which agrarians and officials shared despite their conflicting interpretations of the concept.

However, the collectivization process was decisively shaped by opposing views of the various social and professional groups in the villages. Solidarity among villagers was therefore difficult to achieve. In a similar vein, memories were colored by the concerns and interests of actors (farmers as well as functionaries) in the 1950s, when collectivization directly challenged the traditional values and notions of private property, self-reliance, and family work. Entrenched social hierarchies were demolished, as farmers, peasants, and agricultural laborers had to cooperate on roughly equal terms in the new collectives. Thus, the enforced merger of independent producers triggered long-standing resistance and occasionally violent opposition from villagers. Even after collectivization had officially been accomplished in the GDR in April 1960, farmers felt especially helpless and overwhelmed by agitation groups that had swarmed villages in 1959–60 in order to force farmers to join or create collectives.[2] These confrontations, along with contemporary experiences and past memories, underscored the complex local dynamics of social and economic upheaval collectivization created. The present chapter investigates the proposition that collectivization is to be conceptualized as an ensemble of social practices fuelled by individual and collective agencies. The intention, here, is to shed light on these dynamics by concretely reconstructing and analyzing social relations in rural society in the 1950s and early 1960s.

Previous research on the social and cultural history of collectivization in the East German "Workers' and Peasants' State" has concentrated on certain states and regions. Moreover, most studies have either dealt with the decision-making process at the higher echelons of power or on the issue of repression and peasant resistance to collectivization. These investigations have largely treated both "the state" and "the peasantry" as unitary, monolithic actors, denying or underrating the agency of individual or collective actors. In fact, the crucial role of these actors as historical subjects has been largely neglected. Despite the well-explored notion of a "civil war"[3] between peasants and regime functionaries, few studies have investigated peasant reactions to collectivization, which fundamentally challenged traditional modes of production, rural values, and life styles.

---

[2] For a detailed account of the main problems of collectivization in the GDR, see Jens Schöne's contribution to this volume.

[3] With regard to the collectivization of agriculture and forced grain procurements in the USSR, see Viola, *Peasant Rebels*, 44, 130, 180, 233, 235.

The differentiated responses to collectivization from diverse rural groups in specific local and regional contexts—ranging from steadfast resistance to accommodation and collaboration with the regime—has received little attention in recent historiography, which has grossly neglected the comparative perspective and cross-border entanglements.[4]

Similarly, the conflicts and methods of solidarity among peasants and between peasants and villagers (including local party functionaries and state officials) have been also neglected. Agency on the part of the villagers has thus not been systematically related to party and state policies. According to published investigations, peasants were flexible and able to adapt to new conditions and to even attain some of their immediate aims and economic interests. Ultimately, however, they had to succumb to collectivization. As the framework of national historiography has endured, moreover, the impact of experiences and memories across national borders and cultural boundaries has yet to be reconstructed and explained. As will be demonstrated in this investigation, villagers and officials perceived and interpreted the protracted, multilayered and complex process of collectivization in the GDR through the lens of the preceding violent transformation of agriculture in the Soviet Union. Soviet collectivization served as a foil, along with opposing appropriations of German history, both of which triggered conflicting memories and associations. Moreover, the emerging West German state was an important reference point, both to farmers and leading functionaries of the Socialist Unity Party of Germany (SED). Collectivization, in essence, sealed off the GDR from the Federal Republic and thus prevented rather than promoted the reunification of the two German states, which the SED had officially endorsed until the 1960s.[5]

Following key theoretical and methodological considerations on concepts of agency and memory, this chapter will elaborate on conflicting memories of Soviet collectivization. The third section addresses conflicting conceptions and interpretations of German history which served to legitimize collectivization or to inspire resistance to the process. The concluding section will summarize research findings regarding the role of historical narratives and memories as (ambiguous) resources of agency.

---

[4] Schöne, "Landwirtschaft und ländliche Gesellschaft." On the lack of comparative studies more generally see Kocka, "Bilanz und Perspektiven," 767. For opposing views in the ensuing debate, see Lindenberger and Sabrow, "Zwischen Verinselung;" Bispinck, et al., "Die Zukunft."

[5] See Kleßmann, ed. *The Divided Past*; Kleßmann, *Verflechtung und Abgrenzung*; Heydemann, "Integrale deutsche Nachkriegsgeschichte;" Jarausch, "Die Teile."

This study is based mainly on police and opinion reports from villages of Brandenburg, which was partitioned into the districts (*Bezirke*) of Potsdam, Frankfurt (Oder), and Cottbus in 1952.[6] The northern and eastern territories of Potsdam and Frankfurt (Oder) were characterized by large landownership before the end of the Second World War. In the early postwar years, however, the social and economic structure changed fundamentally. After the 1945 land reform had stripped the estate owners of their property, "new peasants" settled in the countryside, many who struggled to establish their (often miniscule) plots. In 1952–53, "new peasants" were among the first to merge into collectives. By contrast, medium-sized farms dominated in the western regions of the Potsdam *Bezirk* outside Berlin. The former German capital, particularly its western sectors, was an important market for agricultural producers in the Potsdam region, as city-dwellers were faced with a food shortage as late as the early 1950s. In the Cottbus *Bezirk*, on the other hand, many small peasants combined agricultural and industrial work, especially in coal mines. Despite this burden, with which women had to cope as much as men, peasants were slow to collectivize in this district.[7]

Theoretical Framework

Despite their uncontested power, the communist dictatorships that were established in Eastern Europe in the late 1940s and early 1950s could not simply enforce their rule on the public. On the contrary, the party needed the consent and compliance of a significant portion of society beyond their limited group of supporters. They depended on support even if only to legitimize their policies. Local and regional party and state functionaries therefore had to mediate between the aims and directives of the central party leadership and the interests, values, norms, mentalities, and worldviews of the villagers. Moreover, the prevailing (economic and social) conditions and traditions had to be taken into account. Thus, political interventions were adapted and at times even distorted in accordance with unique social, economic and political conditions.[8] Studies of individual and collective agency reconstruct and explain the interrelationships,

---

[6] Schreckenbach and Künzel, "Das Land Brandenburg;" Hanja, "Die Bildung."

[7] Kotsch, *Das Land Brandenburg*, 261–96; Ribbe, "Das Land Brandenburg," 677–726, esp. 515–723; Kotsch, "Die Bezirke Potsdam," 732–40, 743, 746–56.

[8] Fulbrook, *People's State*, esp. 235–49; Fulbrook, "Retheorising," 2002, 280–98. Also see Lindenberger, "Die Diktatur der Grenzen," 34, 39.

encounters, and interchanges between concrete actors in the countryside of various East European states in order to shed light on modes of appropriations, asymmetrical negotiations and local and regional arrangements.

This particular investigation of individual and collective agency in the process of agricultural collectivization highlights the resources of specific actors, their worldviews, mentalities, perceptions, interpretations, and appropriations to specific contexts and conditions, which in turn are at least partially susceptible to agency. This approach, which has been proposed by sociologist Anthony Giddens, emphasizes the appropriation of structures through social action, which individuals and collectives pursue in space and time. Conversely, social action reproduces structures, though not unchangeably. On the contrary, structures serve as resources for individual or collective action and thus provide incentives and options to subjects. New experiences do not only change agency, but also worldviews, interpretations, appropriations as well as methods of communication, representation and linguistic transmission. Remembrances and language can therefore not merely be imposed "from above" as a straightjacket determining agency, but they are flexible and open to creative usages by actors. Stressing embeddedness and "structuration,"[9] this chapter will deal with the transfer of particular resources (in a broad sense) and compensations between individual or collective actors whose power and status depended on their economic, social, cultural, and symbolic "capital," to borrow a term from Pierre Bourdieu.[10] In these networks, actors need to generate social capital above all, as it lends them agency. In particular, actors placed at the "crossroads of social circles" (*Kreuzung sozialer Kreise*, Georg Simmel) gain considerable autonomy *vis-à-vis* structural constraints. Placed at crucial "cutpoints," these key actors mediate between institutionally separated groups. Local elites therefore receive particular attention in studies of agency, which has to be related to experiences and interests as crucial categories of social analysis.[11]

Theories of social and cultural practice offer an overall analytical framework for studies of individual and collective agency. According to these conceptual considerations, actors and structures materialize in inter-

---

[9] Giddens, *Constitution of Society*. Also see Welskopp, "Der Mensch," 156–69; Pred, *Practice and Structure*, 4–11, 195–98; Lamla, *Anthony Giddens*, 45–62.

[10] On Bourdieu, see Wehler, *Die Herausforderung*, 15–44; Reichardt, "Bourdieu für Historiker."

[11] The theoretical considerations are based on Granovetter, "Strength of Weak Ties," 1973; Nolte, "Georg Simmels Historische Anthropologie der Moderne," 239/f, 243, 246; Suter, "Kulturgeschichte des Politischen," 29, 32–34, 52–55.

related practices, which constitute the social as a field. In the GDR, villagers delayed collectivization by disseminating their own versions of recent history (particularly of the USSR), as an attempt to preserve their individual autonomy and personal integrity. Investigations of popular reactions to collectivization therefore highlight these social dynamics unleashed by communist policies.[12]

These studies also underscore the central role of language in social interaction, particularly in asymmetrical power relationships (such as between recalcitrant farmers and superior party functionaries) and conflicts. Thus, opponents of collectivization frequently expressed their disgust and contempt by using highly emotive language that diametrically contrasted to the official doctrines of the communist rulers. Although they did not succeed in stemming the tide of collectivization, these dissenting voices reflected the desire to contradict the official jargon, to preserve their self-respect against political indoctrination and thus to save individual agency in a small-scale rural society.[13]

As a corollary, peasants rejected the profound politicization of rural milieus that the SED leadership promoted. By defining collectivization as "class warfare," First Party Secretary Walter Ulbricht and his lieutenants in the Central Committee sought to expand collectivization to the realm of politics. They conceived the formation of collectives as a plebiscite for or against their rule. By contrast, farmers and peasants insisted on a narrower understanding of politics. Pressured by agitation brigades, they attempted to cast resistance as non-political in order to evade the label of treason. In studies of collectivization, politics can be investigated as a process of conflicting representations that are based on communication between specific actors. It consists of symbolic interaction, though not necessarily on equal terms. In fact, asymmetrical power relations have to be taken into account, as historical studies of dictatorships succinctly demonstrate.[14]

Within this framework, collectivization is conceived as an ensemble of social and cultural practices. Research and investigations based on this

[12] Schatzki, "Introduction," esp. 2f. For an illustrative case study, see Reichardt, "Praxeologie und Faschismus." With regard to the GDR, see Bauerkämper, "Not Dusk, but Dawn."

[13] On language in the GDR, see Schmitt, *Doktrin und Sprache*; Jessen, "Diktatorische Herrschaft." For more general remarks, see Pred, *Lost Words*, 2, 6f.

[14] Frevert, "Neue Politikgeschichte;" Frevert, "Neue Politikgeschichte: Konzepte;" Mergel, "Überlegungen." For a critical assessment, see Weichlein, "Politische Geschichte."

approach deal with the attitudes and conduct of certain individuals and groups involved in collectivization in order to explain the preconditions, resources, dynamics, and impacts of the process. More importantly, this opens the field of study regarding concrete interactions and exchanges among and between peasants, farmers, state and party functionaries, village rural elites, and non-agricultural workers. A detailed study of reactions to collectivization in the GDR, however, reveals that individual and collective action was closely linked to specific experiences and recollections. In fact, the proclaimed "class war" in the countryside gave rise to a vast array of different and even opposing memories involving local actors in a myriad of social practices. SED functionaries strove to impose their politics of history (*Geschichtspolitik*) on the villagers on the one hand[15] and opponents of collectivization disputed the official version of history on the other. The notion of "class war" and in particular the claim that farmers, peasants, and agricultural laborers had "voluntarily" joined collectives clashed with the actual experiences of the bulk of the East German population. In their specific memorial culture, successive collectivization drives were tainted by violence and oppression many villagers associated with Soviet communism. Although these two memorial cultures partially overlapped and cannot be unambiguously related to specific actors in the East German countryside, they can nonetheless be distinguished for analytical purposes.[16]

As this chapter demonstrates, collectivization was largely shaped by memories of the past.[17] Attitudes and modes of conduct *vis-à-vis* the collectivization drive initiated in Eastern Europe in the late 1940s and early 1950s were influenced by the trauma of fascist and authoritarian dictator-

---

15 Bauerkämper, "Not Dusk, but Dawn." The concept of *Geschichtspolitik* refers to attempts to politically exploit history in order to achieve hegemony in public debates; this practice highlights the place of history in politics and the public sphere. See Wolfrum, *Geschichtspolitik*, 25–30. In communist dictatorships, however, competing versions could not be articulated in the official public sphere. In fact, they were confined to small scale, face-to-face communication in local milieus.

16 The concept of "memorial culture" has been proposed in order to study the ambivalent relationship between subjective experiences, historical scholarship and cultural commemorations. See Cornelißen, "Was heißt Erinnerungskultur," esp. 555; Cornelißen, Klinkhammer, and Schwentker, "Nationale Erinnerungskulturen," 12. Also see Hockerts, "Zugänge zur Zeitgeschichte," 17f.

17 Memory and the past are by no means identical concepts. Memory can be conceived as a particular (selective) appropriation of the past that is based on specific worldviews, values, and interests. For a distinction between the two concepts, see Andreas Kablitz, "Geschichte," 233f., 220, 229; Schulze, "Erinnerung per Gesetz," 366.

ships, encounters with the proponents and victims of collectivization in the Soviet Union, and news about more recent transformation processes and experiences of communist policies. Thus, the campaign for building "socialism in the countryside" evoked multilayered memories which were deeply engrained in society's "communicative" and "cultural memory."[18] Contrary to dominant notions of a homogeneous "memorial culture," the clashes of different and even opposing memories in social conflicts merit closer attention and investigation. Moreover, memories of preceding collectivization processes were colored by the experiences actors had gained during the enforcement and appropriation of a wide range of communist agricultural policies. Obviously, these experiences and memories were by no means uniform. On the contrary, the affected subjects appropriated history to express different, often conflicting attitudes and reactions to the SED's agricultural policy. Not only did "individual memory" often clash with "collective memory" and official memorial politics, "individual memory" was also heterogeneous and, as this chapter establishes, can thus be conceptualized as an ensemble of "remembrances."[19]

Following an overview of the political aims, concepts, and strategies of the party leaders, these conflicting memories of villagers and local functionaries will be traced and explained in detail.

CONFLICTING MEMORIES OF THE SOVIET UNION: PEASANTS' PARADISE OR PEASANTS' COMPOUND?

Collectivization of agriculture in the GDR had been preceded by land reform, which was inaugurated by the Soviet occupation authorities and German communists in September 1945. All landed estates greater than 100 hectares and possessions of Nazi *Junkers* were confiscated and partitioned. Small plots of five to ten hectares each were allocated to so-called "new peasants."[20] Yet many "new peasants" were unable to consolidate their possessions. Once collectivization had been officially proclaimed by SED leaders, barren land was integrated into the collectives, or the *Landwirtschaftliche Produktionsgenossenschaften* (LPG). The remaining, economi-

---

[18] For the difference, see Assmann, *Der lange Schatten*, 34, 54.

[19] See the remarks by Dejung, "Oral History," 96, 98, 104f.

[20] "New peasants" designated a group of small landholders in the Soviet zone of occupation and the GDR. They received plots from the commissions that had partitioned the land of expropriated large estate-owners.

cally weak "new peasants" were the first to join the new enterprises. In contrast, independent farmers, who were deeply rooted in local communities, were generally far more reluctant to enter the collectives, wishing instead to preserve their traditional economic autonomy and maintain their social prestige in villages. Many stubbornly resisted attempts by party and state officials to lure or force them into collectives.[21]

After Soviet proposals for the reunification of Germany had foundered in March and April 1952, collectivization was introduced in the GDR.[22] At the Second Party Congress of the SED in July 1952, after most of the officially persecuted *Großbauern* had fled the GDR, many of the recently created "new peasants" abandoned their holdings. Agricultural production therefore fell sharply, and party leaders officially urged peasants to "voluntarily" establish collectives in order to overcome severe shortfalls in East German agriculture. Yet the first collectivization drive, initiated by party commissions, failed in the wake of the popular uprising of June 1953. Organized under strong political pressure, what collectives remained thereafter were increasingly torn apart by internal conflicts. By January 1954, more than 400 farms had dissolved in the GDR. Nevertheless, SED leaders rejected demands from some party functionaries and agronomists who urged the leaders to subsidize individually owned and collective farms, and in 1958 began once again to accelerate the collectivization drive. Starting in the district of Eilenburg in 1959, all farmers were finally pressed to join or establish collectives. In the Socialist Spring of 1960 nearly a half million people were caught up in the process of collectivization, which the *Volkskammer* declared to be completed on April 25, 1960. As new agricultural collectives were hastily established in 1959–60, however, a sizeable number were still fragile and subsequently collapsed. It took the party functionaries and state officials another three years to carry out the forced integration of farmers and peasants into collectives.[23]

---

21 On land reform, see Sandford, *From Hitler to Ulbricht*, 82–118; Bauerkämper, ed., *Junkerland in Bauernhand*; Bauerkämper, *Ländliche Gesellschaft*, 68–122, 239–88.

22 Wolkow, "Die deutsche Frage;" Scherstjanoi, *SED-Agrarpolitik*, 345–424.

23 For detailed accounts of collectivization in the GDR, see Schöne, *Frühling auf dem Lande*, 73–153, 180–234; Bauerkämper, *Ländliche Gesellschaft*, 159–94. For a brief overview, see Steiner, *Von Plan zu Plan*, 115–118. On the revolt of June 1953 in the countryside and its political context, see Witkowski, "Peasants' Revol;" Kluge, "Die verhinderte Rebellion;" Ostermann, *Uprising in East Germany*, XIX–XXXVII, 1–42. See also Jens Schöne's contribution to this volume.

Collectivization in the GDR not only aimed to profoundly transform agriculture but also rural society. In order to mobilize supporters in the countryside, communist rulers presented collectivization in a series of successive campaigns that would involve the wholesale change of elites and establish a new system of agricultural production based on collective or state property in the framework of a planned economy. Thus, SED leaders adopted and pursued policies such as the action "Industrial Workers to the Countryside" that was to transfer workers from industrial plants to agriculture and rural communities. It was initiated in April 1953 and continued until 1963.[24] Party and state functionaries promoted the politicization of social relations in the countryside in order to nourish disagreements and widen divisions between their opponents. Villagers therefore often used collectivization campaigns as a means of settling old scores and to enrich themselves by randomly confiscating property. By staging "class war" in the countryside, communist leaders attempted to enforce the party's conception of political, economic, and social transformation of agriculture and rural society.[25]

"Class war" in the countryside not only mobilized regional and local functionaries but also generated an almost unprecedented shift in the social dynamic of rural society from the late 1940s onwards. As they identified and punished "enemies" of the "Workers' and Peasants' State," officials encouraged the rural population to participate in the suppression of big farmers and in the dissolution of traditional agricultural associations. By promoting collectivization, compliant villagers were able to prove their loyalty to communist rulers. More active supporters expected material rewards and promising careers, hoping party allegiance would lead to influential, prestigious and privileged positions. When pressuring farmers and peasants to join or establish collectives, local and state party functionaries used memories of collectivization in the Soviet Union, which the party glorified as a model. With the support of compliant or enthusiastic villagers, local officials seemed to be able to fulfill the historic mission of the Bolshevik Revolution. As they initiated collectivization in local communities, the functionaries not only demonstrated their power but also appealed to the memories of the emergence of Soviet communism in the countryside from 1929 to 1933. The collectivization that Stalin and Molotov had enforced in the USSR in 1929 was hailed as a major step

[24] Witkowski, "On the Campaign Trail."

[25] Ross, *Constructing Socialism*, 60–66.

towards communist society that allegedly gave birth to communities of selfless villagers cooperating in collectives. By contrast, party and state officials suppressed memories of the persecution and annihilation of the *kulaks*, who had been denounced as "class enemies" in the late 1920s and early 1930s. The official account of collectivization as a huge success, a notion spread by Soviets in the GDR, also nourished the illusion to achieve a historic breakthrough within a few years. Ultimately, this vision reflected the belief in the efficacy of Soviet-style campaign policies.[26]

Yet the official memorial culture of Soviet collectivization did not remain uncontested in the East German countryside. Villagers who rejected or refused to join collectives pointed to the disastrous repercussions of Stalin's violent collectivization campaign in the Soviet Union in 1929. As soldiers or prisoners of war, many Germans—who were already strongly influenced by Nazi propaganda and deep-seated anti-communism—had witnessed starvation and poverty in the countryside. In the village of Werneuchen in *Kreis* Bernau (*Bezirk* Frankfurt/Oder), for instance, farmers argued that they had come to know the despised *kolkhoz* as prisoners of war in the Soviet Union. They rejected collectivization, claiming the system was an economic and social disaster. The decision to abandon collectivization in Poland in October 1956 seemed to confirm this assessment. As late as November 1960, villagers in *Kreis* Liebenwerda (*Bezirk* Cottbus), for instance, argued that a famine would have occurred if collectives had not been dissolved in Poland four years earlier.[27]

Moreover, many former prisoners of war associated collectivization with their personal experiences of hunger. They held the Soviet policy of forced collectivization responsible for severe economic shortages, hunger, famine, and mass migration from the countryside, which they had personally experienced or witnessed. Reports of negative comments about the Soviet collectivization campaign made by Russian forced laborers in Germany during the Second World War were also widely circulated.[28] In

---

[26] Weir, "Der Rausch im Plan," esp. 253f., 258–62; Schöne, *Das sozialistische Dorf*, 84–111. For an instructive introduction to and documentation of the collectivization and dekulakization policies enforced by Stalin and Molotov in the USSR, see Viola, et al., eds., *War Against the Peasantry*, 171–369. A comparative investigation is provided in Bauerkämper, "Vertreibung als Exklusion." Also see Lynne Viola's contribution to this volume.

[27] BLHA, Rep. 730, File 734 (report of February 10, 1960); Rep. 930, File 713/2 (report of November 7, 1960).

[28] BLHA, Rep. 730, File 954 (report of December 24, 1952).

the GDR, these accounts of recent Soviet history reinforced the widespread tendency to identify collectivization with political violence and economic hardship. Refugees from Germany's eastern provinces and territories occupied by the Red Army in 1940 were particularly susceptible to fears of collectivization. Those expelled from Bessarabia, for instance, complained to SED propagandists who tried to convince them to join collectives in *Kreis* Pritzwalk in the autumn of 1952: "You cannot tell us anything, we are coming from *kolkhoz* country ... We have fled from the *kolkhozes* in Russia, and now we are being caught in Germany." Opponents in the East German countryside attempted to uphold their individual autonomy, citing their superior knowledge of collectivization in the USSR and first-hand insights into its disastrous effects. Indeed, farmers and peasants insisted on their agency, which was strongly tied to the freedom to dispose of their property. Deeply influenced by the traditional principle of hereditary succession, they sought to protect their land, buildings, cattle and machinery in order to pass them on to the next generation. Collectivization, however, was imposed "from above" and thus it raised only deep-rooted fears of losing their property but also of being deprived of their birth rights and entrenched cultural traditions. Many villagers feared the loss of their traditional cultural values, anticipated disruptions to their daily lifestyles and resented the unrelenting pressure from communist leaders and functionaries. They complained that the change in agricultural policy had not been announced in the preceding party programs or public statements. They were held accountable for the sudden shift which farmers and peasants saw as coming from party leaders out of touch with village culture and life.[29]

Despite party propaganda, villagers in the GDR did not regard the Soviet Union as a model for agrarian transformation. On the contrary, the history of the USSR was a nightmare which haunted many East German villagers. Memories and stories of violence and oppression as well as the experience of life in the Soviet Union as either soldiers or prisoners of war generated a strong aversion among many East Germans toward the collectives. The system raised the spectre of forced labor and subservience to the increasingly omnipotent party leaders and their regional lieutenants. In popular communicative memory, Soviet collectivization was seen as a threat rather than as a promise. Thus, in the GDR there were two con-

[29] BLHA, Rep. 530, File 1437: 12. Also see BLHA, Rep. 530, File 201: 38f.; Rep. 539, File 762: 99, 119; Rep. 530, Nr. 1421: 37.

tending narratives of the *kolkhoz*: the popular communicative memories and official historiography.[30]

Protest against the forced merger of peasants and farmers occasionally triggered violence, like physical attacks and arson. Villagers frequently threatened to kill party and state functionaries.[31] The collectivization campaign raised memories of atrocities committed by Red Army soldiers during and after the occupation of East Elbian Germany, which seemed to confirm Nazi propaganda about "Asian hordes" advancing into Central Europe. This undermined support for SED leaders, who were widely seen as Soviet puppets. In the 1950s, official remembrance of the war therefore clashed with the memories of the majority of the population in the countryside. On the "Day of Liberation" on May 9, 1954, for instance, when the USSR was glorified in official political folklore, a farmer in the village of Zernsdorf (*Kreis* Königs Wusterhausen) noted ironically: "Well, they have liberated us only from our watches."[32] Villagers felt occupied rather than liberated and associated the advance of the Red Army with destruction, rape and arson. Altogether, large groups of the village population associated collectivization with "Sovietization." As the agitation commissions of the SED pressured independent farmers and peasants to join or establish collectives in the Socialist Spring of 1960, the campaign raised more general fears of social degradation and moral decline.[33]

Collectivization was thus never merely an economic issue—it also touched on notions of honour. The disastrous performance of many GDR collectives in the early 1950s and experiences with collectivization in the Soviet Union reinforced traditional concepts of "backwardness" associated with Eastern Europe. As collectivization in the GDR progressed, resisting farmers and peasants increasingly saw the process as an attempt to impose the Soviet model on Germany. Growing pressure to join collectives evoked feelings of degradation and shame. Deeply rooted stereotypes about the "Asian" threat also resurfaced in the process of collectivization, as rumors about Mongols allegedly taking over expropriated and barren acreage spread in the GDR. Allegations circulated in the villages that hardcore collectivization opponents were being deported to Siberia. Peasants perceived collectivization not only as a threat to their economic

[30] BLHA, Rep. 530, File 1434 (report of March 8, 1960); Rep. 401, File 59: 8.
[31] See, for instance, Rep. 401, File 59: 8.
[32] BLHA, Rep. 530, File 819: 55.
[33] SAPMO-BArch, DY 30/IV 2/5/1359: 86; BAB, DK-1, File 5944: 147; BLHA, Rep. 530, File 732: 5; Wietstruk, *Die sozialistische Umgestaltung*, 20.

freedom, but also as an assault on their self-esteem and autonomy as independent producers and free villagers. Moreover, in the 1950s, fears of inflation evoked memories of the early 1920s, when peasant protests against the controlled economy of World War I peaked into violent confrontations and tensions between rural producers and urban consumers reached a boiling point.[34] Complaints about declining incomes, plummeting production, and reduced productivity were all commonplace in the countryside. Some villagers even expected expropriated large estate owners to return to reclaim their property. Chilling retrospectives on past *Junker* rule, promoted by the government, did less to raise hopes for "better times," than to nourish fears of work in collectives. In September 1952, for example, "new peasants" in *Kreis* Fürstenwalde flatly rejected calls to join collectives because they did not want to once again live like serfs for a landed aristocracy. Opponents of collectivization in the countryside reminded ardent supporters of the "bell," which had been a symbol of the labor regime imposed by the *Junkers* before. By contrast, some members of collectives favorably compared life on the landed estates to the destitution that reigned in the inefficient LPG. These allusions to the old regime were meant to directly challenge the version of history espoused by the new rulers. Obviously, recent history had not been forgotten. On the contrary, conflicting memories of the past served as an incentive for collectivization or as a warning against the agricultural transformation pursued by the SED leadership.[35]

At least for the independent farmers and peasants, agency was largely tied to the freedom to control their possessions. As collectivization abolished individual property, it was seen as a danger to the established values and way of life. In contrast, SED members and supporters of the SED, along with its supporters in the villages, cherished work and life in the collectives and viewed it as liberating. Varying interpretations of the past served to legitimize different models of agricultural production and rural life. Conflicting versions of the recent Soviet history ultimately proved irreconcilable. Even the elaborated politics of history that the SED pro-

[34] BLHA, Rep. 530, File 1429: 72, 80. On the unrest of the early 1920s, see Moeller, *German Peasants*, 116–38; Moeller, "Winners as Losers;" Moeller "Economic Dimensions."

[35] On *Kreis* Fürstenwalde, see BLHA, Rep. 730, File 954 (report of September 30, 1952). Also see BLHA, Rep. 730, File 954 (report of October 8, 1952); Rep. 530, File 956 (report of December 1, 1961). For police and intelligence reports of rumors about Mongols, see, for instance, BLHA, Rep. 530, File 730: 126; Rep. 530, File 1424: 51.

moted proved to be incapable of bridging the gulf between communicative and official memorial narratives.[36]

COMPETING VERSIONS OF GERMAN HISTORY: A HIGHWAY TO DISASTER OR A TREASURE?

In the collectivization campaign in the GDR, clashes over contradictory interpretations of German history proved to be even more important than contending recollections of the Soviet experience. In many rural communities, the political pressure to join or create collectives led to the emergence of opposing memories that profoundly transformed social relations. In fact, the opposition between different memorial cultures and the dynamics of new social conflicts were inextricably intertwined. In many villages, party functionaries, state officials, and supporters of the communist regime evoked memories of authoritarian rule, dictatorial repression, the suppression of the workers' movement and of laborers who toiled on landed estates. From this perspective, which became the official version of the past in the GDR, the communists had finally defeated *Junker* rule in the countryside, abolished Prussian militarism, and given peasants the freedom that they had striven for since the Peasants' War in 1524–25. In the 1950s, SED functionaries and many state officials propagated a polarized view of the past. Whereas the peasants and workers in agriculture were universally seen as the subdued victims of the *Junkers*, large estate owners and big farmers who employed workers were condemned as oppressors. As the vanguard of the urban and landed "proletariat," the communist party was to pursue collectivization as a means of liberating producers from labor exploitation and social inequality. Thus, the pressing legacy of "class rule" in the countryside would finally be obliterated.[37]

After collectivization had been announced in mid-1952, workers and peasants alike were meant to unite in collectives and thus finally resolve divisions between urban and rural society, between industrial workers and farmers in particular, which would bring about the harmonious socialist society so long desired. In many protocols, which were uniformly drafted

---

[36] BLHA, Rep. 401, File 59: 19.

[37] On Marxist condemnations of the *Junkers*, see Reif, "Die Junker," 70f., esp. 524–27. The most influential Marxist account of the Peasants' War has been published by Engels, *Der deutsche Bauernkrieg*. For an overview, see Peter Blickle, "Der Bauernkrieg," 54, 64.

according to central directives, members of newly founded collectives were made to sign declarations glorifying collectivization as the apogee of the continuous and unchanging determination by communists to free peasants from semi-feudal oppression and capitalist exploitation. In the village of Halenbeck in the Brandenburg *Kreis* Ostprignitz, for example, the formation of the local collective, which was named after SED Chairman Walter Ulbricht, was celebrated as the demise of the hated large estate owners and another step toward mobilizing villagers against backwardness that would usher in the utopia of socialist modernity. Members of the collectives at least officially committed themselves to this view of history:

> Liberating us from fascism, the glorious Soviet army under the leadership of the genial military commander J. W. Stalin dispossessed large estate owners and industrial tycoons ... At present, the challenge is to find increasingly powerful means and ways to develop the countryside that has been backward in relationship to the city and industry in all respects ... Free from the unworthy handcuffs of estate ownership, the agricultural laborer and the peasant awaken from the dull sleep of the past centuries. They stretch their limbs and begin to feel their power. Their consciousness has been awakened. Our agricultural laborers and peasants have realized that only the latest agricultural technology for cultivation will enable them to join forces, as the traditional old way of working impedes a further increase in yields.[38]

According to this statement, which announced the formation of the local LPG, land reform liberated peasants and farm laborers from feudal rule, forged a close relationship between the countryside and cities, and laid the ground for collectivization. By joining collectives, rural populations were supposed to benefit from modernization. The rulers of the GDR thus construed a tradition of continuous progress initiated and propelled by the party leadership, mobilizing and educating the rural population. The exaltation of the SED as a vanguard of progress, which was strongly influenced by Marxist determinism, promoted the idea of history as a process of political and social liberation from feudal or bourgeois oppression, accelerating material improvement and raising living standards. This deterministic view was advanced in order to justify the formation of col-

[38] BAB, DK-1, File 5943: 3.

lectives, which the SED leaders propagated as an antidote to the pressing legacies of German history, especially National Socialism.

In their collectivization campaign, party and state functionaries thus contrasted the sufferings of the past against the promise of progress under socialist rule which would bring social harmony and future happiness. Collectivization was to pave the way for a superior mode of production, which would finally supersede capitalist agriculture in a new egalitarian socialist society. In their attempt to delegitimize the Federal Republic of Germany, the SED leaders claimed that the merger of peasants and farmers in collectives had freed the rural population from the pressing burden of German history. In his declaration on the official accomplishment of collectivization in the GDR on April 25, 1960, Ulbricht celebrated collectivization as a liberation movement that would solidify the cherished alliance between workers and peasants. In his speech before the *Volkskammer*, the SED Chairman claimed the party had triumphed over the pressing legacies of German history and paved the way for the bright future of collective work.[39] In the official interpretation, the GDR was the only German state which had left behind a troubled past. This claim of superior historical knowledge was not only directed against the Federal Republic, where private ownership was maintained in the profound agricultural transformation from the 1950s onwards, but was also aimed at silencing proponents of traditional "capitalist" farming. However, leaders tried to decree a view of the past which was derived from Marxist-Leninist dogma and contradicted popular memories—and all versions of history which did not comply with the official view were extinguished.[40]

However, the leadership of the SEP never gained complete control over the past, which remained a strongly contested field. Although the party's conception of history as a succession of "class struggles" and social "formations" that were necessary for a full-fledged communist society rarely met open opposition and rejection in the countryside, serious reservations against this official doctrine persisted among villagers. As police reports indicate, they never ceased equating collectivization with the oppression that they had experienced on landed estates and many peasants directed their anti-communist resentments and stereotypes against

[39] Ulbricht, *Die Bauernbefreiung*, 1161.

[40] On the antagonism between the two German states on collectivization, see Nehrig, "Im Osten proagiert," 14–16; Bauerkämper, *Ländliche Gesellschaft*, 190f; BLHA, Rep. 730, File 732 (report of August 3, 1953); Rep. 730, File 734 (report of February 12, 1960); Rep. 730, File 779 (report of February 14, 1962).

the new collectives. Despite the official rejection and denunciation of National Socialism, moreover, opponents to Stalinist rule subscribed to a "totalitarian view" of modern history. Many villagers thus associated the SED regime with the Third Reich. Occasionally, they even openly interrelated the two dictatorships in the context of the collectivization campaign. As late as 1962, for instance, members of a collective in Crussow (*Kreis* Angermünde) identified their work with life in prison by proposing to seal off the new agricultural enterprise with barbed wire. In fact, some villagers during the forced collectivization of 1960 identified the collectives with concentration camps. A few farmers even tried to provoke SED functionaries by claiming that Marx was no better than Hitler or even that they preferred Nazi rule to Ulbricht's state socialism. The Socialist Spring also evoked memories of Nazi control over agriculture by the corporate state organization, the *Reichsnährstand.* Thus, farmers in the village of Mürow (*Kreis* Bernau) remembered attempts by functionaries of that organization to convince them to sell their holdings and take over a domain in the Ukraine during the Second World War. According to a party report from the *Kreis* Oranienburg (*Bezirk* Potsdam), a chairman of the Peasants' Mutual Aid Association (*Vereinigung der gegenseitigen Bauernhilfe*, VdgB) had emphasized in 1959 that he would resist the formation of collectives as much as the Nazis. Even some local party functionaries raised doubts about the merger of farmers, peasants, and agricultural laborers. Referring to the collapse of Nazism, they predicted an impending upheaval against SED leaders. Equating the SED regime with the Third Reich, villagers directly challenged the legitimacy of the communist dictatorship in the GDR as representing a break with previous German history. Official narratives were openly questioned, not least in order to voice protest against forced collectivization.[41]

Throughout the 1950s, farmers and peasants disputed SED claims of the economic superiority of collectives, which were often so unstable that they deterred rather than encouraged villagers to join them. Disgruntled members of new collectives, moreover, openly rejected the official view that they had voluntarily given up their individual holdings and private property. Thus, they refused to sign declarations about their voluntary entry into the collectives. Some even crossed out the word *freiwillig*

[41] BLHA, Rep. 530, File 891: 145; Rep. 405/15, File 25: 271; Rep. 730, File 732 (report of August 3, 1953); Rep. 730, File 734 (report of February 12, 1960); Rep. 730, File 779 (report of February 14, 1962); Rep. 930, File 713/2 (report of November 7, 1960).

(voluntary) in the statutes which had to be passed before the new enterprises were registered by county councils. Former farmers and peasants also suspended or cancelled their subscription to the *Freier Bauer*, a newspaper with wide circulation among villagers although it was published as an organ of the Peasants' Mutual Aid Association (VdgB). Moreover, the official doctrine of voluntary merger contained a subversive potential which former farmers, peasants, and agricultural laborers of the collectives exploited in order to legitimize their decision to leave or disband their collectives, particularly in the early 1960s. The subjects of the SED's policy in the countryside thus openly rejected official propaganda and explicitly questioned the party's political legitimacy, demonstrating how official politics of history could not so easily surmount strong reservations in communicative memory.[42]

This challenge did not leave official interpretations of German history untouched. In the struggles over collectivization, pressured farmers and peasants disputed the official version of the German past and the memorial culture party leaders espoused. The despair caused by enforced collectivization led many villagers to romanticize rather than condemn Germany's past. In fact, they yearned for the restoration of their prewar lives. Refugees, in particular, demanded to return to their former homes, which they had been forced to leave in the territories beyond the Oder and Neisse. But the native population also projected their fears of and reservations against collectivization, especially on the issue of the GDR's eastern border. In July 1950, the GDR and Poland signed a treaty establishing a new eastern border at the Oder and Neisse river banks, an agreement met with harsh criticism by the public that served as an outlet for frustration over collectivization. Apart from the myth of the Red Army as "liberators" and the glorified alliance with the Soviet Union, which had risen to doctrinal status by the early 1950s, the new territorial order established after the Second World War was a core element of the legitimizing ideology of the SED regime and the GDR as a peace-loving, "anti-fascist" state. In the summer of 1953 and in the autumn of 1956, however, villagers who had been subjected to collectivization openly questioned the new border. Many refugees publicly called for its revision as they longed to return home and reclaim their lost possessions.

---

[42] BAB, DK-1, File 1226: 93; DK-1, File 5878: 315; Osmond, "Kontinuität und Konflikt," 157. On the instability of many collectives in the early 1960s, see Bauerkämper, *Ländliche Gesellschaft*, 479–82.

Many villagers also perceived collectivization as a barrier against a gradual rapprochement with West Germany, casting doubts on the party's claims that the program would promote the reunification of the two German states. Collectivization thus fuelled frustration and resistance over deepening divisions and memories of the heritage of common national culture and history. Many villagers obviously interpreted German history before 1945 not as a disaster and burden but rather as a treasure and source of security. As a last resort, however, opponents of collectivization inevitably held the Soviet Union responsible for widening the gulf between East and West Germans. Altogether, the conflicts over collectivization were marked by a clash of two opposing, mutually exclusive versions of recent history.[43]

Different historical narratives were closely linked to the unprecedented social dynamics in the village communities. On the one hand, the official doctrine of "class war" politicized social relations. By charging opponents of collectivization with political treason, party and state functionaries elevated the merger of farmers, peasants, and agricultural laborers to a plebiscite about state socialism and to an issue of war and peace. On the other hand, supporters of collectivization were exposed to strong pressure in the local communities. Thus, villagers suspected and accused them of striving for posts as chairman of the *kolkhoz*.[44] Political pressure for collectivization not only deepened the gulf between independent agricultural producers and members of collectives but also heightened the existing tensions between established farmers and the "new peasants" as a result of land reform. However, even the latter were divided on collectivization: many "new peasants" who had failed to consolidate their possessions and were incapable of farming their plots tended to join collectives as early as 1952–53, whereas those land recipients who held viable holdings perceived collectivization as a confiscation of their newly-won property. According to an opinion report from the village of Kuhsdorf-Bullendorf (*Kreis* Pritzwalk), for instance, a "new

[43] BLHA, Rep. 530, File 1461 (reports of October 4 and October 6, 1960); Rep. 530, File 1462 (report of May 4, 1960); Rep. 730, File 957 (report of September 22, 1954); Rep. 604/16, File 54: 44. For a detailed investigation into the changing attitude of the SED regime to the new border up to 1950, see Malycha, "'Wir haben erkannt,'' 193–207. On taboos in public discourses in the GDR, see Fulbrook, *German National Identity*, 155.

[44] BLHA, Rep. 730, File 953 (reports of July 7, 1952 and July 10, 1952).

peasant" called the members of the local LPG "traitors and slaves to the *kolkhoz*."[45]

The collectivization of East German agriculture also intensified the antagonistic relationship between refugees and native residents. By January 1946, those expelled from the lost German territories east of the Oder had been allocated 56,676 small land holdings of 7.7 hectares on average in the Soviet zone of occupation. They had received 30 percent of the land that had been distributed to private owners. In Brandenburg alone, 23,234 refugees were given plots. This equalled their share of the population. Yet they were not allotted the necessary cattle and horses, technical equipment, and housing. Many of these "new peasants" were therefore unable to stabilize their new holdings, which in general were not viable. "New peasants" also lacked adequate farming skills and their efforts to increase agricultural production often failed. Those settlers therefore decided to take up new positions that increasingly opened up in industry in the early 1950s and abandoned their land. Many refugees were also among the first to join new collectives that had been launched in mid-1952. This reinforced resentments that the indigenous population harbored against the unwelcome newcomers who differed from traditional villagers in cultural customs, way of life, and religious allegiances. Against the backdrop of widespread xenophobia, desperate refugees seemed to pave the way to collectivization which many established villagers and especially the respected farmers vigorously rejected. Although by no means did all refugee settlers prove incapable of effectively producing and maintaining their holdings, indigenous villagers largely identified the refugees with collectivization. They were therefore marginalized as "foreign" intruders in rural society, both as largely dispossessed refugees and advocates, or even members of collectives.[46]

Most importantly, the process exacerbated conflicts between members of the collectives and independent farmers in the local communities. Violence erupted between these groups in countless villages throughout the collectivization process. Farmers also boycotted the celebrations of the LPG, for example during harvest festivals. Occasionally, the issue even divided the children of the two groups.[47] Moreover, some wives of collective peasants complained that the spouses of established farmers

45 BLHA, Rep. 401, File 59: 193 ("Verräter und Kolchosknechte").

46 Schwartz, *Vertriebene und Umsiedlerpolitik*, 637–654; Meinicke, "Die Bodenreform;" Bauerkämper, *Ländliche Gesellschaft*, 354–68.

47 BAB, DK-1, File 5878: 178; BLHA, Rep. 530, File 730: 8; Rep. 930, File 44: 62.

looked down on them as if they had come from a funeral. In the eyes of the indigenous population these members of collectives seemed to be sad and depressed.[48] Local functionaries were largely incapable of mediating between the opposing groups, and in fact, were often drawn into the social conflicts.[49] Collectivization also divided families and affected gender relations. Farmers and peasants who refused to join collectives frequently referred to objections made by their wives. Many women, however, urged their reluctant husbands to give in to the demands of the agitation commissions, as they expected collectives to alleviate their work burden. Within families, collectivization triggered intergenerational conflicts. Whereas many elderly farmers clung to their cherished property of land, cattle, and agricultural machinery they had inherited from their predecessors, their offspring generally showed a stronger inclination to join or create collectives.[50]

Overall, collectivization undermined established milieus in the countryside and triggered an almost unprecedented shift in the social dynamics within local communities. The conflicts over the politically induced merger of peasants and agricultural laborers into collectives were accompanied by contrasting and opposing memories of the past, which were transmitted by competing narratives. Whereas the communist functionaries propagated the idea of collectivization as the "liberation" of peasants modelled on the successful process in the Soviet Union, large groups of the rural population displayed their disdain and contempt for collectives, associating them with "concentration camps" and the "bell" of the large landowners of the past. Bullied by the agitation groups, peasants and farmers deliberately used emotive language that raised remembrances of the past, thereby challenging the "anti-fascist" claims of the communist rulers who were ultimately unable to bridge the gulf between communicative and cultural memory in the GDR. This divergence also limited the impact of some other "anti-fascist" narratives, such as the cult of the communist resistance warriors of the Spanish Civil War.[51]

---

48 BLHA, Rep. 730, File 970 (report of June 28, 1958): "… sie sehen mich an, als wenn [sic] ich von der Beerdigung komme." Also see SAPMO-BArch, DY 30/IV 2/2023/61: 29.

49 See Lindenberger, "Der ABV im Text," 158; Bauerkämper, *Ländliche Gesellschaft*, 463–76.

50 BLHA, Rep. 530, File 891: 112; SAPMO-BArch, DY 30/IV 2/5/1358: 116.

51 Krammer, "Cult of the Spanish Civil War;" Pritchard, *Making of the GDR*, 100f., 228; Hammerstein, "Schuldige Opfer," 43.

Conclusion: Historical Narratives and Memories as Resources of Agency

The collectivization of agriculture mobilized opposing memories between officialdom and rural society—yet the confrontation was unequal. Whereas party and state officials were able to propagate their version of collectivization, memories that villagers had experienced were sidelined. Nevertheless, resistance to collectivization in the GDR was fuelled by remembrances of the violence that had accompanied the enforced merger of peasants and farmers in the Soviet Union. Moreover, recollections of the inefficiency of collectives in the USSR led East German farmers and peasants to oppose party pressure to abandon their holdings in favor of cooperation. They also spread rumors in small village communities. These activities lent them agency in rural society. Although many dissenters restricted their objections to collectivization policies and did not openly challenge the basic pillars of communist rule in the GDR, they nevertheless posed a significant barrier to the official politics of history.[52]

Collectivization mobilized reservations and popular resentments against the political doctrine of the "leading role" of the Soviet Union, in particular, and fuelled opposition against the glorification of the USSR as a liberating power in official memorial culture. The violence that accompanied the forced merger of farmers in the GDR, especially in 1953 and 1960, was frequently associated to instances of rape and looting that Soviet soldiers had committed in 1944–45. The widespread persecution and oppression in the East German countryside also reminded many East Germans of life as a POW and the stories about *dekulakization* and collectivization that they had heard in the USSR or in Germany. Collectivization thus raised a wide scope of caustic and even contradictory memories. This process was by no means restricted to the GDR. In Hungary, for instance, the suicides of old farmers who had been forced to join collectives in 1954 evoked memories of rape and persecution, thus calling once again the official narrative of the "liberation" of the country by the Soviet Union into question.[53]

Throughout the collectivization process in the 1950s, dissenting views of the past continued to fuel discontent, non-compliance, protest,

[52] In comparative perspective, see Fitzpatrick, *Stalin's Peasants*, 287–96.
[53] Mark, "Remembering," esp. 149.

and opposition toward the politically induced transformation of agriculture and rural society. Yet past memories and experiences resurfaced most clearly in popular memory in times of heightened tension caused by intensified collectivization and political crises, in particular the uprising of 1953 and the temporary thaw following de-Stalinization in 1956–58. During these years, SED leaders proved unable to control the dynamics of long-lasting and more recent memories in the countryside. Although all the grumbling, protest, and revolt did not prevent party and state officials from attempting to impose collectivization on defiant peasants, past memories and experiences shaped the interaction among and between villagers and functionaries. These memories also influenced the modes of appropriation on the part of the diverse actors and led to specific arrangements that by no means conformed to central directives, but lent collectivization its specific combination of both durability and volatility.[54]

As this contribution has demonstrated, the diverse reactions to collectivization were inextricably intertwined with the past experiences of both its instigators and supporters as well of as those individuals and groups affected by the destruction of traditional rural society. Whereas officials and adherents of the regime in rural society used the past as a tool to promote collectivization, the majority of peasants and opponents of the ruling party's policies usually referred to their past experiences to legitimize their opposition to newly founded collectives. Experiences and perceptions of the preceding collectivization campaign in the Soviet Union and interpretations of German history were deeply ingrained in popular communicative memory, which was constituted by social exchange. However, perceptions of collectivization were by no means influenced exclusively by history. Moreover, past experiences are not directly reflected in memories, which are never authentic. Instead memories are shaped by propaganda and popular mentalities which filter perceptions and influence their interpretations.[55]

In the East German countryside, the notion of "class war" was a key point of conflicting memories and competing versions of the past. Yet it had an ambiguous impact on the countryside. While dividing village communities, the campaign for collectivization was continually interpreted and appropriated in ways that the party leaders and state officials had not

[54] This also applies to collectivization in the USSR. See Fitzpatrick, *Everyday Stalinism*, 125–38; Fitzpatrick, *Stalin's Peasants*, 103–51.

[55] See Assmann, *Der lange Schatten*, 25f., 32f.

anticipated. For example, villagers in the GDR exploited calls for "class war" in order to denounce unwelcome neighbors, competitors, and opponents in social and economic conflicts. Collectivization was also frequently used as a means to settle old scores. Altogether, the collectivization drive resulted in an uncontrollable shift in the political and social dynamics in the countryside. Social interaction and economic exchange were marred by profound politicization that ultimately backfired on its supporters. Perceiving any opposition to collectivization as uncompromising hostility to communist rule, party and state functionaries insisted on and reinforced their ideological foundations, and eventually resorted to violence. Although the political radicalization zealously pursued by party functionaries and state officials proved to be self-referential and self-defeating, the campaign intensified social interchange between diverse individual and collective actors.[56]

On a broad scale, memories must be related to their use of history in the process of communication and social exchange. Thus, the role of these types of memories in social action is a promising field of investigation, as this contribution on collectivization in the GDR demonstrates. In village communities and within families, the social dynamics of memories were tied to specific actors utilizing their interpretations of history in order to legitimize their respective behaviour *vis-à-vis* the party policies pursued by the SED. Proponents of collectivization justified their actions by official versions of German history and memorial cultures of the Soviet Union. By contrast, their opponents flatly rejected these historical narratives and attempted to establish an entirely different memorial culture. In the GDR, collectivization was accompanied by social conflicts over the past which were utilized as an incentive or as a barrier to the formation of collectives. In either case, memories and historical narratives were a powerful source of individual and collective action. The area of scholarship on the complex relationship between memory and agency warrants additional investigation in more detail.[57]

[56] Weir, "Der Rausch im Plan," 262f.

[57] See the conceptual deliberations in Schwelling, "Politische Erinnerung."

BIBLIOGRAPHY

ARCHIVAL SOURCES

Brandenburgisches Landeshauptarchiv, Potsdam (BLHA), Rep. 730, File 734 (report of 10 February 1960); Rep. 930, File 713/2 (report of 7 November 1960).

BLHA, Rep. 730, File 954 (report of 24 December 1952).

BLHA, Rep. 530, File 1437: 12. Also see BLHA, Rep. 530, File 201: 38f.; Rep. 539, File 762: 99, 119; Rep. 530, Nr. 1421: 37.

BLHA, Rep. 530, File 1434 (report of 8 March 1960); Rep. 401, File 59: 8.

BLHA, Rep. 530, File 819: 55.

Stiftung Archiv der Parteien und Massenorganisationen im Bundesarchiv, Berlin (SAPMO-BArch), DY 30/IV 2/5/1359: 86; Bundesarchiv Berlin (BAB), DK-1, File 5944: 147; BLHA, Rep. 530, File 732: 5.

Siegfried Wiestruk, Die sozialistische Umgestaltung der Landwirtschaft im Kreis Zossen 1952–1962 (Zossen, 1981).

BOOKS AND ARTICLES

Assmann, Aleida. *Der lange Schatten der Vergangenheit. Erinnerungskultur und Geschichtspolitik*. Munich: Beck, 2006.

Bauerkämper, Arnd, ed. *"Junkerland in Bauernhand?" Durchführung, Auswirkungen und Stellenwert der Bodenreform in der Sowjetischen Besatzungszone*. Stuttgart: Steiner, 1996.

———. "Vertreibung als Exklusion gesellschaftlicher Führungsgruppen. Die Verdrängung der 'Großbauern' in der SBZ/DDR und die Vernichtung der 'Kulaken' in der UdSSR im Vergleich" in Günther Schulz, ed. *Vertriebene Eliten. Vertreibung und Verfolgung von Führungsschichten im 20. Jahrhundert*. Munich: Boldt, 2001, 125–63.

———. *Ländliche Gesellschaft in der kommunistischen Diktatur. Zwangsmodernisierung und Tradition in Brandenburg 1945–1963*. Vienna: Böhlau, 2002.

———. "Collectivization and Memory: Views of the Past and the Transformation of Rural Society in the GDR from 1952 to the Early 1960s." *German Studies Review* 25, no. 2 (2002): 213–25.

———. "Not Dusk, but Dawn: The Cultural Turn and German Social History After 1990." *East Central Europe* 34–35, nos. 1–2 (2007/2008): 39–62.

Bispinck, Henrik, et al. "Die Zukunft der DDR-Geschichte. Potentiale und Probleme zeithistorischer Forschung." *Vierteljahrshefte für Zeitgeschichte* 53, no. 4 (2005): 547–70.

Blickle, Peter. "Der Bauernkrieg" in Peter Wende, ed. *Große Revolutionen in der Geschichte. Von der Frühzeit bis zur Gegenwart*. Munich: Beck, 2000, 49–64.

Cornelißen, Christoph. "Was heißt Erinnerungskultur? Begriff–Methoden–Perspektiven." *Geschichte in Wissenschaft und Unterricht* 54 (2003): 548–63.

———, Lutz Klinkhammer, and Wolfgang Schwentker. "Nationale Erinnerungskulturen seit 1945 im Vergleich" in Christoph Cornelißen, Lutz Klinkhammer, and Wolfgang Schwentker, eds. *Erinnerungskulturen. Deutschland, Italien und Japan seit 1945*, 2nd ed. Frankfurt am Main: Fischer Taschenbuch Verlag, 2004, 9–27.

Dejung, Christof. "Oral History und kollektives Gedächtnis. Für eine sozialhistorische Erweiterung der Erinnerungsgeschichte." *Geschichte und Gesellschaft* 34, no. 1 (2008): 96–115.

Engels, Friedrich. *Der deutsche Bauernkrieg*. Berlin: Internationaler Arbeiter Verlag, 1946.

Fitzpatrick, Sheila. *Stalin's Peasants. Resistance and Survival in the Russian Village After the Collectivization*. Oxford: Oxford University Press, 1994.

———. *Everyday Stalinism. Ordinary Lives in Extraordinary Times: Soviet Russia in the 1930s*. Oxford: Oxford University Press, 1999.

Frevert, Ute. "Neue Politikgeschichte" in Joachim Eibach and Günther Lottes, eds. *Kompass der Geschichtswissenschaft*. Göttingen: Vandenhoeck & Ruprecht, 2002, 152–64.

———. "Neue Politikgeschichte: Konzepte und Herausforderungen" in Ute Frevert and Heinz-Gerhard Haupt, eds. *Neue Politikgeschichte. Perspektiven einer historischen Politikforschung*. Frankfurt am Main: Campus, 2005, 7–26.

Fulbrook, Mary. *German National Identity after the Holocaust*. Cambridge: Polity Press, 1999.

———. "Retheorising 'state' and 'society' in the German Democratic Republic" in Patrick Major and Jonathan Osmond, eds. *The Workers' and Peasants' State: Communism and Society in East Germany under Ulbricht 1945–71*. Manchester: Manchester University Press, 2002.

———. *The People's State. East German Society from Hitler to Honecker*. New Haven: Yale University Press, 2005.

Giddens, Anthony. *The Constitution of Society: Outline of the Theory of Structuration*. Cambridge: Polity Press, 1984.

Granovetter, Mark S. "The Strength of Weak Ties." *American Journal of Sociology* 78, no. 6 (1973): 1360–80.

Hammerstein, Katrin. "Schuldige Opfer? Der Nationalsozialismus in den Gründungsmythen der DDR, Österreichs und der Bundesrepublik" in Regina Fritz, Carola Sachse and Edgar Wolfrum, eds. *Nationen und ihre Selbstbilder. Postdiktatorische Gesellschaften in Europa*. Göttingen: Wallstein, 2008, 39–61.

Hanja, Karl Heinz. "Die Bildung der Bezirke in der DDR ab Mitte 1952." *Zeitschrift für Geschichtswissenschaft* 37, no. 1 (1989): 291–303.

Heydemann, Günther. "Integrale deutsche Nachkriegsgeschichte," *Aus Politik und Zeitgeschichte. Beilage zur Wochenzeitung "Das Parlament,"* B 3/2007 (January 15, 2007): 8–12.

Hockerts, Hans Günther. "Zugänge zur Zeitgeschichte: Primärerfahrung, Erinnerungskultur, Geschichtswissenschaft." *Aus Politik und Zeitgeschichte. Beilage zur Wochenzeitung "Das Parlament,"* B 28/2001 (July 6, 2001): 17f.

Jarausch, Konrad H. "Die Teile als Ganzes erkennen. Zur Integration der beiden deutschen Nachkriegsgeschichten." *Zeithistorische Forschungen* 1, no. 1 (2004): 10–30.

Jessen, Ralph. "Diktatorische Herrschaft als kommunikative Praxis. Überlegungen zum Zusammenhang von 'Bürokratie' und Sprachnormierung in der DDR-Geschichte" in Alf Lüdtke and Peter Becker, eds. *Akten. Eingaben. Schaufenster. Die DDR und ihre Texte. Erkundungen zu Herrschaft und Alltag*. Berlin: Akademie-Verlag, 1997, 57–75.

Kablitz, Andreas. "Geschichte–Tradition–Erinnerung? Wider die Subjektivierung der Geschichte." *Geschichte und Gesellschaft* 32, no. 2 (2006): 220–37.

Kleßmann, Christoph. "Verflechtung und Abgrenzung. Aspekte der geteilten und zusammengehörigen deutschen Nachkriegsgeschichte." *Aus Politik und Zeitgeschichte. Beilage zur Wochenzeitung "Das Parlament,"* B 29/1993 (July 16, 1993): 30–41.

———, ed. *The Divided Past: Rewriting Postwar German History*. Oxford: Oxford University Press, 2001.

Kluge, Ulrich. "Die verhinderte Rebellion. Bauern, Genossenschaften und SED im Umfeld der Juni-Krise 1953 in der DDR" in Wolther von Kieseritzky and Klaus-Peter Sick, eds. *Demokratie in Deutschland. Chancen und Gefährdungen im 19. und 20. Jahrhundert. Historische Essays. Festschrift für Heinrich August Winkler.* Munich: Beck, 1999, 317–35.

Kocka, Jürgen. "Bilanz und Perspektiven der DDR-Forschung. Hermann Weber zum 75. Geburtstag," *Deutschland Archiv* 36, no. 5 (2003): 764–69.

Kotsch, Detlef. "Die Bezirke Potsdam, Frankfurt (Oder) und Cottbus in der DDR (1952 bis 1990)" in Ingo Materna and Wolfgang Ribbe, eds. *Brandenburgische Geschichte,* Berlin: Akademie-Verlag, 1995, 727–94.

———. *Das Land Brandenburg zwischen Auflösung und Wiederbegründung. Politik, Wirtschaft und soziale Verhältnisse in den Bezirken Potsdam, Frankfurt (Oder) und Cottbus in der DDR (1952 bis 1990)*. Berlin: Berlin Verlag, 2001, 261–96.

Krammer, Arnold. "The Cult of the Spanish Civil War in East Germany." *Journal of Contemporary History* 39, no. 4 (2004): 531–60.

Lamla, Jörn. *Anthony Giddens*. Frankfurt am Main: Campus, 2003.

Lindenberger, Thomas. "Der ABV im Text. Zur internen und öffentlichen Rede über die Volkspolizei der 1950er Jahre" in Alf Lüdtke and Peter Becker, eds. *Akten, Eingaben, Schaufenster. Die DDR und ihre Texte: Erkundigungen zu Herrschaft und Alltag.* Berlin: Akademie-Verlag, 1997, 137–66.

———. "Die Diktatur der Grenzen. Zur Einleitung" in Thomas Lindenberger, ed. *Herrschaft und Eigen-Sinn in der Diktatur. Studien zur Gesellschaftsgeschichte der DDR.* Cologne: Böhlau, 1999, 13–44.

——— and Martin Sabrow. "Zwischen Verinselung und Europäisierung: Die Zukunft der DDR-Geschichte." *Deutschland Archiv* 37, no. 1 (2004): 123–27.

Malycha, Andreas. "Wir haben erkannt, daß die Oder-Neiße-Grenze die Friedensgrenze ist. Die SED und die neue Ostgrenze 1945 bis 1951." *Deutschland Archiv* 33, no. 2 (2000): 193–207.

Mark, James. "Remembering: Divided Social Memory and the Red Army in Hungary 1944–1945." *Past and Present* 188, no. 1 (2005): 133–61.

Meinicke, Wolfgang. "Die Bodenreform und die Vertriebenen in der Sowjetischen Besatzungszone" in Arnd Bauerkämper, ed. *"Junkerland in Bauernhand?" Durchführung, Auswirkungen und Stellenwert der Bodenreform in der Sowjetischen Besatzungszone.* Stuttgart: Steiner, 1996, 133–51.

Mergel, Thomas. "Überlegungen zu einer Kulturgeschichte der Politik." *Geschichte und Gesellschaft* 28, no. 4 (October–December 2002): 574–606.

Moeller, Robert G. "Winners as Losers in the German Inflation: Peasant Protest over the Controlled Economy, 1920–1923" in Gerald D. Feldman, et al., eds. *Die deutsche Inflation—eine Zwischenbilanz.* Berlin: de Gruyter, 1982, 255–88.

———. "Economic Dimensions of Peasant Protest in the Transition from Kaiserreich to Weimar" in Robert G. Moeller, ed. *Peasants and Lords in Modern Germany. Recent Studies in Agricultural History*. London: Allen & Unwin, 1986, 140–67.

———. *German Peasants and Agrarian Politics, 1914–1924. The Rhineland and Westphalia*. Chapel Hill: University of North Carolina Press, 1986.

Nehrig, Christel. "Im Osten proagiert, im Westen diffamiert. Die Kollektivierung der Landwirtschaft in der DDR" in Adolf-Grimme-Institut, ed. *Unsere Medien, unsere Republik* 2 (March 1992), 14–16.

Nolte, Paul. "Georg Simmels Historische Anthropologie der Moderne. Rekonstruktion eines Forschungsprogramms." *Geschichte und Gesellschaft* 24, no. 2 (1998): 225–47.

Osmond, Jonathan. "Kontinuität und Konflikt in der Landwirtschaft der SBZ/DDR zur Zeit der Bodenreform und der Vergenossenschaftlichung" in Richard Bessel and Ralph Jessen, eds. *Die Grenzen der Diktatur. Staat und Gesellschaft in der DDR*. Göttingen: Vandenhoeck & Ruprecht, 1996, 137–69.

Ostermann, Christian F. *Uprising in East Germany 1953: The Cold War, the German Question, and the First Major Upheaval Behind the Iron Curtain*. Budapest: CEU Press, 2001.

Pred, Allan. *Practice and Structure. Social and Spatial Transformation in Southern Sweden: 1750–1850*. Cambridge: Barnes and Noble, 1986.

———. *Lost Words and Lost Worlds: Modernity and the Language of Everyday Life in Late Nineteenth-Century Stockholm*. Cambridge: Cambridge University Press, 1990.

Pritchard, Gareth. *The Making of the GDR 1945–53. From Antifascism to Stalinism*. Manchester: Manchester University Press, 2000.

Reichardt, Sven. "Bourdieu für Historiker? Ein kultursoziologisches Angebot an die Sozialgeschichte" in Thomas Mergel and Thomas Welskopp, eds. *Geschichte zwischen Kultur und Gesellschaft. Beiträge zur Theoriedebatte*. Munich: Beck, 1997, 71–93.

———. "Praxeologie und Faschismus. Gewalt und Gemeinschaft als Elemente eines praxeologischen Faschismusbegriffs" in Karl H. Hörning and Julia Reuter, eds. *Doing Culture. Neue Positionen zum Verhältnis von Kultur und sozialer Praxis*. Bielefeld: Transcript, 2004, 129–53.

Reif, Heinz. "Die Junker" in Etienne François and Hagen Schulze, eds. *Deutsche Erinnerungsorte, Vol. 1*. Munich: Verlag C. H. Beck, 2001, 520–36.

Ribbe, Wolfgang. "Das Land Brandenburg in der SBZ/DDR (1945 bis 1952)" in Ingo Materna and Wolfgang Ribbe, eds. *Brandenburgische Geschichte* Berlin: Akademie-Verlag, 1995, 677–726.

Ross, Corey. *Constructing Socialism at the Grassroots: The Transformation of East Germany, 1945–65*. Houndsmill: Palgrave, 2000.

Sandford, Gregory W. *From Hitler to Ulbricht: The Communist Reconstruction of East Germany 1945–46*. Princeton: Princeton University Press, 1983.

Schatzki, Theodore. "Introduction. Practice theory" in Theodore Schatzki, Karin Knorr Cetina, and Eike von Savigny, eds. *The Practice Turn in Contemporary Theory*. London: Routledge, 2001, 1–14.

Scherstjanoi, Elke. *SED-Agrarpolitik unter sowjetischer Kontrolle 1949–1953*. Munich: Oldenbourg, 2007, 345–424.

Schmitt, Dieter. *Doktrin und Sprache in der ehemaligen DDR bis 1989. Eine politikwissenschaftliche Analyse unter Berücksichtigung sprachwissenschaftlicher Gesichtspunkte.* Frankfurt am Main: Lang, 1993.

Schöne, Jens. "Landwirtschaft und ländliche Gesellschaft in der DDR" in Rainer Eppelmann, Bernd Faulenbach, and Ulrich Mählert, eds. *Bilanz und Perspektiven der DDR-Forschung* Paderborn: Schöningh, 2003, 254–59.

———. *Frühling auf dem Lande? Die Kollektivierung der DDR-Landwirtschaft.* Berlin: Links, 2005.

———. *Das sozialistische Dorf. Bodenreform und Kollektivierung in der Sowjetzone und DDR.* Leipzig: Evangelische Verlagsanstalt 2008.

Schreckenbach, Hans-Joachim and Werner Künzel. "Das Land Brandenburg und der brandenburgische Landtag 1945–1952," in Kurt Adamy and Kristina Hübener, eds. *Geschichte der brandenburgischen Landtage. Von den Anfängen 1823 bis zur Gegenwart.* Potsdam: Verlag für Berlin-Brandenburg, 1998, 309–21.

Schulze, Winfried. "Erinnerung per Gesetz oder 'Freiheit für die Geschichte?'" *Geschichte in Wissenschaft und Unterricht* 59, no. 7–8 (2008): 364–81.

Schwartz, Michael. *Vertriebene und "Umsiedlerpolitik." Integrationskonflikte in den deutschen Nachkriegs-Gesellschaften und die Assimilierungsstrategien der SBZ/DDR 1945 bis 1961.* Munich: Oldenbourg, 2004.

Schwelling, Birgit. "Politische Erinnerung. Eine akteurs- und handlungsbezogene Perspektive auf den Zusammenhang von Gedächtnis, Erinnerung und Politik" in Horst-Alfred Heinrich and Michael Kohlstruck, eds. *Geschichtspolitik und sozialwissenschaftliche Theorie.* Stuttgart: Steiner, 2008, 99–121.

Steiner, André. *Von Plan zu Plan. Eine Wirtschaftsgeschichte der DDR.* Munich: Deutsche Verlags-Anstalt, 2004.

Suter, Andreas. "Kulturgeschichte des Politischen—Chancen und Grenzen" in Barbara Stollberg-Rilinger, ed. *Was heißt Kulturgeschichte des Politischen?* Berlin: Duncker & Humblot, 2005, 27–55.

Ulbricht, Walter. *Die Bauernbefreiung in der Deutschen Demokratischen Republik, vol. 2: Juli 1958–Dezember 1960.* Berlin: Dietz, 1962.

Viola, Lynne. *Peasant Rebels under Stalin: Collectivization and the Culture of Peasant Resistance.* Oxford: Oxford University Press, 1996.

———, Viktor Petrovich Danilov, Nikolai Alekseevich Ivnitskii and Denis Kozlov, eds. *The War Against the Peasantry, 1927–1930*, translated by Steven Shabad. New Haven: Yale University Press, 2005.

Wehler, Hans-Ulrich. *Die Herausforderung der Kulturgeschichte.* Munich: Beck, 1998.

Weichlein, Siegfried. "Politische Geschichte" in Stefan Jordan, ed. *Lexikon Geschichtswissenschaft.* Stuttgart: Reclam, 2002, 238–41.

Weir, Todd H. "Der Rausch im Plan. Ursachen und Folgen der Inszenierung von 'Klassenkampf' in der Kollektivierung der DDR-Landwirtschaft 1952–1953." *Deutschland Archiv* 37, no. 2 (2004): 253–63.

Welskopp, Thomas. "*Der Mensch und die Verhältnisse. 'Handeln' und 'Struktur' bei Max Weber und Anthony Giddens*" in Thomas Mergel and Thomas Welskopp, eds. *Geschichte zwischen Kultur und Gesellschaft. Beiträge zur Theoriedebatte.* Munich: Beck, 1997, 173–98.

Wiestruk, Siegfried. *Die sozialistische Umgestaltung der Landwirtschaft im Kreis Zossen 1952–1962.* Potsdam: Zossen, 1981.

Witkowski, Gregory R. "On the Campaign Trail: State Planning and *Eigen-Sinn* in a Communist Campaign to Transform the East German Countryside." *Central European History* 37, no. 3 (2004): 400–22.

———. "Peasants' Revolt? Re-evaluating the 17 June Uprising in East Germany." *German History* 24, no. 2 (2006): 243–66.

Wolfrum, Edgar. *Geschichtspolitik in der Bundesrepublik Deutschland. Der Weg zur bundesrepublikanischen Erinnerung 1948–1990.* Darmstadt: Wissenschaftliche Buchgesellschaft, 1999, 25–30.

Wolkow, Wladimir K. "Die deutsche Frage aus Stalins Sicht (1947–1952)." *Zeitschrift für Geschichtswissenschaft* 48, no. 1 (2000): 20–49.

The Appropriation and Modification of the "Soviet Model" of Collectivization: The Case of Hungary[1]

ZSUZSANNA VARGA

As Cold War conflict intensified toward the end of the 1940s, the efforts to Sovietize Central and Eastern Europe were accelerated, resulting in the large-scale implementation of the Stalinist social, political, and economic model.[2] Despite key differences in the timing and the methods applied in each country, the supremacy of this model was not disputed until Stalin's death.[3] Although criticism of Stalinism first surfaced after his death in 1953, the actual turning point was the Twentieth Congress of the Communist Party of the Soviet Union in February 1956, when Khrushchev delivered his landmark speech denouncing the cult of Stalinism.[4] The resolutions adopted by Soviet party leaders acknowledged the possibility that each country may find its own way to socialism, according to its own special national characteristics. This statement contained an inherent criticism of the dominance of the Soviet model as it was applied across the Eastern Bloc.

---

1 I would especially like to thank Arnd Bauerkämper and Constantin Iordachi for their helpful editorial comments made on earlier versions of this paper. My paper was supported by the János Bolyai Research Scholarship of the Hungarian Academy of Sciences.

2 For more on this period in English, see Berend, *Central and Eastern Europe*; Crampton, *Eastern Europe*; Pittaway, *Eastern Europe 1939–2000*; Swain and Swain, *Eastern*.

3 On the issue of the existence of the Soviet model, I agree with Nigel Swain's interpretation: until Stalin's death the Soviet model was a Stalinist model. However, after 1953 this model changed, it became a "moving target," a changing set of features, due to the Khrushchev's reforms.

4 Fejtő, *A népi demokráciák*, 46–56.

Both the Polish October and the Hungarian Revolution of 1956 had a serious impact on relations between the Soviet Union and its satellites, causing a genuine shock to the socialist system.[5] Stalinist leaders expressed sharp criticism of the policy of de-Stalinization and of the attempts at liberalization that followed. The Chinese leadership, who resented the preeminence of the Soviet Union, was concerned that this development would lead to the weakening of the Communist Party's authority in Eastern Europe. In this tense situation, it was essential for Khrushchev to reconfirm the support of Eastern European communist parties. The Congress of Communist and Workers' Parties in November 1957 provided an ideal opportunity to restore unity and stability to the Socialist Bloc. The resolution adopted there reflected the endurance of Moscow's dogmatic policy—the most obvious sign of which was the party's decision to resume its controversial collectivization program in Eastern European countries.[6] The renewal of this program followed earlier, deeply problematic attempts to implement the program as part of Sovietization at the beginning of the 1950s. Those early campaigns were only partially achieved, and often resulted in severe economic crises and social tensions due to falling food production and problems with the supply chain.[7]

Nigel Swain has identified four different paths of collectivization and socialist agriculture in the Soviet-dominated Central and Eastern Europe: 1) states where the program was abandoned (Poland and Yugoslavia); 2) states which adapted Stalinist model (Romania and Albania); 3) states following a "neo-Stalinist" program (Czechoslovakia, the GDR, and Bulgaria); 4) and the Hungarian case, which constitutes a path of its own.[8] In this volume, József Ö. Kovács examines all three stages of forced collectivization in Hungary (1949–53, 1955–56, and 1959–61), laying the

[5] Fejtő, *A népi demokráciák*, 112–26.

[6] The Moscow declaration contained only general ideological guidelines with no direct instructions as to the acceleration of collectivization. The elimination of capitalist property and placing main means of production under public ownership and a gradual socialist transformation of agriculture were common guidelines of the "building of socialism." Concerning proprietorship the formulation was clear: the fact that the countryside had not yet fully adopted social conditions was a major obstacle in the "building of socialism." The inequalities between industry and agriculture, between towns and countryside, were deemed very harmful and were to be eliminated as soon as possible. See *Az emberiség békéjéért*, 1957, 22–24.

[7] On early collectivization campaigns, see Swain's contribution to this volume.

[8] Swain, "Decollectivization politics." See also his contribution to this volume.

emphasis on the similarities with other socialist countries. In this chapter, I focus mainly on the final phase of the collectivization campaign in Hungary. Taking into account patterns of continuity and discontinuity with the earlier phases, I am searching for possible explanations for the peculiar agricultural development which took place in Hungary after 1959.

I argue that this period is characterized by what can be called a "learning process." What is exactly meant by "learning?" How can historians use this concept? Different scholarly fields—pedagogy, philosophy, policy literature, psychology, and sociology—have approached the "process of learning" from various analytical perspectives. The most useful framework of analysis for my research comes from the field of policy, which since the 1970s has advanced several key theories of "policy learning" that challenge conventional conflict-oriented theories dominant in other disciplines.[9]

It is obvious that political learning encompasses several distinct processes. First, the existing theories differ substantially on the question of who must learn. Some restrict learning to high-level politicians and officials, others extend it to a more general social process involving most members of society. Second, the process also involves an object of learning, that is, something which one is learning about. The existing theories vary on this issue: While all see learning as a general increase in knowledge about policies, some conceive this process in terms of instruments, some in terms of programs, some in terms of policy goals, and some as a combination of these three elements. Finally, there is the question of the effect of learning, i.e. the impact learning has on generating policy change. Again, scholarship is divided on this topic. Some scholars see learning as culminating in organizational change, some as resulting in changes in programs or instruments, and some as resulting in major paradigm shifts in how policy problems are viewed—that is, in which policies are preferred and which are not.

The actors of the present study are decision makers of the Hungarian Socialist Workers' Party (HSWP) on the highest level. In the party-state

[9] These include notions of "political-learning" developed by Hugh Heclo, "policy-oriented learning" developed by Paul A. Sabatier, "lesson-drawing" analyzed by Richard Rose, "social learning" discussed by Peter A. Hall, and "government learning" identified by Lloyd S. Etheredge. There is no shortage of theorization, however, there is a relative lack of empirical work. For a recent summary of different theories, see Nils C. Bandelow's introduction in Maier, Nullmeier, and Pritzlaff, eds., *Politik als*; Dolowitz and Marsh, "Who Learns What."

system functioning in communist Hungary, the learning process I describe refers to the following "leading bodies:" the Political Committee, the Central Committee, and the Secretariat. My investigation is based on documents originating from these bodies as well as from those responsible for producing reports and propositions, e.g. the Agricultural Department and the secretaries of the Central Committee. The second component of the "learning process" is what has been learned. I argue that the object of learning among Hungarian political leaders of the period was how to attain more efficiently their goals in the collectivization campaign. Like in other communist countries, party leadership in Hungary intended to complete collectivization with the least possible social tension and economic recession.

The central questions of this chapter are in what ways did the Hungarian leadership learn from the process of collectivization and from whom. In the first part, I shall examine what HSWP leaders learned from the errors made by the communist regime prior to 1956. The second part will look at what the Hungarian leadership, in preparing for collectivization, had learned from the examples of "fraternal countries." The third part explores a further dimension of "learning," examining what the party leadership had learned from the peasant society. In the conclusion, I will attempt to summarize the characteristics of this complex "learning process." Underlying this study is the key question of whether there indeed was a unique "Hungarian model" of socialist agriculture.

Learning from the Past

In the middle of the 1950s, there was no social group in rural Hungary that did not suffer from the consequences of the Stalinist agrarian policy of the previous years.[10] All farmers were subjected to obligatory deliveries or outstanding taxes. In the worst cases, they even came into conflict with the police or courts. Virtually all landowners were affected by the changes. Some plots had been redistributed via forced land consolidation or the private farmers "voluntarily" offered their land to the state in order to escape party pressure and discriminatory economic penalties. Tens of

[10] For a full discussion of this period, see József Ö. Kovács's contribution to this volume. For more on the subject in both English and Hungarian, see Erdmann, *Begyűjtés*; Lampland, *Object of Labor*; Nagy, "A kulákkérdés;" Szakács, "From Land Reform;" Varga, "Agrarian Development."

thousands were placed on the so-called "*kulak*-list."[11] Many members of collectives intended to leave the collective farms, but were often prevented by a three-year prohibition on leaving according to the Model Charter of 1950.

The unprecedented measures to which to villagers were subjected did not lead to open resistance, but its negative consequences were clearly visible: hundreds of thousands of people abandoned the agrarian sector and 570,000 hectares were left uncultivated. The quantity of agricultural production during the Five-Year Plan (1950–54)—with the exception of the positive year 1951—did not reach the levels of the last prewar year, 1938. In the meantime, tensions grew across the countryside. Recent research suggests that by 1956 despair and hopelessness had permeated rural Hungarian society, and it became increasingly obvious a serious crisis was in the making.[12]

The Hungarian Revolution began on October 23, 1956, with student demonstrations and subsequent armed clashes in Budapest and a number of urban centers. The revolution spread to rural areas within a few days.[13] In nearly every settlement in the country, without any coordination from above, there was revolutionary activity, with villagers establishing new political organizations like national councils, revolutionary committees, and national assemblies, where new leaders were elected and issues discussed. These new forums defined areas of concern villagers expected Imre Nagy's government to address. The urgency of these concerns is demonstrated in the fact that villagers did not compile suggestions and requests, but rather presented the government their demands.

These demands can be divided into two categories. First, there were three basic demands regarding the revolution's fundamental goals that could be found in every village document: 1) Soviet troops must leave the country; 2) the secret police or *Államvédelmi Hatóság*, the despised strong arm of the communist dictatorship, must be disbanded; and 3)

[11] Originally it was a tax-list, which included those wealthy peasants who were forced to pay a special tax (the agricultural development contribution) that was modified in the summer of 1949. Accordingly, a *kulak* was a person whose lands reached or exceeded 14.25 hectares, or whose net cadastral income reached or exceeded 350 gold crowns. Farmers could not be removed from the list, even if they reduced their land to below 14.25 hectares.

[12] See more on this Szakolczai and Varga, eds., *A vidék forradalma*; Szakolczai, ed., *A vidék forradalma*.

[13] Szakolczai, ed., *A vidék forradalma*.

political parties must be allowed to function freely. The second category of demands emanated from the grievances of peasants, which generally called for the wholesale rejection of Stalinist agrarian policies. There were differences in emphasis and the phrasing of these demands, but in essence they were similar.[14] Peasants demanded the government put an end to forced collectivization, return expropriated lands, abolish compulsory deliveries, and reduce taxes.

Farmers continued to articulate their demands even after November 4, when the Soviet army brought an end to the Hungarian Revolution and a Soviet-sponsored Hungarian Revolutionary Workers' and Peasants' Government was established under the leadership of János Kádár.[15] Even though these freshly established village councils did not confront the Kádár government as aggressively as those in the cities, Kádár agreed to meet the farmers' demands.

These demands, which were formulated by farmers between October 26 and November 4, were eventually incorporated within the framework of the socialist system. Even after Kádár took power, farmers did not give up on any of these demands, and sought to improve their situation by continuing the "silent revolution," which would later prove to be a great advantage. In the first half of the 1950s, the majority of farmers stuck to their privately owned land despite repeated collectivization campaigns. This meant that their subsistence did not entirely depend upon the state, unlike that of industrial workers, and they were in a position of keeping the government at bay since the majority of produce came from their land. The Kádár government soon realized that open confrontation with farmers would be politically and economically detrimental, especially with the country's entire industrial production shut down due to a general strike. This situation may explain why the government proved willing to meet farmers' demands and to reform the system's agricultural policy. In November 1956, the Kádár regime issued a statement:

---

[14] See the following examples: "In the interest of the radical change in the situation of working peasantry, we demand changes in delivery-system as well as taxation and the end of forced collectivization." (Kaposvár, October 26, 1956); "A healthy peasant-policy, the abolition of compulsory delivery and changes in the taxation system." (Szakmár, October 28, 1956); "Free farming and an immediate return of landed property as well as equipment and tools unlawfully expropriated from peasants." (Makó, October 29, 1956) in Izsák, Szabó, and Szabó, eds., *1956 plakátjai*, 491, 305, 366.

[15] The new government encountered general dissent, armed opposition, and strikes throughout the country. See Romsics, *Hungary*, 303–11.

> The Revolutionary Workers' and Peasants' Government totally disapproves of the defective agricultural policy of the last years, forcing the organization of cooperatives, forcing farmers to join in, all the procedures that had led to a several-year-long regression of the Hungarian agriculture ... The government is firmly determined to support the farmers' efforts by any possible means to revive agricultural produce and is ready to offer economic aid for cooperatives and individual farmers as well.[16]

The abolition of compulsory deliveries was the most significant measure adopted by the government,[17] and meant one of the seemingly unchangeable elements of the state's planned economy had been eliminated. Forced deliveries had at one time been considered an integral part of the socialist system. Hungary was the first country in the Socialist Bloc to eliminate forced deliveries,[18] a point underscored by Nigel Swain in the final chapter of this volume. He argues that rather than resorting to economic repression, the state established commercial relations with agricultural producers, peasants, and collectives. This paved the way for the introduction of a (limited) free market system to one of the main sectors of the post-1956 Hungarian economy.[19] The end of forced deliveries was soon followed by another significant measure: the permission to quit agricultural collectives, a key demand advanced by farmers during the revolution.

As a result of these measures, two-thirds of the existing collectives were dissolved by January 1957, replaced by several hundred thousand private agricultural units. In general, formerly landed peasants left collectives for independent farming and those who originally had little or no land remained in collective farms.[20] By early 1957, however, both independent and collective farmers largely rejected the system of *kolkhozes* for its lack of alternatives. Farm members wanted the collectives to function differently,

---

16 *Magyar Közlöny*, November 27, 1956. 579–80.

17 *Törvények és rendeletek*, 1957, 62, 68–69, 263–65.

18 Wädekin, *Agrarian Policies*, 1982, 65. See also the chronology.

19 MOL M-KS-288. f. 28/1957/13. ő. e. *Az Élelmezésügyi Minisztérium*, November 19, 1957.

20 The dissolution rate of cooperatives did vary greatly by country regions. The most affected area was the southwestern region (Zala, Somogy, Baranya, and Vas counties), the least affected were Pest, Csongrád, Bács, and Heves counties. MOL M-KS-288.f. 28/1957/11.ő.e. *A Földművelésügyi Minisztérium jelentése*, February 4, 1957.

and abandoned the Stalinist Model Charter.[21] According to a county report from eastern Hungary written in late 1956: "agricultural cooperatives wish to follow the route of free farming and intend to introduce a new order that gives them total autonomy from machine stations and the bank. It is their most sincere suggestion that there should be no fixed patterns or a unified model charter at all, but merely a conceptual framework."[22]

Collective farmers began shaping their own policies, designing more effective forms of producing and regulating their livelihoods according to local needs and circumstances.[23] In many cases, the democratically elected leadership argued with local authorities, especially those in the Communist Party who wanted to enforce the party's supremacy. The primary mission of collective members was to ensure regular and adequate income. Rather than wait for allocation during certain "work units" or "labor days" at the end of the financial year, the collectives decided to offer members shares of produce using a traditional sharecropping or related reward system.[24] This meant that collective members received income throughout the year either as cash or as a share of the produce harvested—and the needs of the members were met before those of the local or state authorities.

As the party reconstructed itself in the wake of 1956, new leadership surveyed the transformation of the country's agriculture system. The radical changes they observed in the surviving collectives generated fierce arguments among party leaders. The more dogmatic elements vehemently attacked initiatives taken by collectives for adapting a capitalist means of production. They argued that it was not possible to build socialism by capitalist methods.[25] At the same time, the reformist faction led by Lajos Fehér argued that by allowing grassroots initiatives, members would become more interested in increasing output and efficiency. This would increase the likelihood of the collective's success, which would in

---

[21] For more on the main features of the Soviet Model Charter of 1935 and its appropriations in Eastern Europe after 1945, see Viola's and Swain's paper in this volume.

[22] HBML XXIII. 2. 4.k. *Jegyzőkönyv a Hajdú-Bihar*, December 4, 1956.

[23] For more detailed information on the history of subsisting or reformed cooperatives after November 1956, see Varga, *Politika*, 17–57.

[24] For some examples of reports which show these tendencies see the documents: MOL M-KS-28/1957/17. ő.e. *Megyei jelentések a termelőszövetkezeti*, 1957.

[25] The advocates of the hard line in agrarian politics came not from the party's Agricultural Department (headed by Lajos Fehér), but from the leadership of the Ministry of Agriculture, lead by Imre Dögei. This is discussed in Adair, "The Agrarian Theses."

turn serve the interests of the state. This group argued that errors in the agrarian policy prior to 1956 and the lessons to be drawn from them had been summarized in a party document entitled "Agrarian Theses of the HSWP" published in July 1957.[26] This document declared that both cooperative and private farming should be supported. It was stressed that the HSWP, in order to improve living standards and to increase the raw material base of industry as well as agrarian exports, intended to create the kind of agrarian policy that would lead to a qualitative and quantitative development in agricultural production.

Learning from "Fraternal Countries"

In the first half of 1957, the HSWP expected Hungarian agriculture would eventually become a multisectored system—a combination of private and collective farming. Distancing itself from the agrarian policy of their predecessors—the Hungarian Workers' Party (HWP)—the HSWP declared both collective and private farming to be a more favorable policy. By the end of 1957, however, the agrarian policy of the HSWP took a sudden turn in response to changes across the socialist region. Party leaders in other countries, acting on directives from Moscow, had reinstituted collectivization campaigns with the aim to complete the process of transforming small-scale farms into large-scale collectives.

Following the Meeting of Communist and Workers' Parties in Moscow in November 1957, the Hungarian party leadership's approach to collectivization had clearly changed.[27] Kádár, in a speech on collectives in mid-December 1957, indicated that the party planned to reintroduce collectivization measures and bring socialism to the countryside. *Népszabadság*, the party's official organ, carried the speech, with the headline: "LET US DEVELOP THE ENTIRE AGRICULTURE—SOCIALISM IS THE ULTIMATE GOAL IN THE VILLAGES, TOO." The contradiction of the title was later resolved in the article by the statement "We support socialist large farms," which was explained as follows:

> Nowadays, some 96% of all peasants, 500 million people in *China* have joined agricultural cooperatives. The same percentage

[26] See Vass and Ságvári, eds., *Az MSZMP*, 102–22.
[27] MOL M-KS-288. f. 5/51. ő.e. *Jegyzőkönyv a PB*, November 23, 1957, Agenda 1.

> is 87% in *Bulgaria,* more than 60% in *Czechoslovakia* and more than 42% in *Romania.* This is the right way! We shall not cheat anyone and *we will not use sticks to force anyone into the cooperatives for that is something we have learnt not to be right. The choice is all yours, which one is better: join a cooperative or return to the old capitalist poverty?*[28] (Italics in original)

The speech was also published in the rural newspaper *Szabad Föld,*[29] and following this, the topic remained on the agenda in the coming years. From early 1958, not a month passed without at least one high-positioned communist politician addressing the issue of collectivization,[30] the speed of "socialist reorganization" emphasized most strongly.

The year 1958 was marked by the party's preparation for collectivization, which included an analysis of experiences in other socialist countries. Party leaders, including Kádár and Prime Minister Ferenc Münnich, travelled abroad to neighbouring states within the Socialist Bloc, visiting nearly all the leaders of their "fraternal parties."[31] Between February and December 1958, the group met leaders in Romania, the GDR, Albania, Bulgaria, and Czechoslovakia. The Soviet Union and Poland, meanwhile, sent high-ranking delegations, led by Khrushchev and Gomułka, to Hungary.[32] Although the purpose was primarily to gather information, these visits were also important for the party's political orientation and a key opportunity for Kádár to establish his footing in the socialist camp.[33]

---

28 *Népszabadság,* December 15, 1957.

29 *Szabad Föld,* December 22, 1957.

30 It is interesting to note that while documents written for inner party circles usually use the expression "collectivization," in statements conceived for the public it is replaced by the "socialist/large-scale reorganization of agriculture." For more on the preparations for the collectivization in the press see the second chapter of my monograph: Varga, *Politika,* 58–81.

31 The Political Committee of the HSWP made a decision on this tight visit in its session held on December 28, 1957. MOL M-KS-288. f. 5/59. ő.e. *Jegyzőkönyv a PB,* December 28, 1957.

32 Nagy, ed., *A magyar külpolitika,* 28–34.

33 These visits were organized by the Department for Foreign Affairs, and the Department for Party and Mass Organizations of the Central Committee of the HSWP, and the documents and reports on preparations serve as valuable sources of scholarly investigation. See MOL M-KS-288. f. 32. cs. and 21. cs.

The most similar production structure to Hungary's agricultural system existed in Poland.[34] Like in Hungary, most Polish collectives were closed after 1956, although Gomułka never relaunched collectivization after that, arguing that the preconditions for such a massive reorganization had not yet emerged in Poland.[35] From the viewpoint of the HSWP, Polish communists also intended to achieve collectivization but lacked a concrete plan to carry it out.[36]

Romania served as a warning to Kádár and his group. In Romania, some provinces in 1957 had been reorganized within a short time, and the collectivization campaign resumed, leading to a breakthrough in the Constanţa region. Full collectivization in that region was accompanied by much violence, and led to peasant unrest and uprising in the immediate post-collectivization period.[37] The Hungarian ambassador to Bucharest, Ferenc Keleti, produced numerous reports on the problems of Romanian collectivization efforts. In October 1957, he reported that in Constanţa, the first and only fully collectivized province, with the best equipment and machinery, that year's wheat crop had been significantly lower than the average wheat harvest in other territories.[38]

Before the Hungarian delegation's official visit to Romania in February 1958, the Department for Foreign Affairs of the CC of the HSWP produced several reports summarizing the local agricultural situation. The Hungarian expert reporting on the agricultural conditions in Romania expressed great concern:

> Even though the acceleration of the socialist reorganization of agriculture yielded significant results in 1957 ... in practice, there were, due to a disregard of the principles of volunteership and progressivity, a great number of abuses and aggressive interventions. This had caused the local population first of all in the provinces of Constanta and Galati, such dissatisfaction, that the party and the government had to take measures to stop forced collectivization.[39]

---

34 MOL M-KS-288. f. 32/1958/14. ő.e. *Jelentés a lengyel*, 1958. See more on this: XIX-J-1-j KÜM TÜK Lengyelország 1945–64, 8. d.

35 See Dariusz Jarosz's contribution to this volume.

36 MOL M-KS-288. f. 5/78. ő.e. *Jegyzőkönyv a PB*, 1958.

37 See Iordachi and Dobrincu's contribution to this volume for a discussion on that collectivization stage. For a detailed analysis, see Iordachi, "Constanta."

38 MOL XIX-J-1-j KÜM TÜK Románia 1945–1964, 27. d.

39 MOL M-KS-288. f. 32/1958/15. ő.e. *Politikai és gazdasági helyzetkép*, 1958.

Initially, the experience observed in Bulgaria and Czechoslovakia was much more positive. Collectivization ended in Bulgaria in the spring of 1958 and around the same time in Czechoslovakia, where some 82 percent of the arable land had been collectivized by late 1958.[40] While the information gathered on Czechoslovakia was carried out primarily through the embassy,[41] Hungarian officials sent high-level delegates to Bulgaria. The first visit was for the Seventh Congress of the Bulgarian Communist Party's celebration of the victory of collectivization in agriculture. The Hungarian delegation was led by Lajos Fehér, who learned the key to Bulgaria's success was the fact that collectivized land had remained private property, with the collective paying its members rent. In his report on the visit he pointed out: "Rent for land is a new economic category without any form of exploitation of cooperative members. This rent does not interfere with the socialist nature of the cooperative. These cooperatives have a flexible structure that allows them permanent development."[42] After the delegation had arrived home, Lajos Fehér assigned his assistant János Keserű to assemble a team of experts to travel to Bulgaria to study the local collectives in more detail.[43]

János Keserű served as the deputy leader of the Agricultural Department of the HSWP and led the delegation sent to Bulgaria. Among the wealth of information gathered, the most significant was the importance of political persuasion in the reorganization of Bulgarian agriculture. One method was to dispatch distinguished party workers and agrarian professionals, who were village natives, to rural areas to rally support for collectivization. Another was to assign members of the Political Committee to manage specified areas or reorganize certain communities in one stretch (which was essential to avoid the consolidation of land-strips). Further methods involved making significant starting capital available to collectives or the flexible management of the size of household plots. Keserű continued:

---

40 See the Gruev and Rychlík chapters in this volume.

41 In the documents of the annual schedule for the embassy of Prague, the Ministry for Foreign Affairs emphasized that the embassy should pay special attention to the situation of the cooperative movement, especially pricing issues, machinery, the social and financial benefits of cooperative members, etc. MOL XIX-J-1-j KÜM TÜK Csehszlovákia, 1945–1964, 27. d.

42 MOL M-KS-288. f. 28/1958/3. ő.e. *Tájékoztatás a Bolgár Kommunista*, 1958.

43 Csete and Erdei, eds., *A szocialista mezőgazdaság*.

> Beside all the good methods we have not concealed the mistakes made by the Bulgarian leaders. One thing we have considered particularly harmful was the occasionally aggressive nature of the methods (pressure from delivery obligations, etc.). This was the aspect that we have cautioned the Hungarian organizers against. The delegation's report laid special emphasis on the dissimilarities between the conditions in Bulgaria and Hungary. In Bulgaria, the party plays a much more important role in the population's everyday life than in Hungary. As a further difference we have pointed out that village communities that include communal labor have, unlike Hungary, a long tradition in Bulgaria.[44]

Keserű also noted how special Hungarian circumstances remained an issue for Hungarian leadership, as party leaders emphasized that "even though experience coming from our friendly socialist nations may be of great use and are definitely worth examining, the conditions in Hungary are still quite unique." This argument would certainly not have been acceptable had it been made prior to the Twentieth Congress of the Communist Party of the Soviet Union. As noted earlier, the final resolution at this congress specified that national characteristics and adaptations in the process of building socialism were justified and acceptable.

The leadership of the HSWP, however, felt pressured to keep pace with the other socialist countries. Yet what had been learned from visits abroad in 1958, combined with Hungary's own experience of the double failure of collectivization in the first half of 1950s, made party leaders cautious. Kádár, in reference to this problem, stated: "Everyone is way ahead of us, even the Albanians. This is true. But what are we supposed to do? Should we make decisions that do not take us any further but eventually even set us back? I believe we must not do this."[45] In April 1958, the Central Committee issued a decree to "elaborate a detailed plan of socialist reorganization" allotting a period of four months for drawing up the scheme that would involve a nationwide discussion.[46] Kádár gave the Agricultural Department specific guidelines for this phase. For example, according to Kádár's directives: "You should by all means make explicit references to some of the best international results and experiences (first of

44 Keserű, *Parasztsorsfordítók*, 88–89.

45 MOL M-KS-288. f. 4/20–21. ő.e. *Jegyzőkönyv a Központi*, 1958. See Örjan Sjöberg's chronology in this volume.

46 MOL M-KS-288. f. 4/16. ő.e. *Jegyzőkönyv a Központi Bizottság*, 1958.

all, examples from Bulgaria and Czechoslovakia). We do not live in isolation and neither do we discuss these matters in such a way ... when do we want to make use of them if not now?"[47]

When the material came under review by the Political Committee in September, leaders decided an additional two months would be necessary to finalize the program.[48] Important factors under debate by party leadership were the preparation, the timing, and the applicable methods of collectivization. Two camps emerged. The leaders of the Ministry of Agriculture—Minister Imre Dögei and Deputy Ministers Mihály Keresztes, András Magyari, Antal Márczis, and Mátyás Szőke—were highly dogmatic, and on every question insisted on reverting to pre-1956 agrarian policy. On the opposing side, the party's Agricultural Department headed by Lajos Fehér vehemently criticized pre-1956 ideas, and in conjunction with the Central Committee, supported reforming the HSWP's agrarian political line in order to avoid past errors and respect peasant interests.[49]

Both camps cited international examples to support their viewpoints. Although they did not share common ground on Soviet agriculture, they agreed it was a successful and exemplary program. However, while the dogmatists considered the strict observance of the *kolkhoz* model and Stalin's Model Charter as the key element of the success of Soviet agriculture, Lajos Fehér and his reformist group held the following interpretation:

> The lifting of rigid restrictions, the extended independence of the *kolkhoz* in the planning process, including the formulation of their own statute, in mechanization, and the implementation of monetary interests in produce became the key elements in the rapid development of Soviet agriculture ... The extended independence of the *kolkhoz* and the increase in monetary interests did not overshadow their responsibility towards the state, in fact, it even came to reinforce it.[50]

---

47 MOL M-KS-288. f. 17/7. ő.e. *Kádár János levele a Mezőgazdasági Osztálynak*, 1958.

48 MOL M-KS-288. f. 5/95–96. ő.e. *Jegyzőkönyv a Politikai Bizottság*, 1958.

49 For more information on the debates between the both sides, see Sipos, "Reform és megtorpanás."

50 MOL M-KS-288. f. 21/1959/28. ő.e. *Jelentés a Szovjetunióban járt*, 1959.

Leaders also turned to China's example, where collectivization ended as early as 1956–57.[51] Dögei and his supporters argued that success in China was a result of the program's reliance mostly on poor peasants and the policy of forcing peasants with medium-sized farms into collectives through taxation pressure. They did not acknowledge that in China a wide range of simple collective forms had been introduced ensuring a more gradual transition. Following a study tour of China, leaders of the National Board of Cooperatives, who shared Lajos Fehér's views, reported:

> We have seen a way of collectivization that is definitely worth considering. The vast majority of peasants had been integrated in lower-grade cooperatives and they were later introduced to the upper-grade ones collectively. This made the consolidation of land-strips more or less unnecessary, since entire communities joined lower- and upper-grade cooperatives collectively ... it is remarkable, that, during the whole process of collectivization, and even now, cooperatives have been granted a considerable level of independence.[52]

In early December 1958, the two camps discussed their opposing platforms in a three-day debate before the Central Committee.[53] On December 7, the Committee sided with the dogmatists, abandoning the reformists' ideas for a more gradual approach to collectivization.

Cross-Border Transfers during the Collectivization Campaign

At the beginning of the collectivization campaign in the winter of 1958–59, just 13 percent of all arable land had been collectivized. By March 1959, this figure had grown to nearly 34 percent, and members of collectives had increased from 200,000 to 500,000 over that three-month period.[54] Thus, the first wave had been quite successful in numerical terms. Moreover, the party was able to collectivize counties (especially

[51] MOL M-KS-288. f. 17/3. ő.e. *A pekingi magyar nagykövetség*, 1958.
[52] MOL M-KS-288. f. 32/1958/15. ő.e. *Jelentés a SZÖVOSZ*, 1958. See also: MOL XIX-J-1-j KÜM TÜK Kína 1945–64. 9. d.
[53] MOL M-KS-288. f. 4/20–21. ő.e. *Jegyzőkönyv a Központi Bizottság*, 1958.
[54] *Mezőgazdaságunk a szocialista*, 1963, 147.

in the western part of the country) where the program had not gone smoothly before. The first campaign of collectivization was evaluated by the Central Committee on March 6, 1959, at which time several participants pointed out that the results were nothing to be ashamed of, even on the international level.[55]

At the same time, farmers resorted to similar defence mechanisms and survival strategies they had developed in previous collectivization drives.[56] Families, for instance, sought means of subsistence outside collectives, sending the best worker to the industrial sector if there was available employment, leaving the wife or an elder member to join the collective. These members worked in the collective only enough to grant them the right to household plots. In the absence of industrial work, the head of the family would seek other opportunities of extra employment, as a day laborer, at construction sites (illegally), or on nearby state farms where pay was given on-site.

Discipline problems in the collective farms worsened. A significant portion of registered members (25 to 60 percent) did little or no work for the collective due to a lack of personal incentives. The average amount of work done by members of collectives decreased each year. The number of work units per farmer family fell from 390 in 1958 to 301 in 1959 and then to 169 in 1960. In addition to this, family members had not been drawn into collective work.[57] As a result, a considerable proportion of peasants continued to exhibit an uncooperative attitude toward joining collectives, although in a different way than before. The entire agricultural sector suffered from the fact that collectives were becoming the scene of passive resistance.

The economic and social problems arising in the agrarian sector eventually became so severe that leadership was compelled to initiate corrective measures. These reforms, however, were preceded by an extensive debate over the different approaches.[58] Kádár believed the suggestion made by Lajos Fehér's reformist group was politically risky. Fehér's camp had questioned one of the most fundamental elements of socialist agriculture, the

---

[55] MOL M-KS-288. f. 4/22. ő.e. *Jegyzőkönyv a Központi Bizottság*, 1959.

[56] This is discussed in Kovács's paper in this volume. For more on the peasants' defense and survival strategies, see Bokovoy, *Peasants and Communists*; Fitzpatrick, *Stalin's Peasants*; Scott, *Weapons of the Weak*; Viola, *Peasant Rebels*.

[57] MOL M-KS-288. f. 17/5. ő. e. *A Központi Statisztikai*, 1960.

[58] For the background to this correction of the HSWP agrarian policy, see Varga, "Impact of 1956."

kolkhoz farm, intending to replace it with a collective that conformed to Hungary's specific conditions.

The Political Committee made every effort to collect as many arguments and examples from the "fraternal countries" as possible, especially Czechoslovakia, where collectivization was completed by 1959.[59] Simultaneously, the Hungarian leaders attempted to find out the opinion of the Soviet Union. 1959 offered many possibilities to do so. The Twenty-First Congress of the Communist Party of the Soviet Union was held in February 1959. Here, the Soviet leadership announced the program of catching up with and even surpassing the achievements of the United States. The Central Committee of the Hungarian Socialist Workers' Party dedicated an entire session to discuss the report on the events of the Twenty-First Congress.[60] In early June, Khrushchev paid a visit to Hungary, and later that month, leaders of socialist countries gathered for a meeting in Bucharest. At the Soviet Union's request, an additional conference on agrarian economic issues was held in Moscow in August.[61] At the end of November, Khrushchev came to Hungary once more in order to attend the Seventh Congress of the HSWP.

Hence, a series of high-level bilateral negotiations were held throughout 1959. At the same time, however, a series of multilateral meetings—including the 13th session of COMECON, held in Budapest July 26–29 and a meeting of communist and workers' parties in Moscow in November—gave Hungarian leadership the opportunity to experience a more extensive international orientation. The series of consultations continued in 1960. At the beginning of February that year, party delegations from across the Eastern Bloc gathered in Moscow for an assembly on agricultural development.[62] Kádár summarized the conference for the Political Committee of HSWP:

> Comrade Khrushchev held a very expressive and explicit speech at the conference and especially at the reception that followed it, on the purpose of providing the population with better supply. Comrade Khrushchev puts it as follows: the masses measure communism and socialism on the degree to which it has improved

59 MOL XIX-J-1-j KÜM TÜK Csehszlovákia 1945–64. 5. d.

60 MOL M-KS-288. f. 5/117. ő.e. *Jegyzőkönyv a Politikai Bizottság*, 1959.

61 Csete, ed., *A moszkvai agrárgazdasági.*

62 The importance of the Moscow meeting is confirmed by the fact that the upper leadership of the HSWP was dealing with it several times in early 1960. MOL M-KS-288. f. 4/29. ő.e.; 5/170. ő.e.; 15/4.ő.e.; 17/5. ő.e.

> their lives. The masses actually consider communism and socialism the form which is secondary to the content. Socialism and communism is a way for them to pursue a better life. If we fail to properly satisfy their [demands], they inevitably lose their faith in socialism and communism itself.[63]

Regarding the sensitive issue of production levels and food supply, Kádár continued:

> The situation in connection with the production of grain is that in our [socialist] camp ... not only can we make no impact on world markets ... but we can supply our own internal needs only with difficulties ... In this connection the opinion was expressed ... that the first duty of every member of the socialist camp is to assure its own grain supplies. In the name of the Soviet delegation, comrade Khruschchev made it absolutely clear to us that they were not willing to take over all the growing (word missing) themselves for an indefinite period of time and offer their grain to every other country.[64]

Khrushchev's statement hit a nerve. Hungary was an important exporter of corn before 1945, but under collectivization the country required corn imports.[65]

The HSWP's suggestion for a viable option was indicated by the title of the first agenda item presented by the Agricultural Department at the following session of the Political Committee: "Proposition Concerning Certain Methods of Remuneration and Distribution of Income Aiming at the Enhancement of the Personal Interestedness of Cooperative Members."[66] If the complicated title obfuscated the proposition's meaning, Kádár was remarkably explicit:

---

63 MOL M-KS-288. f. 4/29. ő.e. *Jegyzőkönyv a Központi Bizottság*, 1960.

64 MOL M-KS-288. f. 4/29. ő.e. *Jegyzőkönyv a Központi Bizottság*, 1960.

65 Between 1934 and 1938 there had been, on average, more than half a million tons of exportable surplus a year. Between 1959 and 1962, an annual average of 227 thousand tons of bread grains had to be imported. See Pető and Szakács, *A hazai gazdaság*, 156–64, 391–408.

66 MOL M-KS-288. f. 5/170. ő. e. *Jegyzőkönyv a Politikai Bizottság*, 1960.

> We have certain principles and measures of proved value that make up the essence of the cooperative movement. This is the work unit ... This is why I cannot agree with the starting point of the material, which declares that the remuneration system due to working units does not properly satisfy the cooperative members' personal monetary interests ... If we issue this, they will say: working units are done for. However, if we say that in many respects, it needs improvement, or that it is, in fact, possible to improve it, we have a quite different starting point.[67]

Although Kádár defended the basic principles of the classic *kolkhoz*—based on Stalin's model, and remuneration based on work units—he took a rather open position in other areas. Ultimately, this meant he would tolerate forms of remuneration that differed from the original Soviet model which had been used during the "capitalist period." These solutions had the advantage of yielding adequate profits without a significant capital investment—and moreover, was an important gesture of compromise with peasants.

Recognizing this dramatic turn in agrarian policy involved serious ideological risks, Kádár sought support from Moscow. In December 1959, at the Seventh Congress of the HSWP, Kádár told Khrushchev about the proposed reforms. Khrushchev responded: "profit-sharing and bonus system have a common result. That is, to incite the individual to reduce the plan. If we could invent a system that would simultaneously raise personal shares and incite workers to fight for an increased plan, then this battle could be won."[68] Khrushchev's latter remark made share cropping acceptable in the end. This form of remuneration, although acknowledged as a feudal and capitalist relic, was nevertheless a key to collectivization's success, as it boosted incentives by allowing members to work for profit. On the other hand, bonus systems tied to a plan's fulfilment kept members interested in a less-ambitious plan, and if the plan proved unrealistic (or unattainable due to bad weather conditions), it immediately reduced the incentives by the members.[69]

After a long debate, the Political Committee decided "that the remuneration of the workers according to the locally proven methods receive

[67] MOL M-KS-288. f. 5/170. ő. e. *Jegyzőkönyv a Politikai Bizottság*, 1960.
[68] MOL M-KS-288. f. 5/170. ő. e. *Jegyzőkönyv a Politikai Bizottság*, 1960.
[69] For more detailed information on different remuneration systems of Hungarian collective farms, see Varga, *Politika*, 51–57.

the widest possible dissemination ... what constitutes the best way of doing this has to be decided by the general meeting of the cooperatives in every case."[70] Although the regime had officially accepted that collective farms could design policies that would advance the profit of their members, the Political Committee issued the decision only to county party committees, adding there was no need "at the present stage for press and broadcast to report on new methods of remuneration." In early 1960, the dogmatic Minister of Agriculture Imre Dögei was replaced by a "progressive" manager of a collective, Pál Losonczi—a more public indication that reforms to Hungary's agrarian policy were underway.[71]

At the Political Committee meeting in February 1960, a majority of the leaders accepted that forms of remuneration which differed from the Soviet model were only a temporary concession. Most local party leaders did not tolerate such reforms, not even as temporary solutions, and adhered to the principles of Stalinist agrarian policy.[72] In response, Lajos Fehér and his group in 1960 undertook an extensive campaign to gather examples from Hungary and abroad in support of a turn in agrarian policy.

Although collectivization just ended in the GDR in the spring of 1960,[73] there were only a few references to the East German example. This finding is worthy of further examination since Hungarian experts in collectivization paid visits to the GDR and the Hungarian Embassy had sent information on agriculture on a regular basis. The main point of reference remained Czechoslovakia. Hungarian delegations were especially interested in problems that had arisen after collectivization had been completed in 1959, as well as the progress of consolidating existing collectives. Ministerial visits became a regular occurrence, and, as documents from the Ministry for Foreign Affairs show, the ministry was expecting information on certain issues from the embassy in Prague,[74] including what methods party leaders used to keep young people in villages, the value of working units and the development of produce in agricultural collectives. Among the observations were the following:

---

70 MOL M-KS-288. f. 5/170. ő. e. *Jegyzőkönyv a Politikai Bizottság*, 1960.

71 *Népszabadság*, January 15, 1960.

72 In a very interesting process, the party leadership began "educating" the local party organizations. See more on the channels of this process: Varga, *Politika*, 82–91.

73 Bauerkämper, *Ländliche Gesellschaft*; Schöne, *Frühling*.

74 MOL XIX-J-1-j KÜM TÜK Csehszlovákia 1945–64. 14. d.

> The comrades from Czechoslovakia informed our delegation—the report of the delegation leader said—that one of the main problems had been the one of manpower ... the widespread adoption of monetary interest is considered an important means of increasing agricultural produce, motivating cooperative members and consolidating cooperative farming. Monetary interest comes across in a variety of forms... The situation after the completion of massive integration required the adoption of a new set of incentives ... The individual cooperatives are granted a great level of independence in the choice of their incentives.[75]

There appeared a number of similar explanations from delegates sent to the Soviet Union, too. The uniformity of the interpretations is supported by the following quotation:

> The lifting of rigid restrictions, the increase of the cooperatives' independence as well as the widespread application of monetary interests became key elements in the rapid development of Soviet agriculture ... Although they considered working units the basic elements, the *kolkhoz* apply the various incentives with a great level of independence.[76]

The emphasis on the variety of local solutions and on the independence of collectives was a significance feature in the debates among the leaders, because not all members of the county and district party apparatus agreed with Lajos Fehér's reformist policies. Furthermore, there appeared a great number of critical references to the Chinese "People's Communes."[77] In the summer of 1960, participants of the Romanian party congress in Bucharest disapproved of how China expressed the supremacy of communes over the *kolkhoz* and collective forms of agriculture. Participants agreed the Chinese should abstain from imposing its own models on other parties.[78] Hungarian sources reported on the problems of Chinese agriculture, with increasing directness:

[75] MOL M-KS-288. f. 17/5. ő.e. *Jelentés a Csehszlovák*, 1960.
[76] MOL M-KS-288.f. 28/1960/30. ő.e. *Jelentés a Szovjetunióban*, 1960; MOL M-KS-28/1960/30. ő.e. *Feljegyzés az SZKP KB mellett*, 1960.
[77] For more on this period, see Fairbank, *China*, 368–81.
[78] MOL M-KS-288. f. 4/35. ő.e. *Jegyzőkönyv a Központi Bizottság*, 1960.

> Wherever we went, everybody kept saying that they consider agricultural communes the most appropriate form that carries the germs of communism. They referred to various data supporting the superiority of communes. They considered the role of monetary interests secondary in the agricultural domain, they also believed that people's ideological and political behaviour and conscience are essential in increasing the produce.[79]

By late 1960, diplomats continued to report on the crisis of agriculture and food supply, adding that "the worsening situation in agriculture was preceded by the establishing of people's communes. Our Chinese comrades have overestimated political work at the expense of economic work. Peasants were discussing political issues for several days and left the harvest untouched or did not cultivate their land properly."[80]

Reports indicated the increasing severity of the Chinese agricultural crisis, documenting how meat rations in Beijing were reduced to a minimum of roughly 100 grams per month, that poultry and eggs were practically unavailable—only the ill and pregnant women were given eggs, and even then, were allotted a maximum of five for the entire duration of their pregnancies.[81] The reports added that despite all the difficulties, Chinese leaders should continue to believe that people's communes were much better models of collective farming than the *kolkhoz* in the Soviet Union or collectives in other "fraternal countries."[82] Therefore, the Hungarian criticism of China was not based on ideology but rather on the fact that China's agrarian policy had endangered the food supply of the entire population. This shows how internal stability and the security of supply remained a high priority for the Hungarian decision makers.

Making Collective Farms Work: Learning from Peasant Practices

In February 1961, the Central Committee of the HSWP issued a communiqué informing the public that a turning point had been reached regarding Hungarian agriculture: "Following industry, Socialist relations in agri-

[79] MOL M-KS-288.f. 21/1960/29. ő.e. *Jelentés a Kínában járt*, 1960.
[80] MOL M-KS-288. f. 17/6. ő.e. *Különböző jelentések*, 1960.
[81] MOL M-KS-288. f. 17/6. ő.e. *Különböző jelentések*, 1960.
[82] MOL M-KS-288. f. 17/6. ő.e. *Különböző jelentések*, 1960.

culture have also become dominant ... around 90% of all arable land in Hungary belongs to the socialist sector. State farms have a share of 14%, agricultural cooperatives some 72% and lower-type cooperatives a share of 3%. Some 3,055 out of a total of 3,273 towns and villages in Hungary are cooperative towns and villages."[83] In 1961 there were 271 state farms, approximately 4,200 collectives, and only 165,000 registered individual farms.[84]

In statistical terms, therefore, collectivization was a success—yet serious problems persisted. Over the next five years (1960–65) agricultural production hardly reached the average yields of 1958–59. There were also supply shortages for several years because collectivization had generated a sharp fall in national livestock herds (excluding sheep) and the condition of livestock deteriorated.[85] Most collectives initially failed to demonstrate the advantages of large-scale production, as they were burdened by shortages of equipment and labor.

Despite the formal success of collectivization and its rapid completion, authorities realized that while they had been able to force people to join collectives, ensuring members would work diligently and conscientiously was a more difficult challenge.[86] In order to promote growth in agricultural produce, the party needed the diligence and ambition of the Hungarian farmers badly, not to mention the buildings and instruments of production used in the household plots. This situation again brought forth pragmatic problem solving that had only been accepted as a temporary concession during the final phase of collectivization. This meant the consideration and toleration of the demands made by the peasants.[87] The way to proceed was learning from the peasants. In order to do this, they had to become acquainted with the local initiatives made by the peasants in detail.

The Agricultural Department of the Central Committee and the Ministry of Agriculture gathered information on local remuneration and work organization initiatives. In addition, two Hungarian Academy of Sciences research institutes, the Agricultural Research Institute and the Institute of

---

83 Vass and Ságvári, eds., *Az MSZMP határozatai*, 497–501.

84 *Mezőgazdaságunk a szocialista*, 1963, 25–27.

85 *Mezőgazdaságunk a szocialista*, 1963, 68–79.

86 Fazekas, *A mezőgazdasági*, 187.

87 The cooperative membership for example won the right to keep more animals on their household plots, to undertake sharecropping, and to receive premiums in kind. For a full discussion of "family labor" and "socialist wage labor" in Hungary's cooperative agriculture, see Swain, *Collective Farms*, 25–50.

Economics, were enlisted to examine more closely the agricultural practices launched by Hungarian farmers.

Each year after analysing the year's productivity, the Minister of Agriculture issued suggestions for possible changes in the remuneration system and distribution of income in collective farms. The minister was therefore able to offer solutions that had not been authorized officially but often had been ongoing practice in the everyday operation of collectives. In early 1961, the Minister of Agriculture announced his first proposals for distributing profits and for the remuneration of the work of its members. He proposed to accept those methods of rewarding and organising work that had been practiced by farmers in the cooperatives although they had not been legally allowed.[88]

The following examples show the unique dynamics of the interactions between the party-leadership and the agricultural collectives.[89] The first case follows how party leaders adjusted their opinions on share cropping. In 1961, this practice moved from illegal to tolerated, at least temporarily. In 1962, sharecropping was legalized after the party acknowledged the practice served the interest of both collective members and the state. Even though the practice was rooted in the exploitive feudal system, party leaders acknowledged it provided the necessary economic incentives and helped boost large-scale production. The second case deals with the practice of cash payments. During the 1950s this form of remuneration belonged to the "forbidden category" because it had little to do with the system of work units. In 1962, cash payments to members became universal, abandoning the dogmatic Stalinist model of remuneration based on work units.

All of this was an important message toward the membership of agricultural cooperatives. The predominance of those forms of remuneration that had been acceptable for members became apparent. This made it possible for members to anticipate the amount of income they would receive for a certain kind of work. Life in the cooperative became more predictable, partly due to the stabilization of household farming plots.

By the time the three-year prohibition to quit the cooperatives was over and members learned that they were unable to retrieve their piece of land, they faced a serious dilemma. And, due exactly to the above mentioned experience, the vast majority of the members remained in the cooperatives. In the background, we may also recognize an interesting sociopsychological

88 *Javaslatok a jövedelemelosztás*, 1961.

89 For more detailed information on this process, see Varga, *Politika*, 82–103.

change. As an indication of this change, let me quote from one of the interviews: "It is hard to relinquish individual farming and this attitude but once we are in a cooperative, this is where we have to make our living and work enough to be able to do so."[90] The gradual change in the members' attitude is also reflected in the statistical data: the number of working units per capita increased as did the involvement of family members.[91]

In taking a more pragmatic line, party leaders faced the problem that local measures concerning remuneration and work organization were largely inconsistent with the Stalinist idea of the *kolkhoz*. Since Hungarian leadership did not wish to confront the basic doctrines of the Soviet model, local reforms and measures had been authorized only in practice without being officially legalized. The practice of authorizing local organizations was an important factor. The administration of the HSWP largely consisted of former members of the Hungarian Workers' Party who strictly adhered to Stalin's original principles, and the idea that the primary form of work organization was the brigade and the main form of socialist remuneration was the system of work units.

Both the mass media and training courses served as tools for changing the dogmatic attitudes of local leaders and to spread more pragmatic agrarian policy. Nevertheless, this could only yield long term results. More importantly, however, a new mediator between the political leadership and the collectives had been developed: the chairmen of the collectives, who had played a very different role than the farm president in 1950s. The farm president was a political cadre usually of worker origin, selected by the party leaders to ensure the collective farm followed the Soviet *kolkhoz* pattern and delivered required produce to the state. During the third wave of collectivization, the party began to accept successful middle-class farmers and so-called *kulak*s as chairmen, a move that put experienced and generally respected local farmers into positions of leadership.[92]

Party leadership also accepted the idea that grassroots initiatives would not weaken or disintegrate collectives. On the contrary, such initiatives tended to strengthen the collective, and the interaction between farmers and local and central leaders meant an increasing number of local initiatives were not only tolerated but openly supported by party officials. This practice gave agricultural collectives increasing room to

90 The author's interview with L. F., January 21, 2002.

91 *Mezőgazdasági Statisztikai*, 1965, 281.

92 The sociological research was dealing with this tendency already from the early 70s. For a sample of this work, see Swain, *Collective Farms*, 114–29.

manoeuvre.[93] Eventually, in the course of the economic reform process of 1966–67, the *de facto* realities in the cooperative sector were given legal recognition.

Conclusion

The 1956 Hungarian Revolution posed a major challenge to the Soviet dominated Eastern European system. The event had a powerful impact on both Hungarian and Soviet decision-makers. After 1956, Hungary became a "high priority" country for Soviet authorities,[94] and was given "special treatment," which often meant a higher level of tolerance. Soviet leadership insisted on internal stability but otherwise showed remarkable flexibility in tactical questions. This was reflected most clearly in Hungary's unique path in terms of its agricultural development. Another factor was the growing importance of agriculture. While the Stalinist economic policy subordinated agriculture to industrial development, Hungarian leadership attempted to compensate for its lack of political legitimacy after 1956 by pursuing policies to raise living standards. This effort depended heavily on food supply. Meanwhile, after the hard-line Stalinist model failed twice in Hungary in the early 1950s, party leaders were compelled to seek an alternative.

The search for a special path to Socialist agriculture can be seen as a "learning process." In 1957, during its first year in power, the HSWP focused on learning from the errors in agrarian policy made by their predecessors. Confronting past failures was characteristic of the political discourses and practices of the time. Moreover, agriculture was the first sector to undergo reforms with the new, publicly announced guidelines.

The need to learn from other communist states became more pressing after 1958. Kádár summarized this necessity as follows: "We should say that we keep an eye on international experience, in general, with special regard to the example of the Czechs and the Bulgarians being the most similar to our own conditions."[95]

---

[93] The tendency mentioned appears with regard to both national and local archival sources. MOL M-KS-288.f. 28/1962/9. ő.e. *Megyénkénti jelentések*; 28/1963/37., 38., 39. ő.e.; 28/1963/13. ő.e. *Jelentés a Földművelésügyi Minisztérium Kollégiumához*.

[94] Rainer, "The Sixties in Hungary."

[95] MOL M-KS-288. f. 5/109. ő.e. *Jegyzőkönyv a PB*, 1958.

The minutes of the Political Committee and of the Central Committee of the HSWP meetings show that most references concern Bulgaria, Czechoslovakia, and—surprisingly—China. These three countries were clearly mentioned as positive examples by party leadership. Although it is difficult to assess the actual impact of these examples, we may assume that the Hungarian leadership followed Bulgaria's example of obligatory rent for land. The course of collectivization in Czechoslovakia confirmed the notion that it was useless to push the organization of "higher"-type collectives in territories with unfavorable physical conditions. China offered a great deal of experience concerning the gradual organization and the role of "lower"-type collectives. This might have led Hungarian professionals to establish these "lower" collectives in certain areas of Hungary.[96] In 1960, party leaders withdrew their support for the Chinese example. Although the Soviet-Chinese conflict played an important part in the reconsideration of the Chinese example, for Hungarian leaders the key factor was China's severe food supply crisis. Providing standard food provisions had been a top priority for the Hungarian party leadership since the 1956. Hence, countries like Romania, and later China, were perceived as negative examples. In high-level party documents, there are relatively few references to Soviet agriculture. This is mainly due to the fact that after 1956, the Soviet leadership did not require other socialist countries to refer to them in every case.

This learning process from neighbouring communist countries and parties was a conscious effort by Hungary's Communist Party. According to Kádár: "Our party is the youngest one among the brotherly parties. As to our work, we may say that we have two critics, who are supporters at the same time: the Hungarian workers and the community of brotherly parties."[97] Hungarian leadership used every opportunity to acknowledge the importance of continuous learning, even at times when it was no longer required. An important factor in this dynamic was the changing relationship between Kádár and Khrushchev, marked initially by interdependence and later by sympathy and even friendship.

In 1960–61, there were changes in the orientation of this "learning process," as party leaders began to learn from peasants as opposed to other states and leaders. Or, as Kádár many times said, "the knowledge of the Political Committee derives from the experience of the masses." The pri-

96 For a local study of "lower" type cooperatives, see Hann, *Tázlár.*

97 MOL M-KS 288. f. 4/14. ö.e. *Jegyzőkönyv a KB*, 1957.

ority of learning from the peasants was not a deliberate decision but rather a result of internal and external pressure.

This "learning process" did not stop at the end collectivization. In fact, the process of learning from peasants continued, and was further supported by scientific research institutes. Likewise, there emerged another change in the direction: Hungarian leadership began to learn from the West, despite the continued rhetorical emphasis on experiences from the other communist countries. Hungary began to import modern technology, including not only machines but also closed-production systems, from the Western countries, especially from the US and the Federal Republic of Germany. This new direction of the "learning process" was motivated by a pragmatic factor, as well: the need for quick food production.

From the beginning of the 1970s onwards, Hungarian agriculture had indeed been able to satisfy the requirements of three different markets: the domestic market, COMECON, and even the capitalist sector. Within COMECON, Hungary's success was a unique achievement. But the "Hungarian agricultural miracle" also became a topic among Western analysts who tended to talk about a specific "Hungarian model." In the 1980s, Hungary's agricultural system became a model for liberal reforms around the Socialist Bloc.[98]

However, the party's cautious and gradual deviation from the Stalinist model was never openly admitted by Hungarian leadership. Kádár and his associates sought to avoid an ideological dispute with Moscow, instead making selective adjustments to the Soviet model based on successful grassroots agrarian practices. Recent research shows that this experience from agricultural practice had a strong affect on market, financial, and commodity conditions, which in turn had a significant stimulating effect on the economic reform process in the 1960s.[99] It helped usher in the New Economic Mechanism (NEM), which was the most far-fetched and theoretically innovative reform of any country in the region outside Yugoslavia.[100]

98 Harcsa, Kovách, and Szelényi, "Hungarian Agricultural Miracle."

99 See more on this: Varga, "Agriculture."

100 For a comparative history of socialist economic reforms, see Boyer, ed., *Zur Physiognomie*.

BIBLIOGRAPHY

ARCHIVAL SOURCES

Magyar Országos Levéltár, Hungarian National Archive, (MOL):

Fond: M-KS-288. Files: 4/14. ő.e. *Jegyzőkönyv a KB 1957. november 29-i üléséről* [Minutes of the CC of HSWP, November 29, 1957]; 4/16. ő.e. *Jegyzőkönyv a Központi Bizottság 1958. április 25-i üléséről* [Minutes of the CC of HSWP, April 25, 1958] Agenda 1. *Jelentés a falusi pártmunka helyzetéről* [Report on the situation of party work in the countryside]; 4/20–21. ő.e. *Jegyzőkönyv a Központi Bizottság 1958. december 5–7-i üléséről* [Minutes of the CC of HSWP, December 5–7, 1958]; 4/22. ő.e. *Jegyzőkönyv a Központi Bizottság 1959. március 6-i üléséről* [Minutes of the CC of HSWP, March 6, 1959]; 4/29. ő.e. *Jegyzőkönyv a Központi Bizottság 1960. február 12-i üléséről* [Minutes of the CC of HSWP, February 12, 1960]; 4/35. ő.e. *Jegyzőkönyv a Központi Bizottság 1960. június 29-i üléséről* [Minutes of the CC of HSWP, June 29, 1960]. Agenda 1: *Beszámoló a román pártkongresszusról* [Report of the party delegation on the Romanian party congress]; 5/51. ő.e. *Jegyzőkönyv a PB 1957. november 23-i üléséről* [Minutes of the PC of HSWP, November 23, 1957] Agenda 1. *Kádár János beszámolója a kommunista pártok moszkvai értekezletéről* [János Kádár's report on the meeting of communist and workers' parties in Moscow]; 5/59. ő.e. *Jegyzőkönyv a PB 1957. december 28-i üléséről* [Minutes of the PC of HSWP, December 28, 1957]; 5/78. ő.e. *Jegyzőkönyv a PB 1958. május 13-i üléséről* [Minutes of the PC of HSWP, May 13, 1958] *Kádár János beszámolója a lengyel párt- és kormány küldöttséggel folytatott tárgyalásokról* [János Kádár's report on the negotiations with the Polish party and government delegation]; 5/95–96. ő.e. *Jegyzőkönyv a Politikai Bizottság 1958. szeptember 23-i üléséről* [Minutes of the PC of HSWP, September 23, 1958]; 5/109. ő.e. *Jegyzőkönyv a PB 1958. december 29-i üléséről* [Minutes of the PC of HSWP, December 29, 1958]; 5/117. ő.e. *Jegyzőkönyv a Politikai Bizottság 1959. február 11-i üléséről* [Minutes of the PC of HSWP, February 11, 1959] *Kádár János beszámolója az SZKP XXI. kongresszusáról* [János Kádár's report on the Twenty-First congress of the Communist Party of the Soviet Union]; 5/170. ő.e. *Jegyzőkönyv a Politikai Bizottság 1960. február 16-i üléséről* [Minutes of the Political Committee of HSWP. February 16, 1960] Agenda 1. *Határozati javaslat a termelőszövetkezeti tagság anyagi érdekeltségének fokozására irányuló különböző munkadíjazási és jövedelemelosztási módszerekkel kapcsolatban* [Proposition concerning certain methods of remuneration and distribution of income aiming at the enhancement of the personal interestedness of cooperative members]; 17/3. ő.e. *A pekingi magyar nagykövetség jelentése Kína mezőgazdasági miniszterének tájékoztatójáról. 1958. február 6* [Report of the Hungarian embassy in Beijing on the speech held by the minister for agriculture of China, February 6, 1958]; 17/5. ő.e. *A Központi Statisztikai Hivatal feljegyzése a paraszti családok és a paraszti népesség számának alakulásáról* [Record of the Central Statistical Office on the Change of the Number of Farmer Families and Agricultural Population in General] November 1960.; 17/5. ő.e. *Jelentés a Csehszlovák Szocialista Köztársaságban járt mezőgazdasági pártmunkás küldöttség tapasztalatairól, 1960. október* [Report on the experience gathered by the agricultural party delegation in Czechoslovakia. October 1960]; 17/6. ő.e. *Különböző jelentések, feljegyzések a kínai helyzetről, 1960. december* [Various reports and notes on the situation in China, December 1960]; 17/7. ő.e. *Kádár János levele a Mezőgazdasági Osztálynak, 1958. július 23.* [A letter from János Kádár to the Agricultural Department, July 23, 1958]; 21. ő.e.; 21/1959/28. ő.e. *Jelentés a Szovjetunióban járt pártmunkás küldöttség tapasztalatairól* [Report on the experience gathered by the party delegation during their trip to the Soviet Union]; 28/1957/11.ő.e. *A Földművelésügyi Minisztérium jelentése a termelőszövetkezeti mozgalom helyzetéről* [Report of the Ministry of Agriculture on the situation of collective farms, February 4, 1957]; 28/1957/13. ő. e. *Az Élelmezésügyi Minisztérium előterjesztése az MSZMP Politikai Bizottságához az új felvásárlási rendszer eredményeiről és tapasztalatairól* [Food Ministry submission to the Political Committee of the HSWP on the achievements and experiences with the new central purchasing system], November 19, 1957; 28/1957/17. ő.e. *Megyei jelentések a termelőszövetkezeti mozgalom helyzetéről* [Country reports on the situation of cooperative movement]; 28/1958/3. ő.e. *Tájékoztatás a Bolgár Kommunista Párt VII. kongresszusának lefolyásáról* [Report on the events of the Seventh congress of the Bulgarian Communist Party]; 28/1960/30. ő.e. *Jelentés a Szovjetunióban járt mezőgazdasági pártmunkás küldöttség tapasztalatairól, 1960. június* [Report on the experience gathered by the agricultural party delegation in the Soviet Union, June 1960]; 28/1960/30. ő.e. *Feljegyzés az SZKP KB mellett működő Mezőgazdasági Osztály helyettes vezetőjével, Kresztyaninov elvtárssal folytatott megbeszélésről. 1960. június 24.* [Report on the consultation with comrade Chrestaninov, deputy leader of the Department for Agriculture of the CC of the Communist Party of the Soviet Union. June 24, 1960]; 28/1962/9. ő.e. *Megyénkénti jelentések a termelőszövetkezetek anyagi ösztönzése területén történt változások hatásairól* [Reports from counties on the impact of the changes in remuneration of cooperatives]; 28/1963/13. ő.e. *Jelentés a Földművelésügyi Minisztérium Kollégiumához a termelőszövetkezetek jövedelemrészesedési és munkadíjazási tapasztalatairól* [Report to the Ministry of Agriculture on the experience of the distribution of income and remuneration in cooperatives]; 28/1963/37., 38., 39. ő.e. *A megyei párt és tanácsi vezetőkkel folytatott megbeszélések a mezőgazdasági termelés és irányítás fontosabb kérdéseiről* [Discussions with county party and local county leaders on the essential issues of agricultural produce and management]; 32/1958/14. ő.e. *Jelentés a lengyel mezőgazdaság jelenlegi helyzetéről, 1958. április* [Report on the actual situation of Polish agriculture. April, 1958]; 32/1958/15. ő.e. *Politikai és gazdasági helyzetkép a Román Népköztársaságról, 1958. január* [Political and economical overview on the People's Republic of Romania. January, 1958]; 32/1958/15. ő.e. *Jelentés a SZÖVOSZ delegáció kínai útjáról, 1958. március* [Report on the trip of the delegation of the National Board of Cooperatives to China, March 1958].

Fond: XIX-J-1-j Külügyminisztérium, Titkos Ügykezelésű Iratok (KÜM TÜK), (Ministry for Foreign Affairs, Classified Documents). Files: Lengyelország 1945–64. 8. doboz. (d.) [Poland, 1945–1964, Box No. 8]; XIX-J-1-j KÜM TÜK Csehszlovákia, 1945–1964. 2. d. [Czechoslovakia, 1945–1964, Box No. 2]; XIX-J-1-j KÜM TÜK Csehszlovákia 1945–64. 5. d. [Czechoslovakia, 1945–1964, Box No. 5]; MOL XIX-J-1-j KÜM TÜK Csehszlovákia 1945–64.

14. d. [Czechoslovakia, 1945–1964, Box No. 14]; XIX-J-1-j KÜM TÜK Kína 1945–64. 9. d. [China 1945–64 Box No. 9]; MOL XIX-J-1-j KÜM TÜK Románia 1945–1964. 27. d. [Romania, 1945–1964, Box No. 27].

Hajdú-Bihar Megyei Levéltár, Hajdú-Bihar County Archive (HBML):
XXIII. 2. 4.k. *Jegyzőkönyv a Hajdú-Bihar megyei Tanács Végrehajtó Bizottságának üléséről* [Minutes of the County Council in Hajdu-Bihar], December 4, 1956.

OTHER SECONDARY MATERIALS

Magyar Közlöny [Hungarian Gazette], issues: November 27, 1956.
Népszabadság [People's Freedom], issues: December 15, 1957; January 15, 1960.
Szabad Föld [Free Land]: December 22, 1957.
Javaslatok a jövedelemelosztás és munkadíjazás formáira a termelőszövetkezetekben [Suggestions for forms of the distribution of income and the system of remuneration in agricultural cooperatives], *Mezőgazdasági Értesítő* (Agricultural Gazette), 1961, no. 2, 9–14.
Törvények és rendeletek hivatalos gyűjteménye, 1956 [Official collection of laws and ordinances, 1956]. Budapest: 1957.

INTERVIEWS

The author's interview with L. F., January 21, 2002.

BOOKS AND ARTICLES

Adair, Bianca L. "The Agrarian Theses and Rapid Collectivization: Accommodation in Hungarian Agriculture, 1956–60." *Journal of Communist Studies & Transition Politics* 17, no. 2 (2001): 131–48.
Az emberiség békéjéért, jövőjéért. A szocialista országok kommunista és munkáspártjainak nyilatkozata, Moszkva, 1957. november 14–16. [For the Peace and Future of Humankind: a Declaration by the Communist and Labor Parties of the Socialist Countries, November 14–16, 1957, Moscow]. Budapest: Kossuth Kiadó, 1957.
Bauerkämper, Arnd. *Ländliche Gesellschaft in der kommunistischen Diktatur. Zwangsmodernisierung und Tradition in Brandenburg 1945–1963.* Vienna: Böhlau, 2002.
Berend, Iván T. *Central and Eastern Europe 1944–1993: Detour from the Periphery to the Periphery.* Cambridge: Cambridge University Press, 1996.
Bokovoy, Melissa K. *Peasants and Communists. Politics and Ideology in the Yugoslav Countryside, 1941–1953.* Pittsburgh: University of Pittsburgh Press, 1998.
Crampton, Richard J. *Eastern Europe in the Twentieth Century and After.* London: Routledge, 1997.
Csete, László, ed. *A moszkvai agrárgazdasági és üzemgazdasági konferencia, Moszkva, 1959. aug. 25–szept. 2* [Conference in Moscow on Agrarian Economy. Moscow, August 25–September 2, 1959]. Budapest: AKI, 1959.
Csete, László and Ferenc Erdei, eds. *A szocialista mezőgazdaság fejlődése Bulgáriában. Három tanulmányút 1959–1960-ban* [The Development of Socialist Agriculture in Bulgaria. Three study-tours in 1959–1960]. Budapest: AKI, 1961.
Dolowitz, David P. and David Marsh. "Who Learns What from Whom: a Review of the Policy Transfer Literature" *Political Studies* 44, no. 2 (1996): 343–57.
Erdmann, Gyula. *Begyűjtés, beszolgáltatás Magyarországon 1945–1956* [Compulsory Deliveries in Hungary, 1945–1956]. Békéscsaba: Tevan Kiadó, 1993.

Fazekas, Béla. *A mezőgazdasági termelőszövetkezeti mozgalom Magyarországon* [The Movement of Agricultural Cooperatives in Hungary]. Budapest: Kossuth Könyvkiadó, 1976.

Fejtő, Ferenc. *A népi demokráciák története.* II. kötet [The History of "People's Democracies," Vol. II]. Budapest: Magvető Kiadó, 1991.

Fitzpatrick, Sheila. *Stalin's Peasants: Resistance and Survival in the Russian Village After Collectivization.* Oxford: Oxford University Press, 1994.

Hann, Chris. *Tázlár: a Village in Hungary.* Cambridge: Cambridge University Press, 1980.

Harcsa, Iván, Imre Kovách, and Iván Szelényi, "The Hungarian Agricultural Miracle and the Limits of Socialist Refoms" in Iván Szelényi, ed. *Privatizing the Land: Rural Political Economy in Post-Communist Societies.* London, New York: Routledge, 1998, 21–42.

Iordachi, Constantin. "Constanţa, the First Collectivized Region: Soviet Geopolitical Interests and National and Regional Factors in the Collectivization of Dobrogea (1949–62)" in Constantin Iordachi and Dorin Dobrincu, eds. *Transforming Peasants, Property and Power: the Collectivization of Agriculture in Romania, 1949–1962.* Budapest: CEU Press, 2009, 103–39.

Izsák, Lajos, József Szabó, and Róbert Szabó, eds. *1956 plakátjai és röplapjai* [Posters and handbills from 1956]. Budapest: Zrínyi, 1991.

Keserű, János. *Parasztsorsfordítók között* [Among agrarian politicians]. Budapest: Napvilág Kiadó, 2007.

Lampland, Martha. *The Object of Labor: Commodification in Socialist Hungary.* Chicago: The University of Chicago Press, 1995.

Maier, Matthias L., Frank Nullmeier, and Tanja Pritzlaff, eds. *Politik als Lernprozess?* Opladen: Leske, Budrich 2003.

Mezőgazdaságunk a szocialista átszervezés idején, 1958–1962 [Our agriculture in the time of Socialist reorganization, 1958–1962]. Budapest: KSH, 1963.

Mezőgazdasági Statisztikai Zsebkönyv, 1965 [Statistical Handbook of Agriculture]. Budapest: KSH, 1965, 281.

Nagy, József. "A kulákkérdés és megoldása az 1948–1953-as években" [The *Kulak* question and its solution, 1948–1953]. *Múltunk* 44, no. 3 (1999): 41–97.

Nagy, Miklós, ed. *A magyar külpolitika, 1956–1989. Történeti kronológia* [Hungarian foreign policy, 1956–1989: Historical chronology]. Budapest: MTA Jelenkorkutató Bizottság, 1993.

Pető, Iván and Sándor Szakács. *A hazai gazdaság négy évtizedének története 1945–1985. I. Az újjáépítés és a tervutasításos irányítás időszaka 1945–1968.* [The History of four decades of domestic economy, 1945–1985, vol. 1: The Period of rebuilding and command economic planning, 1945–1968]. Budapest: Közgazdasági és Jogi Könyvkiadó, 1985.

Pittaway, Mark. *Eastern Europe 1939–2000.* London: Arnold, 2004.

Rainer, János M. "The Sixties in Hungary—Some Historical and Political Approaches" in György Péteri, ed. *Muddling Through in the Long 1960s: Ideas and Everyday Life in High Politics and Lower Classes of Communist Hungary.* Trondheim: Norwegian University of Science and Technology, 2005, 2–26.

Romsics, Ignác. *Hungary in the Twentieth Century.* Budapest: Corvina and Osiris, 1999.

Schöne, Jens. *Frühling auf dem Lande. Die Kollektivierung der DDR-Landwirtschaft.* Berlin: Ch. Links Verlag, 2005.

Scott, James C. *Weapons of the Weak: Everyday Forms of Peasant Resistance.* New Haven: Yale University Press, 1985.

Sipos, Levente. "Reform és megtorpanás. Viták az MSZMP agrárpolitikájáról (1956–1958)" [Reform and Balking. Debates on the agrarian policy of the HSWP, 1956–1958]. *Múltunk* 36, no. 2–3 (1991): 188–97.

Swain, Geoffrey and Nigel Swain. *Eastern Europe since 1945. 3rd ed.* New York: Palgrave Macmillan, 2003.

Swain, Nigel. *Collective Farms which Work?* Cambridge: Cambridge University Press, 1985.

———. "Decollectivization Politics and Rural Change in Bulgaria, Poland and Former Czechoslovakia." *Social History* 32, no. 1 (February 2007): 1–26.

Szakács, Sándor. "From Land Reform to Collectivization (1945–1956)" in Péter Gunst, ed. *Hungarian Agrarian Society from the Emancipation of Serfs (1848) to the Re-privatization of Land (1998).* New York: Columbia University Press, 1998, 257–98.

Szakolczai, Attila and László Á. Varga, eds. *A vidék forradalma, 1956. I. kötet* [Revolution in the Countryside, 1956. Vol. I.]. Budapest: 1956-os Intézet and Budapest Főváros Levéltára, 2003.

Szakolczai, Attila, ed. *A vidék forradalma, 1956. II. kötet.* [Revolution in the countryside, 1956. Vol. II.]. Budapest: 1956-os Intézet and Budapest Főváros Levéltára, 2006.

Varga, Zsuzsanna. *Politika, paraszti érdekérvényesítés és szövetkezetek Magyarországon 1956–1967* [Politics, the Assertion of Agrarian Interests and Cooperatives in Hungary, 1956–1967]. Budapest: Napvilág Kiadó, 2001.

———. "Agriculture and the New Economic Mechanism" in Anssi Halmesvirta, ed. *Hungarologische Beiträge 14.* Jyväskylä: Kopi-Jyvä, 2002, 201–18.

———. "Agrarian Development from 1945 to the Present Day" in János Estók, ed. *History of Hungarian Agriculture and Rural Life, 1848–2004.* Budapest: Argumentum Kiadó and Magyar Mezőgazdasági Múzeum, 2004, 221–52.

———. "The Impact of 1956 on the Relationship between the Kádár Regime and the Peasantry, 1956–66." *Hungarian Studies Review* 34, no. 1–2 (2007): 155–76.

Vass, Henrik and Ágnes Ságvári, eds. *Az MSZMP határozatai és dokumentumai, 1956–1962* [Resolutions and Documents of the Hungarian Socialist Workers' Party, 1956–1962]. Budapest: Kossuth Kiadó, 1973.

Viola, Lynne. *Peasant Rebels under Stalin: Collectivization and the Culture of Peasant Resistance.* Oxford: Oxford University Press, 1996.

Wädekin, Karl-Eugen. *Agrarian Policies in Communist Europe: a Critical Introduction.* The Hague, London: M. Nijhoff, 1982.

Collectivization at the Grass Roots Level: State Planning and Popular Reactions in Bulgaria, Romania, Poland, and the GDR, 1948–1960[1]

GREGORY R. WITKOWSKI

At the end of the Second World War, communist parties throughout Eastern Europe sought to transform the agricultural system by implementing collectivization of agricultural land. State planners and party functionaries proceeded from the notion that massive change could be engendered through the rationalization of production and the elimination of village opposition. Communist plans for the countryside centered primarily on the policy of collectivization, which sought to replace traditional village elites and apply industrial standards to agricultural production—in short, to transform the social and political relationships in the countryside. This essay analyzes the results of this planning and its implementation, focusing on collectivization programs in four countries: Romania, Bulgaria, Poland, and the German Democratic Republic (GDR). In line with recent works on the topic, my investigation rejects models of communism that posit an almost complete control of society by the state, a process

---

1 This essay included a great deal of research into microfilmed archival sources. I want to thank John Kozlovich, who served as my graduate assistant for two years, for completing much of the initial review of the State Department microfilm. I also appreciate the feedback I received from all project participants during meetings in Budapest and Berlin, but especially from Nigel Swain. Finally, the guidance and feedback from the editors of this volume, Constantin Iordachi and Arnd Bauerkämper, have helped this essay tremendously.

most prominently summed up by the theory of totalitarianism.[2] Instead, I argue that (asymmetrical) interactions between the central policy makers and the populace in the periphery shaped the contours of this complex process of political and economic transformation.

This essay builds upon the numerous studies examining peasant resistance, which provide a deep and diversified view of village interactions. I proceed from the notion that state planners had numerous hurdles to their policies but not all of these were the result of principled resistance. In taking this approach, I seek to integrate works that concentrate solely on resistance while differentiating between confrontation, instrumentalization, and societal inflexibility. In doing so, I aim to examine how the state's plans to transform society through collectivization were resisted, adapted, and co-opted by peasants and local leaders in the periphery. I do not conceptualize society as following a strict divide between government and citizenry but rather between the center and the periphery. My own research and other works on collectivization campaigns in Eastern Europe show that state and party representatives at the county, town, and village level were often as willing to co-opt the system as individual citizens. In showing the complicity of some representatives side-by-side with resistance, my aim is to give a fuller view of how communist parties actualized their agrarian policies in diverse states. In this way, my work sees a divide between the center, defined as the policy makers—that is Politburo members and central government officials—and the periphery, which included peasants as well as state and party representatives in the countryside.

After explaining my approach in greater detail, I examine four case studies of postwar collectivization policies, focusing on popular reactions to communist agrarian programs. While no examples can claim to be truly comprehensive, these four provide a diversity that is generally representative of the different experiences throughout Central and Eastern Europe. I first analyze Poland, where peasants overturned collectivization policies, before examining the other ways that peasants on the periphery effected policy in Bulgaria, the GDR, and Romania.

---

2 For more on my own approach to this issue, see Witkowski "On the Campaign Trail." For a critique of the totalitarian model applied to collectivization in Romania and, more generally, in Eastern Europe, see Iordachi and Verdery, "Conclusions."

METHODOLOGICAL APPROACH

For much of the Cold War, contemporary and scholarly study of the Eastern Bloc emphasized the power and capability of the state. Whether employing the terms "totalitarianism" or "Stalinism," these works stressed the ability of those regimes to actualize their power and mobilize support for policies.[3] While scholars continue to rely on this approach, which had a limited rebirth after the fall of communism, my work follows others who challenge this theory.[4] In Soviet studies, this "revisionist" approach rejected the totalitarian model in favor of a bottom up methodology to understanding communism by focusing on the grass roots level. This approach, begun in the 1970s, has continued to influence scholars of the Eastern Bloc.[5] I follow in this tradition by emphasizing the "negotiated" nature of communist dictatorships even while accepting that the state possessed and employed a great deal of power. Numerous studies of peasant resistance to collectivization have shown the variety of strategies pursued at the local level. Collectivization posed a threat not only to peasant livelihoods but also to their identities. Even in the most extreme examples of violent collectivization measures, there was a culture of resistance. This opposition varied from more active displays, like revolts, to more passive forms of resistance, such as slacking and avoiding duties.[6] Peasants did not have a tradition of organized strikes but still possessed specific methods of more sophisticated and muted demonstrations. In contrast to urban workers, peasants more often directed their anger at individual representatives of power than at national leaders or policies. Furthermore, they sought primarily to reclaim public space and reestablish their influence at the local level with little interest in overthrowing the government. They used threats against local officials to secure their demands and additional room for maneuver in their village. Peasants sometimes employed violence, especially in Bulgaria and Romania, but more often limited them-

[3] For an examination of Stalinism and Totalitarianism in the Soviet Union, see Boffa, *Stalin Phenomena*; and Wippermann, *Totalitarismus theorien*.
[4] See, for example, in the German context: Schroeder, *Der SED Staat*.
[5] For more on revisionism see Fitzpatrick, "Revisionism in Retrospect."
[6] For a discussion of definitions of resistance in the collectivization process in the Soviet Union see Viola, "Popular Resistance;" for the classic account of passive resistance, see Scott, *Weapons of the Weak*; on Romania, see Iordachi and Verdery, "Conclusions," esp. 464–65.

selves merely to threats against local representatives of the regime. This approach emerged most naturally from a village culture. First, villagers knew each other; they often had grown up together and had developed relationships over years. The nature of village life also meant that daily interactions, even across class boundaries, were a common occurrence. In this regard, villagers had access to local leaders.[7]

Collective action, which exposed these actors to reprisals by the state, was carried out only infrequently. More often, peasants sought to modify state policies through active accommodation. This adaptation was followed by both peasants and local officials, who served as the mediators between policy makers and the populace.[8] My analysis builds upon Alf Lüdtke's concept of *Eigen-Sinn*, which emphasizes not popular resistance but the totality of individual responses.[9] The term evokes both peasant stubbornness to change and the ways each person individualized the meaning of state policies.[10] In this sense, *Eigen-Sinn* corresponds with some definitions of resistance but seeks to provide a more encompassing approach. My research reveals that local leaders and peasants acted in ways—both compliant and non-compliant with party goals—that the leadership had not anticipated.

In employing this theoretical approach to examine collectivization from the bottom up, I am drawing on my primary research on the GDR.[11] I have found indications that the center/periphery dichotomy which occurred in the GDR also existed in Poland, Bulgaria, and Romania. In this way, this work examines the difficulties central policy makers had in implementing their plans in the countryside.

Reactions to collectivization varied by country and region but exhibited common elements. Despite different topographies and histories, the process in the GDR, Bulgaria, Poland, and Romania was highly influenced by the tumultuous postwar period. All state leaders needed to cope with the destruction of the Second World War, including the dramatic demographic transformations brought about by migration during and after the war and the changing borders in the postwar era. In addition to these

---

7 This opposition fits in with Charles Tilly's definition of premodern resistance. Tilly, *Contentious French*, 391–92.

8 Fitzpatrick, *Stalin's Peasants*. For the GDR, see Last, "Rural Functionaries."

9 The concept of *Eigen-Sinn* was first developed by Alf Lüdtke and was applied to the GDR by Thomas Lindenberger in "Die Diktatur;" Lüdtke, *Eigen-Sinn.*

10 Lindenberger, "Die Diktatur," esp. 23–26.

11 Witkowski, "On the Campaign Trail."

challenges, new governments also faced an often anti-communist and anti-Soviet populace. The new communist governments began their agrarian policy with land reforms, in theory targeted only at Nazi collaborators and large landowners, in order to inaugurate the transformation of the countryside. In Poland, Romania, and Bulgaria, this process was quickly followed by the start of collectivization in 1948–49. In the GDR, collectivization only became official policy in mid-1952. In order to better understand this process throughout Central and Eastern Europe, this essay examines four cases that encompass different historical traditions and geographies, and ultimately, different forms of collective farms.[12]

POLAND

After the Second World War, the borders of Poland were redefined as Allied powers shifted the state's demarcations westward, with the eastern territories becoming part of the Soviet Union, and formerly German territory becoming Polish. These border changes led to mass dislocation of the population on the heels of the war's destruction, which had a great effect on the collectivization process.[13] Germans fled west and the territory left behind was populated with Polish peasants who were quicker to join collectives. As such, it is best to conceive of the Polish periphery as divided into two distinct areas: the areas dispossessed by Germans, where there was a concentration of collective and state farms; and the traditional Polish farmlands, where there was opposition to collectivization from established "old peasants."

The Polish government launched its collectivization efforts in July 1948. It was built on the expropriation of land from land reforms, during which 814,000 new farms were established in the country, many in the former German lands, accounting for four of the six million hectares of redistributed land.[14] Peasants who received these farms were more likely to join a collective. Thus in 1950, over 60 percent of all members of collectives were settlers in these and other territories. An additional 22 percent were farmhands and other formerly landless individuals. In contrast, less than 18 percent were old peasants who had owned land prior

[12] Swain, "Decollectivization."
[13] Adams and Adams, *Men versus Systems*, 100–1.
[14] See Dariusz Jarosz's contribution and the chronology for Poland in this volume.

to the land reforms.[15] Most of the land distributed to settlers and landless peasants had been located in the so-called "new provinces" or other areas vacated by ethnic Germans, and many collective farms were established in these areas. By 1954, these territories accounted for over 75 percent of the collectives, with the new provinces alone equaling almost half.[16] This area clearly led the way in terms of the party's transformation of the countryside.

In some ways, this trend could have been beneficial for the party's plan to initiate collectivization. The area was generally free from the influence of old peasants, who hindered collectivization in other regions. Many of the state farms in these areas were carved from old German *Junker* estates. These large landed estates had traditionally been owned by German nobility and farmed by (often Polish) peasant laborers. They were geared for larger-scale farming and, after the conversion to state farms, continued to function as a single facility. Thus, they provided an ideal opportunity for the state to quickly transform a farm and be successful. While not collective farms, these state farms could have gone a long way to proving the effectiveness of large scale farming under communist leadership. In addition, many of the recipients of land had experience farming, as they were relocated from farms in territories given to the Soviet Union after the Second World War. Unable to bring their equipment with them, these peasants became reliant on the state, and so collectivization may have provided them an opportunity to succeed, which they could not have done otherwise. Those who joined a collective received some protection from failure as they banded together with others in similar situations and distributed risk. They also received economic benefits from state agencies. All of these factors, as well as government pressure, contributed to the growth of collectivization in these areas.

This development of the new provinces created hurdles for the party, too. While displaced peasants from the eastern territories were experienced farmers, they did not have knowledge of the local conditions required for successful agricultural production. Furthermore, they believed that the border with Germany would soon change and their land would once again

[15] *Nowe Drogi*, vol. IV, No. 5, Sept–Oct. 1950, 69, as cited in Korboński, *Politics of Socialist Agriculture*, 177, footnote 62.

[16] The provinces (*Voivodship*) in question were heavily German: Bydgoszcz, Gdańsk, Katowice, and Poznań as well as the new provinces of Olsztyn, Koszalin, Szczecin, Zielona Góra, Wroclaw, and Opole. See Korboński, *Politics of Socialist Agriculture*, 176.

belong to the original owners. As a result, they were more likely to join a collective, as many felt like they had less to lose compared to investing in privately owned farms. But the notion that collective farms were only temporary meant that collective members were not committed to making capital improvements, which hurt the long-term economic performance of these enterprises.[17] Furthermore, the government did not provide them with sufficient tools, draft animals, or equipment to succeed. As such, collectives were less productive than independent farms. Despite collectivizing almost 80 percent of the villages in Szczecin and Wrocław, for instance, the party did not succeed in creating a "socialist countryside" in these regions where peasants viewed collective farms as economically competitive and the best means of production.[18] When General Secretary Gomułka announced that collectivization was voluntary in October 1956, these collectives, unable to compete, quickly dissolved.

In contrast to the "new territories," the other areas of Poland continued to be dominated by small peasants. 88 percent of all farms were ten hectares or less in size.[19] These peasants received high quotas that at times made it difficult for them to succeed. Still they resisted collectivization despite the economic benefits—in terms of lower quotas and taxes—it offered. For instance, by the end of 1955, in areas such as Lodz, Krakow, Warsaw, and Lublin, less than three percent of farmland had been collectivized. While these are the lowest participation rates in all the districts, this is a stark contrast to the level of collectivization achieved in the former German lands.[20]

The Polish government employed varied means of persuasion, including raising taxes and quotas on independent farmers, and employing other repressive measures, to pressure farmers to join collectives. From 1952–55, the regime initiated more than 550,000 legal proceedings against peasants for failing to meet their quotas.[21] In addition, peasants were summoned to meetings that could last until 4AM, where they were pressured to join collectives. Although some gave in, peasants in these areas also deployed various forms of passive resistance to prevent the

[17] Foreign Service Despatch, November 10, 1055, 748.00/11-1055, 3. For the view of a German living in Poland see Foreign Service Despatch, September 13, 1955, 848.20/9-1355, 4.

[18] Foreign Service Despatch, June 4, 1956, 848.20/6-456, 3–4.

[19] Adams and Adams, *Men versus Systems*, 100–1.

[20] Foreign Service Despatch, June 4, 1956, 848.20/6-456, 3.

[21] Jarosz, "Polish Peasants," 72–73.

encroachment of state power. Some sold their livestock and even their land before a census could be conducted to determine the size of their property. They then used their valuables, including jewelry and even foreign hard currency, to buy what they needed locally.[22] In other cases, peasants exploited bad weather to overstate their losses either of livestock or crops so that they could keep more for themselves.[23] While these efforts helped them succeed as independent peasants for four years, in the 1950s they faced increasingly strong pressure to join collectives.

Peasants and farmers tried to delay this process as long as possible. Peasants held out in hope that rumors that collectivization would be ended were true. They reacted to and often spread rumors that collective farms were economically detrimental and tightly controlled. Among the more popular rumors were that collective farms would not provide enough food for peasants to survive, that all members would have to put their babies in state *crèches*, and that collectives would limit what one could purchase, including items like clothing.[24] These rumors became convenient excuses for peasants to avoid or leave collective farms. One can imagine that in this process, there must have been an element of collusion among the local administrators who would have heard these rumors and could have reacted against them. By allowing peasants to use these excuses to reject collectivization, local officials undermined state goals.

Ultimately, peasants in Poland were successful in delaying the process of collectivization. As part of the larger de-Stalinization process in Poland, the new government introduced the possibility of voluntary farming at the Eighth Plenum in October 1956. There was an immediate and massive rejection of collective farming. By March 1957, Polish peasants dissolved 83 percent of all collectives, accounting for 87 percent of all the land farmed by collectives.[25] Ultimately this outpouring of opposition on the periphery and a change by the leadership meant that collectivization never took root in Poland. The center realized the degree of opposition on the periphery and chose not to enforce collectivization. In pursuing this path, Poland became one of two Eastern Bloc states where collectivization was repealed.

[22] Foreign Service Despatch, September 30, 1955, 748.00/9-3055, unpaginated.

[23] Foreign Service Despatch, April 17, 1956, 848.20/4-1756, 2.

[24] Jarosz, "Polish Peasants," 66–67. For more on rumors in collectivization in the Soviet Union, see Viola, "The Peasant Nightmare."

[25] Korboński, *Politics of Socialist Agriculture*, 257–58. Here, cross-reference to Jarosz's chapter on Poland (for more background information on Eighth Plenum).

BULGARIA

In contrast to Poland, Bulgaria followed through with collectivization policies and became the first country in Eastern Europe to complete collectivization after the Second World War. Bulgaria was a predominantly agrarian country, with agriculture accounting for 70 percent of economic production in 1948.[26] Government policy clearly focused on agriculture in a way that party leaders did not in Poland. Of the four countries discussed, Bulgaria had the longest history of agrarian political activism. The Agrarian Union, a Bulgarian party that ruled for a short time in the interwar period, retained strong support even after the Communist Party had cracked down on it. To eliminate this support, Bulgarian leaders resorted to mass violence, which also effected collectivization. Despite these great socioeconomic and historical differences with Poland, Bulgarian collectivization likewise led to peasant resistance and adaptation of state policies.

A failed effort after the First World War to institute land reform had created an agricultural system dominated by small farms, a system that lasted until well into the twentieth century. These holdings made it difficult for some peasants to support their families, but perhaps as importantly for collectivization, this made it difficult to inherit sufficient land. In 1934, for instance, 63 percent of all landholdings were smaller than five hectares, while 26 percent were between 5 and 10 hectares. After the war, these smallholdings remained part of the socioeconomic make-up, with only one percent of farms larger than 30 hectares.[27] Furthermore, in the hills and mountains, it was important to have land in different climatic zones for maximum harvest in varied planting seasons. As a result, peasants often had many smaller parcels (on average 16) that were often widely dispersed.[28] This trend was compounded by two factors. First, unemployment in Bulgarian cities was so great (approximately 20 percent) that after the war, many went to villages to survive.[29] This movement diluted an already small land pool. Second, unlike Poland, there was no large German minority to divest of their land. Because of this, the Bulgarian land reform was smaller in scope and emphasized land nationalization and

---

26 Crampton, *Bulgaria*, 319.

27 Jones and Jankoff, "Agriculture;" 287–88; and Shopov, *Socialist Development*, 5.

28 Creed, *Domesticating Revolution*, 37–38.

29 Crampton, *Bulgaria*, 319

redistribution.[30] This effort, begun even during the Second World War, was followed with a call for collectivization in 1948.

There was a great deal of upheaval brought about by collectivization in Bulgaria. The regime used economic and punitive coercion, including higher quotas and taxes for independent peasants. Peasants outside of the collective system who failed to fulfill their quotas would be subjected to more fines and arrests, but quotas were lower and less rigorously enforced for collective farmers.[31] Before the New Course, state policy also targeted the wealthiest peasants and prevented peasants from selling land to one another so that no one could accumulate wealth. Taxes on richer peasants were designed to take 80 to 90 percent of their income to weaken their economic power. At the same time, the state introduced a law curbing private enterprises, like taverns, which wealthy peasants had generally owned to provide a source of additional income. In doing so, the regime sought to diminish both the economic standing and perhaps more importantly the social influence of wealthier peasants, since taverns especially were common gathering places for all villagers and offered owners the opportunity to influence other peasants with their ideas and conceptions.[32]

State officials also employed physical threats and violence against peasants throughout the collectivization process. Bulgaria arguably had the most violent experience of collectivization throughout Eastern Europe. In a village near Russe in northern Bulgaria, the Security Police killed three peasants who had refused to join a collective by throwing them into the River Danube with millstones around their necks.[33] Although this was an extreme case, the state employed force and punitive measures regularly to implement its collectivization policies. The experience of the son of a *kulak* in the village of Tianovci, west of Vidin, demonstrates the state's approach to those who refused to join a collective. In April 1950, the family was given an order that its freedom of movement would be restricted from 6 am to 6 pm. The following month, the Security Police arrested the son at 10 pm for violating this order and brought him to Vidin, where they interrogated and beat him. The Security Police asked if he would join a collective farm. When he refused, the police told him it would have severe consequences for himself and his father. They impris-

[30] Crampton, *Bulgaria*, 330.

[31] See Gruev in this volume, and "Interrogation of Goran I. K.," January 11, 1951, 769.00/1-1151, 2.

[32] Crampton, *Bulgaria*, 343, 330.

[33] Foreign Service Despatch, May 29, 1956, 848.20/5-2956, unpaginated.

oned the son for twenty days. His release from prison turned out to be short-lived: he was arrested again shortly thereafter, and imprisoned for 15 weeks until he declared he would work for the Communist Party to bring about collectivization.[34] This experience is instructive in that it indicates the regime's willingness to continue to employ increasingly harsh punitive measures. When a short term prison stay did not bring the required change by the son or his father, the police imprisoned the son for a longer period of time.

Peasant reactions to collectivization policies took many forms, including active and passive resistance, as well as co-option and instrumentalization. In 1951, for instance, peasants rose up in the districts of Kula and Vidin and the southern regions of Assenovgrad and Plovdiv. The general fear of reprisals kept much of the population in check, but, as Mikhail Gruev argues in this volume, criticism by central government officials of the harshest measures led some peasants to seek the opportunity to oppose collectivization.[35]

Uprisings in the districts of Plovdiv and Haskovo in 1953 were a successful demonstration of mass peasant resistance. These areas were dominated by tobacco production. Traditionally, the peasants grew the tobacco in the summer and processed, cured, and stored it in the winter so that they remained employed year round. In fall of 1952, the state changed this situation, shipping the harvest to the Soviet Union for processing. The Bulgarian peasants suddenly found themselves unemployed. Unwilling or unable to undertake construction work offered as a replacement, peasant workers took to the streets in May 1953. According to a report, more than 20,000 people marched on the Plovdiv City Hall carrying signs and calling for work and bread. The mayor, who almost certainly had nothing to do with the decision, attempted to placate the crowd. His efforts failed and the peasants hurled sticks and stones. Fearing for his safety, he fled with his wife. The Minister of Heavy Industry, who had previously worked in the tobacco industry in Plovdiv, also made a vain effort to placate the crowd. Peasants drowned out his speech with whistles and jeers before throwing sticks and stones at him. He was unable to complete his speech, and ultimately, the police resorted to firing on the crowd in an effort to quell public resistance.[36]

---

[34] Foreign Service Despatch, June 9, 1952, 769.00/6-952, unpaginated.

[35] Dellin has March 1950 as the date for Kula and Vidin. See also Foreign Service Despatch, March 27, 1951, 769.00/3-2751, unpaginated.

[36] Foreign Service Despatch, September 16, 1953, 769.00/9-1653, 2–3.

Here, there occurred a mixture of public demonstration and traditional peasant attacks. In order to demonstrate, peasants marched on the local figure of authority, the mayor, and placed their demands before him. They hoped to solve the problem locally, but when they could not, they turned to violence. While the police dispersed the crowd, the extent of the protest and the reality of peasant threats would not have been lost upon the mayor. In this fashion, peasants could establish influence over local officials. They were not always successful in achieving their goals but could change local power dynamics through their protests. Without full access to Bulgarian archives, it is difficult to gauge how representative this case was but the report does not indicate that it was an exceptional event, which I see as evidence that other protests of this nature likely occurred in Bulgaria. Peasants also employed many forms of passive resistance. They hid animals from government representatives, understated harvests, and neglected upkeep of machinery owned by the state or collective. Peasants also attacked collective property so that these farms would be less profitable. Breaking tractors or other machinery, and setting fire to farms were common methods of attacking mechanized farming and collective farms. One example of such resistance comes from a teenage peasant from the village of Sinapovo. This boy was the son of a relatively wealthy peasant, who owned 12 hectares of land and 120 sheep. The boy's father was imprisoned because another son had fled the country. During his two-year prison term, the farm was gradually taken by the regime, as the family was unable to meet delivery quotas. One day, while the boy was delivering oxen, he came across a collective in the village of Rozovo and saw the harvested wheat, hay, and grass stacked high near the road. "I was so mad at these collective farms which had caused the ruin of my family that I decided to set fire to all," he later recalled.[37] This was not a singular event. Bulgarian newspapers reported that in one district alone, there were 18 fires on farms in one month between June 18 through July 18, 1951.[38] While fires did occur naturally, the number of fires (more than every other day in that month) indicates peasant resistance was likely the cause of these incidents.

Another reaction, less a form of resistance than it was simply a rejection of collectivization policies, was to abandon farming or encourage chil-

[37] Istanbul News Bureau, November 26, 1955, unpaginated, United States State Department Confidential Files Bulgaria.

[38] Jones and Jankoff, "Agriculture," 304.

dren to move to cities to find work. Others fled the country for Turkey or Yugoslavia. For instance, eleven adults and four children left the small village of Kapitan Andreevo (near the Turkish border) from 1948 to 1950.[39] Given the proximity to the border, this number is perhaps not too surprising even in the face of soldiers posted to the village. What is perhaps more telling, is how villagers seemed to react toward the state's policy. While 160 families joined a collective, many did so in order to prevent persecution as *kulaks*. They wanted to protect their property and prevent the authorities from seizing it. They were not convinced of the benefits of collective farming before joining a collective and became even more dissatisfied with it over time. They complained that not everyone worked hard enough, and that the director and brigadiers showed favoritism.[40]

In an indication that many individuals joined collectives without any dedication to cooperative farming, the newspaper *Dunavska Pravda* ran multiple articles in October 1951 warning that enemies of the state were joining collectives to undermine them, emphasizing that statutes needed to be followed more closely.[41] In Bulgaria, as in other states, people simply refused to farm cooperatively even after joining a collective. They used their membership in the collective to prevent the state from seizing their property but then continued to farm independently as before.

The northwestern town of Zimfirovo, illustrates one way in which peasants instrumentalized collectivization policies. Here there were some who pushed for the forced merger of farms. Because there was insufficient land for all peasants to inherit, a number of newlyweds and even older sons were forced to live with their parents or in-laws and work the land under the orders of the head of the household. These individuals saw collectives as a way to gain some independence from this situation.[42] This decision was likely easier in Bulgaria than other countries because of the long tradition of cooperative farming that stretched from the nineteenth century into the interwar period. The tradition of having two to three large fields in which all families held strips of land, often at some distance from one another, meant that decisions about what to plant had traditionally been decided by the village and not individual peasants.[43] This convention

[39] Interrogation of Nicolai G. K., August 25, 1950, 769.00/8-2550, 1.
[40] Interrogation of Nicolai G. K., August 25, 1950, 769.00/8-2550, 2.
[41] Foreign Service Despatch, October 30, 1951, 769.00/10-3051, 4.
[42] Creed, *Domesticating Revolution*, 57–58.
[43] Crampton, *Bulgaria*, 286.

gave the collectives a sense of legitimacy among some villagers that was absent elsewhere in Eastern Europe. As such, the earliest members of the collective in Zimfirovo were neither the poorest nor the least experienced, as in other countries. In fact, this demographic composition also points to a younger, more energetic collective farmer than the norm.

According to anthropologist Gerald Creed, there was also a snowballing effect in Zimfirovo, so that as more people joined, other family members and friends signed on. Local officials canvassed for new members by starting a parade through the streets, mirroring rituals carried out for marriage or for holy days. This practice lent legitimacy to the cause as both a significant, life-changing event, and a positive experience. In this carnival atmosphere, people had difficulty rejecting collectivization, and their neighbors who joined them.[44] Despite the possibility that those retelling their experiences to Creed may have downplayed the difficulties and struggles they may have had with collectivization, the fact that the collective memory is positive is a significant finding.[45] Here one sees a very small network on the periphery that banded together and followed communist policies. They likely did so less out of a commitment to communist principles and more because of the village social network. In banding together to receive government benefits, they were following their own interests while also allowing younger peasants a voice in the production. As long as the collective did not have to answer to an outside authority, villagers were able to use the process of collectivization to achieve their goals. In fact, it was only when collectives were amalgamated that problems arose. Only within the tight village social network did cooperative farming make sense to these villagers.[46] Thus, on the periphery, common cause among villagers initially co-opted state policy by allowing villagers to create collectives that worked to fulfill their own goals without necessarily following state regulations for how to farm collectively.

While Bulgaria officially declared the socialist transformation of the countryside complete in 1958, the process of collectivization indicates that peasants on the periphery had a great effect on the character of the collectives that were formed. Peasants both rejected and embraced aspects of collectivization. In the face of extreme violence by the police they relented,

---

44 Creed, *Domesticating Revolution*, 66–67.

45 In my own research on East German collectivization, I found that individuals interviewed years after the events tended to minimize any difficulties they encountered, especially in comparison with the archival record.

46 Creed, *Domesticating Revolution*, 74.

but within the village community, peasants rejected, resisted, and adapted the state's policies. In this way, the process of collectivization was a negotiated one between center and periphery that was not finished when the state declared collectivization complete. This theme will be developed further in the next two case studies.

German Democratic Republic

Formed in 1949 shortly after its western counterpart, the GDR continued to be defined by its historical competition with the Federal Republic. The GDR was the last of the four countries to begin collectivization, a result of its unique position in Eastern Europe as part of a divided nation. While the Soviet-occupation government oversaw land reforms, portrayed as de-Nazification, starting in 1945, the call for collectivization officially began in 1952. The state declared the process finished less than eight years later, in 1960, making the GDR the fastest collectivization program of these case studies. This more focused approach to collectivization also led to intense peasant reactions that included participation in an uprising to overthrow the regime, mass emigration to the Federal Republic, as well as other forms of resistance and accommodation mentioned in the case studies above.

Much like in Poland, the East German experience of collectivization was defined by mass transfer of people as borders were reshaped. In addition to the population dislocations caused during the fighting of the Second World War, Germans were forced from territories that became part of Poland and the Soviet Union. These emigrants benefited from the land reforms carried out in the immediate postwar years, receiving parcels of land to help them subsist. In exchange, communist leaders expected to gain support from these "grateful" peasants and hoped to enlist them to overthrow old sociopolitical relations and defend the new order. In East Germany, they even gave peasants a deed to their plots of land in hopes of making them feel more committed to the new communist order.

Although the land reform did not overcome many of the peasants' traditional aversions to communism, it did transform the countryside. By February 1946, the Reform Commissions had seized 2.2 million hectares of land from 6,330 large estates and about 3,000 other farms. An additional 300,000 hectares were confiscated by 1949. Through land reform, over 200,000 peasants, and approximately 90,000 migrants, gained control of land. In the state of Brandenburg, 38 percent of total farmland was distributed, and in Mecklenburg 52 percent.[47] Still, with little farm equipment, weak ties to the community, and often little experience farming, many of these so-called "new peasants" moved on either to cities or to West Germany after a short time in the East German countryside.

It is this mass migration from the GDR that most distinguishes the East German experience of collectivization from the other examples. Because of collectivization, an increasing number of peasants chose to flee the GDR instead of submitting to highly regulated collective farm work. In 1951, a total of 4,343 peasants left the country, and the following year, that figure was 14,141. By the end of 1952, roughly 13 percent of the total agricultural land in the GDR had been abandoned in the previous two years. In 1953, another 37,296 people had emigrated from the GDR.[48] These numbers dropped greatly when pressure to collectivize decreased and jumped again in 1959–60, in reaction to government pressures for full collectivization. Emigration doubled between 1959 and 1960,[49] with 12,634 peasants (and an unidentified number of their family members) leaving the GDR in 1960. The rate of emigration also increased into 1961: more peasants left in the first three months of 1961 than had left in the same period in 1960.[50] Significantly, once someone from a village emigrated, it was more likely that others would follow. The tendency to leave in groups indicates the continuation of strong village connections despite the upheavals of the Second World War and land reform. Because it was a crime to "flee the Republic," emigration too required a certain level of acceptance by local officials. While some fled quickly in the night with few belongings, others divested assets before moving, something that local officials in a village would have known.

Emigration may have ultimately helped quell opposition by draining villages of those inhabitants who clearly opposed collectivization. But there was still substantial resistance among peasants who remained.[51] The first major uprising in the Eastern Bloc occurred in the GDR in 1953. The

---

47 Bauerkämper, "Neue und traditionelle," 82; Sandford, *From Hitler to Ulbricht*, 76.

48 Baring, *Uprising in East Germany*, 14.

49 See *Die Flucht aus der Sowjetzone*, 16–17.

50 Kiesler, Central Committee Division of Agriculture to SED Secretariat, June 1, 1961, "Einschätzung der Republikflucht auf dem Gebiet der Landwirtschaft," SAPMO DY 30 J IV 2/3J 190. West German figures show that although the rate increased in the earlier months, as a whole fewer peasants left the GDR in the first six months of 1961 than in 1960. See *Die Flucht aus der Sowjetzone*, 1961, 17.

51 For exit-voice phenomenon see Pfaff, *Exit-Voice Dynamics*.

June 17 uprising, which included protests in cities and villages throughout the GDR, was eventually suppressed, but only after Soviet military forces were deployed. Peasants formed an important part of the opposition during this uprising, which threatened the stability of the regime. As in cities, most village protests started with economic complaints. These were directed primarily against collectivization, which some referred to as "enslavement." But peasants had a more overtly political agenda. Rural resistance included dissolving collectives, threats and personal attacks against local officials, as well as calls for free elections and changes to political leadership.[52] In the days and weeks before and after the uprising, sources indicate 302 examples of resistance in villages with populations lower than 2,000, where agrarian interests generally predominated.[53] The prevalence of protests in small farming villages demonstrates the intensity of opposition in the countryside, as these peasants knowingly risked injury and imprisonment to show their opposition toward the Communist Party and its policy of collectivization.[54]

After the June 17 uprising in 1953, the rate of agricultural collectivization slowed. From 1955 to 1957 only 32,000 peasants joined an established collective or helped form 450 new ones. The process stagnated as collectives attracted primarily "new peasants" and farmhands who had received plots through the communist land reform. Those peasants who joined collectives were the least successful. They had the poorest quality land and owned less farm equipment and livestock than established peasants.[55] As late as 1957, these groups and their family members accounted for over 70 percent of those who had joined a collective farm.[56] By contrast, more successful peasants generally opposed collectivization. Because of this trend, many collectives were not economically viable while others continued to produce much less efficiently than independent peasants.

---

52 Witkowski, "Peasants' Revolt."

53 Kowalczuk, *17.6.1953*, 2003, 289–90.

54 In Zodel, police arrested a total of 22 people, who received sentences of up to 15 years imprisonment. "Bericht über die Fahrt des Kollegen Sekretär Hofmann nach Görlitz zur Ratssitzung..." June 22, 1953, Barch DO-1/8 361, 141; Brant, *The East German Rising*, 142.

55 For more on the conditions of the new peasants see Bauerkämper, "Die Neubauern."

56 Of the 229,026 members of collectives, 97,341 were former fieldhands or family members and 65,179 were new peasants and family members. See *Statistisches Jahrbuch*, 1958, 368–69.

Despite the government's rhetoric and policies designed to favor collectives, independent farmers and peasants were more productive because they were able to make far greater use of technology than government-supported enterprises. In addition, those peasants who had lived longer in the villages were often protected from the policies of the central authorities by local officials. These administrators often felt a stronger allegiance to their fellow villagers than to the central authorities and frequently supported those peasants who resisted collectivization. For instance in Ottendorf (District Schwerin, in what is currently Mecklenburg) the mayor was friends with a group of residents who did not want to join a collective and therefore offered no support to the collective. In another town in Schwerin, Roggendorf, the mayor backed the claim that collectives could only be formed with more than four farmers even though the law did not stipulate a minimum number of members.[57] While this trend was important in the first wave of collectivization in 1952–53, it remained so even later.[58] The hope for unification, along with the Polish example—where collectivization was halted in 1956—fueled the idea that if they held out long enough, East Germans might be able to avoid collectivization altogether.

Party leadership sought to gain more control by sending loyal workers to the countryside to improve economic performance, strengthen political influence, and break local authority. Between 1951 and 1965 the government recruited more than 100,000 people to live in villages and work on agricultural enterprises for two to five years. Despite successful recruitment, the campaign ultimately did not bring about the socialization of the countryside the party hoped for. Instead, these representatives found conditions in the countryside difficult and traditional production norms too embedded to create the anticipated change. Many returned to their factory jobs before completing their terms in the countryside.[59]

Many peasants did not support collectivization and had joined a collective farm only to escape persecution or gain economic benefits.[60] They had no economic or political interest in sowing their land through forced collaboration—in fact many simply continued to farm independently.[61] In some cases a family would essentially control the collective either by all

[57] Ministry of Agriculture, "Bericht über die Unterstutzung," 12.

[58] RIAS Archiv, DC 004367, April 19, 1960.

[59] Witkowski "Factory to Farm."

[60] For more on the persecution of peasants during collectivization see Wekentin, *Politische Strafjustiz*.

[61] For more on this practice, see Osmond, "Kontinuität und Konflikt."

family members joining the collective, or more often, by simply assuming the leadership of the collective. In the 1950s, many peasants who had been fieldhands for *Großbauern* or larger landowners remained under the influence of these individuals by direct contact, through an intermediary, such as a neighbor, or by mail. The landowner frequently sent packages to his former fieldhands, continuing the paternalistic relationship that existed in some parts of the East German countryside before the formation of the GDR. On the other hand, the landowner expected to get his land back and the secret police worried that threats from these individuals prevented the full collectivization of farmland.[62]

Later in the 1960s, many peasants continued to resist cooperative farming. They joined a collective under coercion during the Socialist Spring of 1960. However, they held out hope that collectivization would be reversed, as it had been in Poland. They felt that both popular resistance and economic benefit—private farms continued to out-produce collective farms through the 1950s—would compel the regime to reverse its policy of collectivization. These collective members formed farms that existed only on paper. They left their property marked as separate, did not hold meetings to organize production, and did not follow the normal statutes of collective farms. They simply continued to farm their own land. The secret police estimated that about 30 percent of the collectives formed during the Socialist Spring in 1960 continued to farm in this manner a year later.[63]

In sum, the interlinked factors in the GDR—including personal ties in the village that often led local officials to side with villagers, the advantages of trading outside the established market system, and the resistance to collectivization policies—limited the power of the center to impose its policies on the periphery. In everyday life, peasants tended to avoid the most oppressive imposition of government policies by not conforming to state goals or by exploiting government policies for their own ends. In the GDR, this non-conformity was supplemented by a non-confrontational and therefore passive form of resistance, namely emigration to the Federal Republic. Hundreds of thousands of East Germans fled across the open border to the West German state before the Berlin Wall was constructed in 1961. This became one of the most powerful forms of resistance in the GDR, and peasants emigrated in proportionally larger numbers than the

[62] "Feindstützpunkte und Feindmethoden," July 31, 1957, BStU, Dok 003557.

[63] "Orientierung über die Lage," BStU, Dok 005157, 4–5. For more on the period after 1961 see Last, *After the Socialist Spring*.

rest of the population. Even the threat of leaving often helped peasants get benefits and accommodations from the state. This aspect of East German peasant resistance is unique to the GDR but in other respects similar to trends of interaction between center and periphery that occurred in other countries. In expressing their opposition, peasants were unable to prevent collectivization, but did shape how it was implemented.

Romania

Romania's collectivization campaign, the longest of the four cases studied here, began in 1949 and officially ended in 1962. Romania was an agrarian country with about two-thirds of its population in rural areas after the Second World War (compared to 14 percent in the GDR and 45 percent in Poland). Of the other countries presented in this article, its economy was most similar to that of Bulgaria.[64] Transforming the countryside was a clear priority of the communist government, a process met with some of the strongest and most aggressive resistance as well. This final case study also indicates the difficulty of implementing central policies in the agrarian periphery, as well as the ways in which peasants avoided or co-opted policies.

Romania traditionally had many holdings too small for anything other than subsistence farming.[65] Despite this agrarian structure, the communist government initiated a land reform in March 1945 in which all estates over 50 hectares were redistributed to peasants. In the fall of 1947, farms of 5 hectares or less made up 73 percent of total farmland in Romania, while those of 5 to 10 hectares in size made up an additional 19 percent.[66] When the government launched its collectivization campaign in 1949, the poorest peasants signed up first. In Rãşcani (Vaslui county) 35 families had 67 hectares of land, 10 plows, and seven oxen.[67] This was not rare, as Agriculture Minister Vasile Vaida admitted: "We see when creating collective farms that in the best cases some 50 to 60 peasant families come to a farm with some 30 carts, a horse, and two cows."[68] As in Poland, the first areas to collectivize were also regions that included land "vacated" by emi-

64 Enyedi, "Changing Face of Agriculture," 364.
65 Keil, *Romania*, 185.
66 Levy, *Ana Pauker*, 97.
67 Levy, "First Wave," 31.
68 As cited in Levy, *Ana Pauker*, 99.

grating ethnic minorities, in this case containing a mixture of ethnicities.[69] The government sought to bring wealthier peasants into socialized agriculture and in 1951 started allowing for the formation of associations in which private property could be maintained. Most peasants chose to join such an association over a collective.

Peasant opposition never reached the mass scale of that in the June 17 uprising in the GDR, but remained constant and included sporadic guerrilla fighting. This aspect of resistance made the agricultural socialization process more dangerous for the regime. In mountain areas, small groups of 20 to 40 people, formed in response to collectivization in 1949, continued to fight until 1960. Their impact should not be overstated as they were poorly armed and relied on villagers for food and shelter in the winter. But they were a constant reminder of opposition to the regime.[70] The second push for collectivization also included peasant unrest. Among the areas of disturbances were Suraia, Vadu Roşca, and Răstoaca.[71] At Fălticeni, 1,000 to 2,000 peasants banded together to resist collectivization in March 1958. When troops arrived, peasants confronted them. The villagers taunted the troops and threw rocks at them. In reaction, the troops opened fire and in the process caused 200 casualties.[72] Despite this show of government force, the region remained politically active and rioted again in 1962.

There were, of course, cases when government or party representatives successfully used force to achieve their mandates. In Vâlcea County, peasants were beaten, had their hands tied behind their backs, and then hung from the ceiling. In a village in Dolj County, the secret state police (Securitate) targeted respected villagers, stripped them naked and held them in a refrigerated storage room for a night.[73] Here they directly attacked the hierarchy in the village community. In stripping these elders of their dignity, they likely hoped to convince others that resistance was futile. These methods were used repeatedly during the early efforts of collectivization. From 1950–52, the Securitate evicted 17,000 families from their households; a total of 89,000 peasants were arrested for resisting collectivization efforts.[74]

---

[69] Foreign Service Despatch, March 19, 1957, 766.00/3-1957, 3.

[70] Deletant, *Communist Terror*, 228.

[71] See the chapter by Iordachi and Dobrincu and the chronology for Romania in this volume.

[72] Telegram USAIRA, April 10, 1958, 766.00/4-1058.

[73] Levy, *Ana Pauker*, 106.

[74] Keil, *Romania*, 204.

These actions sometimes backfired, making the wealthier peasant or *chiabur* the victim and increasing support for him. The Securitate eventually realized this and adjusted its tactics to fit local conditions, sometimes attacking *chiaburi* and other times seeking to integrate them into collectives.[75] The process of attacking the *chiaburi* also illustrates the power that officials on the periphery had in the face of centralized policies. Officially, the criteria for declaring someone a *chiaburi* were the size of his landholdings (above 5 hectares) and the use of paid labor. These criteria were quite malleable as local officials often included those with less land and excluded those with more. As Katherine Verdery reveals, the reasons behind these decisions were multifaceted and she points both to local officials' desires to keep some wealthy peasants as allies in the collectivization process as well as the personal friendships that could develop within a village.[76] What is perhaps more revealing in the context of the center–periphery divide is the way in which wealthy peasants mobilized support from villagers, collecting signatures in their bids to get a reversal of their classification as a *chiabur.* Despite the regime's use of force, peasants rallied behind these wealthier neighbors to support their efforts to oppose state classifications. These support letters reveal how peasants were willing to unite behind a village elder. More importantly, the content of the letters, similar in style to one another and advancing arguments in the language of communism, illustrates that villagers were likely coached by someone familiar with demands of the regime. While it is unclear who coached them, the letters from local officials, which sometimes mirrored the villagers' letters, may indicate complicity there. Regardless, the entire process shows how centralized goals were modified in application as a number of people who clearly were the wealthiest peasants in villages were not classified as *chiaburi.*[77]

In other examples of resistance, peasants worked to hide their land from central authorities. Instead of turning over personal property to the state, they sold their land or slaughtered their animals before the state could take them away. They declared their farms smaller than they were and thus received lower delivery quotas. Peasants "shrunk" their land by claiming smaller measurements or declaring some land insufficiently arable for planting.[78] They pursued similar strategies with livestock, hiding

---

75 Verdery, "Exploiters Old and New," 309.
76 Verdery, "Exploiters Old and New," 320–21.
77 Verdery, "Exploiters Old and New," 307–28.
78 Verdery, *What Was Socialism*, 144.

them in forests during a census. Official numbers of livestock dropped dramatically, partly from peasants slaughtering them but also from hiding them. For example, there was a dramatic change in the number of registered cows: In 1950, authorities registered an average of 15.4 cows per 100 hectares of land; by 1955, this number dropped to 13.9. It increased again in 1965 after collectivization was completed, and all animals accounted for, to 14.2.[79]

By hiding their assets, peasants could produce enough with the "extra" land and livestock to pay the high taxes levied on independent peasants or the quotas for association members. This strategy was based on more than simply peasant guile; it required that local officials look the other way. In the confines of a village, local leaders knew whether land was arable and who owned large livestock. They were complicit in the farce. Officials who registered the land also were culpable. Whether because of personal connections, bribes, or a lackadaisical attitude, these officials registered the acreage declared by peasants without checking it. In one village, a peasant woman had seven different recorded figures for her lands depending on the need of the time. In pursuing their self-interest in this fashion, local officials in the hinterland aligned themselves with the peasants and ultimately undermined government policies. Even after collectivization was completed, officials often were happy to keep the lower acreage totals recorded as it made their yields per hectare appear more impressive.[80]

Peasants used social pressures within villages to resist state policies. This non-violent opposition included similar approaches to those mentioned earlier. Peasants spread rumors, much like in Poland, about the collectives. In Rãşcani, which had a poor collective, the *chiaburi* tried to assert their position of privilege with the collective farmers, declaring that the regime would not last long, that the Americans would shoot collective members when they came, and that the *chiaburi* would take power. In addition to these threats, they spread rumors about what collectivization would mean, declaring that the collective would take everything from them, including their children.[81] In a village in Fãlciu County, *chiaburi* sought to undermine the reputation of collectives among villagers. They therefore paid some poorer peasants to go to the village and beg for money, saying

79 Eckart, *Regionale und strukturelle*, 57.
80 Verdery, *What was Socialism*, 143.
81 Levy, "First Wave," 31.

that they worked at the collective farm and had nothing to eat.[82] In other cases, they allied with priests who declared their opposition to the policy. In Bălţaţi, a religious choir was formed in which the children were told of the negatives of collectivization and stirred up trouble with villagers in the collective, declaring them poor and unsuccessful, according to a report. The end effect of these rumors was, according to one report of communist agitators, that "whatever we get done by day, they demolish by night."[83]

Communist reports complained as well of the resistance that women had to agricultural policies. For instance at a Bucharest market, a peasant woman was told that she could not sell cheese until she completed her state quota. She replied that she needed to buy her son shoes and could not afford to do so with the low prices paid. She was joined in her complaints by other peasant women who apparently brandished their cheese wrapped in cloth and swung it in the air against the controller. He retreated and called on the militia, who when they arrived let the women continue to sell their products.[84] In this case, we see that concerted collective action drove the state agent away, until more amenable local officials, in this case the militia, could be dealt with. It also is an example of the power that women had in their opposition to government policies.[85] The militia essentially ignored the policy in the face of this resistance. This gendered opposition was an important element of resistance to communist policies, as seen here and in the case studies throughout this book. Women were able to resist in ways that men were not. In fact, men often "complained" to state officials that their wives did not want them to join a collective. The wives were likely no more against it than their husbands, and so this approach constituted a form of acceptable resistance to collectivization. While the state ultimately countered this argument by declaring that peasants needed to convince their wives, this strategy bought these peasants more time.[86]

Despite these efforts, collectivization was ultimately forced in Romania, as it had been in Bulgaria and the GDR. Still, the ways in which collectivization was accomplished also reveal the input that peasants, local and regional administrators had on state plans.

---

[82] Levy, *Ana Pauker*, 108.
[83] Levy, "First Wave," 29, 31.
[84] Foreign Service Despatch, October 7, 1957, 766.00/10-757.
[85] Women had a great deal of leeway in their protests in the Soviet Union as well. For more, see Farnsworth and Viola, *Russian Peasant Women*.
[86] Levy, "First Wave," 30.

Conclusion

Communist parties in Eastern Europe sought to transform the hinterland through their policy of collectivization—a process conceived as a matter of changing economic and political structures to create new methods of production and even new village mentalities. These grandiose goals have been linked by some historians of the Soviet Union with the concept of the "garden state" in which the state seeks to order and organize society—weeding out "enemies" such as the *kulaks*, *Grossbauern*, or *chiabui* and reorganizing society, much as a gardener organizes otherwise haphazard, organic growth.[87] In this way, the state could create an organized, efficient and centralized structure that would make society easy to oversee and to govern.

Despite the tremendous power all of these governments possessed, they were unable to actualize their plans and change the village landscape as easily as they had hoped. In all four cases, individual peasants and local officials had interests that conflicted with the government. While it is clear the state possessed more power and forced collectivization throughout most of Eastern Europe (only Poland and Yugoslavia were exceptions), the process took much longer and involved more accommodation than planned. Despite similar power structures, some similarities and differences emerge from this study. Significantly, the most important dynamic in terms of the speed of collectivization appears to have been demographic changes. In all the countries involved, those areas that had recently been vacated by emigrants provided a good starting place for collectivization efforts.[88] Those regions with more established peasants proved more resilient in its opposition to socialized agriculture. In addition to peasant resistance, state planners had to cope with collusion between local officials with villagers. Whether out of a sense of common purpose or out of village loyalty established over generations, local officials often allowed peasants to avoid collectivization and even sided with peasants against state directives. While violence, especially from outsiders, was a threat that forced some to collectivize, local loyalties allowed many peasants to avoid

[87] For more on the concept of the garden state in the Soviet Union see Holquist, "'Information is the Alpha." The concept was originally proposed for the Nazi period in Bauman, *Modernity and the Holocaust*. It resonates as well with James Scott's idea of high modernism, see Scott, *Seeing Like a State*.

[88] See the example of Dobrogea and of the Banat in Iordachi, "'Constanţa;" and Goina, "Never Leave."

or adapt collectivization directives. In these cases, the regimes may have wanted to reorganize the "garden" but with the strong roots that existed in villages this proved difficult. To continue the analogy, the "gardeners" chose to follow a different plan than they were given by the state.

The four case studies also illustrate diversity in state policy and peasant reaction. In the case of Poland, peasants were able to outlast the initial push for collectivization. It was not peasant resistance *per se* that led to the end of collectivization, as state policy changed with a new regime, but without that resistance the policy never would have been changed. In the other locations, such as Bulgaria and Romania, the party responded to resistance with violence. In the GDR, the opportunity for East Germans to flee west to the Federal Republic prevented the regime from employing the harshest methods, although collectivization was forced there as well. These strategies were met with diverse peasant responses. In Romania, peasants supported violent opposition, keeping some bands of armed warriors fed in order to fight the regime for years. While these groups were small in number, they illustrate a specific form of resistance in Romania that was not seen in similar measures elsewhere. In the GDR during the June 17 uprising, peasants in more industrialized areas joined with workers in street demonstrations. They called on officials to overturn collectivization, to oust the communist government, and for free elections in a unified Germany. This form of protest was limited to those areas that had a high population density and a strong working-class tradition. More often, in the GDR and elsewhere, peasants employed less overt forms of protest.

Peasants in Poland, Bulgaria, Romania, and the GDR also employed many similar repertoires of resistance. Their active protest tended to be localized, attempting to challenge a local official in an attempt to reclaim local public space. This localized resistance could be much more successful in avoiding state policies. Passive resistance measures, which did not directly challenge the authority of the state, are seen in all cases in this study and were bolstered by collusion, incompetence, and/or apathy from local officials. Whether by spreading rumors about the perils of collectivization, by hiding production, or by continuing to farm collective farms as individuals, peasants throughout Eastern Europe found ways to not only reject policy initiatives personally but to work against the success of government policies. These efforts had the greatest success in delaying the implementation of collectivization in many ways, and even in revising the goals of collectivization itself. Whereas persecution of *kulaks*, *Grossbauern*, or *chiabui* formed an integral part of the first attempts at collectivization, local allegiances meant that these village elders were often able to redefine themselves and play a role in collective farms. Certainly these farmers also needed to redefine their own roles within the language of the communist regime but many were able to keep leadership roles in the community.

This study of agricultural collectivization in four states reveals that it was a negotiated process by which collectivization occurred, but one in which the communist governments were able to implement the majority of their policies. Peasant resistance and collusion with local officials changed the timeline of collectivization, increasing the costs both domestically and internationally for these regimes. State planners were not able to overcome traditional influences and never saw the gains in efficiency on which they had counted. Nonetheless, for peasants in the GDR, Romania, and Bulgaria resistance ultimately failed, as full collectivization was enforced in these countries. To return to the analogy of the garden state, the state was not able to "weed out" the influence of these traditional elites and in this sense the garden never took the structure that state planners had wanted. On the other hand, these states were able to impose a garden structure on what had previously developed independently. While communist leaders were unable to define completely the nature of the garden, or to work without gardeners, the regime succeeded in establishing a system in which virtually all agricultural production was collectivized or transformed into state farms. In this way, the regime defined the contours of economic and social production.

BIBLIOGRAPHY

ARCHIVAL SOURCES

"Bericht über die Fahrt des Kollegen Sekretär Hofmann nach Görlitz zur Ratssitzung..." June 22, 1953, Bundersarchiv Berlin DO-1/8 361.

"Feindstützpunkte und Feindmethoden in der Landwirtschaft," July 31, 1957, Bundesbeauftragte für die Unterlagen des Staatsicherheitsdienstes, Office of the Federal Commissioner, Berlin, (BStU), ZA, MfS-BdL, Dok 003557.

Ministry of Agriculture. "Bericht über die Unterstutzung der Arbeit der örtliche Organe der Staatsgewalt zur Bildung und Unterstützung der LPG," Barch DK-1 5893.

"Orientierung über die Lage in der Landwirtschaft in der DDR," (BstU), ZA, MfS-BdL, Dok 005157.

Rundfunk im amerikanischen Sektor, Radio in the American Sector, Berlin, (RIAS), DC 004367, Die Zeit im Funk 19.4.60.

Bundesarchiv Berlin, DK-1.

Statistisches Jahrbuch der DDR 1957. Berlin: Dietz Verlag, 1958.

Stiftung Archiv der Parteien und Massenorganisationen im Bundesarchiv, Foundation Archives of Parties and Mass Organizations of the GDR in the Federal Archives, Berlin, (SAPMO-BArch), DY 30 J IV 2/3J 190 (1961).

United States State Department, Confidential Files Bulgaria, 1945–49, 1950–54, 1955–59.

United States State Department, Confidential Files Poland, 1950–54, 1955–59.

United States State Department, Confidential Files Romania 1950–54, 1955–59, 1960–63.

BOOKS AND ARTICLES

Adams, Arthur and Jan Adams. *Men versus Systems: Agriculture in the USSR, Poland and Czechoslovakia.* New York: Free Press, 1971.

Baring, Arnulf. *Uprising in East Germany: June 17, 1953.* Ithaca: Cornell University Press, 1972.

Bauerkämper, Arnd. "Neue und traditionelle Führungsgruppen auf dem Lande." *Berliner Debatte* 4, no. 5 (1995): 79–92.

———. "Die Neubauern in der SBZ/DDR 1945–52" in Richard Bessel and Ralph Jessen, eds. *Die Grenzen der Diktatur.* Göttingen: Vandenhoeck and Ruprecht, 1996, 108–36.

Bauman, Zygmunt. *Modernity and the Holocaust.* Ithaca: Cornell University Press, 1991.

Boffa, Giuseppe. *The Stalin Phenomena,* trans. Nicholas Fersen. Ithaca: Cornell University Press, 1992.

Brant, Stefan. *The East German Rising.* Westport, CT: Praeger, 1957.

Crampton, Richard J. *Bulgaria.* Oxford: Oxford University Press, 2007.

Creed, Gerald. *Domesticating Revolution: From Socialist Reform to Ambivalent Transition in a Bulgarian Village.* University Park: Pennsylvania State University Press, 1998.

Deletant, Dennis. *Communist Terror in Romania: Gheorghiu-Dej and the Police State 1948–65.* New York: St. Martin's Press, 1999.

Die Flucht aus der Sowjetzone und die Sperrmaßnahmen des kommunistischen Regimes vom 13. August 1961 in Berlin. Berlin: Bundesministerium für Gesamtdeutsche Fragen, 1961.

Eckart, Karl. *Regionale und strukturelle Wandlungen in der Landwirtschaft ausgewählter RGW-Lander.* Opladen: Westdeutscher Verlag, 1984.

Enyedi, György. "The Changing Face of Agriculture in Eastern Europe." *Geographical Review* 57, no. 3 (July 1967): 358–72.

Farnsworth, Beatrice and Lynne Viola. *Russian Peasant Women.* Oxford: Oxford University Press, 1992.

Fitzpatrick, Sheila. *Stalin's Peasants: Resistance and Survival in the Russian Village after Collectivization.* Oxford: Oxford University Press, 1994.

———. "Revisionism in Retrospect: A Personal View." *Slavic Review* 67, no. 3 (Fall 2008): 682–704.

Goina, Calin. "'Never Leave 'till Tomorrow What You Can Do Today!' A Case Study of a Model Collective Farm: 'New Life' Santana (Arad Region)" in Constantin Iordachi and Dorin Dobrincu, eds. *Transforming Peasants, Property and Power: The Collectivization of Agriculture in Romania, 1949–1962.* Budapest: CEU Press, 2009, 369–98.

Iordachi, Constantin. "'Constanţa, the First Collectivized Region': Soviet Geo-Political Interests and National and Regional Factors in the Collectivization of Dobrogea (1949–1962)" in Constantin Iordachi and Dorin Dobrincu, eds. *Transforming Peasants, Property and Power: The Collectivization of Agriculture in Romania, 1949–1962.* Budapest: CEU Press, 2009, 103–40.

——— and Katherine Verdery. "Conclusions" in Constantin Iordachi and Dorin Dobrincu, eds. *Transforming Peasants, Property and Power: The Collectivization of Agriculture in Romania, 1949–1962.* Budapest: CEU Press, 2009, 455–72.

Jarosz, Dariusz. "Polish Peasants versus Stalinism" in A. Kemp Welch, ed. *Stalinism in Poland.* New York: St. Matin's Press, 1999, 72–73.

Jones, B.D. and Dimiter Jankoff. "Agriculture" in L. Dellin, ed. *Bulgaria.* New York: Praeger, 1957, 287–312.

Keil, Thomas. *Romania: Tortured Road to Modernity.* New York: Columbia University Press, 2006.

Korboński, Andrzej. *Politics of Socialist Agriculture in Poland: 1945–60.* New York: Columbia University Press, 1965.

Kowalczuk, Ilko-Sascha. *17.6.1953: Volksaufstand in der DDR.* Bremen: Edition Temmen, 2003.

Last, George. "Rural Functionaries and the Transmission of Agricultural Policy: The Case of *Bezirk Erfurt* from the 1960s to the 1970s" in Mary Fulbrook, ed. *Power and Society in the GDR 1961–1979. The "Normalization" of Rule?* New York: Berghahn Books, 2009, 76–101.

———. *After the "Socialist Spring:" Collectivization and Economic Transformation in the GDR.* New York: Berghahn Books, 2009.

Levy, Robert. *Ana Pauker: The Rise and Fall of a Jewish Communist.* Berkeley: University of California Press, 2001.

———. "The First Wave of the Collectivization Campaign: Central Policies and their Regional Implementation, 1949–53" in Constantin Iordachi and Dorin Dobrincu, eds. *Transforming Peasants, Property and Power: The Collectivization of Agriculture in Romania, 1949–1962.* Budapest: CEU Press, 2009, 27–48.

Lindenberger, Thomas. "Die Diktatur der Grenzen: Zur Einleitung" in Thomas Lindenberger, ed. *Herrschaft und Eigen-Sinn in der Diktatur.* Vienna: Böhlau Verlag, 1999, 13–44.

Lüdtke, Alf. *Eigen-Sinn. Fabrikalltag, Arbeitererfahrungen und Politik vom Kaiserreich bis in den Faschismus.* Hamburg: Ergebnisse Verlag, 1993.

Osmond, Jonathan. "Kontinuität und Konflikt in der Landwirtschaft der SBZ-DDR zur Zeit der Bodenreform und der Vergenossenschaftlichung, 1945–61" in Bessel and Jessen, eds. *Grenzen der Diktatur.* Göttingen: Vandenhoeck and Ruprecht, 1996, 137–69.

Pfaff, Steven. *Exit-Voice Dynamics and the Collapse of East Germany.* Durham, N.C.: Duke University Press, 2006.

Sandford, Gregory W. *From Hitler to Ulbricht.* Princeton: Princeton University Press, 1983.

Schroeder, Klaus. *Der SED Staat: Geschichte und Strukturen der DDR.* Munich: Ernst Vögel Verlag, 1998.

Scott, James. *Seeing Like a State: How Certain Schemes to Improve the Human Condition Have Failed.* New Haven: Yale University Press, 1998.

———. *Weapons of the Weak: Everyday Forms of Peasant Resistance.* New Haven, Connecticut: Yale University Press, 1985.

Shopov, Vulkan. *Socialist Development of the Bulgarian Countryside.* Sofia: Sofia Press, 1972.

Swain, Nigel. "Decollectivization, Politics, and Rural Change in Bulgaria, Poland and the former Czechoslovakia." *Social History* 32, no. 1 (February 2007): 1–26.

Tilly, Charles. *Contentious French.* Cambridge, Mass.: Belknap Press of Harvard University Press, 1986.

Verdery, Katherine. *What Was Socialism and What Comes Next?* Princeton: Princeton University Press, 1996.

———. "Exploiters Old and New: Making and Unmaking 'Rich Peasants' in Aurel Vlaicu (Hundedoara Region)" in Constantin Iordachi and Dorin Dobrincu, eds. *Transforming Peasants, Property and Power: The Collectivization of Agriculture in Romania, 1949–1962.* Budapest: CEU Press, 2009, 307–28.

Viola, Lynne. "The Peasant Nightmare: Visions of the Apocalypse in the Soviet Countryside." *Journal of Modern History* 62, no. 4 (Dec 1990): 747–70.

———. "Popular Resistance in the Stalinist 1930s: Soliloquy of a Devil's Advocate" in Lynne Viola, ed. *Contending with Stalinism: Soviet Power and Popular Resistance in the 1930s.* Ithaca: Cornell University Press, 2002, 17–43.

Wekentin, Falco. *Politische Strafjustiz in der Ära Ulbricht.* Berlin: Ch. Links Verlag, 1995.

Wippermann, Wolfgang. *Totalitarismustheorien: Die Entwicklung der Diskussion von den Anfängen bis heute.* Darmstadt: Primus Verlag, 1997.

Witkowski, Gregory R. *Factory to Farm, Farm into Factory: A Case Study of Communist Campaigns and Agricultural Collectivization in the German Democratic Republic.* Buffalo: SUNY Press, 2003.

———. "On the Campaign Trail: Communist Campaigns, State Planning and *Eigen-Sinn* in the Transformation of the East German Countryside." *Central European History* 37, no. 3 (September 2004): 400–22.

———. "Peasants' Revolt? Reevaluating the 17 June Uprising in East Germany." *German History* 24, no. 2 (May 2006): 243–66.

Eastern European Collectivization Campaigns Compared, 1945–1962

NIGEL SWAIN

This contribution represents the culmination of more than three decades of research into the collectivization of agriculture in Eastern Europe and its demise.[1] It is based on my doctoral research into Hungarian collectivization, intermittent monitoring of rural issues in Hungary in the 1980s,[2] information about the collectivization process that emerged from research projects that I led in the 1990s investigating decollectivization in 54 Eastern European villages, and extensive reading of the secondary literature, including the recent publication by Constantin Iordachi and Dorin Dobrincu on Romania, as well as contributions to this volume.[3]

In my youthful naivety I claimed many "unique" features of Hungarian collectivization, which I subsequently discovered existed elsewhere. Equally, as I read accounts of collectivization in other countries, I discovered other scholars making similar claims of their target countries' "uniqueness." We all, it seems, fell into the same trap. We knew "our" country well, and knew that it did not conform to a relatively simplistic understanding of what the collectivization norm was; we assumed that other countries did conform to the norm, and therefore concluded that

---

[1] I am indebted to the editors, Constantin Iordachi and Arnd Bauerkämper, for their perceptive comments on early drafts and to all contributors to this volume for their helpful advice. I was helped writing this contribution by a questionnaire circulated to the project participants (hereafter Questionnaire).

[2] Swain, *Collective Farms*; Swain, "Hungarian Agriculture."

[3] Iordachi and Dobrincu, eds., *Transforming Peasants.*

the experience of "our country" had to be "unique." In the conditions of the Cold War this was not necessarily a manifestation of intellectual laziness. Research into collectivization was not easy, and for many countries of the region it was difficult to get beyond the ideologized claims of the political establishment. But, twenty years after "system change," facile comparisons between one country whose history is known well and a vague socialist norm is unacceptable. Research into collectivization in all of the countries of the region is well advanced, as this volume admirably demonstrates.

This chapter attempts therefore to compare the collectivization processes of the countries covered by this volume, not against an assumed typical model, from which one or another country might partially or temporarily have diverged, but simply against each other. It does not do away with the idea of a model entirely, however, not least because it is impossible to use the comparative method without some sort of model, implicit or explicit, of what the processes being compared entail. Weber developed his notion of the "ideal type" to address this problem.[4] In the case of collectivization, however, we are saved the methodological chore of constructing for ourselves an "ideal type" collective farm because Stalin provided a concrete and prescriptive model in the form of the Model Charter for collective farms of 1935.[5] The political leaders of Eastern Europe faced a common project: to establish rural structures that conformed to the Model Charter.[6] But, in Marx's famous words, they did not do so "under circumstances chosen by themselves, but under circumstances directly encountered, given and transmitted from the past."[7] They could not avoid entanglements with their pasts, the Soviet Union or their fellow Eastern Bloc members. This chapter will attempt to tease out the common features imposed by Stalin's model (and subsequent changes to it), and the entanglements that each country had with its national histories and traditions, Soviet pressure, and one another.

Collectivization in Eastern Europe was a complex process, and the conditions within which it was implemented varied considerably in the

---

[4] Weber, *Economy and Society*, 19–22; Gerth and Mills, *From Max Weber*, 60.

[5] For two rather different English translations of the text of this charter, see Hubbard, *Economics of Soviet Agriculture*, 131–46; and Stuart, *Collective Farm in Soviet*, 199–210.

[6] The various model charters implemented in Eastern Europe are discussed in Wädekin, *Agrarian Policies*, 17, 63–82.

[7] Marx, "Eighteenth Brumaire," 97.

countries covered by this volume. There is a very real danger of failing to "see the wood for the trees." In order to avoid this, this chapter does not follow a chronological approach but rather is divided into two sections. The first of these focuses on the skeleton of collectivization, the key dates relating to the implementation of the Stalin model in the region, and the changes to the nature and content of that model that took place after Stalin's death. The second section addresses the flesh of the national cases, their entanglements with their own history and the experience of their neighbors: the myriad of deviations from, and variants on, the model that such entanglements entailed. This section adopts a thematic approach: certain axes of comparison are selected across which to investigate the national cases. These axes relate first to the context (political and material) in which collectivization was implemented, for the conditions in postwar Eastern Europe differed radically from country to country, and were everywhere dissimilar from those of the Soviet Union in the 1920s and 1930s. They then address the manner and pace with which key policies were applied (policies towards *kulaks* and campaigning generally), as well as responses to them.

A Comparative Framework—Chronology and Model

Chronology

As Lynne Viola's contribution to this volume makes clear, collectivization as a model for the socialist restructuring of agricultural production took its first form in the Soviet Union in the late 1920s and 1930s. By the end of the 1930s it had become the established norm for socialist agriculture, so that when, in late 1939 and 1940, parts of eastern Poland, Bessarabia and the Baltic states were incorporated into the Soviet Union, collectivization in accordance with the Model Charter was implemented almost immediately. But Stalin's wartime alliance with the United States and Great Britain occasioned a change towards more inclusive Popular Front policies, symbolized by the dissolution of the Communist International in May 1943. The communist parties that either came to power or dominated the political scene in immediate postwar Eastern Europe were under instructions to make no mention of collectivization and focus rather on land reform, a policy which could be espoused by most major non-communist parties. In the Baltic states, reintegrated into the Soviet Union, collectivization was more prominent on the agenda, and it was seriously discussed in 1947; once the Paris Peace Treaty had removed some postwar political uncertainties (both domestic and international), it began in the winter of 1947–48. But, in formally independent Eastern Europe, land reforms were implemented and collectivization disappeared from the political agenda, until the Stalin-Tito dispute triggered an abrupt policy change. In May 1948, Stalin and Molotov criticized the Yugoslav leadership amongst other things for referring to the peasants as the "strongest pillar of our state" and for not recognizing the "sharpening class struggle" in the countryside. At the Cominform meeting in Bucharest that June, when Tito's rift with Stalin became public knowledge (and the Berlin Blockade began), the message was clear. Yugoslavia moved quickly to collectivize agriculture, not because Tito wanted to be more Stalinist than Stalin, but, as Melissa Bokovoy argues, because he believed (erroneously as it turned out) that his regime enjoyed widespread peasant support. Yugoslavia embarked on collectivization in January 1949, but the country's communist leadership had doubts after the first year, and abandoned it in 1953.[8]

Although they had not been the objects of Stalin's criticism, the communist leaders of Eastern Europe saw the need to change tack too, a policy shift that was facilitated by the fact that, by this time, internal political opposition, and in particular the peasant parties, had been neutralized. Most countries immediately increased economic pressure on richer peasants; and Bulgaria embarked forthwith on the formation of cooperative farms, such that it had already experienced both a campaign and a softening of the line by June 1949, when campaigns in Czechoslovakia (May, 1949), Romania (March 1949), and Albania (February–March, 1949) were just beginning. Hungary took initial measures in 1948, but also began its first full-scale collectivization campaign in 1949. Poland had formulated a cautious program in September 1948, following the dismissal of Gomułka; and a goal of one percent collectivization by the end of 1949 was announced. The exception was the GDR, not yet an independent state when the Cominform line changed. There Stalin warned against pushing for the socialist transition in the countryside, and collectivization

[8] Swain, "Deciding to Collectivize;" Feest in this volume; Clissold, ed., *Yugoslavia and the Soviet Union*, 1975, 189–90; Bokovoy, *Peasants and Communists*, xv, 86–87.

did not start until June 1952, when he was convinced that there was little chance any more of a united Germany.[9]

The first phase of East European collectivization lasted generally until 1953 when Stalin died. The exception was Bulgaria, which had already experienced a second cycle of intense collectivization in January 1950. This resulted in unrest in both 1950 and 1951 and further collectivization was suspended between 1951 and 1956. The New Course policies of 1953–54 saw a reduction in the pressures on the peasantry everywhere else; even the leaders of the Socialist Unity Party in the GDR, where collectivization had scarcely begun, accepted the changed Soviet line, allowing, for example, peasants who had fled to the West to return without penalties, although this did not prevent the uprising of June 17, 1953.[10]

Collectivization came back onto the agenda in the summer of 1955 as Khrushchev consolidated his power. Hungary and Czechoslovakia (along with China) renewed their campaigns, while Albania followed suit in December 1955 and again in May 1956. Bulgaria's leading functionaries did not respond to Soviet pressure until 1956, but they then continued and were, like Albania, unaffected by the events of the autumn of that year. Poland too, in February 1956, hoped to increase the pace of collectivization over its 1949–55 experience. But collective farms in Hungary began to disband spontaneously in the summer and autumn of 1956, and, after the events of October, their number plunged dramatically.[11] In the wake of October 1956, collectivization was abandoned entirely in Poland. Even in Czechoslovakia and Romania, where there was no political disruption, politicians became concerned about cross border influence, and pressure to collectivize was temporarily relaxed.[12]

As Zsuzsanna Varga notes in her contribution to this volume, it was in Moscow in November 1957 that the decision was taken to renew the pressure to collectivize yet again, thereby initiating the final phase. Collectivization in Czechoslovakia began again in late 1957, but with greater

[9] Swain, *Fate of Peasant Parties*; Chronology of the GDR in this volume; Örjan Sjöberg's contribution to this volume, 382–83; Sjöberg, *Rural Change*, 86–89; Orbán, *Két agrárforradalom*, 71–73, 79–87; Kaliński, "Collectivization," 171; Korboński, *Politics of Socialist*, 1965, 153; Bauerkämper, *Ländliche Gesellschaft*, 127–29; Schöne, *Das Sozialistische Dorf*, 93.

[10] Mihail Gruev's contribution to this volume; Schöne, *Das Sozialistische Dorf*, 113.

[11] Orbán, *Két agrárforradalom*, 147–57; Donáth, *Reform és Forradalom*, 158; Varga, "Impact of 1956," 157.

[12] Myant, *Czechoslovak Economy*, 73; Sjöberg, *Rural Change*, 87–88; Kaliński, *Collectivization*, 179; Rychlík, in this volume; Oprea, "The Final Offensive," 49–80, 64.

commitment in 1958 after Novotný became president. It had achieved 75 percent total land area by 1958, by which time Bulgarian collectivization had reached 92 percent of agricultural territory and was effectively complete. Hungary followed suit in the wake of a Central Committee meeting of December 1958, and announced successful collectivization two years later in February 1961. Collectivization in the GDR, by contrast, was characterized by extreme speed. Progress had been relatively modest in the seven years prior to the advent of Gerhard Grüneberg as Agriculture Secretary for Central Committee, although membership did double from 20 percent to 40 percent between 1953 and 1958. Grüneberg opted for intensely rapid collectivization, resulting in the GDR's Socialist Spring, which completed the process in the first four months of 1960. Full collectivization took considerably longer in Romania, although the completion date was brought forward from 1965 to 1962. It took even longer in Albania where, although the bulk of the lowlands was collectivized by 1960, collectivization in the central and northern mountains was not complete until 1967, and livestock was still being collectivized in the 1980s.[13]

Model

Collectivized agriculture was Soviet-inspired, and in particular Stalin-inspired. It is appropriate to categorize the model, as exemplified in the Model Charter, as "Stalinist" rather than "Soviet" for two reasons. First, key aspects of it were insisted on by Stalin, who, as Viola's contribution to this volume makes clear, was personally involved in developing collectivization policy. Second, after Stalin's death, two further notable developments took place. On the one hand, collectivization policy changed, as will be discussed fully below; on the other, some countries ignored Soviet practice under Stalin's successors, while the Soviet Union itself began to learn from its neighbors as much as the other way around. The nature of the model for collectivization changed: it was no longer concrete Soviet practice, but rather a policy ideal with no concrete equivalent in the real world.

Despite the Model Charter, even during the Stalin years, there were three differences between the Soviet practice of the 1930s and Eastern

[13] Varga in this volume; Orbán, *Két agrárforradalom*, 210, 216, 241–42; Schöne, *Frühling*, 258–60; Schöne, *Das Sozialistische Dorf*, 135; Last, *After the Socialist*, 13; Iordachi, "Constanța," 106; Sjöberg in this volume, 392–94.

European collectivization.[14] The first was related to land ownership. In the Soviet Union, all land was nationalized in November 1917 and leased "in perpetuity" to the peasantry. In the majority of the countries of Eastern Europe, there was no attempt in the 1940s or thereafter either to nationalize the soil or create cooperatively owned land. Land ownership remained formally with the peasants. But there were exceptions. Hungary began creating collective ownership of land following its land law of 1967. In Albania land was *de facto* nationalized as early as 1946, in that all holdings over five hectares were expropriated and the sale or leasing of land was forbidden, but it was only in 1976 that this was recognized in the constitution. In the case of the Baltic states, their incorporation into the Soviet Union ensured the application of the Soviet model of nationalization, followed by leasing in perpetuity, as had been implemented in the areas of interwar Poland that were incorporated into the Soviet Union in 1939.[15]

The second difference concerned the type of violence associated with collectivization. With the exception of the Baltic states, where they did take place, mass deportations did not accompany collectivization. The partial exceptions to this rule were deportations in Romania in 1949 and 1951. In 1949, *moşieri* (landowners with more than 50 hectares who had been exempted from the 1945 reform) were deported throughout the country once their assets had been confiscated. In 1951, groups from ethnic minorities resident in border areas were deported to the Bărăgan plain for either security reasons or because they opposed collectivization.[16]

The third difference from Soviet practice of the 1930s had to do with permissible types of collective farms. Soviet regulations from 1919 had recognized a looser cooperative form, the TOZ, but, with mass collectivization at the end of the 1920s, the *artel'* became the preferred form,[17] the one for which the Model Charters of 1929 to 1935 were produced, and the only one allowed in the Baltic states. In Eastern Europe, with the

[14] For a comparison between the campaign of collectivization in Russia on the one hand and that in post-1945 Eastern Europe on the other, focusing on the status of land property, periodization, the use of violence and anti-*kulak* policies, see also Iordachi and Verdery, "Conclusions," 455–71, esp. 455–59.

[15] Wädekin, *Agrarian Policies*, 14–30; Viola in this volume; Swain, *Collective Farms*, 53–54; Sjöberg in this volume; Dariusz Jarosz, Questionnaire.

[16] Feest in this volume, 81–111; Iordachi in this volume, 266, 271, 278; Vultur, "The Role of Ethnicity," 146–147.

[17] Davies, *Soviet Collective Farm*, 68–115.

exception of Bulgaria, the Soviet practice of the early 1920s was offered in that, initially at least, a choice of simpler forms of cooperatives were presented. Generally these different types of cooperatives were simply designated Type I, Type II, Type III, and Type IV, the latter two being roughly the equivalent of the *artel*.' However, the terms meant slightly different things in each country, and not all countries had all variants. Hungary and the GDR had only three types, for example, and Poland introduced subvariants in 1950 (Types Ia and Ib) because members were reluctant to pool horses and agricultural machinery. Romania only ever had two types, and the preferred type was always the equivalent of the *artel*.' Initially, in 1949, the simpler variants were very loose "tillage associations" (*întovărăşiri*), overseen by the Machine and Tractor Stations, which were established to secure machine services and had no statute of their own. In March 1952, Gheorghiu-Dej, with Stalin's backing, successfully proposed their transformation into Soviet-type TOZs, permanent cooperatives, conceived of as stepping-stones to fully-fledged collectives.[18]

The essential difference of principle between these types consisted in how much activity took place on a collective basis and, crucially, what percentage of rewards was based on labor performed rather than land contributed. In Type III and IV farms (as on the Soviet *artel*'), rewards were almost exclusively based on labor, although members might receive a nominal land rent. In the looser variants, a smaller proportion of rewards came from labor performed, and fewer activities were performed in common. Because communal livestock farming could only realistically take place on the basis of extensive pooled resources (the Soviet TOZs had not pooled livestock of any sort),[19] this, rather than basis of remuneration, became the yardstick for the Type III farm in the GDR in particular, but also in popular conceptions elsewhere. Bulgaria, as noted, only ever advocated a single type of cooperative at any point in time, but its model charter became increasingly like the *artel*,' such that by 1953 only 25 percent of cooperative income could be paid out in rent. Albania abandoned its Yugoslav-inspired model charter from 1946 for a single *artel*'-type one in 1949, and Types I and II were permitted by revised model statutes in 1954. Yugoslavia elaborated four types of cooperative in May 1949 but did not prioritize any of them. Of these, the Type IV included

[18] Wädekin, *Agrarian Policies*, 17, 64; Feest, Questionnaire; Bauerkämper, *Ländliche Gesellschaft*, 509; Last, *After the Socialist*, 89; Kaliński, *Collectivization*, 183; Jarosz in this volume, esp. 121; Levy, "The First Wave," 27, 39–40.

[19] Davies, *Soviet Collective Farm*, 68.

joint ownership of all land aside from the household plot, while the others retained private ownership and different degrees of rental.[20]

Collectivization policy changed after Stalin's death in March 1953, and in particular in the wake of Khrushchev's speech criticizing Stalin at the Soviet Twentieth Party Congress in February 1956, which triggered dramatic challenges that autumn to the regimes in Poland and Hungary. One of the key concepts in the Stalin-model had been "class war in the countryside." The formation of producer cooperatives by and for the rural poor (in alliance with the middle peasantry) went hand in glove with the destruction of the *kulaks*, the actual or potential capitalist exploiters of the rural world. Only two countries had ignored this policy: Yugoslavia and China, both of which had achieved revolutionary transformation without the help of the Red Army, but with significant peasant support. In China, the changed Soviet line of 1948 came before the revolution was complete and while Mao was still dependent militarily on peasant soldiers. Even when in power, his initial focus was on neutralizing the richer peasants politically rather than physically. The Mutual Aid Teams (China's cooperatives of a lower type), established from 1953 onwards (after the death of Stalin), allowed rich peasants to join. Furthermore, in a document dated January 1956 (after the July 1955 call to accelerate collectivization), Mao permitted even "former landlords" and "counter revolutionaries" to join the cooperatives of the higher type. In Yugoslavia, while richer peasants did suffer from higher compulsory deliveries, one of the few issues of substance in the Stalin-Molotov 1948 criticisms of the Yugoslav comrades was the accusation that they had failed to identify class divisions within the peasantry. The Kardelj faction in the leadership continued to support the policy of permitting richer peasants to join cooperatives. In Kardelj's words of November 1947, "The *kulak* is a political concept—a man who does not fulfill his obligation to the state."[21]

With Stalin gone, Eastern European policymakers began to switch without openly acknowledging it to the Yugoslav and Chinese position of

---

20 Bergmann, *Agrarpolitik und Agrarwirtschaft*, 113–14; Fazekas, *A mezőgazdasági*, 1976, 57–59; Kaliński, *Collectivization*, 181–82; Jarosz in this volume, 121–22; Karlik, *Questions of Ownership*; Last, *After the Socialist*, 28–30; Bauerkämper, *Ländliche Gesellschaft*, 168; Schöne, *Das Sozialistische Dorf*, 106–107; Meurs et al., "Bulgaria," 102; Sjöberg in this volume, 393, 401–2; Chronologies for Bulgaria, Yugoslavia, and Albania; Bokovoy, *Peasants and Communists*, 105.

21 Spence, *Search for Modern China*, 521–22; Selden, ed., *People's Republic of China*, 341, 360; Bokovoy, *Peasants and Communists*, 90–91.

divorcing "class war" from collectivization. As part of the New Course, Hungary and Czechoslovakia, for example, both abolished their "*kulak* lists" in 1953. But when the New Course abruptly ended, traditional anti-*kulak* economic pressure shaped the 1955 collectivization in Hungary, and it was only after the political fallout of the following year that significant changes took place. Hungary's revolutionary government abolished compulsory deliveries on October 30, 1956 (a decision already taken in principle by the Agricultural Department of the party's Political Committee on October 22, the day before the revolution broke out).[22]

The GDR had permitted *kulaks* to join cooperatives as early as December 1954 under restricted circumstances, but the measure had little impact, especially following the rejection of a reformist program for collectivization in 1956–57; and the real policy change seems to have been in 1958. In July 1956, Bulgaria permitted *kulaks* to join (but not to become managers of) cooperatives and Poland welcomed "loyal *kulaks*" into its cooperatives. The Czechoslovak literature points more obliquely to the second half of the 1950s as the point when *kulak* policy changed, allowing them even to become chairmen. Hungary's post-1956 policy-determining Agrarian Theses of July 1957, while they stated that a policy of "class peace" would be a mistake, also criticized the "liquidation" methods of the past and resolved even to permit *kulak* employment of labor so long as it was taxed. As early as 1955, a cooperative of the simple type had been established headed by an experienced farmer who had once been on the *kulak* list. But the formal right of *kulaks* to join full producer cooperatives came in a decree of 1959, which even permitted former *kulaks* to lead cooperatives after two years exemplary service, a process which could be accelerated with the agreement of the county administration.[23]

Bulgaria paved the way for making collective agriculture more attractive when it introduced pensions for farm workers from January 1957.[24]

[22] Varga, *Politika*, 22–23; Varga, "The Impact of 1956," 159–60; Myant, *The Czechoslovak Economy*, 72; Orbán, *Két agrárforradalom*, 147–50, 217–56.

[23] Gruev in this volume, 363; Jarosz in this volume, 122; Schöne, *Frühling auf dem Lande*, 178, 221, 208; Last, *After the Socialist*, 8; Rychlík in this volume, 201; Březina, "Kolektivizace zemědělství," 131; "Az MSZMP," 225; Magyar, *Dunaapáti II*, 678.

[24] Gruev in this volume, 363, 372. Hungary introduced them in 1958 and Czechoslovakia in the late 1950s. See Swain, *Collective Farms*, 39; and Rychlík, Questionnaire.

After Khrushchev had condemned the "labor day" in December 1958,[25] and two years later stressed the importance of material incentives, the Hungarian leadership understood the changed realities.[26] As Varga's paper in this volume shows, it was swift to respond to the new environment in which entanglements were possible with fellow Bloc members as well as the Soviet Union. When it sent out delegations in 1958 to learn from the experience of their neighbors, it was particularly impressed by Czechoslovakia's solution to incentives problems, not least because these coincided with the suggestions that were emerging from domestic reformers and grass roots pressure. It became convinced of the importance of material incentives and satisfying consumer needs. As a consequence, a series of minor adjustments made collective farm membership less unattractive in Hungary's final round.[27]

In Albania, which rejected most of Khrushchev's reformism, *kulaks* who no longer constituted a threat were permitted to join cooperatives. In Romania, although compulsory deliveries had been abolished in December 1956 almost immediately after their abolition in Hungary and the focus had shifted to winning influential villagers over to collectivization, the final round of the campaign in 1957 began with violence. This resulted in another temporary pause before complete collectivization in 1958–62. In this context it is not surprising that in Romania, in March 1959, *kulaks* were obliged, rather than permitted, to join cooperatives and donate their land to the state. Despite the trend to make cooperative membership less unattractive, the GDR nevertheless maintained a membership cost unique in the region: the *Inventarbeitrag*. Contributions in kind (livestock, equipment, buildings) upon entry were standard everywhere, but in the GDR members also had to contribute financially: a minimum contribution, commonly 500 marks, was established, against which in-kind con-

[25] Noted in Edvard Kardelj's speech on May 5, 1959, see RFE News and Information, Evaluation and Research, General Desk No. 709. I am grateful to Melissa Bokovoy for supplying this reference.

[26] Varga in this volume, 437–469. Although from 1957 Khrushchev's policy in the Soviet Union was lurching towards his Virgin Lands debacle and an unsuccessful attempt to convert collective farms into state farms, in 1953–55 he had introduced various measures that increased incomes on collective farms, both from the socialist sector and private plots, and had increased farm autonomy. The critical comments on the labor unit should be seen in this light. See Nove, *Economic History of the USSR*, 327–66.

[27] Varga in this volume, esp. 448–50. See also Varga, *Politika*; Varga, "The Impact of 1956;" Swain, *Collective Farms*, 31–34.

tributions were set; the balance was due in cash, although in theory it was to be repaid in the long run.[28]

Perhaps paradoxically, the final phase of collectivization also witnessed the almost universal demise of the looser alternatives to the *artel.*' In Romania, the Type I farm remained a way of exaggerating the commitment to socialist forms until collectivization was completed, but thereafter only the Type III farm, usually referred to as CAP, was permitted. In Albania, reference to other types disappeared until the introduction of farms of a "higher type" in the 1970s. The two countries that changed collectivization policy most radically, Czechoslovakia and Hungary, had also abandoned the looser type of cooperative by the time of the final campaign. In Czechoslovakia, both Types I and II were of little significance by 1953 and had disappeared entirely by 1957. In Hungary, legislation of September 1956 on new simpler forms of cooperatives, together with the July 1957 Agrarian Theses, did away with the "type" terminology and referred rather to "simpler cooperative forms." These flourished between 1956 and 1958 and then declined in number during the final stage of collectivization, but with a new increase in the very last months, apparently, as in Romania, as a concession to achieve collectivization by the required deadline.[29]

In the GDR, however, the Type I cooperatives continued through the 1960s and into the 1970s. The reason for this exceptional development is linked with the GDR's peculiar generosity in making collectivized agriculture attractive. The GDR provided government subsidies to guarantee remuneration levels on the Type III cooperatives, so obviating the problems associated elsewhere with the "remainder principle" aspect of the "labor day."[30] But this expensive policy was only possible (until the sector consolidated) provided a significant number of farmers continued to work according to the Type I model, where remuneration according to labor contribution was less significant.

---

[28] Oprea, "The Final Offensive," 64–65; Iordachi, "Constanţa," 117; Ţârău, "Collectivization Policies," 215; Verdery, "Exploiters Old and New," 321; Chronologies for Romania, Albania, and the GDR; Last, *After the Socialist*, 30; and Swain, "Local experience of the Oder," 145–62, 161.

[29] Iordachi, "Constanţa," 117; Miller, "Law and Propaganda," 81–100, 81; Bergmann, *Agrarpolitik*, 114; Decree of the Council of Ministers, 1.091/1956 (IX.11) MT sz hat; "Az MSZMP," 234–35; Fazekas, *A Mezőgazdasági*, 65, 76, 88, 94, 116, 129, 137.

[30] Last, *After the Socialist*, 59–61. Problems of the "labor day" and "work unit" are discussed in Swain, *Collective Farms*, 42–44, and Varga, *Politika*, 13–14.

Axes of Comparison

Policy context—land reform and resettlement

Although Stalin's policy recipe from the war's end until June 1948 was for the communist parties to promote a land reform which favored the poorer peasants and to postpone collectivization, it is necessary to note that not all parties in the region were "on message" immediately. In the countries of the Balkans that had not relied exclusively on the Red Army to instigate revolutionary change, committed communists passed collectivization measures before switching focus to the land-reform line. Bulgaria, the only country in the region to have any tradition of producer cooperatives, progressed furthest in this regard. Even before the beginning of World War II, there had been 28 producer cooperatives in Bulgaria, and a further 82 were established during the war itself by forces close to the Communist Party. As the war ended, the Communist Party sought to build on this base of 110 cooperatives to promote collectivization in 1944 and early 1945. An act on producer cooperatives was passed in April 1945. But then enthusiasm waned and, following Soviet pressure, the communists began to compete with the Bulgarian Agrarian National Union, the main peasant party, over land reform, which was implemented in March 1946. Committed communists, however, were not enthusiastic about land reform in a country where all knew that its impact would be minimal, and despite the switch in policy, the formation of cooperatives continued. Bulgaria had some 400 producer cooperatives—8 percent of farms—in place by 1948 when collectivization returned to the agenda. Discriminatory compulsory deliveries had been introduced as early as the autumn of 1947 (as in the Baltic states), and measures were taken in the winter and early spring of 1948 to expropriate privately owned large machinery and incorporate collective farms into the quantitative planning structures.[31]

In Yugoslavia, the government passed a series of measures between February and December 1945 to establish an agrarian system that discriminated against private agriculture in favor of the state and cooperative sector. Within these, the model of the peasant work cooperative was

[31] Swain and Swain, *Eastern Europe*, 24–30; Stillman, "Collectivization of Bulgarian," 68; Meurs et al., "Bulgaria," 92; Gruev in this volume, 336–38; Brus, "Postwar Reconstruction," 589; Chronology for Bulgaria.

promoted in March, while the Law on Agrarian Reform and Colonization passed in August. By the end of the year, 31 peasant work cooperatives were in operation, on an impressive 96,000 hectares of land; but the focus then switched to promoting general agricultural cooperatives until 1948. In Albania, cooperative statutes based on the Yugoslav model were adopted in 1946 and thereafter a number of cooperatives were established on the coastal plain.[32]

These exceptions aside, the immediate postwar focus was on land reforms and on winning the poorer peasants over to the communists and away from the peasant parties. In Hungary, for example, land reform was insisted on by Marshal Voroshilov, the head of the Soviet occupation authorities, for both military and political reasons: land reform in the already liberated east would serve as a beacon for peasants in the west still under German occupation. Indeed it was the Communist Party that, under Soviet pressure and against some opposition from old guard communists, watered down the original program of the National Peasant Party. In Czechoslovakia, land reform was part of the politics of attacking the power base of the interwar Agrarian Party, which had been banned from that country's postwar political sphere. Similarly, in Poland and Romania, land reform was a way of countering the influence of the Polish Peasant Party and Romania's National Peasant Party, respectively. In the Soviet Occupation Zone of Germany (the future GDR) it was aimed at mobilizing poor peasant support more generally, there having been no competitive political system since 1933.[33]

In all countries, the major beneficiaries of the land reforms were poorer peasants. But, the extent of rural overpopulation in the interwar years was such that, with the exception of Czechoslovakia, the additional land that poorer peasants received under the reform could not transform them into prosperous peasant farmers. Furthermore, in the context of the aftermath of World War II, land reforms became entangled with population transfers and were rarely ethnically neutral. To a greater or lesser extent, land reform took place at the same time as the expulsion of enemy populations and the resettlement or colonization of the newly vacated areas. In Poland this process was complicated by the fact that the country's borders moved westward at the expense of Germany. Land reform in "old" Poland thus was accom-

[32] Chronologies for Yugoslavia and Albania; Bokovoy, *Peasants and Communists*, 46–47; Sjöberg in this volume, 382–83.

[33] Kenez, *Hungary*, 107–17; Orbán, *Két agrárforradalom*, 25–27; Rychlík, Questionnaire; Korboński, *Politics of Socialist*, 68; Hitchins, *Rumania*, 537–39.

panied by resettlement in the "Regained Territories" vacated by Germans. The majority of the new settlers came from eastern territories now incorporated into the Soviet Union, areas which had experienced collectivization in the 1939–41 period; but others were ethnic Ukrainians who had escaped transfer to Ukraine and inhabited regions where fighting between Ukrainians and Poles had escalated to a civil war. Under Operation Vistula they too were resettled in the west between 1947 and 1948.[34]

In Czechoslovakia, the Beneš Decrees expelled the Sudeten Germans, and associated measures (not the Decrees themselves) rendered the status of Hungarians uncertain. In Yugoslavia, the expulsion of Germans (and some Hungarians) rendered large sections of Vojvodina available for resettlement by poorer populations from Macedonia, Bosnia and Montenegro. Hungary also expelled ethnic Germans from Transdanubia and, in a protracted process, exchanged Slovaks for Hungarians from Czechoslovakia. In Romania, unlike in other Central European countries, the German population was not expelled after the war; yet the properties of ethnic Germans were expropriated under the 1945 land reform, while many Germans had already been taken for labor service in the Soviet Union by the Red Army. Settlers from newly Soviet Bessarabia, northern Bukovina, and Bulgarian southern Dobrogea were moved in, to share the necessarily limited facilities after the Germans returned. (However, some 15 percent of their number did not return, and 50 percent of those who did subsequently emigrated to Germany or Austria.) In Bulgaria, 155,000 ethnic Turks emigrated in 1950 providing opportunities for resettlement. The Soviet Occupation Zone of Germany had an analogous problem to Romania, not the settlement of displaced persons into empty territories, but their incorporation into already populated zones, which were also undergoing land reform.[35]

Colonization and resettlement complicated the dynamics of land reform. Where there was no colonization or resettlement, the beneficiaries of land reform were the rural poor who lacked the human and material capital necessary for independent farming. Throughout the region

---

[34] Brus, "Postwar Reconstruction," 586–96, esp. 594–95; Kersten, *Establishment of Communist Rule*, 392.

[35] Rychlík, Questionnaire; Bokovoy, *Peasants and Communists*, 37–39; Tóth, "The Relocation of the Germans," 253–312; Hitchins, *Rumania*, 1994, 538–39; Cartwright, *Return of the Peasant*, 52–57; Chronologies for Romania and Bulgaria; Southern Dobrudja had been resettled in 1940 under the population exchange that followed the Treaty of Craiova. See Iordachi, "Constanţa," 108–9; and Iordachi in this volume.

(excluding the Baltic states where uniquely there was no post-land reform "land hunger"), these new farmers, formerly landless or from the landed poor, struggled to make a success of private farming and formed the nucleus of early collective farms. Even when, briefly, as in Romania under Ana Pauker and in Poland in 1950, there were government-sponsored or spontaneous measures to keep poorer peasants out of collective farms in order to maintain their economic viability, these groups constituted the majority of applicants.[36] Settlers and colonizers also found their new conditions difficult. The rural poor from eastern Hungary who took over German farms in Transdanubia struggled and were attracted to collectivization. This was equally true of their counterparts from Slovakia who, together with expelled Hungarians, moved into formerly German-occupied border areas where the earliest cooperatives were formed. Romanian settlers, too, were ill-equipped and some were early collectivizers.[37]

But the situation of colonizers and refugees was not identical. Those who had been displaced to a new property rather than invited to resettle in order to improve their lot included people who came from more prosperous backgrounds. Such people had at least the human capital necessary for independent farming, even if material capital in their new environment was lacking or absent. In Poland's "Regained Territories," many of the new inhabitants were not supporters of cooperative farming, and, indeed, the government itself was initially ambivalent. Special Settler Cooperatives were established to help them, cooperatives which were supported by the youth movement of the Peasant Party (commonly known as *Wici*) and did not engage in production. Nevertheless, in the long run it seems that human capital was less important than the shortage of physical resources. Settler Cooperatives formed the basis of the first producer cooperatives established in Olsztyn voivodship in 1949, and agricultural producer cooperatives were located disproportionately in the "Regained Territories" both before and after 1956. Jarosz concludes that, on balance, the most significant divide in the postwar years existed between those who had "old

[36] Feest in this volume; the fact that beneficiaries of the land reform struggled and often joined the cooperatives early is reported by all authors in this volume; Levy, "The First Wave," 28–32; Dobrincu, "Persuasion, Delay and Coercion," 284.

[37] Kovács, "Urbanizáció alulnézetből," 288. She confirms in a personal communication that they both struggled with independent farming and headed the early cooperatives; Průcha, "Economic Development," 60; Myant, *Socialism and Democracy*, 101; Myant, *Czechoslovak Economy*, 52; Vultur, "The Role of Ethnicity," 147; and Goina, "Never Leave 'til Tomorrow," 370.

title" to their land and those whose titles were new, whether because they benefitted from the land reform or because they were refugees.[38] Likewise, although many displaced Germans established new, independent farms and were not in the vanguard to join collectives, many others struggled and joined early, or quickly abandoned agriculture entirely. Although officialdom at the time identified the three logically distinct categories of farmer (old farmers, the beneficiaries of the land reform and displaced persons), in the minds of German peasants too, the divide was two-fold: between old farmers and "settlers," whether the latter were displaced persons or locals new to settled, independent farming.[39]

Collectivization was not without a geo-political dimension. Iordachi's study of the Dobrogea region of Romania, where collectivization was first completed, suggests that there was an overwhelming geo-political imperative to the region's early collectivization. Intense Soviet influence in the region, which imperial Russia had been denied under the 1878 Treaty of Berlin (when it was awarded to Romania), was clearly a factor. The link between geo-political imperatives and ideological forms is confirmed to a certain extent in the Hungarian case. István Orgoványi notes that the creation of a 15-kilometer defensive region along the border with Yugoslavia in 1950 was associated with accelerated collectivization.[40]

Material context: resources for collectivization

The corollary of the fact that the first cooperative farms were formed by the rural poor, who struggled to make a success of private farming, was that additional resources were needed if their farms were to survive. Additional land was therefore provided by the state, either from that unallocated by the postwar land reform, or from new land acquired in later, more radical land reforms. Buildings and other resources were likewise obtained from outside sources in some countries, at the expense of the existing cooperative networks. The need for additional resources was

[38] Kaliński, "Collectivization of Agriculture," 168, 187–91; Korboński, *Politics of Socialist*, 87, 176, 178; Bergmann, *Agrarpolitik*, 206–7; Jarosz in this volume, 119–21.

[39] Bauerkämper, "Transformation," 117–18, 128, 131–32, 137; Swain, "The local experience," 149, 159.

[40] Iordachi, "Constanţa;" Orgoványi, "Buffer Zone" 325.

common, but the sources varied according to national endowments and willingness to challenge existing structures.

In Czechoslovakia, the third land reform of March 1948 allowed authorities to confiscate land of over 50 hectares, adding to the fund for collectivized farming. In Romania, the seizure of *moşieri* land in March 1949 generated resources for collectivization even though the 50 hectare limit was the same as in 1945. In Albania, the maximum holding had been reduced from 20 to five hectares in February 1946, but in Poland there was little need of extra land given the extent of the "Regained Territories." In Hungary, measures in August 1948 put a limit on the total amount of land that rich peasants could jointly own and rent, and ceded the amount above this limit to cooperative groups, the very earliest producer cooperatives. Land for cooperatives was less of an issue in the GDR because of the more or less open border with the West until 1961. Rich peasants could abandon agriculture and migrate to the West. Uncultivated land became such an issue that a new institution, the *Örtlicher Landswirtschaftsbetriebe* (ÖLB), had to be created in 1953 in order to cultivate it, although in the end the local collectives were required to take over the land at considerable cost to their profitability.[41]

Hungary's 1948 attack on rich peasants' land coincided with two decisions aimed at providing them with additional resources by expropriating the cooperative movement. First, in August, it promoted the "land cultivator cooperatives" as general peasant cooperatives. These existed in a formal sense in some 2,500 villages and were encouraged to set up "producer cooperative groups." Then, at the turn of 1948–49, it merged the "land cultivator cooperatives" with the pre-1945 cooperative network, which was similarly encouraged to set up "producer cooperative groups." Likewise, in Czechoslovakia in February 1949, "unified cooperatives" were created when the assets of that country's stronger cooperative movement were passed to the producer cooperatives. Gruev uses the epithet "cuckoo syndrome" to describe similar measures in Bulgaria, which were more extensive because of that country's even more developed cooperative movement. In Poland the role of the Settler Cooperatives in supporting *de facto* collectivization in the Regained Territories has been noted. "Peasant Self-Help" general cooperatives had been established at the end of December 1944 to help smaller peasants, in particular those who benefitted from the

[41] Rychlík in this volume, 188–89; Iordachi in this volume, 257, 281; Sjöberg, *Rural Change*, 85; Orbán, *Két agrárforradalom*, 79; Pető and Szakács, *A hazai gazdaság*, 179; Schöne, *Frühling*, 173; Schöne, *Das Sozialistische Dorf*, 121; Bauerkämper, *Ländliche Gesellschaft*, 176.

land reform. Although by 1948 these had been subordinated to the state planning system, there was no systematic policy to transfer their assets to collective farms.[42] Romania, too, appears to have missed out on this phase of collectivization, which coincided with the early Pauker period, when a softer line directed at creating a smaller, but more viable cooperative sector was pursued. When collectivization policy became more aggressive in June 1950 (while she was on an extended holiday), the moment for a fictitious evolution from an existing cooperative movement was past. Collectivization in the GDR also bypassed this moment because of its late start in 1952. Between 1948 and 1950 all remnants of an independent cooperative movement in the GDR had disappeared.[43]

A final resource issue for the new producer cooperatives related to land consolidation: how to rationalize the scattered holdings of cooperative members in a given village so that their joint venture could hold land in consolidated blocks from which it could realize the benefits of large-scale production. The problem for the non-collectivized peasants was that land offered by local administrators in exchange for such consolidated cooperative land was invariably of lower quality. "Exchange" effectively became a variety of expropriation. The salience of land consolidation differed between countries and periods. It was more significant where both land was in short supply and cooperative and private farming coexisted for a lengthy period. Thus, it was less of an issue in the GDR, where there was bountiful abandoned land and the bulk of collectivization took place in a matter of months. But it figured much more prominently in Hungary, where land was not in surplus and the levels of collectivization fluctuated dramatically. It is not discussed so extensively in the Czechoslovak literature, although Březina refers to exchanges of land of unequal quality, and Czechoslovak accounts use the concept of the "minority cooperative" to describe villages where the majority were not members of cooperatives. It was an issue in Poland and Romania, where it was pursued only hesitantly in the early phase; while, in Bulgaria, it came into prominence in the 1951–56 period when private farming and collective farming coexisted for five years.[44]

---

[42] Pető and Szakács, *A hazai gazdaság*, 179; Orbán, *Két agrárforradalom*, 119; Rychlík in this volume, 190; Gruev in this volume, 351–52; Korboński, *Politics of Socialist*, 103–4; Landau and Tomaszewski, *Polish Economy*, 206.

[43] Levy, "The First Wave," 35–36; Schöne, *Das Sozialistische Dorf*, 95–100.

[44] Orbán, *Két agrárforradalom*, 85, 103–5, 150; Březina, "Kolektivizace," 131; Myant, *Czeschoslovak Economy*, 58; Jarosz in this volume; Levy, "The First Wave," 32–34; Gruev in this volume, 360–62; Meurs et al., "Bulgaria," 96.

Kulaks and kulak policy in the "class war" phase

A curious feature the "class war" period of Eastern European collectivization was the lack of precision concerning who the class enemy was. In theory, of course, *kulak* was a technical term of Marxist theory. *Kulaks* were rural capitalists, those who exploited the labor of others. In countries where a legal definition of *kulaks* emerged, it everywhere made reference to the employment of non-family labor. But many countries then extended the definition to criteria that were easier to identify, such as size of agricultural holding. The exceptions were Yugoslavia, Czechoslovakia, and Romania. In Yugoslavia the term was used primarily as a political tool for those who opposed government policy, as noted above. In Czechoslovakia, a *kulak* was broadly defined as a farmer who owned between eight and 20 hectares of land. In Romania, authorities attempted to formalize the definition in 1952 following the dismissal of Pauker, but the term was never simplified into land holding size and thus always remained open to local interpretation. In Hungary, the upper limit became 14.25 hectares of agricultural land (but less than three hectares of vineyards, or four hectares of fruit), which was essentially the same as in Poland, where it was 14 hectares. In Bulgaria, the limit was somewhat lower, at 10 hectares. In the GDR, it was higher, at 20 hectares.[45]

This lack of consensus was reflected in the terminology used to refer to *kulaks*. In the countries that enthusiastically embraced collectivization, such as Bulgaria, Albania, Hungary, and Yugoslavia—where "Chetnik," "Ustashi," "Cominformist," "papist," "Maček-supporter" or "Mihailović-supporter" were also common—the term *kulak* was used from the start (even if in Bulgaria it was rendered as "kolak" because of its unfamiliarity). Other countries, initially or permanently distanced themselves from Soviet connotations. In Romania and the GDR the local terms for richer peasants, *chiaburi* and *Großbauer* respectively, were used throughout. In Poland and Czechoslovakia there was a change of usage as the "class struggle" intensified. In Poland, *bogacz wiejski* was used in 1948, but *kulak* became the norm after 1949. In Czechoslovakia, initially the term *sedlák*

45 Bokovoy, *Peasants and Communists*, 90, 107; Vomela, "Kolektivizace v Československu," 122; Verdery, "Exploiters Old and New," 264–66; Pető and Szakács, *A Hazai Gazdaság*, 187; Jarosz, Questionnaire; Meurs et al., "Bulgaria," 98; Gruev in this volume, 344–46, 350–51; Schöne, *Frühling*, 68.

was used, followed by the Czech *vesnický boháč* and Slovak *dedinský boháč,* until *kulak* was adopted in 1951.[46]

It was equally unclear initially exactly what the relationship of *kulaks* to the collectivized sector should be. In formal terms, explicit exclusion from membership came after the first initiatives, as failure to collectivize at the required tempo was blamed on *kulak* influence, and the regimes attempted to make *kulaks* pay for the decline in agricultural output that collectivization occasioned. In Hungary, initially, *kulaks* who sold up and joined cooperatives were often welcomed and defended by the local leadership, but the situation changed in 1951 with the introduction of the "*kulak* list" (based on 1948 holdings) and the amendment of the cooperative statute of 1948 in January 1951 to exclude *kulaks* and "exploiters" from membership. The focus switched to unmasking *kulaks* in the membership: four were discovered in Dunaapáti in 1952. In Czechoslovakia, *kulaks* too were allowed initially to sell land and join cooperatives, as they could informally in Poland, but they were excluded from them in 1952. In Romania, *kulaks* could initially do likewise, but anti-*kulak* policies were strengthened after the removal of Pauker in July 1952, although relaxed briefly in the mid 1950s. After 1952, the "*kulak* list" was introduced in Romania and a distinction was introduced between "old *kulaks*" and "new *kulaks*," increasing their overall number by 50 percent. The focus switched there too to unmasking *kulaks* who had already infiltrated the cooperatives—10 were unmasked in one Transylvanian village. In Yugoslavia, larger farmers were explicitly allowed to join cooperatives, but when their behavior transgressed the Yugoslav political definition of *kulak* and they disregarded cooperative rules, there was uncertainty about what to do with "*kulak* infiltrators." In the GDR, *kulaks* could sell animals and machinery before joining, but *kulak* "infiltration" was less of an issue as it became more or less accepted that "middle peasants" could establish their own Type I cooperatives, sometimes out of self-interest rather than commitment to the cooperative cause.[47]

---

[46] Bokovoy, *Peasants and Communists*, 90, 108; Sjöberg, Questionnaire; Gruev in this volume, 344; Project participants, Questionnaires.

[47] Orbán, *Két agrárforradalom*, 95–96; Compare government ordinance 14.000/1948 Korm sz r with Ministry of Agriculture ordinance 18.101/1951 (I.20) FM sz r; Magyar, *Dunaapáti II*, 490–91; Jarosz, Questionnaire; Iordachi and Dobrincu in this volume, 264; Levy, "The First Wave," 35; Iordachi, "Constanţa," 115; Oláh, "Collectivization," 239–45; Ţârău, "Collectivization Policies," 209; Bokovoy, *Peasants and Communists*, 113, 122–24; Schöne, *Frühling*, 223; Bauerkämper, *Ländliche Gesellschaft*, 332, 337, 468, 477.

The initial rounds in the "class war" against *kulaks* were economic. Compulsory deliveries were not a socialist invention. In most cases they had their origins in the wartime economies of Germany (including the occupied Baltic states and Bohemia and Moravia), Hungary, and Bulgaria; and they were simply continued. Poland was unusual in that they were introduced in 1944, abolished in 1946, and only introduced again in 1951 (for grain) and 1952 (for meat, milk, and potatoes). Wartime independent Slovakia had experienced only light-touch measures, but Czech-level deliveries were introduced upon reunification. In Yugoslavia there had been some requisitions since 1944, but the compulsory deliveries system proper was established in the autumn of 1945. In Romania, it was introduced in May 1945 in the context of pressure to meet Soviet reparations payments.[48]

For compulsory deliveries and taxation to become weapons in "class warfare," the burden on certain categories of farmers had to exceed any normal understanding of progressive taxation: tax rates and delivery quotas had to reach penal proportions. This occurred in the Baltic states and Bulgaria in 1947 and elsewhere in 1948. Poland did not have compulsory deliveries at the time, but tax rates were amended in 1949 such that the differential from the top to the bottom rate increased from four to 16 percent, and a compulsory savings scheme for richer peasants was introduced. In Albania, a system of compulsory deliveries was in place by 1946, but the defining criterion was the amount that could be kept by the peasant rather than the quantity to be delivered to the state. In the Soviet Occupation Zone of Germany, differentiation in compulsory deliveries was introduced in 1946, but severe discrimination against *kulaks* waited until 1948/1949. In Hungary, Rákosi's reaction in July 1948 to the changed Cominform line had been that the time might not yet be right for collectivization, but it was for preventing the growth of exploitative elements in the countryside. This prompted a raft of measures restricting the richer peasants: compulsory deliveries were increased by a factor of three, and four on the largest farms. Yugoslavia also launched an economic attack on *kulaks* in June 1948, although it was not aimed at their liquidation. In Romania, while compulsory deliveries had been introduced in 1946, the system of forced requisition of cereals was introduced on July 6, 1948,

48 Rychlík, Questionnaire; Berend and Ránki, *Hungary*, 170; and Meurs et al., "Bulgaria," 96; Bauerkämper, *Ländliche Gesellschaft*, 127; Jarosz, Questionnaire; Bokovoy, *Peasants and Communists*, 43; Chronologies for Yugoslavia, Bulgaria, Hungary, and Czechoslovakia in this volume.

and quotas on milk and meat were introduced in January and December of 1951 respectively. In Czechoslovakia, the economic pressure on richer peasants also began relatively late, in 1950, although the party had decided in principle to collectivize and isolate the rich peasantry in May 1949.[49]

Kulak was quintessentially a political term. If state policy required that more *kulaks* be added to the *kulak* list, then the actors of the local state had to modify the definition, or include on the list "*kulak* stooges" (Hungary) or "sub-*kulaks*" (Poland and Bulgaria) to make the quota. In Hungary in 1949, there were 63,300 farms in the *kulak* category, but 71,600 families on the "*kulak* list," including 21,900 landless peasants and 36,300 whose holdings fell below the proscribed limit. This meant that, at the local level, the term was flexible and negotiated. In Hungary, according to reports of 1951 and 1952, "people are being recorded as *kulaks* who are not *kulaks*," and in Dunaapáti, a peasant whose holding was just under the limit was included by the village notary so he could take his friend off the list in exchange. In Yugoslavia, as noted above, *kulak* was by definition a political term and open to conflicting interpretations. For Romania, where the definition was flexible in principle, Verdery notes numerous examples of fuzziness around the edges to either include or exclude from *kulak* status.[50]

The penal compulsory deliveries and tax rates imposed on the *kulaks* had political as well as economic significance. Economic pressure caused hardship and suffering, but failure to meet economic obligations was a criminal offence and thus opened *kulaks* up to political pressure. They could be prosecuted, or threatened with prosecution, not for being *kulaks*, but for allegedly failing to be good citizens and meet their obligations to the state. In Czechoslovakia and Hungary punishment for politically determined, economic "crimes" could be indefinite. Czechoslovakia introduced resettlement, usually to border areas, in 1951, but it often had a particular sting in the tail. Legislation dating from Austro-Hungarian times permitted those guilty of disorderly conduct to be expelled from their home

49 Landau and Tomaszewski, *The Polish Economy*, 194; Gruev in this volume, 341; Bauerkämper, *Ländliche Gesellschaft*, 127–29; Schöne, *Das Sozialistische Dorf*, 93; Orbán, *Két agrárforradalom*, 71–72, 90; Pető and Szakács, *A hazai gazdaság*, 183–85; Bokovoy, *Peasants and Communists*, 91; Chronologies for Czechoslovakia, Hungary, and Romania; Rychlík in this volume, 190, 194, 196.

50 Project participants, Questionnaire; Pető and Szakács, *A hazai gazdaság*, 187; Orban, *Két agrárforradalom*, 101–2; Magyar, *Dunaapáti II*, 447–48; Verdery, "Exploiters Old and New," 316–22.

villages and prohibited from returning. *Kulaks*, frequently guilty of such "disorderly conduct," could be banished indefinitely from their home village. In Hungary, although there were no large-scale deportations on the Soviet model, some who were classified as *kulaks* were deported to labor camps on the Hortobágy plain and were not initially allowed back to their villages after the camps were shut down.[51]

Punishment of *kulaks* served a further political purpose: it could act as a deterrent to other peasants. An archetypal example of this admonitory function was the trial of the Czech *kulak* Josef Pažout in 1954, which was broadcast over the local radio, taped and preserved for posterity. Show trials were also staged in Bulgaria in 1950 and 1951 in order to make an example of those who opposed collectivization.[52] Given the fuzziness of the definition of *kulak*, few property-owning peasants could feel secure from the threat of imprisonment and penury.

When, in the post-Stalin era, "class war" and its associated anti-*kulak* policies gradually disappeared from the agenda, so too did generalized uncertainty caused by the fuzzy and politicized definition of *kulak*. The fear of arbitrary arrest and instant penury diminished. To this limited, yet significant extent, the post-Stalin climate of collectivization became less hostile for all, not just richer peasants.

Campaigning and techniques of persuasion

Although poorer peasants and those who could not make a success of independent farming had joined cooperatives of their own volition, mass collectivization was not a spontaneous process: peasants had to be persuaded to join. In all countries, and in each phase of collectivization, campaigns were organized by the local party and administrative machines, supported by teams of "agitators" brought in to help convert the peasants to collective farming. These were mostly short-term, ad hoc campaigns, and varied in intensity, both by country and by time period.

Only the GDR developed long-term agitation teams, equivalent to the Soviet 25,000-ers,[53] in the form of the "Industrial workers to the countryside" campaign, which extended for the decade from 1953 to 1963; yet

---

[51] Coulter, "The Hungarian Peasantry," 542; Rychlík in this volume, 198.

[52] Johnston, "Plans to bring prosecutors;" Willoughby, "1950s trial;" Gruev in this volume, 348–49, 357, 359.

[53] Viola, *The Best Sons*.

it curiously does not seem to have been consciously modeled on Stalinist practice. When it came to the Socialist Spring of 1960, the GDR's organizational effort was impressive. Last reports that, in the district of Erfurt, there were 9,724 members of the agitator brigades by March, made up of factory workers and other employees. A common slogan was the implausible, but highly politicized: "He who is against the cooperative is for war," or a supporter of the West German Adenauer government. Elsewhere campaigns were more sporadic. Yet the claim, in Romania, that 25,000 workers were sent to the countryside to promote collectivization in June 1950 is unlikely to have been accidental. Local studies refer to activists involving the entire district council (some of whom had been trained by the regional party organ and were relatively well paid) developing detailed "action plans," which involved intensive two-day training sessions in some instances, followed by three-day visits to villages in a prescribed order, accompanied by theatre shows, singing and dancing. Activists were mobilized in Hungary too, especially in 1955, although the experience in Dunaapáti in 1952 and 1955 suggests that rather little support was given by the district structures to the village party secretaries. In Czechoslovakia too, the 1955 round of collectivization saw the increased use of brigades of agitators and propaganda, although propaganda had been used extensively in earlier rounds. This included the improbable claim, made in the GDR too, that Colorado beetle infestation was an imperialist plot.[54]

Since most peasants had to be persuaded against their will to join collectives, and most were reluctant to be persuaded, it was inevitable that an element of violence was involved. But the degree of violence associated with collectivization was not uniform. *Kulaks*, the "class enemy," were treated with aggression, deprived of their property, arrested on falsified charges, and sent to labor camps, although in the post-Stalin era this threat became somewhat less oppressive.[55] The persuasion techniques used to get poorer peasants to sign up to collective farming included violence too, but

---

[54] Witkowski, "On the Campaign Trail;" Last, *After the Socialist*, 18; Federal Republic of Germany, *The Compulsory Collectivization*, 17; Levy, "The First Wave," 37; Ţârău, "Collectivization," 207–8; Oláh, "Collectivization," 234–35; Dobrincu, "Persuasion, Delay and Coercion," 287–93; Orbán, *Két agrárforradalom*, 150; Magyar, *Dunaapáti II*, 328, 338; Magyar, *Dunaapáti III*, 675–78; Březina, "Kolektivizace," 131; Biolková, "Ohlas zemědělské," 217; Bauerkämper, Questionnaire.

[55] Threats of deportation as a *kulak* were still used in Romania in 1962, but not as a general tactic. See Bodó, "Persuasion Techniques," 360.

were less extreme, if only because there was no "legal" basis for them. Outright brutality, beatings, broken bones and so on were associated with the enthusiastic, early collectivizing states: Bulgaria, Yugoslavia, and probably Albania, although there is no actual evidence in this case. In the general assessment of writers on collectivization in the 1950s, the process was the most violent in Bulgaria, although with the exception of locking people up for more than one night, or sending peasants to a work camp without a trial, the techniques used in Bulgaria were not dissimilar from the standard methods considered below. Meurs reports that "direct physical violence" was rare by the 1950 campaign. Bokovoy, writing on Yugoslavia, also cites examples of physical violence—including brandishing pistols, breaking ribs and knocking out teeth—to enforce compulsory deliveries.[56]

Such extreme brutality does not appear to have been associated with collectivization elsewhere, except for perhaps in Romania. The main techniques used to "persuade" peasants to join farms elsewhere were bullying and threats, and they were quite similar in form. They included teams of cadres interviewing peasants in their houses, often during the evening because peasants disappeared during the day, but preferably inviting them to "discussions" in council offices where the proceedings could better be controlled. Such "discussions" did involve physical beatings on occasion, but more common seems to have been threats and constant pressure—locking peasants up in council offices or police stations to "think things over," refusing to leave peasants' houses until they signed, broadcasting the names of those who had joined or refused to join over loudspeakers, threatening them with imprisonment for failure to meet economic targets, threatening to withdraw their children from school, or prevent a spouse from working in a nearby factory if there was one.[57]

These techniques did not differ between the "class war" and the final phase of collectivization. Persuasion was still necessary in the final phase, and the same techniques of persuasion were used. The West German gov-

---

[56] Sjöberg, *Rural Change*, 86–87 cites only "conflicts." Stillman refers to Bulgaria "alone" using Soviet "shock tactics," see "The Collectivization," 71; Draganova, "Senses of the Private;" Meurs et al., "Bulgaria," 101; Bokovoy, *Peasants and Communists*, 134–40.

[57] Ţârău, "Collectivization," 215; Oláh, "Collectivization," 234–39; Dobrincu, "Persuasion, Delay and Coercion," 291; Jarosz, Questionnaire; Bodó, "Persuasion Techniques," 359–62; Landau and Tomaszewski, *The Polish Economy*, 228–29; Last, *After the Socialist*, 19; Březina, "Kolektivizace," 132. Preventing husbands working in industrial jobs were tactics also used in the Hungarian villages of Dombház and Kissikonda where the author researched in the 1990s.

ernment's report on the GDR's Socialist Spring of 1960, based on interviews with refugees to the West, has examples of all of the techniques listed above. The extensive accounts of Hungary's collectivization from 1960–1961 produced by József Ö. Kovács include examples of never-ending interviews in council offices, threats of violence (sometimes the brandishing of a pistol), and minor acts of violence such as pulling hair, twisting ears, banging heads on tables or rubbing them in the snow, or of ignominy such as making a former *kulak* stand on one leg and support a pencil between his eyebrow and the wall.[58]

Despite the violence of the techniques used, in situations where local activists experienced intense pressure to complete collectivization, the balance of forces was not entirely one-sided. Prospective members could wring out special concessions as conditions for joining. This appears to have been most common in the GDR, where full collectivization had to be completed within months, resulting, for example, in agreements that the first harvest could be made individually, or not all land need be farmed collectively, or more than the permitted number of animals could be kept on household plots, or even that they would receive consumer goods such as TVs and refrigerators if they joined. In Romania, which by 1962 was conscious of its status as laggard and anxious to complete the task, similar examples are observed, such as a special plot for a son, personal use of a horse and cart until a specified date, or a hectare of land to be given to a daughter in the case of her marriage.[59]

Peasant responses

The most common reaction of the peasantry to collectivization campaigns was passive resistance, both with regard to the campaign itself (trying to avoid encounters with the campaigners at all costs) and to participating in collective work once it had been created. Bauerkämper notes more symbolic passive resistance when peasants cancelled subscriptions to the *Free Farmer*, supposedly the paper of the GDR's farmers. But resistance was more active too, throughout the region. In Bulgaria and Yugoslavia, where the campaigns had been most violent, violent protest was also triggered. Bokovoy

58 Kovács in this volume. See also Kovács, "Ekkora gyűlölet;" Federal Republic of Germany, *The Compulsory Collectivization*, 16–19, 29–57.

59 Bauerkämper, *Ländliche Gesellschaft*, 476–78; Bauerkämper Questionnaire; Lățea, "Revolution in Bits and Pieces," 343–47.

gives extensive examples of physical attacks, or threats, against both members of cooperatives and those who met their compulsory deliveries, and of pot shots taken at party secretaries. In Bulgaria, there were peasant revolts in six villages in July 1950, followed by uprisings in the Asenovgrad, Plovdiv, Parvomai regions and elsewhere, while six months later, in the winter of 1950–51, there were more uprisings in the Vidin and Kula regions. In Romania too, collectivization triggered violent backlashes on the part of the peasantry, and not only in the initial "class-war" phase. Iordachi and Dobrincu report uprisings in the summer of 1949, July 1950, and 1958, and Levy documents additional cases of unrest in Romania—in March 1950 and winter 1952. Oprea refers to the personal involvement of Ceauşescu in putting down riots that broke out in three counties in 1960–61.[60]

Public displays of opposition to collectivization were not restricted to those countries where it exploded into peasant revolt. The uprising of June 1953 in the GDR, and the Polish and Hungarian events of October 1956 were fundamentally influenced by collectivization policy and the conditions of rural life. Hungary's influential Petőfi Circle of critical intellectuals and students debated agricultural policy less than a week before the revolution started. Moreover, the spontaneous breakup of Hungarian collectives in the autumn of 1956, against which the authorities decided it was unwise to use "police methods," constituted a peaceful mass rejection of collectivization. Smaller-scale incidents of unrest against collectivization are also recorded, such as demonstrations in a village in Tolna county, where no action was taken because no "opposition elements" participated in them, or 70 acts of arson or sabotage in Hungary in the first half of 1961. More dramatically, in southern Moravia there were illegal groups that destroyed collective farm property and murdered party activists, while in Poland there was open unrest against collectivization in the spring and autumn 1951 and summer 1953.[61]

Other features of the reaction against collectivization are worthy of brief note. In Poland and Slovakia where the Catholic Church was strong, but also in the predominantly Protestant GDR, religion played a role providing support for those who opposed collectivization. In Poland too, but

---

[60] All authors of this volume concur; Bauerkämper, *Ländliche Gesellschaft*, 457; Bokovy, *Peasants and Communists*, 116–20; Gruev in this volume, 357–58, 371; Iordachi and Dobrincu in this volume, 271–76; Levy, "The First Wave," 34–35, 40; Oprea, "The Final Offensive," 71.

[61] Schöne, *Frühling*, 142–53; Varga, *Politika*, 22; Varga, "Földindulás után," 233; Orbán, *Két agrárforradalom*, 158, 220; Rychlík and Jarosz, Questionnaire.

also in Bulgaria, there were examples of women leading opposition to collectivization, as they had done in areas of the Soviet Union. Furthermore, in Poland there were also examples of Luddite attacks on tractors and the machinery associated with collectivization, which had no equivalent elsewhere in the region.[62]

Conclusion

Comparing collectivizations in Eastern Europe is instructive. Clearly the initial decision in favor of collectivization and major changes in collectivization policy were neither local nor autonomous. Stalin initiated the policy as a spin-off from his arguments with Tito. There was probably pressure from Moscow to strengthen the struggle against *kulaks* in 1951, when it was clear collectivization was not progressing as quickly as hoped. Stalin's death and the New Course licensed a relaxation in policy everywhere, and signaled the eventual demise of the "class war" model. Yet hints from Moscow probably prompted the 1955 return to collectivization, and almost certainly a meeting in Moscow in November 1957 initiated the final phase, after the debacle of 1956, which convinced many leaders that some sort of radical break with past policies was necessary.

But, within these parameters, there was national variation as policy became entangled with national contexts. The most striking feature of collectivization in the GDR was the speed of the final phase—the Socialist Spring of 1960. A further peculiarity in the GDR, which had some impact on the dynamics of collectivization in the 1950s, was the relatively open border with the West. As a consequence, probably more land was abandoned by those who chose to flee from "class war" politics, resulting in the creation of a unique institution to cultivate abandoned land, the ÖLB. The GDR also started later than elsewhere, in 1952, although anti-*kulak* policies were commenced according to the norm in 1948.

Bulgaria, Yugoslavia, and Albania initially jumped the collectivization gun in that there were initiatives to implement collectivization immediately after the war, when Stalin's postwar policy called for land reform. Bulgaria remained always in the vanguard of collectivization, followed

---

62 Rychlík and Jarosz, Questionnaire; Bauerkämper, *Ländliche Gesellschaft*, 188, 469; Meurs et al., "Bulgaria," 102; Gruev in this volume. Viola describes women's opposition to Soviet collectivization in "*Bab'i Bunty*;" Jarosz in this volume, 131.

closely by Albania until the end of the 1950s—and neither country was much affected by the events of October 1956. Yugoslavia abandoned collectivization altogether after 1953; but it was rather successful numerically before it did. By the end of 1949, almost 18 percent of arable land in Yugoslavia had been collectivized, more than anywhere else at the time. Poland also abandoned collectivization after 1956, but was never successful numerically, with a maximum of just 11 percent of agricultural land in the collective farm sector in 1955. It has been claimed that "collectivization in Poland was pursued cautiously and without conviction,"[63] and certainly pre-existing cooperative institutions were not hijacked in the way that they were elsewhere. Koenig attributes this to the weakness of the Polish Communist Party at village level, while Korboński and Kaliński stress more the conflicting signals coming from above. There were significant disagreements between high-profile politicians over collectivization policy elsewhere: in Hungary (between Nagy and Rákosi), Czechoslovakia (between Zápotocký and Novotný), and Romania (between Pauker and Dej); but no major politician questioned collectivization itself. In Poland, it was different. Gomułka staked his reputation on his opposition to forced collectivization in 1948, and it was inconceivable that it would continue after his victory in 1956, although the pace of the dissolution of collective farms took even him by surprise.[64]

Hungary experienced the greatest fluctuations in the size of its collectivized sector. Given its traumatic experience and its long-term relative success at making collectivization "work,"[65] there is a logic to identifying Hungary as the source of the less confrontational, more accommodating policies that were to greater or lesser degrees characteristic of the final phase of collectivization. But Varga's contribution to this volume suggests rather entanglements and a cross-fertilization of ideas from Czechoslovakia (which had matched Bulgaria's numerical growth initially, but fallen behind after 1953). Yet arguably, Hungarian policymakers were inclined to "learn" from Czechoslovakia only those lessons that they wanted to hear.

63 Pelczynski, "The Rise and Ebb," 320.

64 Bokovoy, *Peasants and Communists*, 151; see tables give in Bokovoy, *Peasants and Communists*, 121 and Brus, "1950 to 1953," 9. I assume Bokovoy's figures for Yugoslavia are more reliable than those provided by Brus. Even if they are not, Yugoslavia is more collectivized than most; Brus, "1953 to 1956," 52; Koenig, "Collectivization," 109; Korboński, *Politics of Socialist Agriculture*, 139–54, 163, 196, 246–55, 308; Kaliński, "Collectivization," 177.

65 Swain, *Collective Farms*.

The (changing) Chinese case was certainly used by both sets of protagonists, as Varga makes clear. But, if Hungary and Czechoslovakia embraced a less confrontational approach in the later 1950s, Romania and the GDR remained ambivalent. In abolishing compulsory deliveries immediately after Hungary in 1956, Romania embraced elements of reduced confrontation; yet traditional violent policies in 1957 resulted in popular unrest. Nevertheless the final years were characterized by some concessions (personal conditions negotiated on membership, the persistence, although short-term only, of the Type I farm). In the GDR, the Socialist Spring was too impatient to avoid confrontation, but it too permitted personal concessions and the continuation of Type I farms for richer peasants.

With the exceptions of Poland and Yugoslavia (which abandoned the project), Bulgaria (which was always in the vanguard) and Albania (in the vanguard, but trailing significantly behind Bulgaria), the bulk of collectivization took place in the final phase, after 1957. Prior to that date, Czechoslovakia was roughly 50 percent collectivized, but nowhere else was the figure higher than 25 percent. This increased velocity of collective formation in the final phase requires explanation, especially as, with the exception of the abandonment of the "class war" perspective, the techniques used to enforce it remained very much the same. Part of the explanation can probably be attributed to the passage of time and changing realities. The West had not intervened in 1956; the regimes were established for the long term, and they were generating their own sources of wealth and welfare. As the countries industrialized and extended education to the countryside, the threat to deprive family members of employment or education became more onerous. The reality of the late 1950s was that, with continued pressure on private farming, land was no longer a sign of, or vehicle for, prosperity; if anything, it was a burden. Traditional peasant marks of esteem were losing their value. Industrial employment and education, by contrast, were paths out of rural poverty. A decade on from the first attempts to collectivize, there was an increased sense of resignation not only to the fact that collectivization was inevitable if the government felt so strongly about it, but also to that changed world. The village authorities would not take no for an answer, so villagers reluctantly said, yes; but not with conviction or enthusiasm.[66]

---

66 Brus, "1953 to 1956," 52; Verdery, "Exploiters Old and New," 310–11; Lăţea, "Revolution in Bits and Pieces," 340; Goina, "Never Leave 'till Tomorrow," 386; Chelcea, "Here in Reviga," 413; Swain, *Collective Farms*, 26.

The question of how the passive resistance of members could be overcome takes us into the post-1962 era and beyond the scope of this volume. But there are continuities which might briefly be explored. The crux of the matter related to incentives, openness about deviation from ideological orthodoxy (that is to say deviation from the section of the Model Charter that specified rewards based around the "labor day" or "work unit"), and trust of rural communities. Varga, in her contribution to this volume and elsewhere, stresses the importance of the sharecropping debate in Hungary: how a grassroots solution to the "incentives problem" became the subject of an almost decade-long power struggle in the Hungarian leadership between the "agrarian lobby" and the politically orthodox until, by 1967, legislation finally sanctioned grassroots practice.[67] Somewhat less heretical, because it did not connote exploitation in the way that sharecropping did, was the payment of a guaranteed wage for labor performed, rather than the "labor day" or "work unit" which measured a future share in a reward of unknown size. In Hungary, the demise of "class warfare" in the implementation of collectivization was followed, not without political struggle, by ideological compromise: an accommodation to peasant practices, and the creation of a powerful system of material incentives which rejected norms laid down in the Model Charter.

Other countries were not so open about deviating from ideological orthodoxy, but did so nevertheless. Czechoslovakia implemented policies like Hungary's, but seemingly with less political debate. It also, like Hungary, trusted village dwellers sufficiently to abolish piecemeal the system of compulsory deliveries and increase gradually the prices paid to farmers, so allowing wealth to accumulate in the countryside. Although the breakneck Socialist Spring in the GDR was intolerant of compromise, high prices and reduced compulsory deliveries became the norm there too; and the incentive issue was resolved by the country's relative wealth. The state guaranteed the level of the work unit, converting it to essentially a wage from the start. Bulgaria used both regular wages and essentially sharecropping systems as means of remuneration, but agricultural prices were not increased so significantly, and agricultural incomes thus remained relatively low. In compensation, the state, as in Hungary, implemented measures which actively encouraged private household plot production. The other countries discussed so far did the same, but only tacitly.

[67] Varga, *Politika*; Varga, "The Impact of 1956."

In Albania and Romania, however, the still Stalinist political authorities were reluctant to trust rural communities or question openly Model Charter orthodoxy. Prices were not increased, so incomes remained low; and directive planning remained in place, even if compulsory deliveries as such were abolished. Furthermore, household plots were under constant attack. Nevertheless, in Romania at least, below the surface, ideological orthodoxy did not prevail. Parallel with the "work unit" operated the much more popular "accord" system of remuneration, which in essence represented a form of sharecropping under an ideologically more acceptable label. With low agricultural prices and restrictions on household plots, it constituted not so much an incentive as a means of rural survival.

The aftermath of 1989 passed a verdict on these deviations from the Model Charter. In the reformist countries where membership had become associated with adequate or even generous material rewards, there was reluctance to break up collective farms. In Romania and Albania, where membership represented a cost rather than a benefit, there was popular enthusiasm to destroy them. In Poland, which had only ever experienced "class war" collectivization (except in the tiny, genuinely voluntary producer cooperative sector that survived 1956), there was bewilderment at why agrarianists elsewhere could remain so attached to this element of the socialist legacy.[68]

BIBLIOGRAPHY

ARCHIVAL SOURCES

Federal Ministry of All-German Affairs, Government of the Federal Republic of Germany, *The Compulsory Collectivization of Independent Farms in the Soviet Zone of Occupation in Germany.* Bonn: Federal Ministry for All-German Affairs 1960.

BOOKS AND ARTICLES

"Az MSZMP agrárpolitikájának tézisei 1957" [The Theses of the Agricultural Policy of the Hungarian Socialist Workers' Party] in Kálmán Szakács, ed. *Marxizmus és agrárkérdés* [Marxism and the Issue of Agriculture]. Budapest: Tankönyvkiadó, 1974, 218–41.

Bauerkämper, Arnd. "Transformation als Integration? Der Wandel der ländlichen Gesellschaft und die Flüchtlinge im Osten Brandenburgs" in Helga Schultz, ed. *Bevölkerungstransfer und Systemwandel: Ostmitteleuropäische Grenzen nach dem Zweiten Weltkrieg.* Berlin: Berlin-Verlag Spitz, 1999, 105–44.

[68] For a fuller discussion of developments after 1989, see Swain, "A Framework;" and Swain, "Decollectivization Politics."

———. *Ländliche Gesellschaft in der Kommunistischen Diktatur: Zwangsmodernisierung und Tradition in Brandenburg 1945–1963*. Vienna: Böhlau, 2002.

Berend, Iván T. and György Ránki. *Hungary: A Century of Economic Development*. Newton Abbot: David and Charles, 1974.

Bergmann, Theodor. *Agrarpolitik und Agrarwirtschaft sozialistischer Länder*. Saarbücken: Breitenbach, 1979.

Biolková, Jindra. "Ohlas zemědělské politiky v první etapě koleltivizace v Ostravském kraji" in Petr Blažek and Michal Kubálek, eds. *Kolektivizace Venkov v Československu 1948–1960 a Středoevropské Souvislosti*. Prague: Dokořán, 2008, 209–23.

Bodó, Julianna. "Persuasion Techniques and Community Reactions in Corund (the Hungarian Autonomous Region)" in Constantin Iordachi and Dorin Dobrincu, eds. *Transforming Peasants, Property and Power: the Collectivization of Agriculture in Romania, 1949–1962*. Budapest: CEU Press, 2009, 355–68.

Bokovoy, Melissa K. *Peasants and Communists: Politics and Ideology in the Yugoslav Countryside, 1941–1953*. Pittsburgh: University of Pittsburgh Press, 1998.

Březina, Vladimír. "Kolektivizace zemědělství v Československu v letech 1955–1960" in Petr Blažek and Michal Kubálek, eds. *Kolektivizace Venkov v Československu 1948–1960 a Středoevropské Souvislosti*. Prague: Dokořán, 2008, 128–35.

Brus, W. "1953 to 1956: the 'Thaw' and the 'New Course'" in M. C. Kaser and E. A. Radice, eds. *The Economic History of Eastern Europe 1919–1975, Vol. III: Institutional Change within a Planned Economy*. Oxford: Clarendon Press, 1986, 40–69.

———. "Postwar Reconstruction and Socioeconomic Transformation" in M. C. Kaser and E. A. Radice, eds. *The Economic History of Eastern Europe 1919–1975, Vol. II: Interwar Policy, the War and Reconstruction*. Oxford: Oxford University Press, 1986, 564–641.

———. "1950 to 1953: the Peak of Stalinism" in M. C. Kaser and E. A. Radice, eds. *The Economic History of Eastern Europe 1919–1975. Vol. III: Institutional Change within a Planned Economy*. Oxford: Clarendon Press, 1986, 3–39.

Cartwright, A. L. *The Return of the Peasant: Land Reform in Post-Communist Romania*. Aldershot: Dartmouth, 2001.

Chelcea, Liviu. "'Here in Reviga, There Was Nobody to Wage the Class Struggle:' Collectivization in Reviga, Bărăgan Plain (Bucharest Region)" in Constantin Iordachi and Dorin Dobrincu, eds. *Transforming Peasants, Property and Power: the Collectivization of Agriculture in Romania, 1949–1962*. Budapest: CEU Press, 2009, 399–422.

Clissold, Stephen, ed. *Yugoslavia and the Soviet Union 1939–73: A Documentary Survey*. Oxford: Oxford University Press, 1975.

Coulter, Harris L. "The Hungarian Peasantry: 1948–1956." *American Slavic and East European Review* 18, no. 4 (December, 1959): 539–54.

Davies, Robert Williams. *The Soviet Collective Farm, 1929–1930*. London: Macmillan, 1980.

Dobrincu, Dorin. "Persuasion, Delay and Coercion. Late Collectivization in Northern Moldova: The Case of Darabani (Suceava Region)" in Constantin Iordachi and Dorin Dobrincu, eds. *Transforming Peasants, Property and Power: the Collectivization of Agriculture in Romania, 1949–1962*. Budapest: CEU Press, 2009, 275–304.

Donáth, Ferenc. *Reform és forradalom* [Reform and revolution]. Budapest: Akadémiai, 1977.

Draganova, Mariana. "Senses of the Private when Working in Collectives in Socialist Bulgaria." Paper presentation. "Everyday Socialism: States and Social Transformation in Eastern Europe 1945–1965," Conference, London, April 24–26, 2003.

Fazekas, Béla. *A mezőgazdasági termelőszövetkezeti mozgalom Magyarországon* [The Agricultural Collective Farm Movement in Hungary]. Budapest: Kossuth, 1976.

Gerth, H. H. and C. Wright Mills, *From Max Weber: Essays in Sociology*. London: Oxford University Press, 1948.

Goina, Călin. "'Never Leave 'til Tomorrow What You Can Do Today!' A Case Study of a Model Collective Farm: 'New Life' Sântana (Arad Region)," in Constantin Iordachi and Dorin Dobrincu, eds. *Transforming Peasants, Property and Power: the Collectivization of Agriculture in Romania, 1949–1962*. Budapest: CEU Press, 2009, 369–98.

Hitchins, Keith. *Rumania 1866–1947*. Oxford: Oxford University Press, 1994.

Hubbard, Leonard E. *The Economics of Soviet Agriculture*. London: Macmillan, 1939.

Iordachi, Constantin and Dorin Dobrincu, eds. *Transforming Peasants, Property and Power: the Collectivization of Agriculture in Romania, 1949–1962*. Budapest: CEU Press, 2009.

——— and Katherine Verdery, "Conclusions" in Constantin Iordachi and Dorin Dobrincu, eds. *Transforming Peasants, Property and Power: the Collectivization of Agriculture in Romania, 1949–1962*. Budapest: CEU Press, 2009, 455–71.

———. "'Constanţa, the First Collectivized Region:' Soviet Geo-Political Interests and National and Regional Factors in the Collectivization of Dobrogea (1949–1962)" in Constantin Iordachi and Dorin Dobrincu, eds. *Transforming Peasants, Property and Power: the Collectivization of Agriculture in Romania, 1949–1962*. Budapest: CEU Press, 2009, 103–40.

Johnston, Rosie. "Plans to bring prosecutors of wealthy farmers to justice." *Radio Prague*. October 29, 2007. http://www.radio.cz/en/section/curraffrs/plans-to-bring-persecutors-of-wealthy-farmers-to-justice (accessed on December 22, 2012).

Kaliński, Janusz. "Collectivization of agriculture in Poland, 1948–1956," *Acta Poloniae Historica* 57 (1988): 167-201.

Karlik, Jiri. *Questions of Ownership and Use of Land and Other Agricultural Means*. Prague: Ekonomicky Ustav Ceskoslovenske Akademie Ved, 1991.

Kenez, Peter. *Hungary From the Nazis to the Soviets: the Establishment of the Communist Regime in Hungary, 1944–1948*. Cambridge: Cambridge University Press, 2006.

Kersten, Krystyna. *The Establishment of Communist Rule in Poland, 1943–1948*. Berkeley: University of California Press, 1991.

Koenig, Ernest. "Collectivization in Czechoslovakia and Poland" in Irwin T. Sanders, ed. *Collectivization of Agriculture in Eastern Europe*. Lexington: University of Kentucky Press, 1958, 103-26.

Korboński, Andrzej. *Politics of Socialist Agriculture in Poland, 1945–1960*. New York: Columbia University Press, 1965.

Kovács, József Ö. "'Ekkora gyűlölet még nem volt a falukban, mint most.' Szövegek és kommentárok az erőszakos kollektivizálás befejező hullámáról" ["There was never such hatred in our village." Documents and commentaries on the final waves of forced collectivization]. *Századvég* 47, no. 1 (2008): 37–69.

Kovács, Katalin. "Urbanizáció alulnézetből" [Urbanization from below] in József Tóth, ed. *Tér, Idő, Társadalom.* [Space, Time, Society]. Pécs: MTA RKK, 1990, 272–303.

Landau, Zbigniew and Jerzy Tomaszewski. *The Polish Economy in the Twentieth Century,* trans. by Wojciech Roszkowski. London: Croom Helm, 1985.

Last, George. *After the "Socialist Spring:" Collectivization and Economic Transforming in the GDR.* New York: Berghahn Books, 2009.

Lăţea, Daniel. "Revolution in Bits and Pieces: Collectivization in Southern Romania (Craiova Region)" in Constantin Iordachi and Dorin Dobrincu, eds. *Transforming Peasants, Property and Power: the Collectivization of Agriculture in Romania, 1949–1962.* Budapest: CEU Press, 2009, 329–54.

Levy, Robert. "The First Wave of the Collectivization Campaign: Central Policies and Their Regional Implementation (1949–1953)" in Constantin Iordachi and Dorin Dobrincu, eds. *Transforming Peasants, Property and Power: the Collectivization of Agriculture in Romania, 1949–1962.* Budapest: CEU Press, 2009, 27–48.

Magyar, Bálint. *Dunaapáti 1944–1958: Dokumentszociográfia II.* [Dunaapáti 1944–1958: Document-sociography II.]. Budapest: Művelkut. Int., 1986.

———. *Dunaapáti 1944–1958: Dokumentszociográfia III.* (Dunaapáti 1944–1958: Document-sociography III.) Budapest: Művelkut. Int., 1986.

Marx, Karl. "The Eighteenth Brumaire of Louis Bonaparte" in Karl Marx and Frederick Engels, eds. *Selected Works in One Volume.* London: Lawrence & Wishart, 1968, 95–180.

Meurs, Mieke, Veska Kouzhouharova, and Rositsa Stoyanova. "Bulgaria: From Cooperative Village to Agro-Industrial Complex, the Rise and Fall of Collective Agriculture" in Mieke Meurs ed. *Many Shades of Red: State Policy and Collective Agriculture.* Lanham: Rowman & Littlefield, 1999, 87–123.

Miller, Linda. "Law and Propaganda: Rural Land Ownership, Collectivization and Socialist Property in Romania" in Constantin Iordachi and Dorin Dobrincu, eds. *Transforming Peasants, Property and Power: the Collectivization of Agriculture in Romania, 1949–1962.* Budapest: CEU Press, 2009, 81–100.

Myant, Martin. *Socialism and Democracy in Czechoslovakia, 1945–1948.* Cambridge: Cambridge University Press, 1981.

———. *The Czechoslovak Economy 1948–1988: The Battle for Economic Reform.* Cambridge: Cambridge University Press, 1989.

Nove, Alec. *An Economic History of the USSR.* Harmondsworth: Penguin Books Ltd, 1969.

Oláh, Sándor. "Collectivization in the Odorhei District (The Hungarian Autonomous Region)" in Constantin Iordachi and Dorin Dobrincu, eds. *Transforming Peasants, Property and Power: the Collectivization of Agriculture in Romania, 1949–1962.* Budapest: CEU Press, 2009, 229–50.

Orbán, Sándor. *Két agrárforradalom Magyarországon. Demokratikus és szocialista agrárátalakulás 1945–1961* [Two Agrarian Revolutions in Hungary. Democratic and Socialist Agrarian Transformation, 1945–1961]. Budapest: Akadémiai, 1972.

Oprea, Marius. "The Final Offensive: 'The Socialist Transformation of Agriculture' from Slogans to Reality (1953–1962)" in Constantin Iordachi and Dorin Dobrincu, eds. *Transforming Peasants, Property and Power: the Collectivization of Agriculture in Romania, 1949–1962.* Budapest: CEU Press, 2009, 49–80.

Orgoványi, István. "Buffer Zone at the Southern Border: the Southern Border Zone between 1948 and 1956" in Ágnes Tóth, ed. *National and Ethnic Minorities in Hungary, 1920–2001.* New York: Columbia University Press, 2005, 313-53.

Pelczynski, Z. A. "The Rise and Ebb of Stalinism" in R. F. Leslie, ed. *The History of Poland since 1863.* Cambridge: Cambridge University Press, 1980, 299–343.

Pető, Iván and Sándor Szakács. *A hazai gazdaság négy évtizedének története,* 1945–1985 [Economic History of Hungary, 1945–1985]. Budapest: Közgazd és Jogi Kvk., 1985.

Průcha, Václav. "Economic Development and Relations, 1918–89" in Jiří Musil, ed. *The End of Czechoslovakia.* Budapest: CEU Press, 1995, 40–76.

Schöne, Jens. *Frühling auf dem Lande? Die Kollektivierung der DDR-Landwirtschaft.* Berlin: Ch. Links Verlag, 2007.

———. *Das Sozialistische Dorf: Bodenreform und Kollektivierung in der Sowjetzone und DDR.* Leipzig: Evangelische Verlagsanstalt, 2008.

Selden, Mark. ed. *The People's Republic of China: A Documentary History of Revolutionary Change.* New York: Monthly Review, 1979.

Sjöberg, Örjan. *Rural Change and Development in Albania.* Boulder: Westview Press Inc, 1991.

Spence, Jonathan D. *The Search for Modern China.* New York: Norton, 1999.

Stillman, Edmund O. "The Collectivization of Bulgarian Agriculture" in Irwin T. Sanders, ed. *Collectivization of Agriculture in Eastern Europe.* Lexington: University of Kentucky Press, 1958, 67–96.

Stuart, Robert C. *The Collective Farm in Soviet Agriculture.* Lexington, Mass.: Lexington Books, 1972.

Swain, Geoffrey. "Deciding to Collectivize Latvian Agriculture." *Europe-Asia Studies* 55, no. 1 (2003): 39–58.

——— and Nigel Swain, *Eastern Europe since 1945,* 4th ed. Houndmills: Palgrave Macmillan, 2009.

Swain, Nigel. *Collective Farms Which Work?* Cambridge: Cambridge University Press, 1985.

———. "Hungarian Agriculture in the 1980s: Retrenchment Followed by Reform," *Soviet Studies* XXXIX, no. 1 (1987): 24–39.

———. "The local experience of the Oder border region—a Polish and German village compared" in Helga Schultz, ed. *Bevölkerungstransfer und Systemwandel: Ostmitteleuropäische Grenzen nach dem Zweiten Weltkrieg.* Berlin: Berlin-Verlag Spitz, 1998, 145–62.

———. "A Framework for Comparing Social Change in the Post-Socialist Countryside." *Eastern European Countryside* 4 (1998): 5–18.

———. "Decollectivization politics and rural change in Bulgaria, Poland and the former Czechoslovakia." *Social History* 32, no. 1 (2007): 1–26.

———. *The Fate of Peasant Parties during the Socialist Transformation,* paper presented to Session N6 on *From Democracy to Authoritarianism?: Critical Considerations concerning Agrarianism,* World Economic History Conference, Utrecht, August 3–7, 2009.

Țârău, Virgiliu. "Collectivization Policies in the Cluj Region: The Aiud and Turda Districts" in Constantin Iordachi and Dorin Dobrincu, eds. *Transforming Peasants, Property and Power: the Collectivization of Agriculture in Romania, 1949–1962.* Budapest: CEU Press, 2009, 203–28.

Tóth, Ágnes. "The Relocation of the Germans and the Slovakian-Hungarian Population Exchange" in Ágnes Tóth, ed. *National and Ethnic Minorities in Hungary, 1920–2001.* New York: Columbia University Press, 2005, 253–312.

Varga, Zsuzsanna. *Politika, paraszti érdekérvényesítés és szövetkezetek Magyarországon 1956–1967* [Politics, the Assertion of Agrarian Interests and Cooperatives in Hungary, 1956–1967]. Budapest: Napvilág, 2001.

———. "The Impact of 1956 on the Relationship between the Kádár Regime and the Peasantry, 1956–1966." *Hungarian Studies Review* XXXIV, No. 1–2 (2007): 155–176.

———. "'Földindulás után:' a hatalom és a parasztság Magyarországon az 1960-as évek első felében" ['After an Earthquake:' Power and Peasantry in Hungary in the First Half of the 1960s]. *Történelmi Szemle* LI, no. 2 (2009): 231–58.

Verdery, Katherine. "Exploiters Old and New: Making and Unmaking 'Rich Peasants' in Aurel Vlaicu (Hunedoara Region)" in Constantin Iordachi and Dorin Dobrincu, eds. *Transforming Peasants, Property and Power: the Collectivization of Agriculture in Romania, 1949–1962.* Budapest: CEU Press, 2009, 307–28.

Viola, Lynne. "*Bab'i Bunty* and Peasant Women's Protest During Collectivization." *Russian Review* 45, no. 1 (1986): 23–42.

———. *The Best Sons of the Fatherland: Workers in the Vanguard of Soviet Collectivization.* Oxford: Oxford University Press, 1987.

Vomela, Martin. "Kolektivizace v Československu a její politické motivy" in Petr Blažek and Michal Kubálek, eds. *Kolektivizace Venkov v Československu 1948–1960 a Středoevropské Souvislosti.* Prague: Dokořán, 2008, 119–27.

Vultur, Smaranda. "The Role of Ethnicity in the Collectivization of Tomnatic/Triebswetter (Banat Region) (1949–1956)" in Constantin Iordachi and Dorin Dobrincu, eds. *Transforming Peasants, Property and Power: the Collectivization of Agriculture in Romania, 1949–1962.* Budapest: CEU Press, 2009, 141–64.

Wädekin, Karl-Eugen. *Agrarian Policies in Communist Europe: A Critical Introduction.* Dordrecht: Martinus Nijhoff, 1982.

Weber, Max. *Economy and Society: An Outline of Interpretive Sociology.* Berkeley: University of California Press, 1978.

Willoughby, Ian. "1950s trial of 'Czech kulak' subject of radio documentary." *Radio Prague.* September 29, 2006. http://incentraleurope.radio.cz/ice/article/83695 (accessed on December 22, 2012).

Witkowski, Gregory R. "On the Campaign Trail: State Planning and Eigen-Sinn in a Communist Campaign to Transform the East German Countryside." *Central European History* 37, no. 3 (2004): 400–22.

About the Authors

Arnd **BAUERKÄMPER** is Professor of Modern European History at the *Freie Universität Berlin.* He studied history and English at the universities of Bielefeld, Oxford and Göttingen. Selected publications: *Die "radikale Rechte" in Großbritannien. Nationalistische und faschistische Bewegung vom späten 19. Jahrhundert bis 1945.* Göttingen: Vandenhoeck & Ruprecht Verlag, 1991; *Ländliche Gesellschaft in der kommunistischen Diktatur. Zwangsmodernisierung und Tradition in Brandenburg 1945–1963.* Cologne: Böhlau Verlag, 2002; *Die Sozialgeschichte der DDR.* Munich: Oldenbourg Wissensch Verlag, 2005; *Der Faschismus in Europa 1918–1945.* Stuttgart: Reclam Verlag, 2006; *Das umstrittene Gedächtnis.* Paderborn: Schöningh, 2012. Co-editor of: *Demokratiewunder. Transatlantische Mittler und die kulturelle Öffnung Westdeutschlands 1945–1970.* Göttingen: Vandenhoeck & Ruprecht Verlag, 2005 (together with Konrad H. Jarausch and Marcus Payk); *Zwischen Fürsorge und Seelsorge. Christliche Kirchen in den europäischen Zivilgesellschaften seit dem 18. Jahrhundert.* Frankfurt am Main: Campus Verlag, 2009 (together with Jürgen Nautz); *Durchhalten! Krieg und Gesellschaft im Vergleich 1914–1918.* Göttingen: Vandenhoeck & Ruprecht Verlag, 2010 (together with Elise Julien).

Melissa K. **BOKOVOY** is Associate Professor of History, University of New Mexico. Author of *Peasants and Communists: Politics and Ideology in the Yugoslav Countryside, 1941–1953.* Pittsburgh: University of Pittsburgh Press, 1998. Co-editor of: *Sharing the Stage: Biography and Gender in Western Civilization,* 2 vols. Boston: Houghton Mifflin, 2003 (with Jane Slaughter); *State-Society Relations in Yugoslavia, 1945–1992.* New York: St. Martin's Press, 1997 (with Jill Irvine and Carol Lilly).

Dorin **DOBRINCU** is Research Fellow, "A. D. Xenopol Institute of History," Iaşi, and Associate Professor, Department of History, Alexandru I. Cuza University, Iaşi. Areas of specialization: history of Romania, particularly the Second

World War, the Holocaust, the communist period, the memory of fascism and communism; church-state relations, religious nationalism, the communist repression of the Church, and the history of Romanian protestant communities. Editor of: *Listele morţii. Deţinuţi politici decedaţi în sistemul carceral din România potrivit documentelor Securităţii, 1945–1958* [Death lists. Political prisoners deceased in the prison system of Romania according to Securitate documents, 1945–1958]. Iaşi: Polirom, 2008; *Proba Infernului. Personalul de cult în sistemul carceral din România potrivit documentelor Securităţii, 1959–1962* [The test of inferno. Clerics in the prison system in Romania according to Securitate documents]. Bucharest: Scriptorium, 2004. Co-editor of: *Refugiaţii polonezi în România, 1939–1947. Documente din Arhivele Naţionale ale României/Polscy uchodźcy w Rumunii 1939–1947. Dokumenty z Narodowych Archiwów Rumunii* [Polish Refugees in Romania. Documents from the National Archives of Romania], bilingual edition (Romanian-Polish), Warsaw-Bucharest: Arhivele Naţionale ale României, Instytut Pamięci Narodowej Komisja Ścigania Zbrodni przeciwko Narodowi Polskiemu, IICCMER, 2013, vol. I-II; *Istoria comunismului din România* [History of Communism in Romania], vol. I: *Documente. Perioada Gheorghe Gheorghiu-Dej (1945–1965)* [Documents. Gheorghe Gheorghiu-Dej period]. Bucharest: Humanitas, 2009; vol. II: *Documente Nicolae Ceauşescu (1965–1971)* [Documents Nicolae Ceauşescu (1965–1971)] Iaşi: Polirom, 2012 (with Mihnea Berindei and Armand Goşu); *Raportul Final al Comisiei Prezidenţiale pentru Analiza Dictaturii Comuniste din România* [Final Report of the Presidential Commission for the Study of the Communist Dictatorship in Romania]. Bucharest: Monitorul Oficial, 2006; 2nd, rev. ed., Bucharest: Humanitas, 2007 (with Vladimir Tismăneanu and Cristian Vasile); *Ţărănimea şi puterea. Procesul de colectivizare a agriculturii în România (1949–1962)* [The peasantry and the power. The process of collectivization of agriculture in Romania (1949–1962)]. Iaşi: Polirom, 2005 (with Constantin Iordachi).

David **FEEST** is Research Associate at the Berlin-Hohenschönhausen Memorial, Germany. He specializes in modern Russian and Baltic history. Selected publications: *Zwangskollektivierung im Baltikum. Die Sowjetisierung des Estnischen Dorfes 1944–1953.* Cologne, Weimar, Vienna: Böhlau Verlag, 2007. Editor of: *Von den Restgütern zu den Sowchosen in Estland 1939–1953. Dokumentensammlung.* Unter Mitarbeit von Karsten Brüggemann. Münster: Lit Verlag, 2010. Co-editor of: *Imperiale Herrschaft in der Provinz. Repräsentationen politischer Macht im späten Zarenreich.* Cologne: Campus Verlag, 2008 (with Jörg Baberowski and Christoph Gumb); and *Dem Anderen begegnen. Eigene und fremde Repräsentationen in sozialen Gemeinschaften.* Cologne: Campus Verlag, 2008 (with Jörg Baberowski and Maike Lehmann).

Michail **GRUEV** is Associate Professor at Sofia University "St. Kliment Ohridski." His research interests focus on the recent history of Bulgaria, ethnic and cultural issues, minorities, and historical anthropology. Selected publications: *Между петолъчката и полумесеца. Българите мюсюлмани и политическият режим 1944–1959* [Between the red star and the crescent: Muslim Bulgarians and the political regime, 1944–1959]. Sofia: Kota, 2003; *"Възродителният процес." Мюсюлманските общности и комунистическият режим: политики, реакции* *и последици* [The "Revival Process." Muslim communities and the Communist regime: Politics, reactions, and consequences]. Sofia: Ciela, 2008 (co-authored with Alexei Kalionski); *Преорани слогове. Колективизация и социална промяна в Българския северозапад (40-те–50-те години на XX век)* [Reploughed boundaries: Collectivization and Social Change in Bulgarian Northwest (40s and 50s of the XX century)]. Sofia: Ciela, 2009.

Constantin **IORDACHI** is Associate Professor and co-director of Pasts, Center for Historical Studies, at the Central European University, Budapest (CEU). He is specializing in the comparative history of Central and Southeastern Europe, with a focus on nationalism, fascism, and communism. Author of: *Charisma, Politics and Violence: The Legion of the "Archangel Michael" in Interwar Romania.* Trondheim: Trondheim Studies on East European Cultures & Societies, 2004; *Citizenship, Nation- and State-Building: The Integration of Northern Dobrogea in Romania, 1878–1913.* Carl Back Papers in Russian and East European Studies, University of Pittsburgh, 2002. Editor of: *Redobândirea cetățeniei române: Perspective istorice, comparative și aplicate/Reacquiring Romanian Citizenship: Historical, Comparative and Applied Perspectives,* bilingual edition, Bucharest: Curtea Veche, 2012; "Fascism in East Central and Southeastern Europe: A Reappraisal," Special Issue, *East-Central Europe,* vol. 37 (2010), no 2–3; *Comparative Fascist Studies: New Perspectives.* London: Routledge 2009, 2010; "Reconceptualizing the Social: Writing Social History in East Central Europe," Special Issue, *East-Central Europe,* 35 (2008) 1–2. Co-editor of: *Hungary and Romania Beyond National Narratives: Comparisons and Entanglements.* Berlin: Peter Lang, 2013 (together with Anders Blomqvist and Balazs Trencsenyi); *Transforming Peasants, Property and Power: The Process of Land Collectivization in Romania, 1949–1962.* Budapest: CEU Press, 2009 (together with Dorin Dobrincu); *Holocaust in Romania and the Trans-Dnister region: Historical and Comparative Perspectives.* Bucharest: Curtea Veche, 2004 (together with Viorel Achim).

Dariusz **JAROSZ** is Professor of History at the University of Warsaw. He is specializing in the social history of Poland after 1945, including relationships with Western countries. Selected publications: *Polityka władz komunistycznych w Polsce w latach 1948–1956 a chłopi* [The policy of the Communist authorities in Poland in the years 1948–1956 and the peasants]. Warszawa: Wydawnictwo DiG, 1998; *Polacy a stalinizm 1948–1956* [The Poles and Stalinism 1948–1956]. Warszawa: Instytut Historii PAN, 2000; *The Wawrzecki Affair: Meat in Social History of Poland (1944–1970).* Toruń: Wydawnictwo Marszałek, 2004; *The Organization of Leisure in Poland 1945–1956.* Warsaw, 2003; "Transformation of Polish Villages before and after 1989," in Shingo Minamizuka, ed., *The Transformation of the Systems of East-Central Europe Rural Societies before and after 1989.* Kecskemet, 1996, 19–27; "Polish Peasants versus Stalinism, Stalinism in Poland, 1944–1956" in A. Kemp-Welch, ed., *Selected Papers from the Fifth World Congress of Central and East European Studies.* London: Macmillan Press, 1999, 59–77.

József Ö. **KOVÁCS** is Professor of History, Károli Gáspár University of the Reformed Church in Hungary, Budapest. His research interests focus on social history, rural society, and the history of the communist rule in Hungary. Selected publications: *Zsidók a Duna-Tisza közén. Társadalomtörténeti esettanulmányok, XVIII–XIX.század* [Jews in the Region between the Danube and Tisza Rivers in the 18th–19th centuries]. Kecskemét: Kecskeméti Lapok, 1996; *Az újkori német társadalomtörténet útjai* [Ways of German Social History]. Debrecen: Csokonai, 2004; "Arbeiterexistenz in Ungarn nach 1956. Einige Schnittpunkte der Mikro- und Makrogeschichte" in Peter Hübner, Christoph Kleßmann, Klaus Tenfelde, eds. *Arbeiter im Staatssozialismus: Ideologischer Anspruch und soziale Wirklichkeit.* Cologne: Böhlau, 2005, 319–345; "The 'Socrealist' Contexts of the Collectivization of Campaign" [A kollektivizálás "szocreál" kontextusai Magyarországon (1948–1953)]. *Aetas—Journal of History and Related Disciplines*, 24 (2009) 4: 32–45; "'Sűrített népnevelő.' A kollektivizálás tapasztalattörténetei 1958–1959" ['Condensed Mass Educator:' Narratives of Collectivization Experiences]. *Korall—Journal of Social History* 10, no. 36 (2009): 31–54; "Documents from the Period of the Collectivization." *Korall—Journal of Social History* 10, no. 36 (2009): 160–78; "Istoria socială a colectivizării din Ungaria," *Arhivele Totalitarismului* 64–65, no. 3–4 (2009): 68–98; «Соцреаль»-ные контексты коллективизации в Венгрии (1948–1953 гг.), ed., И.В. Крючков, *Ставрополь* 2009, 3: 202–23; "Importanţa violenţei şi interpretarea acesteia." *Arhivele Totalitarismului* 68–69, no. 3–4 (2010): 60–79.

Jan **RYCHLÍK** studied history and ethnology at the Faculty of Art of Charles University. In 1985 he earned his PhD in ethnology and folklore at St. Kliment Ochridski University in Sofia, Bulgaria. In 1998 he earned a PhD in history at the Historical Institute of Czech Academy of Sciences. Since 1992, he is teaching history at the Faculty of Arts of the Charles University in Prague and he is simultaneously working as a senior researcher in the Masaryk Institute and Archives of the Czech Academy of Sciences. He was appointed assistant professor (docent) in 1997 and full time professor of Czech and Slovak history in 2003. Since 2006 he is also teaching the general history of 19th and 20th centuries at the Faculty of Arts of St. Cyril and Methodius University in Trnava, Slovakia. Rychlík specializes in Central European and Balkan history. His main publications include: *R. W. Seton-Watson and his Relations with the Czechs and Slovaks. Documents 1906–1951*, 2 vols. Prague: Ústav T. G. Masaryka; Martin: Matica Slovenská, 1995, 1996 (edited together with T. D. Marzik and M. Bielik); *Etnos i folklor* [Ethnos and Folklore]. Sofia: Vezni, 1997 (in Bulgarian); *Češi a Slováci ve 20. století. Československé vztahy,* 2 vols: 1914–1945, 1945–1992 [Czechs and Slovaks in the 20th Century. Czech-Slovak Relations 1914–1945, 1945–1992]. Bratislava: Academic Electronic Press, 1997, 1998; *Dějiny Bulharska* [History of Bulgaria]. Prague: Lidové noviny, 2000, 2nd ed. 2002; *Rozpad Československa* [The Split of Czechoslovakia, 1989–1992]. Bratislava: Academic Electronic Press, 2002; *Dějiny Makedonie* [History of Macedonia]. Prague: Lidové noviny, 2003, (together with M. Kouba); *Dějiny Chorvatska* [History of Croatia]. Prague: Lidové noviny, 2007, (together with Milan Perenčević). *Mezi Vídní a Cařihradem. Utraření balkánských národů* [Between Vienna and Istanbul. Building up of Balkan Nations I.]. Prague: Historica, 2009 (together with Jan Pelikán, Nadja Danova and Krstju Mančev).

Jens **SCHÖNE** is Land Deputy Commissioner for Stasi files in Berlin and a lecturer at the *Humboldt-Universität zu Berlin.* He is specializing in the history of the German Democratic Republic, with a focus on political, social and economic questions. His publications include: *Frühling auf dem Lande? Die Kollektivierung der DDR-Landwirtschaft.* Berlin: Christoph Links Verlag, 2005; *Das sozialistische Dorf. Bodenreform und Kollektivierung in der Sowjetzone und DDR.* Leipzig: Evangelische Verlagsanstalt, 2008; and *The Peaceful Revolution. Berlin 1989/90—The Path to German Unity.* Berlin: Berlin Story Verlag, 2009.

Örjan **SJÖBERG** is Professor of Economic Geography at the Stockholm School of Economics, Sweden. He received his PhD in human geography at Uppsala University with a dissertation entitled *Urbanization under Central Planning: the Case of Albania* (1991), and has since focused on economic geography and economic development. This includes work initiating the landscape of priorities approach to inter-urban and intra-urban change under central planning: see *Urban Studies* (1999); also articles written together with Michael Gentile, in: *Europe-Asia Studies* (2006); *Annals of the Association of American Geographers* (2010); and *Journal of Historical Geography* (2010); and several contributions on rural development, migration and post-socialist economic reform. For the most part, the geographical focus has been Southeastern Europe, the Baltic states, and Southeast Asia. His publications on Southeastern Europe include: *Rural Change and Development in Albania.* Boulder, CO: Westview, 1991; *Economic Change in the Balkan States: Albania, Bulgaria, Romania and Yugoslavia.* London: Pinter, 1991 (co-edited with Michael L. Wyzan); and "A chequered history: political culture and conflict in Albanian economic reform" (co-authored with Marta Muço) in Lundahl, Mats, and Michael Louis Wyzan, eds., *The Political Economy of Reform Failure.* Routledge, London 2005. His articles on Albania have also appeared in *South Eastern Europe Journal of Economics* (2005); and *Journal of Ethnic and Migration Studies* (2006).

Nigel **SWAIN** is Lecturer in the Department of History at the University of Liverpool. His research has been on the sociology and social history of post war Hungary, and latterly Eastern Europe more generally, with a focus on rural economy and society. Selected publications: *Collective Farms which Work?* Cambridge: Cambridge University Press, 1985; *Hungary: The Rise and Fall of Feasible Socialism.* London: Verso, 1992; *Eastern Europe since 1945* (with Geoffrey Swain). London: Macmillan, 1993, 1998, 2003 and 2009; "Collective farms which work" in T. Shanin (ed), *Peasants and Peasant Societies* (2nd ed.). London: Blackwell, 1987, Penguin, 1988, 111–113; "A framework for comparing social change in the post-socialist countryside," *Eastern European Countryside* 4 (1998): 5–19; "Agricultural restitution and cooperative transformation in the Czech Republic, Hungary and Slovakia." *Europe-Asia Studies* 51, no. 7 (1999): 1199–1219; "From kolkhoz to holding company: a Hungarian agricultural producer cooperative in transition," *Journal of Historical Sociology* 13, no. 2 (June 2000): 142–171. Editor of: *Hungary: A Decade of Economic Reform.* London: George Allen and Unwin, 1981 (with P. G. Hare and H. K. Radice).

Zsuzsanna **VARGA** is Associate Professor at the Department of Modern Hungarian History of Eötvös Loránd University, Budapest. She earned her MA in History (1993) at Kossuth Lajos University, Debrecen, and her PhD in Economic History (1998). Varga's research interests and publications focus on the history of socialist agriculture in Hungary, but she incorporates her research findings within a Central-Eastern European comparative framework. Varga is a member of the Committee of Agrarian History and Sociology of the Hungarian Academy of Sciences. Publications: *Politika, paraszti érdekérvényesítés és szövetkezetek Magyarországon 1956–1967* [Politics, the assertion of agrarian interests and cooperatives in Hungary between 1956 and 1967]. Budapest: Napvilág Kiadó, 2001; "Agrarian development from 1945 to the present day" in János Estók, ed., *History of Hungarian Agriculture and Rural Life, 1848–2004.* Budapest: Argumentum Publishing House, Museum of Hungarian Agriculture, 2004, 221–94; "The Impact of 1956 on the Relationship between the Kádár Regime and the peasantry, 1956–66." *Hungarian Studies Review*, 34, nos. 1–2. (2007): 155–76; "The Agrarian Elite in Hungary before and after the Political Transition" in Christoph Boyer, Friderike Sattler eds., *European Economic Elites. Between a New Spirit of Capitalism and the Erosion of State socialism.* Berlin: Dunckler & Humblot, 2009, 223–50.

Lynne **VIOLA** is Professor at the University of Toronto. She is specializing in twentieth century Russian history, focusing on political and social history. Her research interests include women, peasants, political culture, and Stalinist terror. Viola is the author of over thirty articles and four books: *The Best Sons of the Fatherland: Workers in the Vanguard of Soviet Collectivization.* New York: Oxford University Press, 1987; *Peasant Rebels Under Stalin: Collectivization and the Culture of Peasant Resistance.* New York: Oxford University Press, 1996; *The War Against the Peasantry, 1927–1930.* New Haven, CT: Yale University Press, 2005; and *The Unknown Gulag: The Lost World of Stalin's Special Settlements.* Oxford: Oxford University Press, 2007. Editor or co-editor of: *A Researcher's Guide to Sources of Soviet Social History in the 1930s.* Armonk, N. Y.: M. E. Sharpe, 1990 (with Sheila Fitzpatrick); *Russian Peasant Women.* New York: Oxford University Press, 1992, (with Beatrice Farnsworth); *Kollektivizatsiia i krest'ianskoe soprotivlenie na Ukraine: noaibr' 1929-mart 1930* [Collectivization and Peasant Resistance in Ukraine, November 1929–March 1930]. Vinnitsia: Logos, 1997 (with Valerii Vasil'ev); *Riazanskaia derevnia v 1929–1930 gg.: khronika golovokhruzheniia* [The Riazan Countryside in 1929–1930: A Chronicle of Dizzyiness]. Moscow: ROSSPEN, 1998 (with S. Zhuravlev, T. McDonald, and A. Mel'nik); *Tragediia sovetskoi derevni 1927–37: dokumenty i materialy* [The Tragedy of the Soviet Countryside, 1927–37: Documents and Materials] in 5 volumes. Moscow: ROSSPEN, 1999–2003 (with V. P. Danilov and R. T. Manning); and *Contending with Stalinism: Soviet Power and Popular Resistance in the 1930s.* Ithaca: Cornell University Press, 2002.

Gregory R. **WITKOWSKI** is Associate Professor of Philantropic Studies at the Indiana University Lilly Family School of Philanthropy. He is a sociocultural historian, whose work concentrates on the everyday life of historical actors and focuses on issues such as power and resistance, identity and difference, and modernization and tradition. His first research project, *Workers to the Countryside! Mass*

Mobilizations in Communist East Germany 1949–65, is an examination of the communist use of campaigns as part of their five-year plans. He argues that these mass mobilizations were more than a propaganda tool; they were an integral part of communist planning and policy implementation. His second project examines East German identity formation through the lens of charitable giving for the developing world. His articles have appeared in *Central European History* and *German History* and in numerous edited volumes.

Index

D

E

F

I

J

K

N

S

For Product Safety Concerns and Information please contact our EU representative GPSR@taylorandfrancis.com Taylor & Francis Verlag GmbH, Kaufingerstraße 24, 80331 München, Germany

Batch number: 10703950

Printed by Printforce, the Netherlands